Progress in Mathematics

Volume 190

Series Editors

H. Bass
J. Oesterlé
A. Weinstein

Progress in Mathematics

Volume 190

Arno van den Essen

Polynomial Automorphisms

and the Jacobian Conjecture

Springer Basel AG

Author:

Arno van den Essen
Department of Mathematics
University of Nijmegen
Toernooiveld 1
6500 GL Nijmegen
The Netherlands

2000 Mathematics Subject Classification 14Exx, 34Dxx, 13Pxx; 14E09, 14E07, 34D20, 14E20

A CIP catalogue record for this book is available from the Library of Congress, Washington D.C., USA

Deutsche Bibliothek Cataloging-in-Publication Data

Essen, Arno /van den:
Polynomial automorphisms and the Jacobian Conjecture / Arno van den Essen. - Basel ; Boston ; Berlin : Birkhäuser, 2000
(Progress in mathematics ; Vol. 190)
ISBN 978-3-0348-9567-5 ISBN 978-3-0348-8440-2 (eBook)
DOI 10.1007/978-3-0348-8440-2

Originally published by Birkhäuser Verlag, Basel, Switzerland in 2000
Softcover reprint of the hardcover 1st edition 2000
Printed on acid-free paper produced of chlorine-free pulp. TCF ∞

ISBN 978-3-0348-9567-5

9 8 7 6 5 4 3 2 1

For Sandra and Raïssa

Contents

Preface

It was in December 1985, when I first heard about the Jacobian Conjecture. I was in Paris to give a series of lectures on Algebraic Microlocalization at the Institute Henri Poincaré. At that time my main interest was the study of $\mathfrak{D}$-modules with regular singularities using algebraic methods. In my audience was a mathematician from Togo, his name Kossivi Adjamagbo (Pascal). He was working on a thesis concerning non-commutative determinants extending earlier work of Dieudonné from 1943. The main point of his interest in my lectures was that, using algebraic microlocalization, it is possible to embed a large class of non-commutative domains in a skew-field (this is not always possible for arbitrary non-commutative domains). After my first lecture he came to my office in Paris VI and invited me and my wife Sandra for dinner. During this dinner he explained to us that he had been working for three days and nights (without interruption!) on a famous conjecture in algebraic geometry called the Jacobian Conjecture and that he was close to a solution (this would happen many times later; at almost every occasion we met he came up with a new "solution"). However it was in June 1986, during another visit to Paris, that he explained the Jacobian Conjecture to me in more detail. That's how it all started for me. The Jacobian Conjecture originated from a paper [179] by Ott-Heinrich Keller in 1939 entitled "Ganze Cremona-Transformationen". In this paper he formulated the following question, now known as "Keller's problem": "*given polynomials* $F_1, \ldots, F_n \in \mathbb{Z}[X_1, \ldots, X_n]$ *such that* $\det JF = 1$, *where* JF *denotes the Jacobian matrix* $\left(\frac{\partial F_i}{\partial X_j}\right)_{1 \le i, j \le n}$, *does it follow that each* X_i *can be expressed as a polynomial in* $F_1, \ldots, F_n$ *with coefficients in* $\mathbb{Z}$?".

In the introduction of his paper Keller motivates his question as follows: "*Für manche Integritätsbereiche lässt sich eine lineare Basis* $A = (a_1, \ldots, a_n)$ *angeben, und jedes Element des Integritäts-bereiches lässt sich als Linearkombination* $\sum \lambda_i a_i$ *mit ganzzahligen* λ_i *darstellen. Durch eine unimodulare Lineartransformation erhalten wir eine neue Basis, und umgekehrt hängt jede andere Basis B mit A vermöge einer unimodularen Lineartransformation zusammen. Wir haben also eine vollständige Übersicht über die möglichen Basissystemen.*

Nun gibt es Integritätsbereiche, die keine lineare Basis, wohl aber ein Basis $(x_1, x_2, \ldots, x_n)$ *besitzen, mit deren Hilfe sich jedes Element als Polynom* $\sum \lambda_{i_1,\ldots,i_n} x_1^{i_1} \ldots x_n^{i_n}$ *mit ganzzahligen* $\lambda_{i_1,\ldots,i_n}$ *darstellen lässt. Ein Beispiel sind die symmetrische Funktionen; fur sie sind die symmetrischen Grundfunktionen eine solche Basis. Es ergibt sich die Aufgabe, auch hier eine Übersicht alle möglichen Basissysteme und die vermittelnden Transformationen zu gewinnen*".

"For many integral domains there exists a linear basis $A = (a_1, \ldots, a_n)$ such that every element of this integral domain can be written as a linear combination $\sum \lambda_i a_i$ with integral coefficients. By a unimodular linear transformation we get a new basis and conversely every other basis B is related to A by means of a unimodular linear transformation. So we have a complete overview of all possible basis systems.

Now there exist integral domains which do not have a linear basis, which do however have a basis such that every element can be written as a polynomial $\sum \lambda_{i_1,\ldots,i_n} x_1^{i_1} \ldots x_n^{i_n}$ with integral coefficients. An example is formed by the symmetric functions; the elementary symmetric functions form such a basis. The problem arises here also of getting an overview of all possible basis systems and their relating transformations".

At the end of his introduction Keller writes: "*Mir scheint die Frage einer Untersuchung sehr zu lohnen, Sie scheint jedoch bereits im ebenen Fall sehr schwierig zu sein*".

"It seems to me that it will be worthwhile to investigate this question, however even in the plane it seems to be very difficult."

Indeed Keller was right. Today this question is still open in all dimensions ≥ 2, in spite of the efforts of many mathematicians. Nevertheless the search for a solution of the Jacobian Conjecture has led to a stream of new results concerning polynomial automorphisms. This book will be an account of many of these new results, as I have seen them develop during the last twelve years. It is my sincere hope that this book will bring joy to the reader and that it will stimulate several people to join the growing group of "Jacobians".

Acknowledgements

It is a great pleasure for me to thank everyone who directly or indirectly has contributed to this book. However some people I wish to mention in particular. In the first place Pascal Adjamagbo, who with his everlasting energy, enthusiasm and stream of ideas pursuaded me into this beautiful subject and whose warm friendship and hospitality made every meeting a special occasion. Next I want to thank Harm Derksen, whose many extremely simple and elegant ideas have had a great influence on the contents of this book. Then Engelbert Hubbers, whose computer skills have been indispensable during the preparation of the book and whose computational talents have heavily contributed to our fruitful collaboration. Gary Meisters for his many ideas, conjectures and friendship. Armengol Gasull for giving me a strong motivation to finish this book at the right moment and Andrzej Nowicki for supplying me with his short and elementary proof of the Abhyankar-Moh theorem. Charles Cheng, Stefan Maubach and Peter van Rossum for reading several parts of the text and for giving ideas to improve the readability of it. A special word of thanks to Andrew Campbell, who sent me at least twenty pages full of detailed suggestions to improve the text. The book would have been a lot less readable without all his help!

Then there is my family, my wife Sandra and our daughter Raïssa. Sandra typed a large part of the text and corrected patiently thousands of mistakes in the various drafts. Raïssa's patience during my many hours of typing was more than admirable. The joy, friendship and warmth they bring into my life every day cannot possibly be expressed in words !

Finally I would like to thank the various institutions where I stayed or which I visited during the preparation of this book. First of all the University of Nijmegen and Universitat Autònoma de Barcelona where I gave a seminar on parts of this book during the first six month of 1998, the Centre de Recerca Matematica(CRM), The Max-Planck Institut at Bonn, the Nicholas Copernicus University (Torún), the Jagiellonian University (Krakow), the Banach Center (Warsaw), the University of Turino (Italy), Oakland University, the University of Nebraska at Lincoln and, of course Université Paris VI where I was a guest uncountably many times !

Nijmegen, December 1999 Arno van den Essen

Introduction

This introduction begins by giving the reader some feeling for the kind of questions which will be considered in the book. This is followed by an overview of the whole book and a brief description of its chapters.

For the sake of simplicity we start with polynomial maps over a field k. In Chapter 1 we will consider the more general situation of polynomial rings over an arbitrary commutative ring. So let k be an arbitrary field and let $k[X] := k[X_1, \ldots, X_n]$ denote the polynomial ring in the variables $X_1, \ldots, X_n$ over k. A **polynomial map** is a map $F = (F_1, \ldots, F_n) : k^n \to k^n$ of the form

$$(x_1, \ldots, x_n) \to (F_1(x_1, \ldots, x_n), \ldots, F_n(x_1, \ldots, x_n))$$

where each F_i belongs to $k[X]$. Such a polynomial map is called **invertible** if there exists a polynomial map $G = (G_1, \ldots, G_n) : k^n \to k^n$ such that $X_i = G_i(F_1, \ldots, F_n)$ for all $1 \leq i \leq n$, i.e. G is a left inverse of F. It is not difficult to show (1.1.6) that if a polynomial map F has a left inverse G, which is also a polynomial map, then G is also a right inverse of F. So F is invertible in the sense of morphisms of algebraic varieties.

Warning. Suppose that k is a *finite* field and $F : k^n \to k^n$ a polynomial map. If there exists a polynomial map $G : k^n \to k^n$ such that $G(F(x)) = x$ for all $x \in k^n$, then F need not be invertible (Exercise 1).

Invertible polynomial maps correspond one-to-one with k-automorphisms of the polynomial ring $k[X]$ given by

$$F \longrightarrow F^* : g \mapsto g(F).$$

So describing invertible polynomial maps from k^n to k^n is the same as describing k-automorphisms of $k[X]$. The group of k-automorphisms of $k[X]$ will be denoted by $\mathrm{Aut}_k\, k[X]$. If $F : k^n \to k^n$ is an invertible polynomial map we will call the set $F_1, \ldots, F_n$ a **coordinate system** of $k[X]$. Furthermore, a polynomial $f \in k[X]$ is called a **coordinate** if there exist $F_2, \ldots, F_n$ in $k[X]$ such that $f, F_2, \ldots, F_n$ is a coordinate system of $k[X]$. The name coordinate system is justified by the following useful and easy lemma.

Lemma 1 *Let $F : k^n \to k^n$ be a polynomial map. Then F is invertible if and only if $k[X_1, \ldots, X_n] = k[F_1, \ldots, F_n]$.*

Proof. If F is invertible with inverse $G = (G_1, \ldots, G_n)$ then $X_i = G_i(F_1, \ldots, F_n)$ for each $1 \leq i \leq n$. So $X_i \in k[F_1, \ldots, F_n]$ for each i, implying $k[X_1, \ldots, X_n] = k[F_1, \ldots, F_n]$. Conversely, if $k[X_1, \ldots, X_n] = k[F_1, \ldots, F_n]$, then $X_i \in k[F_1, \ldots, F_n]$ for each i, so $X_i = G_i(F_1, \ldots, F_n)$ for some G_i. So F is invertible. □

This lemma answers one of the first questions one always asks: "do there exist non-linear polynomial automorphisms?" It is interesting to remark that this question

appears more or less explicitly in the Scottish Book: Mathematics from the Scottish. Café [205] as problem 79 (Mazur, Orlicz), namely:
"If $F = (F_1, \ldots, F_n) : \mathbb{C}^n \to \mathbb{C}^n$ is a one-to-one polynomial map whose inverse is also a polynomial map, is each F_i of degree one?". An "Addendum" states "Trivial" and gives the answer "no" by writing the following examples: let $1 \leq i \leq n$ and $a = a(X_1, \ldots, \hat{X}_i, \ldots, X_n) \in k[X_1, \ldots, \hat{X}_i, \ldots, X_n]$ (here $\hat{X}_i$ means omit X_i). Then by Lemma 1 the polynomial map

$$E = (X_1, \ldots, X_{i-1}, X_i + a, X_{i+1}, \ldots, X_n) \tag{2}$$

is invertible since one readily verifies that the components of E generate the k-algebra $k[X]$. In fact the inverse is the map

$$E' = (X_1, \ldots, X_{i-1}, X_i - a, X_{i+1}, \ldots, X_n).$$

A map as in (2) is called an **elementary** polynomial map. Making finite compositions of such elementary maps and elements of the **affine subgroup** $\mathrm{Aff}(k, n)$, i.e. the group of automorphisms F such that $\deg F_i = 1$ for all i, we get more invertible maps. These maps are called **tame**. The subgroup of $\mathrm{Aut}_k\, k[X]$ consisting of the tame automorphisms is called the **tame subgroup** and is denoted by $T(k, n)$. This leads us to the following question.

Tame Generators Problem. Is $\mathrm{Aut}_k\, k[X] = T(k, n)$? In other words is every invertible polynomial map $F : k^n \to k^n$ tame?

It turned out that the answer to this problem is yes if $n = 2$: this result is known as the Jung-van der Kulk theorem (see 5.1.11 and 5.3.6). However if $n \geq 3$ the problem is still open. The general belief is that the answer is no. In fact there are several candidate counterexamples. One of them, denoted σ, was constructed by Nagata in 1972

$$\begin{aligned} \sigma(X) &= X - 2(XZ + Y^2)Y - (XZ + Y^2)^2 Z, \\ \sigma(Y) &= Y + (XZ + Y^2)Z, \\ \sigma(Z) &= Z. \end{aligned}$$

Conjecture. (Nagata) σ is not tame!

So in contrast with the Jung-van der Kulk theorem we have no structure theorem in dimension ≥ 3. On the other hand there is a beautiful result due to Harm Derksen (unpublished) which asserts that, for any dimension $n \geq 3$, the tame group is generated by the affine subgroup and only one non-linear map, namely

$$\varepsilon = (X_1 + X_2^2, X_2, \ldots, X_n)$$

This result is explained in Chapter 5, §2.

Another important problem which forms the motivation for much of the research concerning polynomial automorphisms is the

Invertibility Problem. Given a polynomial map $F : k^n \to k^n$. How can one decide if F is invertible?

To find a necessary condition is rather easy, namely suppose that F is invertible with inverse G. Then $G(F(X)) = X$, where $X = (X_1, \ldots, X_n)$, and hence by the chain rule

$$(JG)(F(X))JF(X) = I.$$

So taking determinants we get

$$\det JG(F(X)) \det JF(X) = 1.$$

Since both determinants are elements of $k[X]$ and the only units of $k[X]$ are non-zero constants, we conclude that $\det JF \in k^*$. So we get

(3) If $F : k^n \to k^n$ is invertible, then $\det JF \in k^*$.

The condition "$\det JF \in k^*$" is called the **Jacobian Condition** and a polynomial map $F : k^n \to k^n$ satisfying the Jacobian Condition is called a **Keller map**. Now one can ask: is the converse of (3) true, i.e. is a Keller map invertible? The answer is no, if k is a field of characteristic $p > 0$. To see this take $n = 1$ and let $F(X) = X - X^p \in k[X]$. Then obviously $\det JF = 1$, however F is not injective since $F(0) = F(1) = 0$. So we have to assume that k has characteristic zero. In that case it is easy to see that if $n = 1$ the condition $F'(X) \in k^*$ implies that $F(X) = \lambda X + \mu$, for some $\lambda \in k^*$ and $\mu \in k$. Hence F is invertible with inverse $G(X) = \lambda^{-1}X - \lambda^{-1}\mu$. So indeed if $n = 1$ the condition $\det JF \in k^*$ is both necessary and sufficient for F to be invertible. However if $n \geq 2$ we do not know if the converse of (3) is true. In fact this problem is known as the famous

Jacobian Conjecture. Let k be a field of characteristic zero. If $F : k^n \to k^n$ is a Keller map, then F is invertible.

As described in the preface the analogous question with $\mathbb{Z}$ instead of k was first formulated by Keller in 1939. On the other hand it will be shown in the next section that Keller's original question is equivalent to the Jacobian Conjecture. Therefore some authors refer to the Jacobian Conjecture as "Keller's Problem". Recently the Jacobian Conjecture appeared as Problem 16 on a list of 18 famous open problems in the paper "Mathematical problems for the next century" by Steve Smale, [270].

To obtain a solution for the Invertibility Problem one can proceed in two ways, either by adding some necessary conditions which together with the Jacobian Condition are sufficient, or by finding a completely new approach to describe invertibility. Indeed several additional conditions were found which together with the Jacobian Condition guarantee invertibility. One such condition is assuming that $k[X]$ is integral over $k[F] := k[F_1, \ldots, F_n]$ (2.2.16). Another one is assuming that the field extension $k(F) \subset k(X)$ is Galois (2.2.16). A completely new way to decide invertibility was found by the author in 1986, based on results from the theory of Gröbner bases. In contrast with the possible "Jacobian Conjecture criterion" this new criterion works for all fields and not only decides if a given polynomial map $F : k^n \to k^n$ is invertible, but in case F is invertible it also computes the inverse (3.2.1).

Using this result one can obtain a remarkable result concerning polynomial automorphisms, first found by McKay and Wang in [208], 1988, namely that a polynomial automorphism $F = (F_1, \ldots, F_n) : k^n \to k^n$ is completely determined by its so-called **face polynomials,** i.e. the n^2 polynomials $F_i(X_j = 0)$. Furthermore given these polynomials one can reconstruct the original map F (3.2.7). Another remarkable result concerning polynomial automorphisms is the following: let k be an algebraically closed field of characteristic zero. If $F : k^n \to k^n$ is injective, then F is a polynomial automorphism. Using this result one can reformulate the Jacobian Conjecture as follows: write $F'(z) := \det JF(z)$ for all $z \in k^n$. Then the Jacobian Conjecture is equivalent to the following "Rolle-type" formulation: If $F(a) = F(b)$ for some $a \neq b$ in k^n, then $F'(z) = 0$ for some $z \in k^n$.

Polynomial automorphisms play a role in several problems. For example the Jung-van der Kulk theorem implies that every group action of a reductive group on affine 2-space is linearizable (9.4.8). In particular every polynomial automorphism $F : \mathbb{C}^2 \to \mathbb{C}^2$ of finite order, i.e. $F^p = I$ for some $p \in \mathbb{N}$, is **linearizable**, i.e. there exists a polynomial automorphism $\varphi : \mathbb{C}^2 \to \mathbb{C}^2$ such that $\varphi^{-1}F\varphi$ is a linear map. We do not know if such a result is true for $\mathbb{C}^n$ if $n \geq 3$. In particular we do not even know if a polynomial automorphism of order 2, i.e. satisfying $F^2 = I$, is linearizable if $n \geq 3$. In fact an affirmative answer would imply that the Cancellation Problem has an affirmative answer too.

The **Cancellation Problem** asks if the following is true: let V be a non-singular affine algebraic variety of dimension $d \geq 1$ and suppose that $V \times \mathbb{C} \simeq \mathbb{C}^{d+1}$ as algebraic varieties; does it follow that $V \simeq \mathbb{C}^d$ as algebraic varieties? Only for the cases $d = 1$ and $d = 2$ is it known that the answer is affirmative. The case $d \geq 3$ remains open. We refer the reader to Chapter 9, §4 for more details.

Another example where polynomial automorphisms played a prominent role is in the solution of a famous conjecture in the field of dynamical systems: the **Markus-Yamabe Conjecture**, also called the **global-asymptotic stability Jacobian Conjecture** which was formulated by Markus and Yamabe in [201], 1960, in an attempt to globalize a classical (local) result on asymptotic stability due to Lyapunov. The conjecture asserts the following: let $F : \mathbb{R}^n \to \mathbb{R}^n$ be a C^1-map with $F(0) = 0$ and such that for each $x \in \mathbb{R}^n$ all eigenvalues of $JF(x)$ have negative real part; then 0 is a **global attractor** of the autonomous system of differential equations

$$y'(t) = F(y(t)), \tag{4}$$

i.e. every solution of (4) is defined for all $t > 0$ and tends to 0 if t tends to infinity. In the case $n = 2$ the conjecture was proved by Gutierrez [147] and Fessler [123] independently in 1993. The polynomial case had already been proved in 1988 by Meisters and Olech in [212]. However in 1995 the author in collaboration with Cima, Gasull, Hubbers and Mañosas ([54]) showed that even the polynomial version of the

Markus-Yamabe Conjecture is false for all $n \geq 3$, by giving an explicit polynomial automorphism F satisfying the hypothesis of the conjecture, but such that the system (4) has a solution which tends to infinity if t tends to infinity! (See Chapter 8)

Of course many topics considered in this book are not mentioned in this section. However the results described above certainly give some flavor of the subject entitled "polynomial automorphisms" and hopefully will motivate the reader to continue reading.

Exercises

1. Let $k = \mathbb{F}_3$. Let $F : k \to k$ be given by $F(x) = x + x^3$. Show that F is bijective and that F is not invertible.
2. Show that the map $F \longrightarrow F^* : g \to g(F)$ gives a one-to-one correspondence between polynomial maps of k^n and endomorphisms of $k[X]$. Furthermore show that this map induces a one-to-one correspondence between invertible polynomial maps of k^n and polynomial automorphisms of $k[X]$.
3. Show that if f is a coordinate of $k[X]$, then $\frac{\partial f}{\partial X_1}, \ldots, \frac{\partial f}{\partial X_n}$ is a **unimodular row,** i.e. there exist polynomials a_i such that $1 = a_1 \frac{\partial f}{\partial X_1} + \cdots + a_n \frac{\partial f}{\partial X_n}$.

   [Use formula (3) above.]
4. Show that Nagata's σ is invertible by computing its inverse.
5. Show that the polynomial map $F : \mathbb{Q} \to \mathbb{Q}$ defined by $F(x) = x^3$ is injective but not surjective.
6. Show that the polynomial map $F : \mathbb{R} \to \mathbb{R}$ defined by $F(x) = x^3$ is bijective but that its inverse is not a polynomial map.
7. A polynomial automorphism $F : k^n \to k^n$ is called **triangular** or **of triangular form** if $F_i \in k[X_i, X_{i+1}, \ldots, X_n]$ for each $1 \leq i \leq n$.

   (i) Show that $F_i = \lambda_i X_i + a_i$, with $\lambda_i \in k^*$ and $a_i \in k[X_{i+1}, \ldots, X_n]$.
   (ii) Show that each triangular map is tame.

Outline of the book

The contents are arranged as follows. The book is divided in three Parts, "Methods", "Applications" and "Appendices". The first five of these appendices should supply sufficient background for reading the text. The last appendix "Special examples and counterexamples" is added since I experienced that many of these examples very often can be used as counterexamples to other problems as well.

In Chapter 1 several basic tools used throughout the book are introduced: the formal inverse function theorem, the clearing map, derivations, the exponential of a derivation and locally nilpotent derivations. They all turn out to play a crucial role in many problems related to polynomial automorphisms.

Chapter 2 is devoted to the interplay between polynomial automorphisms and (locally nilpotent) derivations. For example it is shown how derivations can be used to prove that the degree of the inverse of a polynomial automorphism F is bounded by $(\deg F)^{n-1}$ and that the Jacobian Conjecture holds in the "proper" case and the "Galois" case.

Chapter 3 describes various invertibility criteria and formulae for polynomial automorphisms. In particular we give an algorithm which decides if a given R-algebra homomorphism between finitely generated R-algebras is an isomorphism and which computes the inverse in case it exists. Here R is any finitely generated k-algebra and k an arbitrary field. This result is used to study face polynomials and to give a criterion which decides if a given $(n-1)$-tuple of polynomials can be extended to a polynomial automorphism.

Chapter 4 is centered around a beautiful result of Cynk and Rusek which asserts that every injective endomorphism of an algebraic set V over an algebraically closed field is automatically an automorphism ! In Chapter 5 we study the tame automorphism group of a polynomial ring with coefficients in a commutative $\mathbb{Q}$-algebra. In case R is a domain we show that in dimension 2 the tame automorphism group is the free amalgamated product of the affine subgroup and the de Jonquières subgroup. If $n \geq 3$ it is proved that the tame automorphism group is generated by the affine subgroup and one non-linear automorphism. Furthermore we study embeddings of algebraic varieties and give an elementary proof of the Abhyankar-Moh theorem.

In Chapter 6 we describe the stabilization principle and show how it can be used to reduce the study of the Jacobian Conjecture to special classes of polynomial mappings. At the end of the chapter we discuss a useful pairing between cubic homogeneous mappings and Drużkowski mappings. Chapter 7 is mainly devoted to the study of a large class of polynomial mappings having nilpotent Jacobian matrix. This class gives rise to a large class of polynomial automorphisms whose importance comes from the fact that it can be used as a test class for various open problems concerning polynomial automorphisms.

In Chapter 8, the first chapter of the application part of the book, some problems from the field of dynamical systems are discussed such as the LaSalle problem and the Markus-Yamabe Conjecture. Using results from Chapter 7, counterexamples to these problems in dimensions ≥ 3 are given. In Chapter 9 we show how results on polynomial automorphisms can be used to study group actions on affine varieties and we discuss the interplay between group actions, polynomial automorphisms and locally nilpotent derivations. We will study the rings of invariants for reductive group actions and G_a-actions. In the first case we describe Derksen's algorithm to compute the ring of invariants, and in the second case we give a concrete G_a-action on $\mathbb{C}^7$ whose ring of invariants is not finitely generated.

Finally in Chapter 10 we return once more to the Jacobian Conjecture. We describe several possible approaches to study the conjecture. So we do not give a complete overview of all results known. For example, concerning the two-dimensional case we discuss some classical results using a (possibly) new approach inspired by an unpublished manuscript of Zurkowski. Concerning the n-dimensional case we describe approaches via positive characteristic and using $\mathfrak{D}$-modules.

Part I
Methods

1 Preliminaries

Polynomial maps have of course been common tools in algebraic geometry for many years. But more recently, motivated by the study of some tempting conjectures, the subject started to become a subbranch of affine algebraic geometry. Surprising new results were obtained, several conjectures were made and proofs of existing results were simplified etc. In this chapter we start developing some of the main tools which play a crucial role throughout this book. The first of these is the formal inverse function theorem, the algebraic analogue of the local inverse function theorem of calculus. For example we use it to describe the so-called Lefschetz-principle and the clearing map; both are techniques to reduce the study of polynomial maps over commutative $\mathbb{Q}$-algebras to polynomial maps over such special rings as $\mathbb{Z}$ or $\mathbb{C}$.

Another tool which we will use frequently is the concept of a derivation. The kernel of a derivation on a polynomial ring turns out to contain much information on problems related to polynomial automorphisms. Particularly important are the so-called locally nilpotent derivations. They will be extensively studied and some practical algorithms to describe their kernels and cokernels are given in the last section of this chapter.

1.1 The formal inverse function theorem and its applications.

One of the most important and remarkable results of linear algebra is the fact that an $n \times n$ matrix over a commutative ring R is invertible if and only if its determinant is invertible in the ring R. This result forms the starting point of several well-known results in analysis: let $U \subset \mathbb{R}^n$ be an open set and $F : U \to \mathbb{R}^n$ a C^1-map. Then for each point p in U one has the local inverse function theorem, asserting that F is locally invertible in a neighborhood of p, i.e. with inverse a C^1-map, if and only if the determinant of the Jacobian matrix at p is non-zero, i.e. $\det JF(p) \neq 0$. We get a similar result if we start with an analytic map $F : U \to \mathbb{C}^n$, where $U \subset \mathbb{C}^n$ is an open neighborhood of a point p in U and assume that $\det JF(p) \neq 0$.

In this section we give an algebraic analogue of these results: the formal inverse function theorem. This result will play a crucial role in the study of polynomial maps. The remainder of this section is used to demonstrate this statement by giving several applications of the formal inverse function theorem to polynomial maps.

Throughout this section we have the following notations: R denotes an arbitrary commutative ring, R^* the group of its units and $R[[X]] := R[[X_1, \ldots, X_n]]$ denotes the ring of formal power series in n variables over R. By $\mathfrak{m}$ we denote the ideal in $R[[X]]$ generated by $X_1, \ldots, X_n$. If $0 \neq f \in R[[X]]$ we can write

$$f = f_{(r)} + f_{(r+1)} + \cdots$$

where $f_{(r)} \neq 0$ and each $f_{(i)}$ is a homogeneous polynomial in $X_1, \ldots, X_n$ of degree i. So if $f_{(i)} \neq 0$, then $f_{(i)}$ belongs to $\mathfrak{m}^i$ but not to $\mathfrak{m}^{i+1}$. In particular f belongs to $\mathfrak{m}^r$ but not to $\mathfrak{m}^{r+1}$. Consequently

$$\cap_{r\geq 1} \mathfrak{m}^r = (0). \tag{1.1.1}$$

Finally, if $F = (F_1, \ldots, F_n) \in R[[X]]^n$, with $F(0) = 0$, i.e. $F_i(0) = 0$ for all i and $G = (G_1, \ldots, G_n) \in R[[X]]^n$, then we define the composition $G \circ F$ to be the n-tuple $(G_1(F_1, \ldots, F_n), \ldots, G_n(F_1, \ldots, F_n))$.

Now we are able to formulate the main result of this section.

Theorem 1.1.2 (Formal Inverse Function Theorem). *Let $F = (F_1, \ldots, F_n) \in R[[X]]^n$ with $F(0) = 0$ and $\det JF(0) \in R^*$. Then there exists $G \in R[[X]]^n$ such that $G(0) = 0$ and $G \circ F = X$. Furthermore, such a G is uniquely determined and it also satisfies $F \circ G = X$.*

The proof is based on

Lemma 1.1.3 *Let $F = (F_1, \ldots, F_n) \in R[[X]]^n$ with $F(0) = 0$ and $\det JF(0) \in R^*$. Then there exists $g \in R[[X]]$ with $g(F_1, \ldots, F_n) = X_1$.*

Proof: Composing F by the inverse of the linear part (which exists because $\det JF(0) \in R^*$), we may assume that $F_i \equiv X_i \pmod{\mathfrak{m}^2}$ for all i. Put $h_1 := X_1$. Then

$$h_1(F_1, \ldots, F_n) = F_1 \equiv X_1 + h_2 \pmod{\mathfrak{m}^3}$$

where h_2 is a polynomial homogeneous of degree 2 or $h_2 = 0$. Now observe that $h_2 \equiv h_2(F_1, \ldots, F_n) \pmod{\mathfrak{m}^3}$ since each $F_i \equiv X_i \pmod{\mathfrak{m}^2}$. So

$$h_1(F_1, \ldots, F_n) - h_2(F_1, \ldots, F_n) \equiv X_1 \pmod{\mathfrak{m}^3}.$$

Write

$$h_1(F_1, \ldots, F_n) - h_2(F_1, \ldots, F_n) \equiv X_1 + h_3 \pmod{\mathfrak{m}^4}$$

where h_3 is zero or a homogeneous polynomial of degree 3, and observe again that $h_3 \equiv h_3(F_1, \ldots, F_n) \pmod{\mathfrak{m}^4}$. Continuing in this way we get polynomials $h_1, h_2, \ldots$, where h_i is homogeneous of degree i or zero, such that

$$(h_1 - h_2 - \cdots - h_r)(F_1, \ldots, F_n) \equiv X_1 \pmod{\mathfrak{m}^{r+1}} \tag{1.1.4}$$

for all $r \geq 2$. Now define

$$g := h_1 - \sum_{i=2}^{\infty} h_i \in R[[X]].$$

Since obviously $h_i(F_1, \ldots, F_n) \equiv 0 \pmod{\mathfrak{m}^{r+1}}$ if $i \geq r + 1$, it follows from 1.1.4 that

$$g(F_1, \ldots, F_n) \equiv X_1 \pmod{\mathfrak{m}^{r+1}}, \text{ for all } r \geq 2.$$

So

$$g(F_1, \ldots, F_n) - X_1 \in \cap \mathfrak{m}^{r+1} = (0) \text{ by } 1.1.1.$$

Whence $g(F_1, \ldots, F_n) = X_1$. □

Proof of Theorem 1.1.2: As before we may assume that $F_i \equiv X_i \pmod{\mathfrak{m}}$ for all i. By 1.1.3 there exists $G = (G_1, \ldots, G_n)$ with $G \circ F = X$ and $G_i \equiv X_i \pmod{\mathfrak{m}^2}$ for all i. Now apply 1.1.3 to the n-tuple $G = (G_1, \ldots, G_n)$. So there exists $F' \in R[[X]]^n$ with $F' \circ G = X$. From $G \circ F = X = F' \circ G$ it follows that $F' = F$. So G is the left and right inverse of F, hence it is uniquely determined. □

In the remainder of this section we give several applications of 1.1.2. Therefore we first extend the notion of invertibility defined in the introduction to polynomial maps over arbitrary commutative rings. A **polynomial map** over R is by definition an element of $R[X]^n$.

Definition 1.1.5 Let $F = (F_1, \ldots, F_n) \in R[X]^n$. We say that F is **invertible over** R or that $F_1, \ldots, F_n$ is a **coordinate system of** $R[X]$ if $R[F_1, \ldots, F_n] = R[X_1, \ldots, X_n]$, in other words if there exist $G_1, \ldots, G_n \in R[X]$ such that $X_i = G_i(F_1, \ldots, F_n)$ for each i.

Again, as in the introduction, we get a one-to-one correspondence between the $F \in R[X]^n$ which are invertible over R and the R-automorphisms of $R[X]$ given by

$$F \longrightarrow F^* : g \rightarrow g(F).$$

Proposition 1.1.6 *If $F \in R[X]^n$ is invertible, then $G = (G_1, \ldots, G_n)$ as in* 1.1.5 *is uniquely determined and satisfies $F \circ G = X$.*

Proof: Put $\tilde{F} := F - F(0)$ and $\tilde{G} := G(X + F(0))$. Then $\tilde{F}$ and $\tilde{G}$ are as in 1.1.2. So $\tilde{G}$ is uniquely determined by $\tilde{F}$ and satisfies $\tilde{F} \circ \tilde{G} = X$. The desired result follows easily. □

As another application of 1.1.2 we investigate invertibility under base change. Therefore let η be the nilradical of R, i.e. the set of nilpotent elements of R and $\varphi : R \rightarrow S$ a ring homomorphism. If $F \in R[X]^n$, then F^φ denotes the element of $S[X]^n$ obtained by applying φ to the coefficients of the F_i.

Proposition 1.1.7 (Invertibility under base change). *Let $\varphi : R \rightarrow S$ be a ring homomorphism with* $\ker \varphi \subset \eta$. *Let $F \in R[X]^n$ with* $\det JF(0) \in R^*$. *Then F is invertible over R if and only if F^φ is invertible over S.*

Proof: If F is invertible over R, then obviously F^φ is invertible over S. The proof of the converse is a consequence of the following two lemmas. The details are left to the reader (Exercise 1). □

Lemma 1.1.8 *Let $R \subset S$ be commutative rings and $F \in R[X]^n$ such that* $\det JF(0) \in R^*$. *If F is invertible over S, then F is invertible over R.*

Proof: We may assume that $F(0) = 0$. Since $R \subset S$ we have $R[X] \subset R[[X]] \subset S[[X]]$. By 1.1.2 F has an inverse G in $R[[X]]^n$. Furthermore F has an inverse H

in $S[X]^n$. Since both G and H belong to $S[[X]]^n$ and the inverse of F considered over S is unique by 1.1.2, we get $G = H$. So this inverse belongs to $R[[X]]^n \cap S[X]^n = R[X]^n$. □

For $F \in R[X]^n$ denote by $\overline{F} \in \overline{R}[X]^n$ the element obtained by reducing the coefficients of the F_i modulo η. Here $\overline{R} := R/\eta$.

Lemma 1.1.9 *Let $F \in R[X]^n$. Then F is invertible over R if and only if $\overline{F}$ is invertible over $\overline{R}$.*

Proof: Assume that $\overline{F}$ is invertible over $\overline{R}$. Then $X_i = G_i + \sum_{j=1}^{s_i} c_{i,j}a_{i,j}(X)$ for some $G_i \in R[F], a_{i,j} \in R[X], s_i \in \mathbb{N}$ and $c_{i,j} \in \eta$. Let N be the ideal of R generated by all $c_{i,j}$. So $N^e = 0$ for some $e \in \mathbb{N}, e \geq 1$.We get

$$R[X] \subset R[G] + NR[X] \subset R[F] + NR[X]$$

whence

$$R[X] \subset R[F] + N(R[F] + NR[X]) \subset R[F] + N^2R[X].$$

Similarly

$$R[X] \subset R[F] + N^rR[X], \text{ for all } r \geq 1.$$

Since $N^e = 0$ we get $R[X] \subset R[F]$, whence $R[X] = R[F]$. □

Now we will show how 1.1.7 can be used to relate Keller's original problem to the Jacobian Conjecture. Therefore we first generalize the Jacobian Conjecture. Arguing as in the introduction we get

(1.1.10) If $F \in R[X]^n$ is invertible over R, then $\det JF$ is a unit in $R[X]$.

Conversely we get

(1.1.11) **(Generalized) Jacobian Conjecture.** $JC(R, n)$. Let R be a commutative ring contained in a $\mathbb{Q}$-algebra and $F \in R[X]^n$. If $\det JF \in R[X]^*$, then F is invertible over R.

So Keller's problem is the special case $R = \mathbb{Z}$ and we get the Jacobian Conjecture by taking for R a field of characteristic zero. Now the surprising point is that the Jacobian Conjecture over $\mathbb{C}$ implies the Generalized Jacobian Conjecture, i.e.

Proposition 1.1.12 *Let $n \geq 1$. If $JC(\mathbb{C}, n)$ is true, then $JC(R, n)$ is true for any commutative ring R which is contained in a $\mathbb{Q}$-algebra.*

The proof of 1.1.12 uses a very useful technique to reduce problems over domains of characteristic zero to problems over $\mathbb{C}$. This technique is called the **"Lefschetz-principle"**. The basic tool is the following lemma, whose proof is left to the reader (Exercise 2).

Lemma 1.1.13 (Embedding lemma). *Let $\mathbb{Q} \subset \mathbb{Q}(a_1, \ldots, a_n)$ be a finitely generated field extension. Then there exists an isomorphism $\varphi : \mathbb{Q}(a_1, \ldots, a_n) \simeq L \subset \mathbb{C}$, where L is a subfield of $\mathbb{C}$.*

To demonstrate the "Lefschetz-principle" we show how $JC(\mathbb{C}, n)$ implies $JC(R, n)$ for all domains of characteristic zero.

Lemma 1.1.14 *Let $n \geq 1$. If $JC(\mathbb{C}, n)$ is true, then $JC(R, n)$ is true for any domain R of characteristic zero.*

Proof: Let R be a domain of characteristic zero and $F \in R[X]^n$ with $\det JF \in R[X]^*$. Since R is a domain we have $R[X]^* = R^*$. So $d := \det JF \in R^*$ and hence $d^{-1} \in R$. Let R_0 be the subring of R generated over $\mathbb{Z}$ by d^{-1} and the coefficients of F. By 1.1.13 we get an embedding $\varphi : R_0 \to \mathbb{C}$. Since $\det JF \in R_0^*$ we get $\det JF^\varphi \in \mathbb{C}^*$. Hence F^φ is invertible over $\mathbb{C}$ since we assume that $JC(\mathbb{C}, n)$ is true! So by 1.1.7 F is invertible over R_0 and hence over R. □

Proof of Proposition 1.1.12: i) First we show that it suffices to prove $JC(S, n)$ for all $\mathbb{Q}$-algebras S: namely if $R \subset S$, where S is a $\mathbb{Q}$-algebra and $F \in R[X]^n$ with $\det JF \in R[X]^*$, then obviously $\det JF \in S[X]^*$ and hence, assuming $JC(S, n)$, F is invertible over S. Since $\det JF(0) \in R^*$ (for $\det JF \in R[X]^*$) it follows from 1.1.8 that F is invertible over R.

ii) So by i) we may assume that R is a $\mathbb{Q}$-algebra. Now we will show that we may also assume that R is Noetherian: for if $F \in R[X]^n$ with $\det JF \in R[X]^*$ let R_0 be the $\mathbb{Q}$-subalgebra of R generated by the coefficients appearing in F and in $(\det JF)^{-1} (\in R[X])$. Then R_0 is a finitely generated $\mathbb{Q}$-algebra, hence Noetherian. Obviously $F \in R_0[X]^n$ and $\det JF \in R_0[X]^*$. So, assuming $JC(R_0, n)$ since R_0 is Noetherian, we get that F is invertible over R_0 and hence over R.

iii) So we may assume that R is a Noetherian $\mathbb{Q}$-algebra. Let $F \in R[X]^n$ with $\det JF \in R[X]^*$ and let $\wp$ be a prime ideal of R. Then $\overline{R} := R/\wp$ is a domain of characteristic zero (since it is a $\mathbb{Q}$-algebra) and obviously $\det J\overline{F} \in \overline{R}[X]^*$. So by 1.1.14 $\overline{F}$ is invertible over $\overline{R}$, i.e. $\overline{R}[X_1, \ldots, X_n] = \overline{R}[\overline{F}_1, \ldots, \overline{F}_n]$. So

(1.1.15) $$R[X] \subset \wp R[X] + R[F], \text{ for every prime ideal of } R.$$

Since R is Noetherian there are only finitely many minimal prime ideals of R, say $\wp_1, \ldots, \wp_r$, and we have $\eta = \wp_1 \cap \cdots \cap \wp_r$. From 1.1.15 we deduce that

$$\begin{aligned} R[X] \subset \wp_1 R[X] + R[F] \subset \wp_1(&\wp_2 R[X] \\ &+ R[F]) + R[F] \subset \wp_1\wp_2 R[X] + R[F]. \end{aligned}$$

Continuing in this way we get

(1.1.16) $$R[X] \subset \wp_1 \ldots \wp_r R[X] + R[F] \subset \eta R[X] + R[F].$$

So if $\varphi : R \to R/\eta$ is the canonical map, then it follows from 1.1.16 that F^φ is invertible over R/η. Then the desired result follows from 1.1.7.

(1.1.17) **Warning.** If R is not contained in a $\mathbb{Q}$-algebra, the condition $\det JF \in R[X]^*$ need not imply that F is invertible over R. This can be seen from the following example: take $n = 1$, $R = \mathbb{Z}[Y]/(2Y)$ and $F(X) = X - aX^2$, where $a = \overline{Y}$. Then

$F'(X) = 1$ (since $2a = 0$) and the uniquely determined formal inverse of F is given by the formula

$$G(X) = X + \sum_{k=1}^{\infty} a^{2^k-1} X^{2^k}.$$

Since $a^m \neq 0$ for all $m \in \mathbb{N}$ we see that $G(X) \notin R[X]$, so F is not invertible over R!

Now we will show that the strongest possible converse of 1.1.12 is true, i.e.

Theorem 1.1.18 *If there exists a commutative ring R contained in a $\mathbb{Q}$-algebra such that $JC(R, n)$ is true for all $n \geq 1$, then $JC(\mathbb{C}, n)$ is true for all $n \geq 1$ and hence $JC(R', n)$ is true for all n and all commutative rings R' contained in a $\mathbb{Q}$-algebra.*

Proof: Observe that R has no $\mathbb{Z}$-torsion since it is contained in a $\mathbb{Q}$-algebra. So R contains $\mathbb{Z}$. Suppose now that there exists some n such that $JC(\mathbb{C}, n)$ is not true. Then by 1.1.19 there exists an $N(> n)$ variable counterexample with integer coefficients and Jacobian determinant 1. Since $\mathbb{Z} \subset R$ we have $F \in R[X]^N$. From $\det JF = 1$, it then follows by 1.1.8 that F is invertible over $\mathbb{Z}$, hence over $\mathbb{C}$, a contradiction. □

Proposition 1.1.19 (Connell, van den Dries). *If for some $n \geq 2$ there is a counterexample to $JC(\mathbb{C}, n)$, then for some $N > n$ there is a counterexample $F : \mathbb{C}^N \to \mathbb{C}^N$ with* det $JF = 1$ *and integer coefficients.*

To prove this result we develop two fundamental tools which play a crucial role in the study of polynomial maps.

(1.1.20) **The clearing map**

Let R be a commutative ring and $F \in R[[X]]^n$ with $F(0) = 0$. Let $F = F_{(1)} + F_{(2)} + \cdots$ be its homogeneous decomposition and T a new variable. The **clearing map** associated to F is by definition

$${}^T F := F_{(1)} + T F_{(2)} + T^2 F_{(3)} + \cdots \in R[T][[X]]^n.$$

Viewing everything inside $R[T, T^{-1}][[X]]^n$ one readily verifies

(1.1.21) $${}^T F = T^{-1} F(TX).$$

Finally for each $r \in R$ we put

$${}^r F := ({}^T F)_{|T=r},$$

i.e. the element of $R[[X]]^n$ obtained from ${}^T F$ by making the substitution $T := r$. The basic properties of the clearing map are given in the next lemma, whose proof is left to the reader (Exercise 7).

Lemma 1.1.22 i) *If $F, G \in R[[X]]^n$ with $F(0) = G(0) = 0$, then ${}^T(F \circ G) = {}^T F \circ {}^T G$. In particular $F \circ G = X$ if and only if ${}^T F \circ {}^T G = X$.*

ii) $\det J_X({}^T F) = \det(J_X F)(TX)$.

iii) *Assume that R is a domain, $0 \neq r \in R$ and $F \in R[X]^n$. Then*

a) *F is invertible over R if and only if ${}^rF$ is invertible over R.*
b) $\det JF \in R^*$ *if and only if* $\det J({}^rF) \in R^*$.

The name "clearing map" is explained by the following result.

Proposition 1.1.23 (Clearing denominators). *Let R be a domain, K its quotient field and $F \in K[X]^n$ with $F(0) = 0$, $F_{(1)} \in R[X]^n$ and $\det JF \in R^*$. Then there exists $0 \neq r \in R$ with ${}^rF \in R[X]^n$ and $\det J({}^rF)$ $(= \det JF) \in R^*$.*

Proof: Choose $0 \neq r \in R$ such that $rF_{(i)} \in R[X]^n$ for all $i \geq 2$. Then ${}^rF \in R[X]^n$ and from 1.1.22 ii) we get $\det J({}^rF) = \det JF(rX) = (\det JF)(rX) \in R^*$, since $\det JF \in R^*$. □

(1.1.24) **A Galois descent**

Let k be an infinite field, $k \subset K$ a finite Galois extension, $a_1, \ldots, a_m$ a k-basis of K and denote by $\sigma_0, \ldots, \sigma_{m-1}$ the k-automorphisms of K. Let $F = (F_1, \ldots, F_n) : K^n \to K^n$ be a polynomial map. Denote by $\alpha : k^m \to K$ the map defined by $\alpha(y_1, \ldots, y_m) = y_1a_1 + \cdots + y_m a_m$. The obvious extension $(\alpha, \ldots, \alpha) : (k^m)^n \to K^n$, which we also denote by α, is clearly bijective. So, conjugating F with α we get the map $F^\alpha := \alpha^{-1}F\alpha : k^{mn} \to k^{mn}$. Since k is infinite this map is a polynomial map given by mn polynomials $G_{1,1}, \ldots, G_{m,1}, \ldots, G_{1,n}, \ldots, G_{m,n} \in k[Y_1, \ldots, Y_m]$. Obviously

(1.1.25) $F : K^n \to K^n$ is injective if and only if $F^\alpha : k^{mn} \to k^{mn}$ is injective.

Our main aim will be to show that

(1.1.26) $\det JF \in K^*$ if and only if $\det JF^\alpha \in k^*$.

To prove this result we need some notation: for each $p = \sum c_{s_1,\ldots,s_n} X_1^{s_1} \ldots X_n^{s_n} \in K[X_1, \ldots, X_n]$ we define

$$p^{\sigma_i} := \sum \sigma_i(c_{s_1,\ldots,s_n}) {X_{in+1}}^{s_1} \ldots {X_{in+n}}^{s_n} \in K[X_{in+1}, \ldots, X_{in+n}].$$

Furthermore put $F^{\sigma_i} := (F_1^{\sigma_i}, \ldots, F_n^{\sigma_i})$. This gives a polynomial map

$$\overline{F} := (F^{\sigma_0}, F^{\sigma_1}, \ldots, F^{\sigma_{m-1}}) : K^{mn} \to K^{mn}$$

satisfying

(1.1.27) $$\det J\overline{F} = \det JF^{\sigma_0} \ldots \det JF^{\sigma_{m-1}}.$$

Since by Dedekind's lemma ([280], Lemma 9.1) the σ_i are linearly independent over K, it follows (Exercise 8) that

(1.1.28) $$\det \sigma_i(a_j) \in K^*$$

and hence that $L : K^{mn} \to K^{mn}$ defined by

$$L_{in+j} := \sigma_i(a_1)Y_{1,j} + \cdots + \sigma_i(a_m)Y_{m,j}$$

for all $0 \le i \le m-1$, all $1 \le j \le n$ is an invertible linear map. Furthermore, since $k \subset K$ we can view each $G_{i,j}$ as a polynomial in $K[Y_1, \ldots, Y_m]$ and hence we can view F^α as a polynomial map from K^{mn} to K^{mn}.

Lemma 1.1.29 $F^\alpha = L^{-1}\overline{F}L$.

Proof: Write G instead of F^α. From the definition of F^α we get

$$\begin{aligned} &F_j(a_1Y_{1,1} + \cdots + a_mY_{m,1}, \ldots, a_1Y_{1,n} + \cdots + a_mY_{m,n}) \\ &= a_1G_{1,j} + \cdots + a_mG_{m,j} \end{aligned}$$

where $G = (G_{1,1}, \ldots, G_{m,1}, \ldots, G_{1,n}, \ldots, G_{m,n}) \in k[Y]^{mn}$. Applying σ_i gives

$$F_j^{\sigma_i}(L) = \sigma_i(a_1)G_{1,j} + \cdots + \sigma_i(a_m)G_{m,j}$$

so $(\overline{F} \circ L)_{in+j} = (L \circ G)_{in+j}$ for all i, j, which implies the desired equality. □

As an immediate consequence of 1.1.27 and 1.1.29 we get

Proposition 1.1.30 *Notation as above. Then* $\det JF \in K^*$ *if and only if* $\det JF^\alpha \in k^*$ *and in this case* $\det JF^\alpha = \Pi_i\sigma_i(\det JF)$.

Now we are able to give the proof of 1.1.19. We will use that an injective polynomial map from $\mathbb{C}^n$ to $\mathbb{C}^n$ is automatically invertible. This result will be proved in Chapter 4.

Proof of Proposition 1.1.19: i) Suppose that $f : \mathbb{C}^n \to \mathbb{C}^n$ is a counterexample to $JC(\mathbb{C}, n)$. By the remark above f is not injective, so making a linear coordinate change we may assume that $f(1, 0, \ldots, 0) = f(0, \ldots, 0)$. Furthermore, replacing f by $f - f(0)$ and replacing f by $Jf(0)^{-1}f$ we may assume that $f = X + f_{(2)} + \cdots$, $\det Jf = 1$ and $f(1, 0, \ldots, 0) = (0, \ldots, 0)$. Replace the non-zero coefficients of the monomials of degree ≥ 2 appearing in f by new variables Y_s. Then the determinant of the Jacobian becomes 1 plus some polynomial p in $\mathbb{Z}[X, Y]$. Each monomial X^i appearing in p has some coefficient c_i in $\mathbb{Z}[Y]$. Furthermore the condition $f(1, 0, \ldots, 0) = (0, \ldots, 0)$ can also be expressed as a set of polynomial equations $d_j = 0$, where $d_j \in \mathbb{Z}[Y]$. By the properties of f there exists a solution of the system $c_i = 0, d_j = 0$ in $\mathbb{C}^M$, where M is the number of variables Y_s. Hence the ideal generated by the c_i and d_j in $\overline{\mathbb{Q}}[Y]$ does not contain 1 ($\overline{\mathbb{Q}}$ is the algebraic closure of $\mathbb{Q}$ in $\mathbb{C}$). So by the Nullstellensatz the system has a solution in $\overline{\mathbb{Q}}^M$.

ii) From i) it follows that there exists a finite Galois extension $\mathbb{Q} \subset K$ and a non-injective polynomial map $F : K^n \to K^n$ of the form $F = X + F_{(2)} + \cdots$ with $\det JF = 1$. By 1.1.30 and 1.1.29 we get a non-injective map $F^\alpha : \mathbb{Q}^{mn} \to \mathbb{Q}^{mn}$ with $F^\alpha(0) = 0$, $JF^\alpha(0) = I$ and $\det JF^\alpha = 1$ ($m := |K : \mathbb{Q}|$). Then the desired result follows from 1.1.23, observing that the map obtained after clearing denominators is still non-injective (use 1.1.21).

The next application of the formal inverse function theorem concerns polynomial maps over a field. So let k be an *arbitrary* field and $F = (F_1, \dots, F_n) \in k[X]^n$. By $\bar{k}$ we denote an algebraic closure of k.

Proposition 1.1.31 *If* $\det JF \neq 0$ *in* $k[X]$, *then* $F_1, \dots, F_n$ *are algebraically independent over* k *and* $k(F) \subset k(X)$ *is a finite field extension.*

Proof: i) We will show that $F_1, \dots, F_n$ are algebraically independent over $\bar{k}$ and hence over k. Therefore we view F_i in $\bar{k}[X]$. So we may assume that $k = \bar{k}$. Hence there exists $x \in k^n$ with $\det JF(x) \neq 0$. By replacing F by $F(X+x) - F(x)$ we may assume that $F(0) = 0$ and $\det JF(0) \neq 0$. By the formal inverse function theorem there exists $G \in k[[X]]^n$ with $F_i(G) = X_i$ for all i. Now suppose that $h \in k[X]$ satisfies $h(F_1, \dots, F_n) = 0$. Then $0 = h(F_1(G), \dots, F_n(G)) = h(X_1, \dots, X_n) = h$. So $F_1, \dots, F_n$ are algebraically independent over k.

ii) From i) we get $\operatorname{trdeg}_k k(F) = n = \operatorname{trdeg}_k k(X)$. So $\operatorname{trdeg}_{k(F)} k(X) = 0$, i.e. $k(X)$ is algebraic over $k(F)$, whence $k(X)$ is finite over $k(F)$. □

Theorem 1.1.32 (Adjamagbo, Derksen, van den Essen). *Let k be any field and* $F \in k[X]^n$. *If* $\det JF(x) \neq 0$ *for all* $x \in k^n$, *then the cardinality of the fibers of F is bounded by* $|k(X) : k(F)|$, *the degree of the field extension* $k(F) \subset k(X)$.

Proof: i) It suffices to show that $\#F^{-1}(0) \leq |k(X) : k(F)|$, for if $y \in k^n$ define $\tilde{F} := F - y$ and observe that $\tilde{F}^{-1}(0) = F^{-1}(y)$, $J\tilde{F} = JF$ and $k(\tilde{F}) = k(F)$.

ii) Let $a \in F^{-1}(0)$. Put $F_\star(X) := F(X + a)$. Then $F_\star(0) = F(a)$ and $\det JF_\star(0) = \det JF(a) \neq 0$. So by the formal inverse function theorem there exists $G_a \in k[[Y]]^n$ with $G_a(0) = 0$ and $F_\star(G_a(Y)) = Y$, i.e. $F_i(a + G_a) = Y_i$, for all i. Since $\det JF \neq 0$, it follows from 1.1.31 that $k[F]$ is a polynomial ring in n variables over k, so we get a ring homomorphism $\psi : k[F] \to k[[Y]]$ by sending F_i to Y_i for each i. Finally denote by ϕ_a the ring homomorphism from $k[X]$ to $k[[Y]]$ sending X_i to $(a + G_a)_i$, the i-th component of $a + G_a$. Observe that ϕ_a is injective, for if $h(X) \in k[X]$ satisfies $h(a + G_a) = 0$, then $h((a + G_a)(F_\star)) = 0$, so $h(a + X) = 0$, i.e. $h(X) = 0$ and that ϕ_a is an extension of ψ, since $F_i(a + G_a) = Y_i$, for all i.

iii) Put $S := k[F]\backslash\{0\}$. Then $S^{-1}k[F] = k(F)$. Furthermore since $k[X] = k[F][X]$ and $k(F) \subset k(X)$ is a finite field extension (by 1.1.31) we get that $S^{-1}k[X] = k(F)[X]$ is a field, hence equal to $k(X)$. Consequently localizing with respect to S we get embeddings $\psi : k(F) \to k((Y))$ and $\phi_a : k(X) \to k((Y))$, where again ϕ_a is an extension of ψ. Since $a + G_a = a' + G_{a'}$ if and only if $a = a'$ (for $G_a(0) = G_{a'}(0)$), we get $N := \#F^{-1}(0)$ different embeddings of $k(X)$ which extend ψ. So $N \leq |k(X) : k(F)|$, as desired. □

Corollary 1.1.33 *If* $\det JF(x) \neq 0$ *for all* $x \in k^n$ *and* $k(X) = k(F)$, *then F is injective.*

Corollary 1.1.34 (Flatness of Keller maps). *Let k be any field and* $F \in k[X]^n$. *If* $\det JF \in k^*$ *then*

i) $k[F] \subset k[X]$ *is flat.*

ii) $k(F) \cap k[X] = k[F]$.

Proof: i) We view F in $\bar{k}[X]^n$. Obviously $\det JF \in \bar{k}^*$. So $\det JF(x) \neq 0$ for all $x \in \bar{k}^n$, whence $\#F^{-1}(x) \leq |\bar{k}(X) : \bar{k}(F)|$ for all $x \in \bar{k}^n$. Then by D.2.1 it follows that $F : \bar{k}^n \to \bar{k}^n$ is flat, i.e. $\bar{k}[F] \subset \bar{k}[X]$ is flat. Since $k[F] \subset \bar{k}[F]$ is flat it follows from the transitivity of flatness, D.1.3 ii), that $k[F] \subset \bar{k}[X]$ is flat. Since $k[X] \subset \bar{k}[X]$ is faithfully flat it follows from D.1.3 iii) that $k[F] \subset k[X]$ is flat.

ii) The second statement follows readily D.1.7. □

Corollary 1.1.35 (Keller's theorem). *Let R be a domain with quotient field k and $F \in R[X]^n$. If $\det JF \in R^*$ and $k(F) = k(X)$, then $R[F] = R[X]$, i.e. F is invertible over R.*

Proof: By 1.1.34 we get

$$k[F] = k(F) \cap k[X] = k(X) \cap k[X] = k[X].$$

So F is invertible over k. Then apply 1.1.8. □

(1.1.36) **Warning.** If k is not algebraically closed, then the condition $\det JF(x) \neq 0$ for all $x \in k^n$ need not imply that $k[F] \subset k[X]$ is flat. To see this take $k = \mathbb{R}$ and $F : \mathbb{R}^3 \to \mathbb{R}^3$ given by

$$F(X_1, X_2, X_3) = (X_1, X_2 - X_1X_3, X_3 + X_1X_2).$$

Then $\det JF = 1 + X_1^2$, so $\det JF(x) \neq 0$ for all $x \in \mathbb{R}^3$. Furthermore F has a rational inverse G given by

$$G(X_1, X_2, X_3) = (X_1, (X_2 + X_1X_3)/(1 + X_1^2), (X_3 - X_1X_2)/(1 + X_1^2)).$$

So $\mathbb{R}(X) = \mathbb{R}(F)$. Consequently, if $\mathbb{R}[F] \subset \mathbb{R}[X]$ is flat, then it follows from D.1.7 that $\mathbb{R}[F] = \mathbb{R}(F) \cap \mathbb{R}[X] = \mathbb{R}(X) \cap \mathbb{R}[X]$, which means that F has a polynomial inverse, a contradiction. So $\mathbb{R}[F] \subset \mathbb{R}[X]$ is not flat.

Exercises for §1.1

1. Supply the details of the proof of 1.1.7. [Decompose φ into the maps $\varphi : R \to \varphi(R)$ and the inclusion map $i : \varphi(R) \to S$ and consider the map $\varphi(R) \simeq R/\ker\varphi \to R/\eta$.]
2. Prove the embedding Lemma 1.1.13. [If the extension $\mathbb{Q} \subset \mathbb{Q}(a_1, \dots, a_n)$ is algebraic, use the primitive element theorem; if $\operatorname{trdeg}_{\mathbb{Q}} \mathbb{Q}(a_1, \dots, a_n) \geq 1$, split the extension into a transcendental and an algebraic one and use the fact that $\mathbb{C}$ has infinite transcendence degree over $\mathbb{Q}$.]
3. Let $R[T]$ be a polynomial ring in one variable over a ring R. Consider the following statement:

   $AM(R)$: if $f(T), g(T) \in R[T]$ satisfy $R[f(T), g(T)] = R[T]$, then $\deg f(T)$ divides $\deg g(T)$ or $\deg g(T)$ divides $\deg f(T)$.

   Show that if $AM(\mathbb{C})$ is true, then $AM(R)$ is true for any domain R of characteristic zero. [Remark: $AM(\mathbb{C})$ is the so-called **Abhyankar-Moh theorem**, see Chapter 5, §4.]

4. Give a direct proof of the first part of 1.1.31. [Suppose $h(F_1, \ldots, F_n) = 0$ and use the chain rule.]
5. Give an example of a polynomial map $F : \mathbb{R} \to \mathbb{R}$ with $\det JF(x) \neq 0$ for all $x \in \mathbb{R}$ and such that $\#F^{-1}(x) < |\mathbb{R}(X) : \mathbb{R}(F)|$ for all $x \in \mathbb{R}$.

   Give an example of a polynomial map $F : \mathbb{F}_p \to \mathbb{F}_p$ with $\det JF = 1$ and such that $\#F^{-1}(x) = |\mathbb{F}_p(X) : \mathbb{F}_p(F)| > 1$, for some $x \in \mathbb{F}_p$.
6. (A special case of the Jacobian Conjecture). Let $F = X + H \in R[X]^n$, where each component H_i of H is either zero or homogeneous of the same degree $d \geq 2$. Suppose that $(JH)^2 = 0$.

   (i) Prove that $\det JF \in R[X]^*$.

   (ii) Prove that F is invertible with inverse $G = X - H$.

   [Denote by X^t the transpose of the row matrix X. Let E be the Euler operator $\sum_{i=1}^n X_i \frac{\partial}{\partial X_i}$. Put $N := JH$. Show that $H^t = d^{-1}N.X^t$ and deduce

   (1) $F^t = (I + d^{-1}N)X^t$.
   Let G be the formal inverse of F, i.e.

   (2) $F \circ G = X$.
   Deduce from (1) and (2) that

   (3) $G^t = (I - d^{-1}N(G))X^t$.
   Apply E to (2) and deduce that

   (4) $E(G)^t = (I - N(G))X^t$.
   Deduce from (3) and (4) that $E(G) - dG = (1-d)X$. Finally, writing $G = X + \sum_{i \geq 2} G_i$ its homogeneous decomposition, show that $G = X + G_d$ and conclude that $G_d = -H$.]
7. Prove (1.1.22). [To prove i) and ii) use 1.1.21. For iii) a): if ${}^rF$ is invertible over R then $\det JF(0) \in R^*$, so by the formal inverse function theorem F has a formal inverse G. So by i) ${}^rF \circ {}^rG = X$. Since ${}^rF$ is invertible the unique formal inverse ${}^rG$ is polynomial, hence so is G.]
8. Prove (1.1.28). [$\det \sigma_i(a_j) = 0$ if and only if the rows of the matrix are linearly dependent over K. Since the elements $a_1, \ldots, a_m$ form a k-basis of K this is equivalent to the existence of $c_1, \ldots, c_m \in K$, not all zero such that $c_1\sigma_1 + \cdots + c_m\sigma_m = 0$.]
9. (Formanek) Let $F = (F_1, \ldots, F_n) \in k[X_1, \ldots, X_n]^n$ with $\det JF \in k^*$ and assume that $\operatorname{char} k = 0$. Suppose there exists $F_{n+1} \in k[X_1, \ldots, X_n]$ such that $k[F_1, \ldots, F_{n+1}] = k[X_1, \ldots, X_n]$. We prove that F is invertible over k.

   (i) Define $\alpha: k[U_1, \ldots, U_{n+1}] \to k[X_1, \ldots, X_n]$ by $\alpha(U_i) = F_i$, for all i. Show that α is surjective and $\ker \alpha = (f)$ for some f with $\deg_{U_{n+1}} f \geq 1$.

   [Use 1.1.31].

   (ii) Choose $T_1, \ldots, T_n$ such that $\alpha(T_i) = X_i$. Show that there exist $R_1, \ldots, R_{n+1}$ in $k[U_1, \ldots, U_{n+1}]$ such that $U_i = F_i(T_1, \ldots, T_n) + R_i f$, for all $1 \leq i \leq n+1$.

(iii) Differentiate the equations in ii) with respect to $U_1, \ldots, U_{n+1}$ and deduce that

$$I_{n+1} = \begin{pmatrix} JF & & * \\ & & \vdots \\ & & * \\ \ldots & * & * \end{pmatrix} \begin{pmatrix} & & * \\ & * & \vdots \\ & & * \\ \ldots & * & c \end{pmatrix}, \text{ where } c := \alpha\left(\frac{\partial f}{\partial U_{n+1}}\right).$$

(iv) Deduce from iii) that $c \in k^*$. [Use Cramers rule]

(v) Deduce that $F_{n+1} = -c^{-1} f_0(F_1, \ldots, F_n)$ and hence that F is invertible over k. ($f_0 = f(U_{n+1} = 0)$).

1.2 Derivations

One of the main tools in the study of polynomial automorphisms is the concept of a derivation (see Chapter 2 for some applications). In this section we recall some of the basic properties of derivations on a ring (see also [190], [203] and [227] for more details) and give some fundamental results concerning derivations on polynomial rings over a field.

Let A be any ring (for the moment A need not be commutative) and R a commutative ring.

Definition 1.2.1 A derivation on A is an additive map $D : A \to A$ satisfying Leibniz' rule, i.e.

$$D(a+b) = D(a) + D(b) \text{ and } D(ab) = aD(b) + D(a)b, \text{ for all } a, b \in A.$$

Example 1.2.2 Let $A := R[X] = R[X_1, \ldots, X_n]$. Then the usual partial derivative $\frac{\partial}{\partial X_i}$ is a derivation on A. From Leibniz's rule we obtain (Exercise 2)

$$D^n(ab) = \sum_{i=0}^{n} \binom{n}{i} D^i(a) D^{n-i}(b), \text{ for all } a, b \in A, \text{ all } n \geq 1. \tag{1.2.3}$$

The set of all derivations on A we denote by Der A. Furthermore if A is an R-algebra via a ring homomorphism $f : R \to A$, we say that D is an **R-derivation** if $D \circ f = 0$. The set of all R-derivations on A we denote by $\mathrm{Der}_R A$. Since $1.1 = 1$ we get $D(1) = D(1) + D(1)$, so $D(1) = 0$ for all $D \in \mathrm{Der}\, A$. Hence, viewing A as a $\mathbb{Z}$-algebra we get $\mathrm{Der}\, A = \mathrm{Der}_{\mathbb{Z}} A$. If D and D' belong to $\mathrm{Der}_R A$ then one easily verifies that the **(Lie) bracket**

$$[D, D'] := DD' - D'D$$

also belongs to $\mathrm{Der}_R A$, and in fact with this bracket $\mathrm{Der}_R A$ forms a **Lie algebra** (Exercise 4). Furthermore, if A is commutative $\mathrm{Der}_R A$ is a left A-module with the obvious addition and multiplication by elements of A. The proof of 1.2.4 below is left to the reader (Exercise 5):

Lemma 1.2.4 *Let G be a generating set for the R-algebra A and $D \in \mathrm{Der}_R\, A$. Then D is completely determined by the elements $D(g)$, $g \in G$.*

Proposition 1.2.5 *Notations as in* 1.2.2. *Then* $\mathrm{Der}_R\, R[X]$ *is a free $R[X]$-module with basis $\partial_1, \ldots, \partial_n$ and $[\partial_i, \partial_j] = 0$ for all i, j. Here ∂_i denotes the partial derivative with respect to X_i.*

Proof: i) Let $D \in \mathrm{Der}_R\, R[X]$. Put $D_0 := D - \sum D(X_i)\partial_i$. Then the derivation D_0 is zero on each R-generator X_i, so $D_0 = 0$ by 1.2.4, i.e. $D = \sum D(X_i)\partial_i$.

ii) If $\sum a_i \partial_i = 0$ for some $a_i \in R[X]$, then applying this derivation to X_i gives $a_i = 0$.

iii) Put $D_0 := [\partial_i, \partial_j]$. Then this derivation is zero on each X_i, hence $D_0 = 0$, as desired. □

Remark 1.2.6 We will show in Chapter 2 that for any $\mathbb{Q}$-algebra R the Jacobian Conjecture $JC(R, n)$ is equivalent to the statement that the n-tuple $\partial_1, \ldots, \partial_n$ is essentially the only commutative $R[X]$-basis of $\mathrm{Der}_R\, R[X]$.

Proposition 1.2.7 (Extension of derivations). *Let A be a commutative ring and $D \in \mathrm{Der}\, A$.*

i) *Let $S \subset A$ be a multiplicatively closed subset. Then D induces a derivation $\tilde{D}$ on $S^{-1}A$. If A has no S-torsion then $\tilde{D}$ is the unique extension of D to $S^{-1}A$. In particular every derivation on a domain has a unique extension to its quotient field.*

ii) *Let I be a proper ideal of A (i.e. $I \neq A$) and $\hat{A}$ the I-adic completion of A. Then D induces a derivation $\hat{D}$ on $\hat{A}$. If the I-adic topology is Hausdorff, i.e. $\cap I^n = (0)$, then $\hat{D}$ is the unique extension of D to $\hat{A}$. In particular if A is a Noetherian domain, D has a unique extension to its completion $\hat{A}$.*

iii) *Let $K \subset L$ be a separable algebraic field extension and $D \in \mathrm{Der}\, K$. Then D can be extended uniquely to a derivation $\tilde{D}$ on L.*

For the proof we refer to Exercise 6 (or [190] or [227]).

An immediate consequence of 1.2.7 is

Corollary 1.2.8 *Let $K \subset L$ be a separable field extension and $D \in \mathrm{Der}_K\, L$. If $x \in L$ is algebraic over K then $D(x) = 0$.*

Proof: Apply 1.2.7 iii) to the field extension $K \subset K(x)$. □

Using 1.2.8 several times we will deduce the useful.

Proposition 1.2.9 *Let k be a field of characteristic zero and $F_1, \ldots, F_n$ elements of $k[X] := k[X_1, \ldots, X_n]$. Then* $\mathrm{rk}\, JF = \mathrm{trdeg}_k\, k(F)$.

Proof: By induction on n. The case $n = 1$ is obvious, so let $n \geq 2$. Let $r := \operatorname{trdeg}_k k(F)$. If $r = 0$, then each F_i is algebraic over k and hence belongs to k. Then the result follows easily. So let $r \geq 1$.

i) The case $r = n$. So we need to show that $\det JF \neq 0$. Therefore we define a derivation D on $k[X]$ as follows

(1.2.10) $$D(g) := \det J(F_1, \ldots, F_{n-1}, g), \text{ for all } g \in k[X].$$

Obviously $D(F_i) = 0$ for all $1 \leq i \leq n-1$ and $D(F_n) = \det JF$. So we need to show that $D(F_n) \neq 0$. Therefore suppose that $D(F_n) = 0$. Then $D = 0$ on $k(F_1, \ldots, F_n)$ and hence on $k(X)$ by 1.2.8, since $r = n$ implies that $k(F) \subset k(X)$ is algebraic. So $D(X_i) = 0$ for all i, i.e. $\det J(F_1, \ldots, F_{n-1}, X_i) = 0$. So

$$\det J_{X_1,\ldots,\hat{X}_i,\ldots,X_n}(F_1, \ldots, F_{n-1}) = 0.$$

Now viewing $F_1, \ldots, F_{n-1}$ as polynomials in $k(X_i)[X_1, \ldots, \hat{X}_i, \ldots, X_n]$ it follows by the induction hypothesis that $F_1, \ldots, F_{n-1}$ are algebraically dependent over $k(X_i)$ for all i. Since $F_1, \ldots, F_n$ and hence $F_1, \ldots, F_{n-1}$ are algebraically independent over k, we deduce that each X_i is algebraic over $k(F_1, \ldots, F_{n-1})$, whence $n = \operatorname{trdeg}_k k(X) = \operatorname{trdeg}_k k(F_1, \ldots, F_{n-1}) = n-1$, a contradiction. So $D(F_n) \neq 0$ and the case $r = n$ is proved.

ii) The case $r < n$. Then we may assume that $F_1, \ldots, F_r$ are algebraically independent over k, that F_i is algebraic over $k(F_1, \ldots, F_r)$ for each $i > r$ and that $F_1, \ldots, F_r, X_{r+1}, \ldots, X_n$ is a transcendence basis of $k(X)$ over k. So by i) we deduce that $\det J(F_1, \ldots, F_r, X_{r+1}, \ldots, X_n) \neq 0$, whence $\det J_{X_1,\ldots,X_r}(F_1, \ldots, F_r)$ is not equal to zero. So the $r \times r$ minor obtained from JF by deleting the last $n - r$ rows and columns is non-zero. So $\operatorname{rk} JF \geq r$.

iii) By ii) it suffices to show that each $r+1 \times r+1$-minor of JF equals zero. Therefore let $i = (i_1, \ldots, i_{r+1})$ with $1 \leq i_1 < i_2 < \cdots < i_{r+1} \leq n$ and $j = (j_1, \ldots, j_{r+1})$ with $1 \leq j_1 < j_2 < \cdots < j_{r+1} \leq n$. Put

$$m(i, j) := \det J_{X_{j_1},\ldots,X_{j_{r+1}}}(F_{i_1}, \ldots, F_{i_{r+1}}).$$

Let K be the subfield of $k(X)$ generated over k by all X_i with $i \notin \{j_1, \ldots, j_{r+1}\}$ and view $F_{i_1}, \ldots, F_{i_{r+1}}$ as polynomials in $K[X_{j_1}, \ldots, X_{j_{r+1}}]$. Since $\operatorname{trdeg}_k k(F) = r$ these polynomials are algebraically dependent over k and hence over K. So, if $r < n-1$ then $m(i, j) = 0$ follows from the induction hypothesis. Finally, if $r = n-1$ then $m(i, j) = \det J(F_1, \ldots, F_n) = D(F_n)$ (where D is as in i)). Since $F_1, \ldots, F_{n-1}$ are algebraically independent over k, F_n is algebraic over $k(F_1, \ldots, F_{n-1})$ and so by 1.2.8 $D(F_n) = 0$, as desired. □

(1.2.11) The exponential of a derivation

In the remainder of this section A denotes a **$\mathbb{Q}$-algebra**. One of the most useful tools in the study of derivations is the so-called **exponential map** associated to a derivation: let D be a derivation on A and consider the formal power series ring $A[[T]]$ in one variable T over A. Extend D to a derivation $\tilde{D}$ on $A[[T]]$ by the formula

$$\tilde{D}\left(\sum a_i T^i\right) := \sum D(a_i) T^i.$$

We write again D instead of $\tilde{D}$. Now define

$$\exp TD : A[[T]] \to A[[T]] \text{ by the formula}$$

$$\exp TD(g) := \sum_{p=0}^{\infty} \frac{1}{p!} D^p(g) T^p, \text{ for all } g \in A[[T]].$$

One easily verifies that

$$\frac{d}{dT}(\exp TD(a)) = \exp TD(Da), \text{ for all } a \in A, \tag{1.2.12}$$

$$\exp TD(a)_{|T=0} = a, \text{ for all } a \in A. \tag{1.2.13}$$

The importance of the map $\exp TD$ comes from

Proposition 1.2.14 $\exp TD : A[[T]] \to A[[T]]$ *is a ring automorphism.*

Proof: i) For each $p \geq 0$ D^p is additive, hence so is $\exp TD$.
ii) Let $f, g \in A[[T]]$, then

$$\begin{aligned}
\exp TD(f)\exp TD(g) &= \left(\sum_{p=0}^{\infty} \frac{1}{p!} D^p(f) T^p\right)\left(\sum_{q=0}^{\infty} \frac{1}{q!} D^q(g) T^q\right) \\
&= \sum_{n=0}^{\infty}\left(\sum_{p+q=n} \frac{1}{p!q!} D^p(f) D^q(g)\right) T^n \\
&= \sum_{n=0}^{\infty} \frac{1}{n!} D^n(fg) T^n \quad \text{(by (1.2.3))} \\
&= \exp TD(fg).
\end{aligned}$$

iii) Finally one easily verifies that $\exp(-TD)\exp(TD)$ is the identity map on $A[[T]]$, which concludes the proof. □

From now on we assume that A is commutative. To give a first application of 1.2.14 we show

Proposition 1.2.15 (Seidenberg). *Let A be a Noetherian domain, K its quotient field and $\tilde{A}$ the integral closure of A in K. Let D be a derivation on A and $\tilde{D}$ its unique extension to K. Then $\tilde{D}(\tilde{A}) \subset \tilde{A}$.*

Proof: Let $x \in \tilde{A}$. Since x is integral over A, $A[x]$ is an A-module of finite type contained in K, say $A[x] = \sum_{j=1}^{N} A\frac{b_j}{a_j}$ for some $N \geq 1$, $a_j, b_j \in A$ and each $a_j \neq 0$. Put $a := a_1 \ldots a_N$. Then $a\frac{b_j}{a_j} \in A$ for all j, hence $aA[x] \subset A$. So

$$ax^i \in A, \text{ for all } i \geq 0. \tag{1.2.16}$$

Let $\phi : K \to K[[T]]$ be the restriction of $\exp T\tilde{D}$ to K. Then by 1.2.14 we get

$$\phi(a)\phi(x)^i \in A[[T]], \text{ for all } i \geq 0. \tag{1.2.17}$$

So expanding $a\phi(a)(\phi(x) - x)^n$ we see that this element belongs to the ring $A[[T]]$ since all terms in the expansion are of the form $ax^i\phi(a)\phi(x)^j$, then use 1.2.16 and 1.2.17. Since $a\phi(a) = a^2 + T(\ldots)$ and $(\phi(x) - x)^n = D(x)^n T^n + T^{n+1}(\ldots)$ we conclude that $a^2 D(x)^n \in A$, for all $n \geq 0$. So the A-module $A[D(x)]$ is contained in the finite A-module $A.\frac{1}{a^2}$, which is Noetherian since A is Noetherian. So $A[D(x)]$ is a finite A-module, which implies that $D(x) \in \tilde{A}$. □

(1.2.18) **The kernel of a derivation**

To conclude this section we collect some results concerning the kernel of a derivation on a polynomial ring over a field. The importance of the study of these kernels comes from the fact that, as we shall see later, several problems concerning polynomial automorphisms can be transformed into problems concerning kernels of certain derivations. Before we restrict our attention to derivations on polynomial rings over a field, let us first introduce some notations and formulate a rather general (easy) lemma.

Let D be a derivation on A. The kernel of $D : A \to A$ will be denoted by A^D or $\ker(D, A)$. One easily verifies that A^D is a subring of A (resp. subfield of A in case A is a field), called the **ring** (resp. **field**) **of constants of** D.

Lemma 1.2.19 *If A is a domain, then A^D is integrally closed in A.*

Proof: Let $0 \neq a \in A$ and suppose that a is integral over A^D. So there exist $n \in \mathbb{N}$ and $c_0, \ldots, c_{n-1} \in A^D$ such that

$$a^n + c_{n-1}a^{n-1} + \cdots + c_1 a + c_0 = 0. \tag{1.2.20}$$

Choose n minimal with this property and apply D to 1.2.20. Since $c_i \in A^D$ for each i we get

$$(na^{n-1} + (n-1)c_{n-1}a^{n-2} + \cdots + c_1)D(a) = 0.$$

Since A is a domain, the minimal choice of n implies that $D(a) = 0$, i.e. $a \in A^D$. □

Now we turn to derivations on polynomial rings over a field. One of the fundamental questions concerning such derivations is the following. Let D be any k-derivation on $k[X]$.

Question 1.2.21 *Is $k[X]^D$ a finitely generated k-algebra?*

This question was first answered in the negative by H. Derksen in [75]. There he relates it to the following famous finiteness problem.

Problem 1.2.22 (Hilbert's 14-th problem). Let k be any field and L a subfield of the field $k(X_1, \ldots, X_n)$ containing k. Is $L \cap k[X_1, \ldots, X_n]$ a finitely generated k-algebra?

This problem will be discussed in detail in Chapter 9, §6 (see also Exercise 13). For the moment it suffices to say that it was solved in the negative by Nagata in 1958 in case $n = 32$. Using this example Derksen constructed a derivation on $\mathbb{C}[X_1, \ldots, X_{32}]$ whose kernel equals $L \cap \mathbb{C}[X_1, \ldots, X_{32}]$, where L is the field found by Nagata. On the other hand, Zariski proved that Hilbert's 14-th problem has an affirmative answer in the following special case. We refer to [306] for a proof.

Theorem 1.2.23 (Zariski). *If* $\text{trdeg}_k\, L \leq 2$, *then the answer to Hilbert* 14 *is affirmative.*

As a consequence we get

Corollary 1.2.24 (Nagata, Nowicki). *Let k be a field of characteristic zero. If $0 \neq D$ is a k-derivation on $k[X]$, then $k[X]^D$ is a finitely generated k-algebra if $n \leq 3$.*

Proof: Let $Q(k[X]^D)$ denote the quotient field of $k[X]^D$. Then we get $\text{trdeg}_k\, Q(k[X]^D) \leq n - 1$. Namely if $\text{trdeg}_k\, Q(k[X]^D) = n$, then $Q(k[X]^D) \subset k(X)$ is separable algebraic and since D is zero on $Q(k[X]^D)$ it follows from 1.2.8 that $D = 0$ on $k[X]$, a contradiction. Since obviously

$$k[X]^D = Q(k[X]^D) \cap k[X]$$

the result follows from 1.2.23. □

If we further restrict to $n = 2$ we have a much stronger result.

Theorem 1.2.25 (Nagata, Nowicki). *Let k be a field of characteristic zero and $0 \neq D \in \text{Der}_k\, k[X_1, X_2]$. Then $k[X_1, X_2]^D = k[f]$, for some $f \in k[X_1, X_2]$.*

The proof of this result is based on the following beautiful result of Zaks. See also the notes at the end of this chapter.

Theorem 1.2.26 (Zaks). *Let k be a field, char $k = 0$, $n \geq 1$ and $k \subseteq R \subseteq k[X_1, \ldots, X_n]$ an integrally closed ring of dimension 1. Then $R = k[f]$ for some $f \in k[X]$.*

Proof: We use induction on n. i) The case $n = 1$. By Exercise 13 $R \subset Q(R) \cap k[X_1] = k[u]$ for some $u \in k[X_1]$. Write $u = r/s$ with $r, s \in R$, $s \neq 0$. Since $s \in R \subseteq k[u]$, we can write $s = p(u)$ for some $p(T) \in k[T]$. Hence $up(u) - r = 0$, which shows that u is integral over R. So $u \in R$ (R is integrally closed). Consequently, $k[u] \subset R (\subset k[u])$, whence $R = k[u]$.

ii) Now let $n \geq 2$. Choose $a \in k$ and let $\pi_a : k[X_1, \ldots, X_n] \to k[X_1, \ldots, X_{n-1}]$ be the substitution map sending X_n to a. Put $\phi_a := \pi_a|_R$. Then $M_a := \ker \phi_a = R \cap (X_n - a)$ is a prime ideal in R. If $M_a = 0$, then $R \cong \phi_a(R) \subseteq k[X_1, \ldots, X_{n-1}]$ and the result follows from the induction hypothesis. So suppose $M_a \neq 0$ for all $a \in k$. Then M_a is a maximal ideal (since dim $R = 1$). So $\phi_a(R) \cong R/M_a$ is a field contained in $k[X_1, \ldots, X_n]$. Since $k[X_1, \ldots, X_n]^* = k^*$, we get $\phi_a(R) = k$. So for each $f \in R$ there exists $c \in k$ with $f - c \in M_a \subset (X_n - a)k[X]$. So $f - c = (X_n - a)g$ for some $g \in k[X]$. In particular $\frac{\partial f}{\partial X_i} = (X_n - a)\frac{\partial g}{\partial X_i}$ for all

$1 \le i \le n-1$. So $X_n - a$ divides $\frac{\partial f}{\partial X_i}$ for all $a \in k$ and all $1 \le i \le n-1$. Hence $\frac{\partial f}{\partial X_i} = 0$ for all $1 \le i \le n-1$, i.e. $f \in k[X_n]$. Consequently $R \subset k[X_n]$ and we are done by i). □

Proof of Theorem 1.2.25: Put $R := k[X_1, X_2]^D$ and $s := \text{trdeg}_k\, Q(R)$. Then $s \le 1$. If $s = 0$ then each element of R is algebraic over k and hence $R = k$, since the only elements of $k(X_1, X_2)$ which are algebraically dependent over k are the elements of k. So $R = k[f]$ with $f = 1$ for example. Now assume $s = 1$. Then R has dimension 1. By 1.2.19 R is integrally closed in $k[X_1, X_2]$ and hence in $k(X_1, X_2)$. So R is integrally closed. Now apply 1.2.26.

Remark 1.2.27 i) Recently in his master's thesis [36] Joost Berson generalized theorem 1.2.25 to the case that k is a UFD. For a proof of this result we refer to Exercise 12.

ii) If $n \ge 3$ and $0 \ne D \in \text{Der}_k\, k[X]$ such that $k[X]^D$ is a finitely generated k-algebra, then $k[X]^D$ need not be generated by $n-1$ elements: in fact everything is possible as can be seen from the following result (see [233] for a proof).

Proposition 1.2.28 (Nowicki, Strelcyn). *Let k be a field of characteristic zero, $n \ge 3$ and $r \ge 0$. Then there exists a k-derivation D on $k[X]$ such that the minimal number of generators of $k[X]^D$ is equal to r. If $1 \le r < n$ take $D = X_{r+1}\partial_{r+1} + \cdots + X_n\partial_n$ and if $r \ge n$ take $D = X_1\partial_1 + X_2\partial_2 + (r-n+2)X_3\partial_3$.*

Remark 1.2.29 Recently it has been shown by Daigle and Freudenburg in [72] that for any integer $r \ge 3$ there exists a triangular derivation (see 1.3.17) D on $k[X_1, \ldots, X_4]$ (char $k = 0$) such that $r \le \#\ker D \le r+1$.

Exercises for §1.2

1. Let A be any ring and $a \in A$. Show that the map $b \to [a, b] := ab - ba$ is a derivation on A. Such a derivation is called an **inner derivation of** A.
2. Prove 1.2.3 [Use induction on n.]
3. Let A be commutative. Show that $D(a^n) = na^{n-1}D(a)$ for all $a \in A$, $n \ge 1$ and $D \in \text{Der}\, A$. Deduce that if A has characteristic $p > 0$, then D^p is a derivation on A.
4. Show that $\text{Der}_R\, A$ with the bracket [,] introduced in §3 is an ***R*-Lie algebra**, i.e. satisfies the rules

   (i) [,] is R-bilinear.
   (ii) $[D, D] = 0$ for all $D \in \text{Der}_R\, A$.
   (iii) $[D_1, [D_2, D_3]] + [D_2, [D_3, D_1]] + [D_3, [D_1, D_2]] = 0$ for all D_1, D_2, D_3 in $\text{Der}_R\, A$ (the Jacobi identity).

5. Prove 1.2.4 [Show by induction on n that for each n-tuple $g_1, \ldots, g_n$ in G the element $D(g_1 \cdots g_n)$ can be expressed in terms of the elements g_i and $D(g_j)$.]
6. Complete the proof of 1.2.7. [For ii) use that $D(I^n) \subset I^{n-1}$, so D is continuous with respect to the I-adic topology. Also use that $\cap I^n = 0$ if A is a Noetherian domain by Krull's intersection theorem, see [22] or [203].]

7. Let A be a **filtered ring**, i.e. a ring equipped with an ascending chain $\cdots A_{n-1} \subset A_n \subset A_{n+1} \subset \cdots$ of additive subgroups satisfying $1 \in A_0$, $A = \cup A_n$ and $A_n A_m \subset A_{n+m}$ for all $n, m \in \mathbb{Z}$. Let $D \in \operatorname{Der} A$ be such that there exists $k \in \mathbb{Z}$ with $DA_n \subset A_{n+k}$ for all $n \in \mathbb{Z}$. Let $\operatorname{gr} A := \oplus A_n/A_{n-1}$ be the **associated graded ring of** A. Show that the map $\operatorname{gr} D : gr A \to \operatorname{gr} A$ defined by

$$\operatorname{gr} D(a + A_{n-1}) := Da + A_{n+k-1}, \text{ for all } a \in A_n$$

is a derivation on $\operatorname{gr} A$.
8. (Nagata, Nowicki) Let k be a field of characteristic $p > 0$ and D a k-derivation on $k[X]$.

   (i) Show that $k[X_1^p, \ldots, X_n^p] \subset k[X]^D$ and deduce that the ring of constants $k[X]^D$ is equal to $k[X_1^p, \ldots, X_n^p, f_1, \ldots, f_s]$ for some $f_i \in k[X]^D$.
   (ii) It is shown in [228], Proposition 4.2 that the analogue of 1.2.25 is true if $p = 2$, i.e. $k[X_1, X_2]^D = k[X_1^2, X_2^2, f]$ for some $f \in k[X_1, X_2]^D$. Show that if $p > 2$ and $D = X_1\partial_1 + X_2\partial_2$ then there does not exist $f \in k[X_1, X_2]^D$ with $k[X_1, X_2]^D = k[X_1^p, X_2^p, f]$.

   [Observe that $X_1^{p-1}X_2$ and $X_1X_2^{p-1}$ belong to $k[X_1, X_2]^D$. Assume that f exists. Then these two elements are of the form $u(f)$ resp. $v(f)$ where $u(T)$ and $v(T)$ belong to $k[X_1^p, X_2^p][T]$. Hence $-X_1^{p-2}X_2 = u'(f)\frac{\partial f}{\partial X_1}$, $X_1^{p-1} = u'(f)\frac{\partial f}{\partial X_2}$ and $X_2^{p-1} = v'(f)\frac{\partial f}{\partial X_1}$. Deduce that $u'(f) = cX_1{}^{p-2}$ for some $c \in k^*$ and derive a contradiction by observing that $D(X_1^{p-2}) \neq 0$ $(p > 2)$].
9. Let A be a domain with quotient field $Q(A)$. Let k be a field of characteristic zero and assume that A is a k-algebra. Let $0 \neq D \in \operatorname{Der}_k A$. Show that if $\operatorname{trdeg}_k Q(A) = n \geq 1$, then $\operatorname{trdeg}_k Q(A^D) \leq n - 1$. [Remark: in the next section we describe a class of derivations with $\operatorname{trdeg}_k Q(A^D) = n - 1$.]
10. Let k be a field of characteristic and $f, g \in k[X, Y]$ algebraically dependent over k. Show that $f = p(h)$, $g = q(h)$ for some $p, q \in k[T]$ and $h \in k[X, Y]$. [By 1.2.9 rank $J(f, g) < 2$, so $\det J(f, g) = 0$. Put $D := g_Y\partial_X - g_X\partial_Y$. So $D(f) = 0$. Apply 1.2.25.]
11. Show that Proposition 1.2.9 does not hold if k has positive characteristic.
12. (Berson) Let R be a UFD such that $Q(R)$ has characteristic 0 and let D be a non-zero R-derivation on $R[X, Y]$. We are going to show that $R[X, Y]^D = R[f]$ for some $f \in R[X, Y]$.

   (i) Let $K = Q(R)$. Show, using 1.2.25, that there exists $f \in R[X, Y]$ with $f(0, 0) = 0$ and primitive, i.e. the gcd of all its coefficients equals 1, such that $K[X, Y]^D = K[f]$.
   (ii) Let $0 \neq P \in R[X, Y]^D$ be primitive with $P(0, 0) = 0$. Show that $P = h(f)$ for some $h \in K[T]$ with $h(0) = 0$.

(iii) Show that, in order to prove that $h \in R[T]$, we may assume that R is a discrete valuation ring. [Use that R is the intersection of all $R_\wp$, where $\wp$ runs through all height 1 prime ideals of R and that each $R_\wp$ is a discrete valuation ring.]

(iv) Prove that $h \in R[T]$ in case R is a discrete valuation ring. [Let $\mathfrak{m}$ be the maximal ideal Rt of R. Choose n minimal such that $h_*(T) := t^n h(T) \in R[T]$. If $n \geq 1$, then $\bar{h}_*(\bar{f}) = 0$ gives a contradiction.]

(v) Deduce that $R[X, Y]^D = R[f]$.

13. (Hilbert 14 in case $n = 1$). Let $k \subset L \subset k(X_1)$ be fields. Prove that $L \cap k[X_1] = k[u]$ for some $u \in k[X_1]$. [Suppose $k \subsetneq L \cap k[X_1]$. By Lüroth's Theorem $L = k(u)$, where $u = a/b$, $a, b \in k[X_1]$, $b \neq 0$, $\deg a \geq \deg b$ and g.c.d.$(a, b) = 1$. Let $g(X_1) \in (L \cap k[X_1]) \setminus k$ and write $g = P(u)/Q(u)$ with $P, Q \in k[T]$, $Q \neq 0$ and g.c.d.$(P, Q) = 1$. Deduce from $Q(u)g = P(u)$ and $\deg_X g > 0$ that $b \in k^*$. Finally use that $rP + sQ = 1$ for some $r, s \in k[T]$ and deduce that $Q(u) \in k[X_1]^* = k^*$.]

1.3 Locally finite derivations

In this section we consider a special class of derivations, the so-called locally finite derivations. The importance of these derivations was formulated by the author in [102] as follows: "Locally nilpotent and locally finite derivations are hidden in many problems. However once their presence is revealed, information about these derivations can give valuable information on the problem under consideration." This statement will be amply demonstrated throughout this book.

The main aim of this section is to collect some of the basic facts concerning these derivations. In particular we show that locally finite derivations admit a Jordan decomposition, i.e. they can be written as a sum of a semisimple and a locally nilpotent derivation which commute. First we recall some generalities from ring theory, see [141] Chapter 3 or [190] Chapter XVII for more details.

Semisimple modules

Let A be a (not necessarily commutative) ring. An A-module M is called a **simple** A-module if it contains only trivial submodules, i.e. 0 and M. An A-module M is called **semisimple** if it is a sum of simple A-submodules, or equivalently if it is a direct sum of simple submodules. Furthermore one can show ([141], Chapter 3, Proposition 3.2 or [190], Chapter XVII, §2).

(1.3.1) An A-module M is semisimple if and only if for every A-submodule N of M there exists an A-submodule N' of M such that $M = N \oplus N'$.

Consequently we have

(1.3.2) A submodule and a factor module of a semisimple module are semisimple.

Finally we recall that a ring A is **semisimple** if all left A-modules are semisimple. One can show ([141], Chapter 3) that this is equivalent to the fact that the left A-module A is semisimple and also to the same statement with "left" replaced by "right".

Example 1.3.3 i) A field is a semisimple ring.

ii) If G is a finite group and k a field such that the order of the group does not divide the characteristic of k, then the **group ring** (Exercise 18) $k[G]$ is semisimple (this is **Maschke's theorem**, [190] Chapter XVIII, Theorem 1.2).

Remark 1.3.4 For the structure theory and many other characterizations of semisimple rings we refer to [141], Chapter 3 and [190], Chapter XVII.

Locally finite linear maps

Let k be a field, V a possibly infinite dimensional k-vector space and $l : V \to V$ a k-linear map.

Definition 1.3.5 i) l is called **locally finite** (resp. **locally nilpotent**) if for every $v \in V$ the k-vector space generated by the elements $l^i(v)$, $i \geq 0$ is finite dimensional (resp. if for every $v \in V$ there exists a positive integer n such that $l^n(v) = 0$).

ii) l is called **semisimple** if V is a semisimple $k[l]$-module, where $k[l]$ is the k-subalgebra of $\text{End}_k V$ generated by l and the identity map 1_V.

The following result will be needed in the proofs below.

Lemma 1.3.6 *Let $l : V \to V$ be locally finite. Then l is semisimple if and only if for every finite dimensional linear subspace W of V satisfying $l(W) \subset W$, the restriction $l_{|W}$ of l to W is semisimple.*

Proof: ($\to$) follows from 1.3.1. Conversely, assume that for each finite dimensional l-invariant subspace W of V, $l_{|W}$ is semisimple. It suffices to show that each $v \in V$ is contained in a sum of simple $k[l]$-submodules of V. Therefore observe that l is locally finite, so v is contained in the finite dimensional invariant subspace W generated by the elements $l^i(v)$, $i \geq 0$. Since by hypothesis $l_{|W}$ is semisimple, W is a sum of simple $k[l_{|W}]$-submodules. Consequently W is a sum of simple $k[l]$-submodules. □

Corollary 1.3.7 *Let $\bar{k}$ be an algebraic closure of k. Let $\bar{l} : \bar{k} \otimes_k V \to \bar{k} \otimes_k V$ be the induced $\bar{k}$-linear extension of l to $\bar{V} := \bar{k} \otimes_k V$. Then l is semisimple if and only if $\bar{l}$ is semisimple.*

Proof: First observe that the statement is well-known if V is finite dimensional. Then use 1.3.6 and the fact that every submodule of $\bar{V}$ is of the form $\bar{k} \otimes_k W$, for some submodule W of V. □

Proposition 1.3.8 (Jordan decomposition for locally finite linear maps). *Let V be a k-vector space and $l \in \text{End}_k V$ locally finite. Then there exist a semisimple linear map l_s on V and a locally nilpotent linear map l_n on V such that $l = l_s + l_n$ and $[l_s, l_n] = 0$. This decomposition is unique.*

Proof: Since each $v \in V$ is contained in a finite dimensional l-invariant subspace W of V, because l is locally finite, it suffices to define l_s and l_n on finite dimensional l-invariant subspaces. So let $W \subset V$ be l-invariant and finite dimensional. Then it is well-known that $f := l_{|W}$ has a unique Jordan decomposition $f = f_s + f_n$ and both f_s and f_n are polynomials in f. We define $l_s(w) := f_s(w)$ for all $w \in W$

and $l_n := l - l_s$. So $l_n(w) = f_n(w)$. We need to show that if W' is another l-invariant finite dimensional subspace of V, then $f_{s|Z} = f'_{s|Z}$, where $Z := W \cap W'$ and $f' := l_{|W'}$. To see this it suffices to show that $f_{s|Z} = (l_{|Z})_s$ (arguing similarly for W' we get $f'_{s|Z} = (l_{|Z})_s$). Therefore observe that both W and W' are l-invariant. So Z is f-invariant and hence f_s and f_n-invariant, since both are polynomials in f. Restricting the Jordan decomposition of f to Z gives $l_{|Z} = f_{|Z} = f_{s|Z} + f_{n|Z}$. By 1.3.6 $f_{s|Z}$ is semisimple and obviously $f_{n|Z}$ is nilpotent. Also $[f_{s|Z}, f_{n|Z}] = 0$. The uniqueness of the Jordan decomposition of $l_{|Z}$ implies that $f_{s|Z} = (l_{|Z})_s$ and $f_{n|Z} = (l_{|Z})_n$ as desired. So l_s and l_n are well-defined. Furthermore, by 1.3.6, l_s is semisimple and obviously l_n is locally nilpotent. The remaining properties follow readily from the corresponding properties in the finite dimensional case. □

Proposition 1.3.9 *Notations as in* 1.3.8. *Then* i) $\ker l = \ker l_s \cap \ker l_n$.

ii) *For every k-subspace W of V we have: W is l-invariant if and only if W is both l_s and l_n-invariant.*

Proof: The inclusion "$\supset$" in i) and the arrow ($\leftarrow$) in ii) are obvious. Let $v \in V$. Since l_s and l_n commute there exists a finite dimensional k-subspace V_0 of V containing v and invariant under l, l_s and l_n. The restrictions of l, l_s, l_n to V_0 we denote by $l_0, l_{s,0}, l_{n,0}$ respectively. Observe that $l_{s,0}$ is semisimple by 1.3.6 and $l_{n,0}$ is nilpotent. Since obviously $l_{n,0}$ commutes with $l_{s,0}$, $l_0 = l_{s,0} + l_{n,0}$ is the Jordan decomposition of the linear map l_0. Since there exist polynomials $p(T), q(T) \in k[T]$ with $p(0) = q(0) = 0$ and such that $l_{s,0} = p(l_0)$ and $l_{n,0} = q(l_0)$ i) follows easily: if $l(v) = 0$, then $l_0(v) = 0$, so $l_s(v) = l_{s,0}(v) = p(l_0)v = 0$. Similarly one proves that $l_n(v) = 0$ and the arrow ($\rightarrow$) in ii). □

Remark 1.3.10 Assume that k is algebraically closed and that l is locally finite. Then obviously the simple $k[l]$-modules are one-dimensional. So l is semisimple if and only if V is a direct sum of one-dimensional eigenspaces, i.e. $V = \oplus V_\lambda$, where each V_λ is spanned by an eigenvector v_λ satisfying $l(v_\lambda) = \lambda v_\lambda$.

Locally finite derivations and the Jordan decomposition

Definition 1.3.11 A k-derivation D on a k-algebra A is called **locally finite** (resp. **locally nilpotent**, resp. **semisimple**) if D viewed as a k-linear map is locally finite (resp. locally nilpotent, resp. semisimple).

Observe that each locally nilpotent derivation is locally finite. Also a semisimple derivation is locally finite (1.3.12). To check if a given derivation is locally finite (resp. locally nilpotent, resp. semisimple) the following result is often useful.

Proposition 1.3.12 *Let G be a generating set for the k-algebra A and D a k-derivation on A. Then D is locally finite (resp. locally nilpotent) if and only if for each $g \in G$ the vector space generated by the elements $D^i g, i \geq 0$ is finite dimensional (resp. $D^n g = 0$ for some $n \in \mathbb{N}$). Furthermore, D is semisimple if and only if each $g \in G$ is contained in a D-stable finite dimensional vector space W of A such that $D_{|W}$ is semisimple.*

Proof: i) The locally finite and locally nilpotent statements are left to the reader (Exercise 1). To prove the semisimplicity criterion extend D uniquely to a $\bar{k}$-derivation on $\bar{A} := \bar{k} \otimes_k A$. Denote this extension by $\bar{D}$. By 1.3.7, D is semisimple if and only if $\bar{D}$ is semisimple. From this it is not difficult to verify that we may assume that k is an algebraically closed field.

ii) So let k be algebraically closed. First assume that D is semisimple. Then by 1.3.10, $A = \oplus A_\lambda$, where $A_\lambda = kv_\lambda$ and $Dv_\lambda = \lambda v_\lambda$. Consequently D is locally finite and each $g \in G$ is contained in a finite dimensional D-invariant subspace W of A. By 1.3.6, $D_{|W}$ is semisimple. To prove the converse it is sufficient to show that, if v_1 is contained in a finite dimensional D-invariant subspace W_1 of A such that $D_{|W_1}$ is semisimple and v_2 is contained in a finite dimensional D-invariant subspace W_2 of A such that $D_{|W_2}$ is semisimple, then $v_1 v_2$ is contained in a finite dimensional D-invariant subspace W of A such that $D_{|W}$ is semisimple. To see this, observe that since $D_{|W_1}$ is semisimple there exist a basis $(e_1, \ldots, e_n)$ of W_1 consisting of eigenvectors of $D_{|W_1}$ and a basis $(f_1, \ldots, f_m)$ of W_2 consisting of eigenvectors of $D_{|W_2}$. One readily verifies that each non-zero element $e_i f_j$ of A is an eigenvector of D, and hence the eigenspace W generated by the elements $e_i f_j$ is D-invariant and $D_{|W}$ is semisimple. □

By 1.3.8 we can decompose each locally finite derivation D uniquely as $D = D_s + D_n$ satisfying $[D_s, D_n] = 0$. In the next proposition we will show that D_s and D_n are k-derivations on A.

Proposition 1.3.13 (Jordan decomposition for locally finite derivations). *Let D be a locally finite k-derivation on A. Then D_s and D_n are k-derivations on A.*

Proof: Arguing as in the proof of 1.3.12 we may assume that k is algebraically closed. Since $D_n = D - D_s$ it suffices to prove that D_s is a derivation on A. Since D is locally finite it follows from the finite dimensional Jordan decomposition that $A = \oplus M_\lambda$, where M_λ is the set of $a \in A$ satisfying $(D - \lambda)^p a = 0$ for some $p \in \mathbb{N}$. Furthermore we have

(1.3.14) $$D_s a = \lambda a, \text{ for all } a \in M_\lambda.$$

Since D_s is linear it suffices to show that $D_s(vw) = vD_s(w) + D_s(v)w$ for all $v \in M_\lambda$ and all $w \in M_\mu$. So let $v \in M_\lambda$ and $w \in M_\mu$. Then we get $vw \in M_{\lambda+\mu}$ (for if $(D - \lambda)^n v = 0$ and $(D - \mu)^m w = 0$, then it is not difficult to verify that $(D - (\lambda + \mu))^{n+m}(vw) = 0$). Consequently we deduce from 1.3.14

$$D_s(vw) = (\lambda + \mu)vw = (\lambda v)w + v(\mu w) = D_s(v)w + vD_s(w)$$

which concludes the proof. □

Locally nilpotent derivations

In the remainder of this section we consider only locally nilpotent derivations. In fact we first extend the notion of a locally nilpotent derivation a little, namely in 1.3.11

above we assumed that A is a k-algebra where k is a field. To define a locally nilpotent derivation we do not need this assumption. So from now on A is an R-algebra, where R is any commutative ring, and we say that a derivation D on A is **locally nilpotent** if for every $a \in A$ there exists a positive integer n such that $D^n a = 0$.

Example 1.3.15 The derivation $\frac{\partial}{\partial X_i}$ on $R[X]$ is locally nilpotent.

As in 1.3.12 we get

Proposition 1.3.16 *Let A be generated as an R-algebra by some set G and D an R-derivation on A. Then D is locally nilpotent if and only if for every $g \in G$ there exists $n \in \mathbb{N}$ with $D^n g = 0$.*

As a consequence of this result we obtain a large class of locally nilpotent derivations on $R[X]$, namely

Corollary 1.3.17 *Let D be a so-called* ***triangular derivation*** *on $R[X]$, i.e. a derivation of the form*

$$D = a_1(X_2, \ldots, X_n)\partial_1 + \cdots + a_{n-1}(X_n)\partial_{n-1} + a_n\partial_n \tag{1.3.18}$$

where $a_i \in R[X_{i+1}, \ldots, X_n]$ for all $1 \le i \le n-1$ and $a_n \in R$. Then D is locally nilpotent on $R[X]$.

Proof: By induction on n. The case $n = 1$ is obvious. So let $n \ge 2$ and write $D = a_1\partial_1 + D_0$. Then $D_0 \in \operatorname{Der}_R R[X_2, \ldots, X_n]$ and it has the form 1.3.18, so by the induction hypothesis D_0 is locally nilpotent on $R[X_2, \ldots, X_n]$. Now observe that $Dg = D_0 g$ for all $g \in R[X_2, \ldots, X_n]$ and that $DX_i \in R[X_2, \ldots, X_n]$ for all i. Consequently

$$D^{m+1}(X_i) = D^m(D(X_i)) = D_0^m(D(X_i))$$

which is zero for large m, since D_0 is locally nilpotent on $R[X_2, \ldots, X_n]$ and $DX_i \in R[X_2, \ldots, X_n]$. □

An element $s \in A$ is called a **slice of** D if $D(s) = 1$. If a slice exists, it is obviously uniquely determined up to an element of A^D. Locally nilpotent derivations do not always have a slice, for example $X_2\partial_1$ on $\mathbb{C}[X_1, X_2]$ has no slice. However locally nilpotent derivations which do have a slice have some very pleasant properties. Before describing them, let us first introduce some notations. From now on we assume: A is **commutative**. For each $a \in A$ we denote by $\pi_a : A[T] \to A$ the substitution homomorphism defined by

$$\pi_a(g(T)) = g(a) \text{ for all } g(T) \in A[T].$$

Now observe that if we extend D to $A[T]$ by defining $D(T) = 0$, then

(1.3.19) D is locally nilpotent on A if and only if $\exp TD(A) \subset A[T]$.

So if D is locally nilpotent on A, then $\varphi := \exp TD$ is a ring automorphism of $A[T]$. For each $a \in A$ we denote the composition $\pi_a \circ \varphi : A[T] \to A$ by φ_a.

Lemma 1.3.20 *Let D be locally nilpotent on A and $s \in A$ (arbitrary!). Then each element $a \in A$ is a polynomial in s of the form*

$$a = \sum_{i=0}^{\infty} \frac{1}{i!} \varphi_{-s}(D^i(a)) s^i.$$

Proof: Let $a \in A$. Then $\varphi(a) \in A[T]$, say $\deg_T \varphi(a) = N$. So

$$\begin{aligned} a &= \exp(-TD)\exp(TD)(a) = \exp(-TD)\varphi(a) \\ &= \sum_{i=0}^{N} \frac{1}{i!} \exp(-TD)(D^i(a))T^i \end{aligned}$$

where we used 1.2.14 and the fact that $\exp(-TD)(T^i) = T^i$ for all i. Finally substitute $T := s$. □

Proposition 1.3.21 *Let D be a locally nilpotent derivation having a slice s in A, where A is a $\mathbb{Q}$-algebra. Then A is a polynomial ring in s over A^D, i.e. $A = A^D[s]$ and $D = \frac{d}{ds}$ on A.*

Proof: Let $a \in A$. Then $D(s) = 1$ implies

$$D\varphi_{-s}(a) = D\left(\sum \frac{(-1)^i}{i!} D^i(a) s^i\right) = 0.$$

So

$$\varphi_{-s}(a) \in A^D, \text{ for all } a \in A. \tag{1.3.22}$$

Hence by 1.3.20 every element of A is a polynomial in s with coefficients in A^D. Finally, applying D one readily verifies that if $\sum a_i s^i = 0$ with $a_i \in A^D$, then all $a_i = 0$. So $A = A^D[s]$ is a polynomial ring in s over A^D and $D = \frac{d}{ds}$. □

Corollary 1.3.23 *Let A be an R-algebra (and R be a $\mathbb{Q}$-algebra). Let $D \in \mathrm{Der}_R A$ be a locally nilpotent derivation on A having a slice s in A. Then $A^D = \varphi_{-s}(A)$. In particular if G is a generating set for the R-algebra A, then $\varphi_{-s}(G)$ is a generating set for the R-algebra A^D.*

Proof: By 1.3.22 $\varphi_{-s}(A) \subset A^D$. Conversely, if $a \in A^D$ then $a = \exp TD(a)$, whence $a = \varphi_{-s}(a) \in \varphi_{-s}(A)$. So $A^D = \varphi_{-s}(A)$. □

(1.3.24) **Locally nilpotent derivations on a domain**

From now on we assume that A is a **domain containing** $\mathbb{Q}$. In 1.3.32 below we collect some of the most important properties of locally nilpotent derivations on a domain. The main technique used in the proofs is to study such derivations by localizing A at a suitable multiplicatively closed subset such that the extended derivation 1.2.7 has a slice in the localized ring. Then using 1.3.21 we deduce information for the original derivation on A. Before we can prove the main result 1.3.32, let us explain this localization procedure.

Let $0 \neq D$ be a locally nilpotent derivation on A. Then there exists an element $a \in A$ with $D(a) \neq 0$. Let $n \geq 1$ be the smallest positive integer such that $D^{n+1}(a) = 0$. Put $p := D^{n-1}(a)$. Then $d := D(p) \neq 0$ and $D^2(p) = 0$. So

$$d \in A^D. \tag{1.3.25}$$

Such an element p is called a **preslice of** D. Now put

$$\tilde{A} := A[d^{-1}] \text{ and } s := \frac{p}{d}. \tag{1.3.26}$$

By 1.2.7 the derivation D can be extended uniquely to a derivation on $\tilde{A}$ and one easily verifies that this extension, which we also denote by D, is locally nilpotent on $\tilde{A}$. The point is now that

$$D(s) = 1, \tag{1.3.27}$$

i.e. D has a slice in $\tilde{A}$, and furthermore one easily verifies that

$$\tilde{A}^D = A^D[d^{-1}]. \tag{1.3.28}$$

So by 1.3.21 we get

$$\tilde{A} = A^D[d^{-1}][s] \text{ and } D = \frac{d}{ds} \text{ on } \tilde{A}. \tag{1.3.29}$$

Since $d \in A^D$ by 1.3.25, we get from 1.3.26 and 1.3.29 that

$$Q(A) = Q(A^D)(s) \tag{1.3.30}$$

with s transcendental over $Q(A^D)$. Finally, to "descend" certain properties we use

$$A \cap Q(A^D) = A^D \tag{1.3.31}$$

which follows readily from the definition of A^D.

After these preparations we are now ready to prove

Proposition 1.3.32 *Let k be a field of characteristic zero contained in A and $0 \neq D$ a locally nilpotent k-derivation on A.* i) *If* $\text{trdeg}_k\, Q(A)$ *is finite, then* $\text{trdeg}_k\, Q(A^D) = \text{trdeg}_k\, Q(A) - 1$.

ii) *(Eigenvalue property). If $Da = \lambda a$ for some $0 \neq a \in A$ and $\lambda \in A$, then $\lambda = 0$.*

iii) *A^D **is factorially closed in** A, i.e. if $a, b \in A$, both non-zero, satisfy $ab \in A^D$, then a and b belong to A^D.*

iv) *Let $a \in A$ be such that $Da \neq 0$. Then $\rho(D, a) :=$ the smallest power of D which annihilates a, is equal to $|Q(A) : Q(A^D)(a)| + 1$.*

Proof: i) It follows immediately from 1.3.30. To see ii) observe that by 1.3.29 we can write $a = f(s)$ and $\lambda = g(s)$, with $f(T), g(T)$ in $A^D[d^{-1}][T]$. Since $D = \frac{d}{ds}$ on $\tilde{A}$, the equation $Da = \lambda a$ gives $f'(s) = g(s)f(s)$. So looking at the degrees of these polynomials in s we deduce that $f'(s) = 0$ and hence $\lambda = g(s) = 0$, since $f(s) = a \neq 0$ and A is a domain. To see iii) assume that $ab \in A^D$ with $a, b \in A$. Write again $a = f(s)$ and $b = g(s)$. Then the hypothesis $D(ab) = 0$ becomes $(f(s)g(s))' = 0$. So both $f(s)$ and $g(s)$ belong to $A^D[d^{-1}]$ and by hypothesis to A, so by 1.3.31 to $A \cap Q(A^D) = A^D$, as desired.

To see iv) write $a = f(s) \in Q(A^D)[s]$, say with $\deg_T f(T) = d \geq 1$. Since $D = \frac{d}{ds}$ on $Q(A^D)[s]$ we get $\rho(D, a) = d + 1$. Finally observe that

$$|Q(A) : Q(A^D)(a)| = |Q(A^D)(s) : Q(A^D)(f(s))| = d.$$

□

Corollary 1.3.33 *Notations as in* 1.3.32. *If* $\operatorname{trdeg}_k Q(A) = 1$, *then* $A = A^D[s]$ *where* $Ds = 1$ *and* $k \subset A^D$ *is an algebraic field extension. In particular* $A = k[s]$, *a polynomial ring in one variable over* k, *if* k *is algebraically closed.*

Proof: By 1.3.32 $\operatorname{trdeg}_k Q(A^D) = 0$, so each element of A^D is algebraic over k, hence $k \subset A^D$ is an integral extension, so A^D is a field. Hence d is a unit in $A^D \subset A$. Consequently $s = p/d \in A$, i.e. D has a slice in A. Then apply 1.3.21. □

Corollary 1.3.34 *Let* $0 \neq D$ *be any derivation on* A *and* $0 \neq a \in A$ *Then* aD *is locally nilpotent if and only if* D *is locally nilpotent and* $a \in A^D$.

Proof: i) If $a \in A^D$ then one easily verifies that

(1.3.35) $$(aD)^n(b) = a^n D^n(b), \text{ for all } b \in A, \text{ all } n \geq 1.$$

So if D is locally nilpotent, then so is aD.

ii) Conversely if aD is locally nilpotent on A, then 1.3.32 ii) and the obvious equality $(aD)(a) = D(a)a$ imply that $D(a) = 0$ i.e. $a \in A^D$. Since A is a domain it follows from 1.3.35 that D is locally nilpotent. □

Corollary 1.3.36 *Let* $0 \neq D$ *be a locally nilpotent derivation on* A. *Then* $(A^D)^* = A^*$ *and every irreducible element of* A^D *is irreducible in* A. *In particular if* A *is a UFD then so is* A^D.

Proof: This follows directly from 1.3.32 iii) and Exercise 6. □

Proposition 1.3.37 (Vasconcelos). *Let* $A \subset B$ *be an integral extension where* B *is a domain and* $\mathbb{Q} \subset A$. *If* D *is a derivation on* B *such that* $DA \subset A$ *and the restriction* $D_{|A}$ *of* D *to* A *is locally nilpotent, then* D *is locally nilpotent on* B.

Proof: i) Write D instead of $D_{|A}$ and put $S := A^D \backslash \{0\}$. Localize the extension $A \subset B$ with respect to S, i.e. we get the integral extension $S^{-1}A \subset S^{-1}B$, and extend the respective derivations to $S^{-1}A$ resp. $S^{-1}B$. Obviously the derivation on $S^{-1}A$ is locally nilpotent. Now the point is that by 1.3.29 we get that $S^{-1}A = Q(A^D)[s]$ with s transcendental over $Q(A^D)$ and that $D = \frac{d}{ds}$ on $S^{-1}A$. So if we put $K := Q(A^D)$ we may assume that $A = K[T]$ and $D = \frac{d}{dT}$.

ii) So let $A = K[T]$ and $D = \frac{d}{dT}$. Let $b \in B$. Since b is integral over $K[T]$, hence algebraic over $K(T)$, it follows that $Db \in K(T, b)$ whence $D^i b \in K(T, b)$ for all $i \geq 1$. Since each $D^i b \in B$ and B is integral over $K[T]$ it follows that $B_0 := K[T][b, Db, D^2b, \ldots]$ is contained in the integral closure C of $K[T]$ in $K(T, b)$. Since, as is well-known, C is a finite $K[T]$-module, so is B_0. It remains to see that D is locally nilpotent on the finite $K[T]$-module B_0, whose quotient field $K(T, b)$ is finite over $K(T)$. Using the Lefschetz principle we may replace K by $\mathbb{C}$. Then the result follows from the following lemma. □

Lemma 1.3.38 *Let $\mathbb{C}(T) \subset L$ be a finite field extension and B be a subring of L containing $\mathbb{C}[T]$. Let D be the unique extension of $\frac{d}{dT}$ to L and assume that $DB \subset B$. If B is a finite $\mathbb{C}[T]$-module then $B = \mathbb{C}[T]$.*

Proof:. Let $b \in B$. Since B is finite over $\mathbb{C}[T]$, b must satisfy a differential equation of the form

(1.3.39) $$\frac{d^n}{dT}b + a_{n-1}(T)\frac{d^{n-1}}{dT}b + \cdots + a_1(T)\frac{d}{dT}b + a_0(T)b = 0$$

with $a_i(T) \in \mathbb{C}[T]$ for all i. So viewing b as a function in T, it follows from elementary theory of linear differential equations in one complex variable that b is analytic on the domain where all a_i are analytic, i.e. on $\mathbb{C}$. So b is entire. Also b is algebraic over $\mathbb{C}(T)$. But then it is well-known that $b \in \mathbb{C}[T]$ (see for example [28]). □

(1.3.40) Locally nilpotent derivations on polynomial rings over a field

From now on we assume that k is a field of characteristic zero. In Theorem 1.2.25 we saw that in case $n = 2$, the kernel of any non-zero derivation D on $k[X]$ is generated by one element. On the other hand we know by 1.2.28 that if $n \geq 3$, $k[X]^D$ is in general not generated by $n - 1$ elements. However if D is locally nilpotent, we have the following non-trivial result. We refer to [217] for a proof.

Theorem 1.3.41 (Miyanishi). *Let D be a non-zero locally nilpotent derivation on $k[X_1, X_2, X_3]$. Then* $\ker D = k[f, g]$ *for some f, g which are algebraically independent over k.*

Using this result Daigle showed in [68] that any locally nilpotent derivation on $k[X_1, X_2, X_3]$ is a multiple of a so-called Jacobian derivation. More precisely

Proposition 1.3.42 (Daigle). *Let D be as in Theorem* 1.3.41. *Then there exists $a \in \ker D$ such that $D = aD_1$, where D_1 is the Jacobian derivation given by*

$$D_1(h) := \det J(f, g, h), \textit{for all } h \in k[X_1, X_2, X_3].$$

Furthermore, D_1 is locally nilpotent and irreducible, i.e. the g.c.d. of the elements $D_1(X_i)$, $i = 1, 2, 3$ equals 1.

If $n \geq 4$, we saw at the end of the previous section that there exists a locally nilpotent derivation whose kernel is not generated by $n-1$ elements. However by 1.3.32 the kernel of any locally nilpotent derivation D on $k[X]$ contains $n-1$ algebraically independent elements. This enabled Makar-Limanov in [200] to prove the following useful result.

Theroem 1.3.43 (Makar-Limanov). *Let D be any non-zero locally nilpotent derivation on $k[X]$. Let $f_1, \ldots, f_{n-1}$ be $n-1$ algebraically independent elements of* $\ker D$ *and D_1 the* **Jacobian derivation** *defined by*

$$D_1(h) := \det J(f_1, \ldots, f_{n-1}, h), \textit{for all } h \in k[X].$$

Then there exist non-zero elements $a, b \in \ker D$ such that $aD = bD_1$. In particular D_1 is locally nilpotent.

This result should be compared with the weaker statement in Exercise 7 at the end of this section.

Now we restrict to the case $n = 2$. Our aim is to give complete answers to the following main problems in the theory of locally nilpotent derivations.

(1.3.44) **Classification Problem**. Describe all locally nilpotent derivations on $k[X]$.

(1.3.45) **Recognition Problem**. Give a criterion which decides if a given derivation on $k[X]$ is locally nilpotent.

Let us start with the classification problem in the case $n = 2$. The main ingredient in the classification Theorem 1.3.48 is the following result.

Proposition 1.3.46 *Let $0 \neq D = a\partial_1 + b\partial_2$ be a locally nilpotent derivation on $k[X_1, X_2]$ with* $\gcd(a, b) = 1$. *Then D has a slice in $k[X_1, X_2]$.*

Proof: i) It is left to the reader to verify that we may assume that k is algebraically closed (consider $\bar{k} \otimes_k k[X_1, X_2]$, extend D and check that if $\gcd(a, b) = 1$ in $k[X_1, X_2]$, then also in $\bar{k}[X_1, X_2]$ etc.)

ii) Choose a preslice p in $A := k[X_1, X_2]$ with minimal degree. We claim that $D(p) \in k^*$. Hence $s = p/D(p) \in k[X_1, X_2]$ is a slice. So let us assume that $D(p) \notin k^*$. Then there exists an irreducible factor $q \in k[X_1, X_2]$ which divides $D(p)$. So

$$D(p) \in (q). \tag{1.3.47}$$

Since $D(p) \in A^D$ it follows from 1.3.32 iii) that $q \in A^D$. So we can consider the derivation $\tilde{D} : A/(q) \to A/(q)$, which is obviously locally nilpotent. Now observe that $\tilde{D} \neq 0$ for otherwise $DX_i \in (q)$, for $i = 1, 2$, contradicting $\gcd(a, b) = 1$. So by 1.3.33 $\ker \tilde{D} = k$. From 1.3.47 we get $\tilde{D}(p + (q)) = 0$, so $p + (q) = \lambda + (q)$, for some $\lambda \in k$, i.e. $p - \lambda = hq$ for some $h \in k[X_1, X_2]$. Finally one easily verifies that

$D^2(h) = 0$, since $D^2(p) = 0$ and $D(q) = 0$, and that $D(h) \neq 0$, since $D(p) \neq 0$ and $D(q) = 0$. However $p - \lambda = hq$ implies that $\deg h < \deg p$, contradicting the minimality of p. □

Now we are able to prove

Theorem 1.3.48 (Rentschler). *Let $0 \neq D$ be a locally nilpotent derivation on $k[X_1, X_2]$. Then there exists $\varphi \in \mathrm{Aut}_k\, k[X_1, X_2]$ and $f(X_2) \in k[X_2]$ such that $\varphi^{-1} D \varphi = f(X_2) \frac{\partial}{\partial X_1}$.*

Proof: Let $g = \gcd(a, b)$. Write $a = g\tilde{a}$ and $b = g\tilde{b}$. So $D = gD_0$, where $D_0 = \tilde{a}\partial_1 + \tilde{b}\partial_2$. It follows from 1.3.34 that D_0 is locally nilpotent and $D_0(g) = 0$. Put $k[X] := k[X_1, X_2]$. It follows from 1.3.46 that $D_0(F_1) = 1$ for some $F_1 \in k[X]$, so $k[X] = k[X]^D[F_1]$ by 1.3.21. Since $k[X]^D = k[F_2]$ for some $F_2 \in k[X]$, by 1.2.25, we get $k[X] = k[F_1, F_2]$, hence $F := (F_1, F_2)$ is invertible and φ defined by $\varphi(h) = h(F_1, F_2)$ for all $h \in k[X]$ is a polynomial automorphism of $k[X]$. Since $g \in k[X]^D = k[F_2]$ we can write $g = f(F_2)$, for some $f(T) \in k[T]$. Finally consider the derivation $D_1 := \varphi^{-1} D \varphi = \varphi^{-1} g D_0 \varphi$. Then one easily verifies that $D_1(X_1) = f(X_2)$ and $D_1(X_2) = 0$. So $D_1 = f(X_2)\partial_1$, as desired. □

Remark 1.3.49 In [250] Rentschler obtained a stronger result: in fact he proved that the automorphism φ in 1.3.48 is tame. Using this result he deduced that every element of $\mathrm{Aut}_k\, k[X_1, X_2]$ is tame. We refer the reader to Chapter 5, §1,where Rentschler's proof is given.

(1.3.50) **Warning.** If $n \geq 3$, a result similar to that in 1.3.46 is not true: for example the triangular derivation $X_2\partial_1 + X_3\partial_2$ obviously has no slice, nevertheless $\gcd(X_1, X_2) = 1$. A similar remark holds for 1.3.48; it is shown in Chapter 9 that if $n \geq 3$, there exist locally nilpotent derivations D on $k[X]$ which are not triangulable, i.e. there exists no $\varphi \in \mathrm{Aut}_k\, k[X]$ such that $\varphi^{-1} D \varphi$ is of triangular form (1.3.18).

Now let us consider the Recognition Problem. Before we concentrate on the two dimensional case we need one more general result. Let $D = \sum_{i=1}^n a_i \partial_i$ be any derivation on $k[X_1, \ldots, X_n]$, $n \geq 1$. Then the **divergence of** D, denoted $\mathrm{div}(D)$, is the element $\sum_i \frac{\partial a_i}{\partial X_i}$.

Proposition 1.3.51 *If D is locally nilpotent on $k[X]$, then* $\mathrm{div}(D) = 0$.

Proof: By 1.2.14 and 1.3.19, $\exp TD : k[X, T] \to k[X, T]$ is a polynomial automorphism. So using formula (3) in the introduction, we deduce that

$$j(X, T) := \det J_{X_1, \ldots, X_n, T}(\exp TD) = 1.$$

In particular the coefficient of T in the T-expansion of $j(X, T)$ is equal to zero. Since $\exp TD(X_i) = X_i + D(X_i)T + \cdots$ we get

$$J_{X_1, \ldots, X_n, T}(\exp TD) = I_{n+1} + \begin{pmatrix} \left(\frac{\partial D(X_i)}{\partial X_j}\right)_{1 \leq i, j \leq n} & * \\ 0 & 0 \end{pmatrix} T + \cdots$$

which implies that the coefficient of T in the T-expansion of $j(X,T)$ equals $\sum_i \frac{\partial D(X_i)}{\partial X_i} = \operatorname{div}(D)$. So $\operatorname{div}(D) = 0$. □

Now we are able to prove

Theorem 1.3.52 *Let $0 \neq D = a_1\partial_1 + a_2\partial_2 \in \operatorname{Der}_k k[X_1, X_2]$. Put*

$$d := \max_{i,j} \deg_{X_i} a_j$$

Then D is locally nilpotent if and only if $D^{d+2}X_i = 0$ for $i = 1, 2$.

Proof: i) By 1.3.12 we get ($\leftarrow$). Conversely, let D be locally nilpotent. Then by 1.3.51 $\operatorname{div}(D) = 0$, so $\frac{\partial a_1}{\partial X_1} + \frac{\partial a_2}{\partial X_2} = 0$, whence by 1.3.53 below there exists $f \in k[X_1, X_2]$ such that $a_1 = \partial_2 f$ and $a_2 = -\partial_1 f$. So $D = \partial_2 f \partial_1 - \partial_1 f \partial_2$ and hence $Df = 0$. Obviously $f \notin k$ (otherwise $D = 0$), so f is transcendental over k.

ii) If $\partial_1 f = 0$, then $f = f(X_2)$, $D = f'(X_2)\partial_1$ and the claim follows easily. Similarly if $\partial_2 f = 0$. So assume that both $\partial_1 f$ and $\partial_2 f$ are non-zero. Then $DX_i \neq 0$ for $i = 1, 2$ and we have

$$|k(X_1, X_2) : k(f)(X_1)| = \deg_{X_2} f \text{ and } |k(X_1, X_2) : k(f)(X_2)| = \deg_{X_1} f.$$

Now observe that

$$|k(X_1, X_2) : k(f)(X_i)| \geq |k(X_1, X_2) : Q(k[X_1, X_2]^D)(X_i)|$$

for $i = 1, 2$. Then by 1.3.32 iv) we get $D^{N+1}(X_i) = 0$ for $i = 1, 2$, where $N = \max(\deg_{X_1} f, \deg_{X_2} f)$. Finally observe that

$$d = \max \deg_{X_i} a_j = \max \deg_{X_i} f - 1 = N - 1.$$

So $N = d + 1$, which completes the proof. □

Lemma 1.3.53 *Let R be a $\mathbb{Q}$-algebra and $b_1, \ldots, b_n \in R[X] := R[X_1, \ldots, X_n]$ such that $\partial_i(b_j) = \partial_j(b_i)$ for all i, j. Then there exists $f \in R[X]$ such that $b_i = \partial_i f$ for all i.*

Proof: Let $b_i = \sum c_\alpha{}^{(i)} X^\alpha$, $X^\alpha = X_1^{\alpha_1} \ldots X_n^{\alpha_n}$. Then the hypothesis implies

$$\frac{1}{\alpha_i} c^{(i)}{}_{\alpha_1,\ldots,\alpha_{i-1},\alpha_i-1,\alpha_{i+1},\ldots,\alpha_n} = \frac{1}{\alpha_j} c^{(j)}{}_{\alpha_1,\ldots,\alpha_{j-1},\alpha_j-1,\alpha_{j+1},\ldots,\alpha_n}$$

for all $\alpha_i, \alpha_j \geq 1$ and all $1 \leq i, j \leq n$. Now put

$$f := \sum_{\alpha_1,\ldots,\alpha_n \geq 0} a_\alpha X^\alpha$$

where

$$a_\alpha = \frac{1}{\alpha_i} c^{(i)}{}_{\alpha_1,\ldots,\alpha_{i-1},\alpha_i-1,\alpha_{i+1},\ldots,\alpha_n}, \text{ if } \alpha_i \geq 1 \text{ for some } i$$

and $a_{0,\ldots,0} = 0$. One easily verifies that f has the desired property. □

Remark 1.3.54 Lemma 1.3.53 is a special case of the Poincaré lemma which asserts that the complex of differential forms is exact.

(1.3.55) **Comment** i) The power $d+2$ given in 1.3.52 is sharp: for example the triangular derivation $X_2^d\partial_1 + \partial_2$ satisfies $D^{d+1}(X_1) = d!$. So indeed $d+2$ is the smallest power which annihilates both X_1 and X_2.

ii) More generally, let $d \geq 1$ and $D = \sum a_i\partial_i$ a triangular derivation (1.3.18) with $\max_i \deg a_i = d$, then it can be shown (Exercise 13) that $D^{N+1}(X_i) = 0$ for all i, where

$$N := \sum_{k=0}^{n-1} d^k. \tag{1.3.56}$$

To conclude this section we mention (without proof) a useful formula due to Nowicki. Let R be a $\mathbb{Q}$-algebra and D a derivation on $R[X]$. We define elements $B_0, B_1, \ldots$ in $R[X]$ by the formula

$$\det J_X(\exp TD) = \sum_{p=0}^{\infty} \frac{1}{p!} B_p T^p.$$

Theorem 1.3.57 (Nowicki). *Let $d := \operatorname{div} D$. Then $B_0 = 1$, $B_1 = d$ and $B_{p+1} = dB_p + D(B_p)$, for all $p \geq 0$.*

Corollary 1.3.58 *If $d \in R$, then $B_p = d^p$ for all $p \geq 1$. In particular if $d = 0$ then* $\det J_X(\exp TD) = 1$.

Exercises for §1.3

1. Prove 1.3.12. [Use 1.2.3.]
2. Let $D = X_2^2\partial_1 + X_3^2\partial_2 + \partial_3$. Compute $k[X_1, X_2, X_3]^D$.
3. Give a direct proof of 1.3.32 ii) using the map $\exp TD$. [Apply $\exp TD$ to the equality $Da = \lambda a$ and compare T-degrees.]
4. Give a direct proof of 1.3.32 iii) using the map $\exp TD$.
5. Let A be a domain, $S \subset A\backslash\{0\}$ a multiplicatively closed subset of A. Let $D \in \operatorname{Der} A$ and consider the extension $\tilde{D}$ on $S^{-1}A$.

   (i) Show that $\tilde{D}$ is locally nilpotent on $S^{-1}A$ if and only if D is locally nilpotent on A and $S \subset A^D$.
   (ii) If $S \subset A^D$ show that $(S^{-1}A)^{\tilde{D}} = S^{-1}A^D$ and $S^{-1}A^D \cap A = A^D$.

6. Let A be a subring of a domain B. Suppose that A **is factorially closed in** B, i.e. if $b_1b_2 \in A$, where $b_1, b_2 \in B$, then $b_1, b_2 \in A$. Show that

   (i) $A^* = B^*$.
   (ii) A is integrally closed in B.
   (iii) Every irreducible element of A is irreducible in B.

(iv) If B is a UFD, then A is a UFD.
(v) $B \cap Q(A) = A$.

7. Let A be a domain containing $\mathbb{Q}$. Let D_1, D_2 be locally nilpotent derivations on A such that $A^{D_1} = A^{D_2}$. Show that there exist $a_1, a_2 \in A\backslash\{0\}$ such that $a_1 D_1 = a_2 D_2$. [Use 1.3.29.]
8. (Daigle) Let k be a field of characteristic zero and B a finitely generated k-domain. Let $A \subset B$ be a subalgebra and put $S := A\backslash\{0\}$. Show that the following statements are equivalent.

(i) A is the kernel of some locally nilpotent derivation D on B.
(ii) $Q(A) \cap B = A$ and $S^{-1}B = Q(A)[T]$ a polynomial ring in one variable over $Q(A)$.

[For $ii) \to i)$ write $B = k[b_1, \ldots, b_r]$ and consider $\frac{d}{dT}$ on $Q(A)[T]$. Since $\frac{d}{dT}(b_i) \in S^{-1}B$ for all i, there exists $s \in A\backslash\{0\}$ such that $s\frac{d}{dT}(B) \subset B$. Show that $D := s\frac{d}{dT}$ satisfies $A = B^D$ and D is locally nilpotent on B.]
9. Put $V := kX_1 + \cdots + kX_n \subset k[X]$, where k is any field. Let $D := \sum a_i \partial_i$, with $a_i \in V$ for all i. Let $D_{|V}$ be the k-linear endomorphism of V obtained by restriction of D to V.

(i) Show that D is locally finite.
(ii) Show that D is locally nilpotent if and only if $D_{|V}$ is a nilpotent linear map on V.

10. Let k be a field of characteristic zero and $D = a_1 X_1 \partial_1 + \sum_{i=2}^{n} a_i \partial_i$, where $a_i \in k[X]$ for all i and $a_1 \neq 0$. Show that D is not locally nilpotent on $k[X]$. [Use 1.3.32 ii).]
11. Let k be a field of characteristic zero.

(i) Show that $\cap k[X]^D = k$, where D runs through all locally nilpotent derivations of $k[X]$.
(ii) Show that the k-subalgebra generated by $\cup k[X]^D$ equals $k[X]$, where D runs through all non-zero locally nilpotent derivations of $k[X]$.

12. Let k be a field of characteristic zero. Let A be either the coordinate ring of the unit circle, i.e. the ring $k[X_1, X_2]/(X_1^2 + X_2^2 - 1)$ or the coordinate ring of the "standard" elliptic curve, i.e. the ring $k[X_1, X_2]/(X_2^2 - X_1^3 + X_1)$.

Show that in both cases $D = 0$ is the only locally nilpotent k-derivation on A, i.e. both the unit circle and the "standard" elliptic curve do not admit a non-trivial G_a-action, see Chapter 9. [Use 1.3.33.]
13. Let R be any commutative ring and $d \geq 1$. Put $N := \sum_{j=0}^{n-1} d^j$.

(i) Let $D = X_2^d \partial_1 + \cdots + X_n^d \partial_{n-1} + \partial_n$ on $R[X]$. Show that $D^N X_1 \neq 0$ and $D^{N+1} X_i = 0$ for all i.

(ii) Let $D = \sum a_i \partial_i$ be a triangular derivation with $\max \deg a_i = d$. Show that $D^{N+1} X_i = 0$ for all i.

14. (Nowicki) Let k be a field of characteristic 0 and $D = \partial_X + (1 + XY)\partial_Y$ on $A := k[X, Y]$.

 (i) Show that if $Da = \lambda a$ for some $0 \neq a \in A$ and $\lambda \in A$, then $\lambda = 0$.
 (ii) Show that D is not locally nilpotent.

15. Let D be a derivation on $k[X]$.

 (i) Show that D is locally finite if and only if for every $g \in k[X]$ there exists a positive integer N such that $\deg D^p g \leq N$ for all $p \geq 1$.
 (ii) Let $a_i \in k[X_{i+1}, \ldots, X_n]$, for each $1 \leq i \leq n-1$. Show that the derivation $a_1\partial_1 + \cdots + a_{n-1}\partial_{n-1} + X_n\partial_n$ is locally finite.

16. Let $D = (2X_2 + X_3 + X_2^2)\partial_1 + 3X_3^2\partial_2 + X_3\partial_3$. Compute D_s and D_n. [Let W be the smallest D-invariant k-subspace of $k[X]$ containing all X_i. Compute the Jordan form of $D_{|W}$. Then $D_s = \sum D_s(X_i)\partial_i$, where $D_s(X_i) = (D_{|W})_s(X_i)$.]
17. Let $\varphi \in \mathrm{Aut}_k\, k[X]$. Show that if D is locally finite (resp. locally nilpotent, resp. semisimple) then the same holds for $\varphi^{-1} \circ D \circ \varphi$. [Use 15 i) and the uniqueness of the Jordan decomposition.]
18. Let R be any ring and G a group (or a semigroup), written multiplicatively. Then one can form the **(semi-) group ring**

$$A = R[G] := \oplus_{g \in G} Rg.$$

 Elements of A are finite formal sums of the form $\sum_{g \in G} r_g g$. The multiplication in A is induced by the multiplication in G, i.e.

$$\left(\sum_g a_g g\right)\left(\sum_h b_h h\right) = \sum c_k k$$

 where $c_k = \sum a_g b_h$, with the summation over all pairs $(g, h) \in G \times G$ such that $gh = k$.

 (i) Show that $R[\mathbb{N}]$ is isomorphic to the polynomial ring $R[X_1]$ and more generally $R[\mathbb{N}^n] \simeq R[X_1, \ldots, X_n]$.
 (ii) Show that the group of units of $\mathbb{Z}[C_5]$ contains non-trivial units, i.e. units different from g or $-g$ with $g \in C_5$, where C_5 denotes the cyclic group of order 5.

 [Let x be a generator of C_5. Consider $u := 1 - x^2 - x^3$ and $v := 1 - x - x^4$.]

1.4 Algorithms for locally nilpotent derivations

Throughout this section k is a field of characteristic zero, $A = k[a_1, \ldots, a_n]$, a finitely generated k-algebra without zero-divisors and D a non-zero locally nilpotent k-derivation on A. The main aim of this section is to describe three algorithms concerning locally nilpotent k-derivations. The first algorithm, the so-called **kernel-algorithm**, computes generators for A^D, in case A^D is a finitely generated k-algebra. As we shall see in Chapter 9, this ring need not be a finitely generated k-algebra. On the other hand, if we don't know a priori that for a given locally nilpotent derivation D on A the ring A^D is finitely generated, we can conclude it is if our algorithm terminates after a finite number of steps! Based on the kernel algorithm is the so-called **image membership algorithm**, which decides if a given element $a \in A$ belongs to $D(A)$. Furthermore if a does belong to the image of D, then the algorithm gives an element b in A such that $a = D(b)$.

Finally we discuss the so-called **nilpotency criterion** which gives a partial solution to the still open problem: how to decide if a given derivation on A is locally nilpotent? Our criterion gives a positive solution under the assumption that we know sufficiently many algebraically independent elements in the kernel of D. In several practical situations this hypothesis is satisfied. See for example the extendibility algorithm in Chapter 3, §2. We will use the notations and results of §3. By $p \in A$ we denote a preslice of D, i.e. $d := D(p) \neq 0$ and $D^2(p) = 0$. So $d \in A^D$. Put $\tilde{A} := A[d^{-1}]$. Then $A \subset \tilde{A}$ and the extension of D to $\tilde{A}$, which we also denote by D is again locally nilpotent. Furthermore the element $s := d^{-1}p \in \tilde{A}$ satisfies $D(s) = 1$.

(1.4.1) **An algorithm to compute the kernel of a locally nilpotent derivation**

We begin by recalling several results of §3. By 1.3.23 we have

(1.4.2) $$\tilde{A}^D = k[b_1, \ldots, b_n][d^{-1}]$$

where $b_i := \varphi_{-s}(a_i)$, for all i. Furthermore by 1.3.28 we have

(1.4.3) $$\tilde{A}^D = A^D[d^{-1}].$$

Consequently there exist natural numbers $e_i \geq 0$ such that $c_i := d^{e_i} b_i \in C := A^D$, for all i. So if we put

(1.4.4) $$C_0 := k[c_1, \ldots, c_n, d]$$

then it follows from 1.4.2 and 1.4.3 that

(1.4.5) $$C_0 \subset C \subset C_0[d^{-1}].$$

Now for every $m \geq 1$ we define C_m inductively as the subalgebra of A generated by the elements $h \in A$ satisfying $dh \in C_{m-1}$. Since $d \in C_0$ it follows, by induction on m, that $C_0 \subset C_m$, for all $m \geq 1$ and hence that $C_{m-1} \subset C_m$ for all $m \geq 1$. By induction on m we also see that $C_m \subset C$: the case $m = 0$ is obvious. So let $m \geq 1$ and assume that $C_{m-1} \subset C$. If $h \in C_m$ satisfies $dh \in C_{m-1} \subset C$, then $D(dh) = 0$, so $dD(h) = 0$ $(D(d) = 0)$. Since A is a domain and $d \neq 0$ we deduce that $D(h) = 0$, i.e. $h \in C$.

The key to the kernel-algorithm is given by (the first two points) of

Proposition 1.4.6 i) *For every $m \geq 1$, C_m is a finitely generated k-subalgebra of C.*

ii) *If C is a finitely generated k-algebra, then $C = C_r$ for some $r \in \mathbb{N}$.*

iii) *If there exists $r \in \mathbb{N}$ such that $C_r = C_{r+1}$, then C is a finitely generated k-algebra. More precisely $C = C_r$.*

Proof: i) We use induction on m. So let $m \geq 1$ and assume that $C_{m-1} = k[f_1, \dots, f_l]$. Let J be the set of $P \in k[Y] := k[Y_1, \dots, Y_l]$ such that $P(f) := P(f_1, \dots, f_l) \in Ad$. This is an ideal of $k[Y]$ and hence of finite type. Let $P_1, \dots, P_s$ be its generators. So $P_i(f) = g_i d$, for some $g_i \in A$, which by definition belongs to C_m (since obviously $P_i(f) \in C_{m-1}$). Now we claim

$$C_m = k[f_1, \dots, f_l, g_1, \dots, g_s]$$

which clearly implies i). It remains to show "$\subset$". Therefore let $h \in C_m$ with $dh \in C_{m-1}$. Then $dh = P(f)$ for some $P \in k[Y]$. In other words $P(f) \in Ad$, i.e. $P = \sum_{i=1}^{s} a_i(Y) P_i$ for some $a_i(Y)$ in $k[Y]$. Consequently $P(f) = \sum a_i(f) P_i(f)$. Since $P_i(f) = g_i d$ and $P(f) = dh$ we conclude that $dh = d \sum a_i(f) g_i$, whence

$$h \in \sum k[f] g_i \subset k[f_1, \dots, f_l, g_1, \dots, g_s]$$

which proves the claim.

ii) Now assume that C is a finitely generated k-algebra, say $C = k[I_1, \dots, I_n]$ for some $I_j \in C$. Since by 1.4.5 $C \subset C_0[d^{-1}]$ there exists $r \in \mathbb{N}$ with $d^r I_j \in C_0$ for all j. So $d^{r-1} I_j \in C_1$, whence $d^{r-2} I_j \in C_2$ etc. Continuing in this way we find that $I_j \in C_r$ for all j. Hence $C \subset C_r \subset C$ i.e. $C = C_r$ as desired.

iii) Finally assume $C_r = C_{r+1}$. Observe that 1.4.5 implies that $C = \cup_{j \geq 0} C_j$. So the last statement of the proposition follows if we can show for all $j \geq 0$

(1.4.7) If $C_j = C_{j+1}$, then $C_{j+1} = C_{j+2}$.

To see 1.4.7, assume $C_j = C_{j+1}$ and let $h \in C_{j+2}$ with $dh \in C_{j+1} = C_j$. Then $h \in C_{j+1}$. So $C_{j+2} \subset C_{j+1}$. Consequently $C_{j+1} = C_{j+2}$ since we already know that $C_{j+1} \subset C_{j+2}$. □

Now we are able to describe the kernel-algorithm. Following the proof of 1.4.6 we therefore need to answer the following two questions

Question 1: *Knowing $C_{m-1} = k[f_1, \dots, f_l]$, how can we compute $g_1, \dots, g_s$ in A such that $C_m = k[f_1, \dots, f_l, g_1, \dots, g_s]$?*

Question 2: *How can we decide when $C_r = C_{r+1}$?*

The second question can be answered immediately by the membership algorithm (C.2.3). To answer the first question we need to construct generators $P_1, \ldots, P_s$ of the ideal J. Now observe that if we put $\bar{A} := A/Ad$ and denote by $\bar{f}_i$ the residue class of f_i in $\bar{A}$, then

$$J = \{P \in k[Y] | P(\bar{f}_1, \ldots, \bar{f}_l) = 0 \text{ in } \bar{A}\}.$$

So applying the relation algorithm (C.2.2) we find generators $P_1, \ldots, P_s$ of J. So we know that $P_i(f) = dg_i$ for some $g_i \in A$. It remains to show how we can compute g_i from the element P_i. Let A be given as $k[X_1, \ldots, X_n]/I$ where I is the ideal in the polynomial ring $k[X] := k[X_1, \ldots, X_n]$ generated by the elements $H_1, \ldots, H_t$. Let $f_i = F_i + I, d := d^* + I$ for some $F_i, d^* \in k[X]$. Then $P_i(f) \in dA$ means that

$$P_i(F) = B_i d^* + \sum_{j=1}^{t} G_{j,i} H_j, \text{ for some } B_i, G_{j,i} \in k[X].$$

To find the coefficient B_i of d^* one can use well-known results from Gröbner basis theory: compute a Gröbner basis $(S_1, \ldots, S_p)$ of the ideal $(d^*, H_1, \ldots, H_t)$ in $k[X]$. Write

$$P_i(F) = \sum_{j=1}^{p} L_{j,i} S_j, \; L_{j,i} \in k[X]. \tag{1.4.8}$$

Express each S_i as a $k[X]$-linear combination of $d^*, H_1, \ldots, H_t$ and substitute these expressions in 1.4.8. Then one finds the coefficients B_i of d^*.

Remark 1.4.9 If A is a polynomial ring, the above argument can be simplified: namely from $P_i(f) = dg_i$ one immediately obtains that $g_i = P_i(f)/d$.

(1.4.10) **The image membership algorithm**

From now on a will be a fixed non-zero element of A. So there exists $m \geq 0$ such that $D^m(a) \neq 0$ and $D^{m+1}(a) = 0$. Under the assumption that A^D is a finitely generated k-algebra, we will now describe an algorithm which decides if a is of the form $D(b)$ for some $b \in A$ and computes such an element b if it exists. Assume that $A^D = k[f_1, \ldots, f_l]$. With the notations as above we put

$$b' := \sum_{i=o}^{m} \frac{(-1)^i}{(i+1)!} D^i(a) s^{i+1}. \tag{1.4.11}$$

Then one easily verifies that

$$D(b') = a. \tag{1.4.12}$$

So if we put

$$g := d^{m+1} b' \tag{1.4.13}$$

then $g \in A$, using that $s = p/d$. Write

(1.4.14) $$a = a^* + I,\ d = d^* + I,\ g = G + I \text{ and } f_i = F_i + I$$

for some a^*, d^*, G, F_i in $k[X]$ and let J be the ideal in $k[X, Y] := k[X_1, \dots, X_n, Y_1, \dots, Y_l]$ defined by

$$J = (Y_1 - F_1, \dots, Y_l - F_l, d^{*m+1}, H_1, \dots, H_t).$$

Choose on the set of terms of $k[X, Y]$ an admissible ordering such that $X_i > Y^\alpha$, for all i, α and let B be a Gröbner basis of J with respect to this ordering. Finally let $\tilde{G}$ be the normal form of G with respect to B, i.e. the result obtained by reduction of G modulo B.

Proposition 1.4.15 *With the notations above we get: $a \in D(A)$ if and only if $\tilde{G} \in k[Y]$. Furthermore, if $\tilde{G} \in k[Y]$ then $b := (g - \tilde{G}(f))/d^{m+1} \in A$ and satisfies $D(b) = a$.*

Proof: i) Suppose that $a = D(b)$ for some $b \in A$. Then by 1.4.12 $D(b - b') = 0$, whence by 1.4.13 $g - d^{m+1}b \in A^D$. So $g - d^{m+1}b = H(f_1, \dots, f_l)$, for some $H(Y) \in k[Y]$. Consequently if we write $b = b^* + I$ for some $b^* \in k[X]$ then by 1.4.14 we get

$$G - d^{*m+1}b^* - H(F_1, \dots, F_l) \in I$$

whence $G - H(Y) \in J$ (since $H(F) \equiv H(Y) \pmod{J}$). Since $G - \tilde{G} \in J$ we deduce that $\tilde{G} - H(Y) \in J$. Then it follows that the leading coefficient of $\tilde{G}$ has degree smaller than or equal to that of $H(Y)$, so $\tilde{G} \in k[Y]$.

ii) Conversely, if G reduces to $\tilde{G}$ (modulo B), then

$$G = \tilde{G} + \sum_i a_i(X, Y)(Y_i - F_i) + \sum_j c_j(X, Y)H_j + e(X, Y)d^{*m+1}$$

for some $a_i, c_j, e \in k[X, Y]$. Then making the substitutions $Y_i := F_i$ for all i and reducing the obtained equations modulo I, we get the following equality in A

(1.4.16) $$g = \tilde{G}(f) + bd^{m+1}, \text{ where } b := e(X, F) + I \text{ in } A.$$

From 1.4.16 we obviously get $b = (g - \tilde{G}(f))/d^{m+1}$. Finally substituting 1.4.13 in 1.4.16 and applying D gives $d^{m+1}D(b') = D(b)d^{m+1}$. Then $D(b') = a$ (by 1.4.12) and the fact that A is a domain imply that $a = D(b)$. □

To conclude this section we give a partial answer to the recognition problem 1.3.45. So from now on D denotes a non-zero k-derivation on A, i.e. we do not assume it is locally nilpotent. Let $t := \text{trdeg}_k\, Q(A)$. So $t \geq 1$ since $D \neq 0$. If D is locally nilpotent on A, then it follows from 1.3.32 i) that there exist $x_1, \dots, x_{t-1} \in A^D$ which are algebraically independent over k. Conversely, let D be any k-derivation on A and suppose that we can find $t - 1$ elements $x_1, \dots, x_{t-1}$ in A^D which are algebraically independent over k. Then we are able to decide if D is locally nilpotent. More precisely

(1.4.17) **Partial nilpotency criterion.** Suppose we can find $x_1, \ldots, x_{t-1}$ in A^D algebraically independent over k. Put

$$N := \max_i \{ |Q(A) : k(x_1, \ldots, x_{t-1})(a_i)| , D(a_i) \neq 0\}.$$

Then N is finite and D is locally nilpotent on A if and only if $D^{N+1}(a_i) = 0$ for all i.

Proof: i) Since D is non-zero there exist an a_i such that $D(a_i) \neq 0$. Then a_i is transcendental over $Q(A^D)$ (by 1.2.8) and hence over the subfield $k(x_1, \ldots, x_{t-1})$. So $|Q(A) : k(x_1, \ldots, x_{t-1})(a_i)|$ is finite, implying that N is finite. Obviously 1.3.12 implies the implication ($\leftarrow$). To see the converse just observe that

$$|Q(A) : k(x_1, \ldots, x_t)(a_i)| \geq |Q(A) : Q(A^D)(x_i)|$$

and apply 1.3.32 iv). □

Exercises for §1.4

1. Let D be the derivation $W\partial_U + Z\partial_V + 2V\partial_X$ on $k[X, Y, Z, U, V, W]$. Compute the kernel of D.
2. (Winkelmann) Let $D = (1 + YU - ZW)\partial_X + W\partial_Y + U\partial_Z$ on $A := k[X, Y, Z, U, W]$.

   (i) Compute ker D and show that this ring is regular.
   (ii) Show that $D(X)$, $D(Y)$, $D(Z)$ generate the unit ideal in A.
   (iii) Show that D has no slice in A. [Show that 1 does not belong to Im D.]

3. Let $D = a_1\partial_{Y_1} + a_2\partial_{Y_2}$ with $a_1, a_2 \in k[X_1, \ldots, X_n]$ and $\gcd(a_1, a_2) = 1$. Show that $k[X, Y]^D = k[X_1, \ldots, X_n, a_2Y_1 - a_1Y_2]$.
4. Let $D = X\partial_S + Y\partial_T + Z\partial_U$ on $A := k[X, Y, Z, S, T, U]$.

   (i) Compute ker D.
   (ii) Put $a_0 := 1$. Compute $a_1 \in A$ with $Da_1 = Xa_0$, $a_2 \in A$ with $Da_2 = YZa_1$ and $a_3 \in A$ with $Da_3 = XYZa_2$.

Notes

Nagata's conjecture was first published in [224]. Proposition 1.1.12 was obtained by Bass, Connell and Wright in the paper [29]. Theorem 1.1.32 appeared in a preprint of the University of Nijmegen in 1992, but remained unpublished since the submitted version was lost! Proposition 1.1.19 was obtained by Connell and van den Dries in [63]. In their proof they used what we called the clearing map, which was originally introduced in [30] by Bass and Wright to study localization in K-theory. See also [62]. The clearing map returned without being mentioned explicitly in the proofs of the reduction theorems of Bass, Connell and Wright in [29] and of Drużkowski in [93]. See also Chapter 6 where these theorems are discussed. Also it appears in Derksen's paper [76] concerning the degree of the inverse of a polynomial automorphism (Chapter 2). Flatness of Keller maps 1.1.34 was first observed by Wang in [289]. Its consequence, Keller's theorem, goes back to Keller's 1939 paper [179], which formed the origin of the Jacobian Conjecture. Exercise 9 at the end of section one is Theorem 1 in the paper [125] of Formanek. Proposition 1.2.15 appeared in the paper [265] of Seidenberg. The fundamental result 1.2.25 was given by Nagata and Nowicki in [228] as well as 1.2.24. For a survey of derivations the reader should consult Nowicki's excellent thesis [231]. This is a great source of information! Theorem 1.2.26 was proved by Zaks in [305]. He also treated the case where char $k > 0$ and showed the same result in case R is a Dedekind ring. Proposition 1.3.21 is often attributed to Wright in [293], however this result already appeared in a paper [137] by Gabriel and Nouazé some 14 years earlier! The useful equality in iv) of 1.3.32 originates from the paper [83] of Deveney and Finston and the basic result 1.3.37 goes back to Vasconcelos in [286]. The kernel algorithm described in §4 is due to the author in [105]. The image membership algorithm is inspired by an unpublished note of Derksen in which he describes an algorithm similar to the one given in §4, to compute a slice of a locally nilpotent derivation on a polynomial ring. Finally, the partial nilpotency criterion was given by the author in [103] in relation with the extendibility problem (Chapter 3).

2 Derivations and polynomial automorphisms

In Chapter 1 we encountered several polynomial automorphisms, in particular the elementary ones and the Nagata example. In § 1 of this chapter, we describe a uniform way to construct both examples, namely as so-called exponential automorphisms, i.e. automorphisms of the form $\exp D$ where D is a locally nilpotent derivation on the polynomial ring $k[X]$, see Examples 2.1.9 and 2.1.10. Since in case R is not a field there exist non-tame automorphisms of $R[X]$ of the form $\exp D$, this will be proved in Chapter 5, it is natural to replace the tame generators conjecture by a weaker one, the so-called exponential generators conjecture, which asserts that for any $\mathbb{Q}$-algebra R, every R-automorphism of $R[X]$ is a finite product of exponential automorphisms and elements of $\mathrm{Aff}(R, n)$. In §1 we prove the exponential conjecture for the so-called nilpotency subgroup of $\mathrm{Aut}_R\, R[X]$, i.e. the set of all automorphisms of the form $(X_1 + g_1, \ldots, X_n + g_n)$, where each g_i is a nilpotent element of $R[X]$. In fact we show that each such automorphism is an exponential automorphism!

In §2 we investigate the Jacobian condition, i.e. $\det JF \in R[X]^*$. To such a polynomial map F we associate n pairwise commuting derivations $\frac{\partial}{\partial F_1}, \ldots, \frac{\partial}{\partial F_n}$ which form an $R[X]$-basis of $\mathrm{Der}_R\, R[X]$. Conversely we show that every $R[X]$-basis $D_1, \ldots, D_n$ of $\mathrm{Der}_R\, R[X]$ which is "commutative', i.e. satisfies $D_i D_j = D_j D_i$ for all i, j, comes from a polynomial map F satisfying $\det JF \in R[X]^*$, i.e. there exists such an F with $D_i = \frac{\partial}{\partial F_i}$ for all i. We give several applications of these derivations. For example if k is a field of characteristic zero we show that the Jacobian Conjecture is true in case $k[F] \subset k[X]$ is finite or if $k(F) \subset k(X)$ is a Galois extension. Also in §3 we use the $\frac{\partial}{\partial F_i}$ to show that $\deg F^{-1} \leq (\deg F)^{n-1}$ for any $F \in \mathrm{Aut}_k\, k[X]$. Furthermore we discuss a result which relates the degree of the inverse of a polynomial automorphism satisfying $\det JF \in R^*$ with the Jacobian Conjecture (2.3.6). Various other equivalent formulations of the Jacobian Conjecture in terms of the derivations $\frac{\partial}{\partial F_i}$ are given (2.2.9–2.2.12).

2.1 Locally nilpotent derivations and polynomial automorphisms

In the introduction of this book we described Nagata's example σ and left it to the reader to verify that this map is invertible. In this section we will show how locally nilpotent derivations can be used to construct a large class of polynomial automorphisms, including Nagata's example and all elementary automorphisms.

So let A be any $\mathbb{Q}$-algebra, for the moment A need not be commutative, and D a locally nilpotent derivation on A. If $a \in Z(A)$, the center of A, then the substitution map $\pi_a : A[T] \to A$ given by $\pi_a(g(T)) = g(a)$ is a ring homomorphism, called the substitution homomorphism. So taking $a = 1$, the composition $\pi_1 \circ \exp TD$ is a ring homomorphism from $A[T]$ to A, which by restriction to A gives a ring endomorphism of A, denoted by $\exp D$. So

$$\exp D(a) = \sum_{p \geq 0} \frac{1}{p!} D^P(a), \text{ for all } a \in A.$$

The sum is finite since D is locally nilpotent. In fact $\exp D$ is a ring automorphism with inverse $\exp(-D)$. So we get

Proposition 2.1.1 *If D is a locally nilpotent derivation on A, then* $\exp D$ *is a ring automorphism of A with inverse* $\exp(-D)$.

Automorphisms of the form $\exp D$, where D is a locally nilpotent derivation on A are called **exponential automorphisms** of A. To prove the main result of this section, 2.1.13, we will need a criterion to decide if a given endomorphism of A is an exponential automorphism. To describe this result we introduce some terminology. Let $f : A \to A$ be any map. To f we associate the map $E : A \to A$ given by $E := f - 1_A$. If f is additive, so is E. Also one easily verifies that f is a homomorphism if and only if

$$E(ab) = aE(b) + E(a)b + E(a)E(b), \text{ for all } a, b \in A$$

or equivalently if E is an f-**derivation** on A, i.e.

(2.1.2) $$E(ab) = E(a)f(b) + aE(b), \text{ for all } a, b \in A.$$

Suppose that f is an exponential automorphism, say $f = \exp D$ with D a locally nilpotent derivation on A. Then it follows readily from the definitions that the $\mathbb{Q}$-linear map E is locally nilpotent. Now we will show that the converse is true as well. For a less computational proof we refer to Exercise 6 at the end of this section.

Proposition 2.1.3 *Let $f : A \to A$ be a ring homomorphism. Then f is an exponential automorphism of A if and only if E is locally nilpotent. Furthermore, if E is locally nilpotent, then the map $D : A \to A$ defined by*

$$D(a) = \sum_{i \geq 1} (-1)^{i+1} \frac{E^i(a)}{i}, \text{ for all } a \in A$$

is a locally nilpotent derivation on A and $f = \exp D$.

Proof: i) To prove the first part of the proposition suppose that E is locally nilpotent. To see that f is surjective let $a \in A$. Since $E^n(a) = 0$ for some $n \geq 0$, we have that $(f - 1_A)^n = 0$. Hence the binomial expansion gives that $(-1)^n a$ is a sum of terms of the form $m_i f^i(a)$, where each m_i is an integer and $1 \leq i \leq n$. So $a \in f(A)$. Consequently f is surjective. The same expansion also shows that $f(a) = 0$ implies that $a = 0$, i.e. f is injective.

ii) To prove the second statement of the proposition we need to show that D is a derivation on A. This will be done in 2.1.4 below. Since E is locally nilpotent it follows readily that D is locally nilpotent as well. Finally, since $D = \log(1_A + E)$ we get that $\exp D = 1_A + E = f$, which completes the proof. □

Lemma 2.1.4 *Let f be a ring homomorphism of A and E the corresponding f-derivation of A. If E is locally nilpotent, then D, as in* 2.1.3, *is a derivation on A.*

Proof: From 2.1.2 one deduces easily by induction on n that

$$E^n(ab) = \sum_{k=0}^{n} \binom{n}{k} E^k(a) f^k(E^{n-k}(b)).$$

Now let $a, b \in A$. It follows, replacing f by $E + 1_A$, that $D(ab)$ is equal to

$$\sum_{n\geq 1} \frac{(-1)^{n+1}}{n}\left[\sum_{k=0}^{n}\binom{n}{k}E^k(a)(E+1_A)^k(E^{n-k}(b))\right]$$

(2.1.5)

$$= \sum_{n\geq 1} \frac{(-1)^{n+1}}{n}\left[\sum_{k=0}^{n}\binom{n}{k}E^k(a)\left[\sum_{j=0}^{k}\binom{k}{j}(E^{j+n-k}(b)\right]\right].$$

This expression is a sum of terms of the form $rE^p(a)E^q(b)$, with r a rational number and $p, q \geq 0$. One easily verifies that the sum of all such terms with $p = 0$ and $q \geq 1$ equals $aD(b)$ and the sum of all such terms with $p \geq 1$ and $q = 0$ equals $D(a)b$. So to prove that D is a derivation it suffices to prove that for each pair $p, q \geq 1$ the coefficient of $E^p(a)E^q(b)$ in 2.1.5 is zero. So fix such a pair $p, q \geq 1$. By symmetry we may assume that $p \leq q$. Then the coefficient of $E^p(a)E^q(b)$ in 2.1.5 is given by

$$\sum_{n=q}^{p+q} \frac{(-1)^{n+1}}{n}\binom{n}{p}\binom{p}{q+p-n}.$$

So, since $\frac{1}{n}\binom{n}{p} = \frac{1}{p}\binom{n-1}{p-1}$ it suffices to prove that

$$\sum_{n=q}^{p+q}(-1)^{n+1}\binom{n-1}{p-1}\binom{p}{q+p-n} = 0.$$

or equivalently that

$$\sum_{n=0}^{p}(-1)^{n+q+1}\binom{n+q-1}{p-1}\binom{p}{p-n} = 0$$

This last equality follows from 2.1.6 below, observing that $\binom{p}{p-n} = \binom{p}{n}$. □

Lemma 2.1.6 *Let $n \geq 1$, $0 \leq k < n$ and m an integer ≥ 0. Then*

$$\sum_{p=0}^{n}(-1)^p\binom{k+m+p}{k}\binom{n}{p}=0. \tag{2.1.7}$$

Proof: The case $k = 0$ is well-known and the case $n = 1$ is obvious. So we may assume that $k \geq 1$ and $n \geq 2$. Using induction on k and m and the formula

$$\binom{k+m+p}{k}=\binom{k+m+p-1}{k-1}+\binom{k+m+p-1}{k}$$

we can reduce the proof to the case $m = 0$. In this case the sum in 2.1.7 can be written as

$$\sum_{p=0}^{n-1}(-1)^{p+1}\binom{n-1}{p}\binom{k+p+1}{k}+\sum_{p=0}^{n-1}(-1)^p\binom{n-1}{p}\binom{k+p}{k} \tag{2.1.8}$$

where we used that $\binom{n}{p}=\binom{n-1}{p-1}+\binom{n-1}{p}$. If $k < n-1$ both summands in 2.1.8 are zero by induction on n so we need to consider the case $k = n-1$. In this case 2.1.8 becomes

$$\begin{aligned}
&-n+\sum_{p=1}^{n-1}(-1)^{p+1}\binom{n-1}{p}\binom{n+p}{n-1}+\sum_{p=1}^{n-1}(-1)^p\binom{n-1}{p}\binom{n+p-1}{n-1}\\
&=(-1+n)+\sum_{p=1}^{n-1}(-1)^{p+1}\binom{n-1}{p}\binom{n+p-1}{n-2}\\
&=\sum_{p=0}^{n-1}(-1)^{p+1}\binom{n-1}{p}\binom{n+p-1}{n-2}
\end{aligned}$$

which is zero by induction on n. □

From now on we will restrict our attention to the polynomial ring $R[X] := R[X_1, \ldots X_n]$ where R is any commutative $\mathbb{Q}$-algebra. First we compute some instructive examples.

Example 2.1.9 Let $1 \leq i \leq n$, $a \in R[X_1, \ldots, \hat{X}_i, \ldots, X_n]$ and $D = a\frac{\partial}{\partial X_i}$. So by 1.3.15 and 1.3.34, D is locally nilpotent. One easily verifies that

$$\begin{aligned}
\exp D(X_j) &= X_j, \textit{ if } j \neq i,\\
\exp D(X_i) &= X_i + a.
\end{aligned}$$

So $\exp D = (X_1, \ldots, X_{i-1}, X_i + a, X_{i+1}, \ldots, X_n)$, i.e. we recover all elementary polynomial automorphisms!

Example 2.1.10 Let $D_0 := -2X_2\partial_1 + X_3\partial_2$ on $R[X_1, X_2, X_3]$. Then D_0 is locally nilpotent by 1.3.17. Furthermore $a := X_1X_3 + X_2^2$ satisfies $D_0(a) = 0$, so by 1.3.34 $D := aD_0$ is locally nilpotent. Also one easily verifies that $DX_1 = -2aX_2$, $D^2X_1 = -2a^2X_3$, $D^3X_1 = 0$, $DX_2 = aX_3$, $D^2X_2 = DX_3 = 0$. So we get

$$\begin{aligned}\exp D(X_1) &= X_1 - 2aX_2 - a^2X_3,\\ \exp D(X_2) &= X_2 + aX_3,\\ \exp D(X_3) &= X_3.\end{aligned}$$

Hence $\exp D = \sigma$, Nagata's example described in Chapter 1.

With these examples in mind let us reflect a moment on the Tame Generators Conjecture. As we remarked in Chapter 1 we do not know if the Nagata automorphism σ is tame. On the other hand in Chapter 5, §1 we will show that there do exist non-tame automorphisms in $\mathrm{Aut}_R\, R[X_1, X_2]$ as soon as R is a domain which is not a field. More precisely we show: if R is a domain which is not a field and $0 \neq c$ a non-unit in R, then for $a := cX_1 + X_2^2$ the map $\tilde{\sigma}$ defined by

$$\begin{aligned}\tilde{\sigma}(X_1) &= X_1 - 2aX_2 - a^2c,\\ \tilde{\sigma}(X_2) &= X_2 + ac\end{aligned}$$

is a non-tame R-automorphism of $R[X_1, X_2]$. However one easily verifies that $\tilde{\sigma} = \exp D$, where D is the locally nilpotent derivation

$$D = a\left(-2X_2\frac{\partial}{\partial X_1} + c\frac{\partial}{\partial X_2}\right).$$

All these examples suggest replacing the Tame Generators Conjecture, which as the example $\tilde{\sigma}$ shows, can at most be true for a field, by Conjecture 2.1.11 below. Recalling that the affine subgroup of $\mathrm{Aut}_R\, R[X]$, denoted $\mathrm{Aff}(R, n)$, consists of all automorphisms F such that $\deg F_i = 1$ for all i, we formulate

(2.1.11) **Exponential Generators Conjecture.** Let R be any commutative $\mathbb{Q}$-algebra. Then $\mathrm{Aut}_R\, R[X]$ is generated by $\mathrm{Aff}(R, n)$ and the exponential automorphisms of $R[X]$.

Now we will provide some evidence for this conjecture. More precisely we will show that all elements of the so-called nilpotency subgroup of $\mathrm{Aut}_R\, R[X]$ are exponential automorphisms! As a consequence we show that it suffices to prove the conjecture for reduced rings.

Definition 2.1.12 The **nilpotency subgroup** of $\mathrm{Aut}_R\, R[X]$, denoted by $N(R, n)$, consists of all F of the form

$$(X_1 + g_1, \ldots, X_n + g_n)$$

where each g_i is a nilpotent element of $R[X]$ (or equivalently belongs to $\eta R[X]$, where η is the nilradical of R).

Indeed it follows from 1.1.9 that each such map belongs to $\mathrm{Aut}_R\, R[X]$. The relevance of understanding $N(R,n)$ will be explained in the last section of this chapter in connection with the Jacobian Conjecture.

Proposition 2.1.13 (van den Essen). *If $F \in N(R,n)$, then F is an exponential automorphism.*

Proof: i) Using 2.1.3 we need to show that the F-derivation E associated to F is locally nilpotent. So let $a \in R[X]$. We must prove that $E^p(a) = 0$ for some $p \geq 1$. Therefore replacing R by the subalgebra of R generated by all coefficients appearing in a and F, we may assume that R is Noetherian and hence that $\eta^m = 0$ for some positive integer m.

ii) Now we show that $E^m(R[X]) = 0$. Therefore let $h \in R[X]$. Since each $g_i \in \eta R[X]$ the same holds for $E(h) = h(X_1+g_1, \ldots, X_n+g_n) - h(X_1, \ldots, X_n)$. So $E(R[X]) \subset \eta R[X]$. Since E is R-linear, applying E to the previous inclusion gives $E^2(R[X]) \subset \eta^2 R[X]$. Repeating this argument we finally arrive at $E^m(R[X]) \subset \eta^m R[X] = 0$, as desired. □

Corollary 2.1.14 *It suffices to prove the exponential conjecture for reduced rings.*

Proof: i) Let $\overline{R} := R/\eta$. So $\overline{R}$ is reduced. If $\ell \in \mathrm{Aff}(\overline{R}, n)$ one readily verifies that $\ell = \overline{L}$ for some $L \in \mathrm{Aff}(R,n)$. Here $\overline{L}$ is obtained from L by reducing the coefficients of L modulo η. Furthermore it follows from 2.1.15 below that if $f = \exp d$ for some locally nilpotent derivation d on $\overline{R}[X]$, then there exists a locally nilpotent derivation D on $R[X]$ with $\overline{D} = d$ and $F := \exp D$ satisfies $\overline{F} = f$.

ii) Now suppose that F is an R-automorphism of $R[X]$. Assuming the exponential conjecture for reduced rings, we deduce that $\overline{F}$ is a product of affine automorphisms and exponential ones. By i) each of these factors in this product comes from an affine, respectively exponential, automorphism of $R[X]$, i.e. is obtained by reducing these modulo η. Denoting the corresponding product in $\mathrm{Aut}_R\, R[X]$ by F_*, we get that $\overline{F} = \overline{F_*}$ and hence that $\overline{F_*^{-1} \circ F} = X$, i.e. $F_*^{-1} \circ F \in N(R,n)$. So it is an exponential automorphism, say $\exp D_0$, and hence $F = F_* \circ \exp D_0$. Then the desired result follows from the definition of F_*. □

Lemma 2.1.15 *If d is a locally nilpotent derivation on $\overline{R}[X]$, then there exists a locally nilpotent derivation D on $R[X]$ such that $\overline{D} = d$.*

Proof: i) For each i choose $a_i \in R[X]$ such that $\overline{a_i} = d(X_i)$ and let $D = \sum a_i \partial_i$. As in the proof of the previous proposition, replacing R by the subalgebra of R generated by all monomials appearing in the a_i, we may assume that R is Noetherian and that $\eta^e = 0$ for some positive integer e.

ii) Now let $g \in R[X]$. The nilpotency of d implies that $D^m(g) \in \eta R[X]$ for some positive integer m, i.e. $D^m(g)$ is a finite sum of elements of the form $c_i g_i$, with $c_i \in \eta$ and $g_i \in R[X]$. Similarly there exists a positive integer m_2 such that $D^{m_2}(g_i) \in \eta R[X]$ for all i, whence $D^{m+m_2}(g) \in \eta^2 R[X]$. Repeating this argument we finally get that $D^M(g) \in \eta^e R[X] = 0$, for some $M \in \mathbb{N}$, as desired. □

Finally we mention an interesting result of Alev and Chamarie [18], which gives some support for the truth of Conjecture 2.1.11. Instead of considering polynomial rings over a commutative $\mathbb{Q}$-algebra we consider the so-called quantized polynomial rings over $\mathbb{C}$.

Definition 2.1.16 Let $q = (q_{ij}) \in M_n(\mathbb{C})$ be such that $q_{i,i} = 1$ for all i and $q_{i,j} = q_{j,i}^{-1}$ for all $i \neq j$. Then $\mathbb{C}[X]_{(q)}$, the **quantized polynomial ring over** $\mathbb{C}$, is the $\mathbb{C}$-algebra generated by $X_1, \ldots, X_n$ with relations $X_i X_j = q_{i,j} X_j X_i$ for all i, j. If all $q_{i,j} = 1$ we get the usual polynomial ring over $\mathbb{C}$. So we say that $\mathbb{C}[X]_{(q)}$ is a deformation of $\mathbb{C}[X]$. In case $n = 2$ the ring $\mathbb{C}[X]_{(q)}$ is called the **$\mathbb{C}$-algebra of the quantum-plane** $X_1 X_2 = q X_2 X_1$.

Theorem 2.1.17 (Alev, Chamarie). *Assume that $q_{i,j} = q$ for all $i < j$ and that q is not a root of unity. Then* i) *If $n \neq 3$, $\mathbb{C}[X]_{(q)}$ has no non-zero locally nilpotent derivations and* $\operatorname{Aut}_{\mathbb{C}} \mathbb{C}[X]_{(q)}$ *consists of linear automorphisms.*

ii) *If $n = 3$, the only locally nilpotent derivations on $\mathbb{C}[X]_{(q)}$ are of the form λD with $\lambda \in \mathbb{C}$ and $DX_1 = DX_3 = 0$, $DX_2 = X_1 X_3$. Furthermore* $\operatorname{Aut}_{\mathbb{C}} \mathbb{C}[X]$ *is generated by the linear automorphisms and the elements* $\exp(\lambda D)$.

Exercises for §2.1

1. Let σ be as in Chapter 1. Compute the inverse of σ using locally nilpotent derivations.
2. Let R be any commutative $\mathbb{Q}$-algebra, $r_1, r_2 \in R$ and $f(T) \in R[T]$. Put $D := f(r_1 X_1 + r_2 X_2)(-r_2 \partial_1 + r_1 \partial_2)$. Show that D is locally nilpotent on $R[X_1, X_2]$ and compute $\exp D$.
3. Show that the following polynomial map F is invertible and compute its inverse
$$F = (X_1 - X_4 d, X_2 + X_3 d, X_3, X_4), \text{ where } d = X_3 X_1 + X_4 X_2.$$
[This example is known as **Anick's example**. It is also a candidate non-tame polynomial automorphism over $\mathbb{C}$.]
4. (Furter) Let R be a commutative $\mathbb{Q}$-algebra, d an integer ≥ 2 and $F = (X_1 + H_1, X_2 + H_2) \in R[X_1, X_2]^2$, where each H_i is either zero or a homogeneous polynomial of degree d.

   Show that if $\det JF = 1$, then F is invertible with inverse $G = X - H$. [The Jacobian condition gives $\operatorname{Tr} JH = 0$ and $\det JH = 0$. So by 1.3.53 $H_1 = -\partial_2 P$ and $H_2 = \partial_1 P$ for some $P \in R[X_1, X_2]$. Consequently
$$\partial_1^2 P \partial_2^2 P - (\partial_1 \partial_2 P)^2 = 0.$$
   Put $D := H_1 \partial_1 + H_2 \partial_2$ and show, using the Euler operator, that $D^2 X_i = 0$ for $i = 1, 2$. So $F = \exp D$ etc.]
5. (Furter) Let $R = \mathbb{C}[T]/(T^2)$. Put $\varepsilon := \overline{T} \in R$. Let
$$F = (X_1 + \varepsilon X_1^3, (1 - 3\varepsilon X_1^2) X_2 + X_1^2).$$
   (i) Write F as a product of elementary automorphisms.
   (ii) Compute F^{-1} and observe that $\deg F^{-1} = 4$.

6. Let A be a commutative ring and $D : A \to A$ a locally nilpotent $\mathbb{Q}$ linear map.
   (i) Show that D is a derivation on A if and only if $\exp nD : A \to A$ is a ring homomorphism for all $n \in \mathbb{N}$.
   (ii) Use i) to give another proof of the second part of 2.1.3.

2.2 Derivations and the Jacobian Condition

Let R be a commutative ring and $R[X] := R[X_1, \ldots, X_n]$. Let $F := (F_1, \ldots, F_n) \in R[X]^n$ satisfy the Jacobian Condition, i.e.

$$\det JF \in R[X]^*. \tag{2.2.1}$$

To such an F we can associate an n-tuple of R-derivations, denoted $\frac{\partial}{\partial F_1}, \ldots, \frac{\partial}{\partial F_n}$, as follows

$$\begin{pmatrix} \frac{\partial}{\partial F_1} \\ \vdots \\ \frac{\partial}{\partial F_n} \end{pmatrix} = ((JF)^{-1})^t \begin{pmatrix} \partial_1 \\ \vdots \\ \partial_n \end{pmatrix}. \tag{2.2.2}$$

The following lemma justifies the notation $\frac{\partial}{\partial F_i}$.

Lemma 2.2.3 *Let F satisfy* 2.2.1 *and $\frac{\partial}{\partial F_i}$ be as in* 2.2.2. *Then* i) $\frac{\partial}{\partial F_i}(F_j) = \delta_{ij}$, *for all i, j.*
ii) $\frac{\partial}{\partial F_1}, \ldots, \frac{\partial}{\partial F_n}$ *is an $R[X]$-basis of* $\mathrm{Der}_R\, R[X]$.
iii) $[\frac{\partial}{\partial F_i}, \frac{\partial}{\partial F_j}] = 0$, *for all i, j.*

Proof: i) Let $1 \leq i \leq n$. Applying 2.2.2 to F_i and transposing gives

$$\left(\frac{\partial}{\partial F_1}(F_i), \ldots, \frac{\partial}{\partial F_n}(F_i)\right) = (\partial_1(F_i), \ldots, \partial_n(F_i))(JF)^{-1}.$$

Since $(\partial_1(F_i), \ldots, \partial_n(F_i))$ is the i-th row of JF we get that $(\frac{\partial}{\partial F_1}(F_i), \ldots, \frac{\partial}{\partial F_n}(F_i))$ is the i-th row of $JF.JF^{-1} = I$, which gives i).
ii) follows immediately from 2.2.2 since $((JF)^{-1})^t \in Gl_n(R[X])$.
iii) Let $1 \leq i, j \leq n$. Put $D := [\frac{\partial}{\partial F_i}, \frac{\partial}{\partial F_j}]$. Then $D = \sum a_i \frac{\partial}{\partial F_i}$ for some $a_i \in R[X]$ (by ii)). From i) it follows that $DF_i = 0$ for all i, whence $a_i = DF_i = 0$. So $D = 0$, as desired. □

In fact the properties ii) and iii) of 2.2.3 characterize the n-tuples $(\frac{\partial}{\partial F_1}, \ldots, \frac{\partial}{\partial F_n})$ coming from F's satisfying 2.2.1. More precisely

Proposition 2.2.4 (Nowicki). *Let R be a $\mathbb{Q}$-algebra and $D_1, \ldots, D_n$ a* ***commutative*** *$R[X]$-basis of* $\mathrm{Der}_R\, R[X]$ *(i.e. $[D_i, D_j] = 0$ for all i, j). Then there exists $F = (F_1, \ldots, F_n) \in R[X]^n$ with* $\det JF \in R[X]^*$ *such that $D_i = \frac{\partial}{\partial F_i}$ for all i.*

Proof: Since $D_1, \ldots, D_n$ is an $R[X]$-basis of $\mathrm{Der}_R\, R[X]$ we get

$$(2.2.5) \qquad \begin{pmatrix} \partial_1 \\ \vdots \\ \partial_n \end{pmatrix} = B \begin{pmatrix} D_1 \\ \vdots \\ D_n \end{pmatrix}$$

with

$$(2.2.6) \qquad B = (b_{ij}) \in Gl_n(R[X]), \text{ i.e. } \det B \in R[X]^*.$$

Now write the equations $[\partial_i, \partial_j] = 0$ in terms of the b_{ij} and the derivations D_j and use the hypothesis $[D_i, D_j] = 0$, for all i, j. This gives

$$\sum_p \left(\sum_q b_{iq} D_q(b_{jp}) \right) D_p = \sum_p \left(\sum_q b_{jq} D_q(b_{ip}) \right) D_p.$$

Equating the corresponding coefficients of the D_p, using that $D_1, \ldots, D_n$ is an $R[X]$-basis of $\mathrm{Der}_R\, R[X]$, we get

$$\sum_q b_{iq} D_q(b_{jp}) = \sum_q b_{jq} D_q(b_{ip})$$

for all p, or equivalently

$$\sum_q b_{iq} D_q \begin{pmatrix} b_{j1} \\ \vdots \\ b_{jn} \end{pmatrix} = \sum_q b_{jq} D_q \begin{pmatrix} b_{i1} \\ \vdots \\ b_{in} \end{pmatrix}.$$

In other words

$$\partial_i \begin{pmatrix} b_{j1} \\ \vdots \\ b_{jn} \end{pmatrix} = \partial_j \begin{pmatrix} b_{i1} \\ \vdots \\ b_{in} \end{pmatrix}$$

for all i, j. Then by 1.3.53 we get

$$B^t = \left(\partial_1 \begin{pmatrix} F_1 \\ \vdots \\ F_n \end{pmatrix}, \ldots, \partial_n \begin{pmatrix} F_1 \\ \vdots \\ F_n \end{pmatrix} \right)$$

for some $F_i \in R[X]$. So $B^t = JF$ and hence by 2.2.5 and 2.2.6 we get

$$\begin{pmatrix} D_1 \\ \vdots \\ D_n \end{pmatrix} = ((JF)^t)^{-1} \begin{pmatrix} \partial_1 \\ \vdots \\ \partial_n \end{pmatrix}$$

with $\det JF \in R[X]^*$. □

(2.2.7) **Reformulations of the Jacobian Conjecture**

The above remarks enable us to give several reformulations of the Jacobian Conjecture in terms of the derivations $\frac{\partial}{\partial F_i}$. The first of these reformulations is a consequence of 2.2.4 (the details are left to the reader, Exercise 1).

Corollary 2.2.8 *Let $n \geq 1$ and R a $\mathbb{Q}$-algebra. Then $JC(R, n)$ is equivalent to (2.2.9) $(\partial_1, \dots, \partial_n)$ is, apart from a polynomial coordinate change, the only commutative $R[X]$-basis of* $\mathrm{Der}_R\, R[X]$.

To give a second reformulation of the Jacobian Conjecture we make some preparations. Let $F \in R[X]^n$ satisfy 2.2.1. Then $\det JF(0) \in R^*$, so if $F(0) = 0$ the inverse function theorem implies that there exists a uniquely determined formal power series $G = (G_1, \dots, G_n) \in R[[V_1, \dots, V_n]]^n$ satisfying $G(0) = 0$ and $G(F) = X$, i.e. $X_i = G_i(F_1, \dots, F_n)$ for all i. So $R[[X]] = R[[F]]$. Furthermore the derivations $\frac{\partial}{\partial F_i}$ can be extended uniquely to $R[[F]]$, using 1.2.7. After these preparations we get

Proposition 2.2.10 *Let $F \in R[X]^n$ satisfy* 2.2.1. *Then F is a polynomial automorphism if and only if $\frac{\partial}{\partial F_i}$ is locally nilpotent on $R[X]$ for all i.*

Proof: If F is invertible, then $R[X] = R[F]$ and the result is obvious. Conversely, assume that each $\frac{\partial}{\partial F_i}$ is locally nilpotent on $R[X]$. We may assume that $F(0) = 0$. Let $g \in R[X]$ and $1 \leq i \leq n$. Then $(\frac{\partial}{\partial F_i})^p g = 0$ for some $p \in N$. So viewing g as, formal power series in $R[[F]]$ it follows that g is a polynomial in F_i with $\deg_{F_i} g \leq p - 1$. This holds for all i, so $g \in R[F]$ implying that $R[X] = R[F]$. □

Let us give one more reformulation of the Jacobian Conjecture in terms of the derivations $\frac{\partial}{\partial F_i}$. Therefore let R be a $\mathbb{Q}$-algebra. Observe that if $F = (F_1, \dots, F_n) \in R[X]^n$ is invertible, then $R[X] = R[F]$. So we get

$$\ker\left(\frac{\partial}{\partial F_n}, R[X]\right) = \ker\left(\frac{\partial}{\partial F_n}, R[F]\right) = R[F_1, \dots, F_{n-1}].$$

So if *JC(R, n)* is true, then the following conjecture is true as well.

(2.2.11) **Kernel Conjecture *(KC(R, n))*.** If $\det JF \in R[X]^*$, then

$$\ker\left(\frac{\partial}{\partial F_n}, R[X]\right) = R[F_1, \dots, F_{n-1}].$$

Conversely we have

Proposition 2.2.12 *KC(R, n+1) implies JC(R, n), for all $n \geq 1$.*

Proof: Let $F = (F_1, \dots, F_n) \in R[X]^n$ satisfy 2.2.1. We may assume that $F(0) = 0$. Put $\tilde{F} := (F, X_{n+1}) \in R[X, X_{n+1}]^{n+1}$. Then $\det J\tilde{F} = \det JF \in R[X]^* \subset R[X, X_{n+1}]^*$. So by *KC(R, n+1)* we get

$$\ker\left(\frac{\partial}{\partial \tilde{F}_{n+1}}, R[X_1, \dots, X_{n+1}]\right) = R[F_1, \dots, F_n]. \tag{2.2.13}$$

Now we will show

(2.2.14) $$\frac{\partial}{\partial \tilde{F}_{n+1}} = \partial_{n+1} \text{ on } R[X, X_{n+1}].$$

Let us assume 2.2.14 for a moment. Then 2.2.13 implies

$$R[X] = \ker\left(\frac{\partial}{\partial_{n+1}}, R[X, X_{n+1}]\right) = \ker\left(\frac{\partial}{\partial \tilde{F}_{n+1}}, R[X, X_{n+1}]\right) = R[F],$$

i.e. F is invertible. So it remains to prove 2.2.14. Therefore observe that $\frac{\partial}{\partial \tilde{F}_{n+1}}$ and ∂_{n+1} coincide on $R[\tilde{F}]$, since they coincide on X_{n+1} and each F_i. Hence their extensions to $R[[\tilde{F}]]$ also coincide. However by 1.1.2 $R[[\tilde{F}]] = R[[X, X_{n+1}]]$, so $\frac{\partial}{\partial \tilde{F}_{n+1}}$ and ∂_{n+1} coincide on $R[[X, X_{n+1}]]$ and hence on $R[X, X_{n+1}]$, which proves 2.2.14. □

So if the Jacobian Conjecture is true, then in particular $R[X]^{\frac{\partial}{\partial F_n}}$ should be a finitely generated R-algebra. As we saw already in section two of Chapter 1, for arbitrary derivations D on $\mathbb{C}[X]$, $\mathbb{C}[X]^D$ need not be a finitely generated $\mathbb{C}$-algebra. On the other hand we have

Proposition 2.2.15 *If k is a field of characteristic zero, then KC(k, 2) is true, i.e.*

$$\ker\left(\frac{\partial}{\partial F_2}, k[X_1, X_2]\right) = k[F_1].$$

Proof: By 1.2.25 $\ker(\frac{\partial}{\partial F_2}, k[X_1, X_2]) = k[f]$ for some $f \in k[X_1, X_2]$. Since $\frac{\partial}{\partial F_2}(F_1) = 0$ we deduce that $F_1 = g(f)$ for some non-zero polynomial $g(T) \in k[T]$. Apply $\frac{\partial}{\partial F_1}$ to this equation and observe that $\frac{\partial}{\partial F_1} k[f] \subset k[f]$, since $\frac{\partial}{\partial F_1}$ and $\frac{\partial}{\partial F_2}$ commute (by 2.2.3 iii)). Consequently $1 = g'(f)\frac{\partial}{\partial F_1}(f)$ in $k[f]$. So $g'(f) \in k^*$, i.e. $F_1 = g(f) = \lambda f + \mu$ for some $\lambda, \mu \in k$, $\lambda \neq 0$. Consequently $k[f] = k[F_1]$. □

To conclude this section we restrict our attention to polynomial rings over a field k of characteristic zero. We will show how the derivations $\frac{\partial}{\partial F_i}$ play a prominent role in the proof of the following "classical" result.

Theorem 2.2.16 *Let k be a field of characteristic zero and $F \in k[X]^n$ satisfying* 2.2.1. *Then there is equivalence between*

i) *F is invertible.*
ii) *$k[F] \subset k[X]$ is finite, i.e. $k[X]$ is a finite $k[F]$-module.*
iii) *$k(F) \subset k(X)$ is a Galois extension.*

Proof: i)→ii) and i)→ iii) are obvious. So let us assume that $k[F] \subset k[X]$ is finite. Then by 1.3.37 the derivations $\frac{\partial}{\partial F_i}$, which are obviously locally nilpotent on $k[F]$, are locally nilpotent on $k[X]$. Then by 2.2.10 F is invertible. So it remains to prove iii) → i). The proof is based on the following general result. □

Lemma 2.2.17 *Let $A \subset B$ be integral domains whose quotient fields K, resp. L, form a finite Galois extension satisfying $B \cap K = A$. Let D be a derivation on L such that $DK \subset K$ and $DB \subset B$. If B is integrally closed, then $D\overline{A} \subset \overline{A}$ where $\overline{A}$ is the integral closure of A in L.*

Proof: i) Let $G := Gal(L/K)$ and $g \in G$. Since $DK \subset K$, the derivation gDg^{-1} coincides with the derivation D on K and hence on L by 1.2.7. So $gD = Dg$ for all $g \in G$.

ii) If $x \in L$ is integral over A then $g(x)$ is integral over A, since g fixes all elements of $A (\subset K)$. So $g(\overline{A}) \subset \overline{A}$.

iii) If $x \in \overline{A}$, then x is integral over A hence over B. Since B is integrally closed this implies that $x \in B$. So $\overline{A} \subset B$.

iv) From i), ii), iii) and the hypothesis $DB \subset B$ we deduce

$$gD(\overline{A}) \subset Dg(\overline{A}) \subset D(\overline{A}) \subset DB \subset B.$$

v) Finally to see that $D\overline{A} \subset \overline{A}$ let $x \in \overline{A}$ and put

$$P(X) := \Pi_{g \in G}(X - g(D(x))).$$

Since the coefficients of $P(X)$ are invariant under G we get $P(X) \in K[X]$. Furthermore it follows from iv) that $P(X) \in B[X]$, so $P(X) \in K[X] \cap B[X] = A[X]$. Since $D(x)$ is a zero of $P(X)$, which is monic, it follows that $D(x)$ is integral over A, i.e. $D(x) \in \overline{A}$. □

Proof of Theorem 2.2.16: (finished) Let $1 \le i \le n$. Apply Lemma 2.2.17 to $A := k[F]$, $B := k[X]$ and $D := \frac{\partial}{\partial F_i}$. So $\frac{\partial}{\partial F_i}\overline{A} \subset \overline{A}$, whence by 1.3.37 $\frac{\partial}{\partial F_i}$ is locally nilpotent on $\overline{A}$ for all i. Since B is integrally closed we have $\overline{A} \subset B = k[X]$. So arguing as in Proposition 2.2.10 we get $\overline{A} \subset A$, i.e. $\overline{A} = A$. Since $Q(\overline{A}) = Q(B)$ we deduce that $Q(A) = Q(B)$. Then $A = B$ follows from Keller's theorem. □

Exercises for § 2.2

1. Give a proof of 2.2.8. [Use that if φ is an endomorphism of $R[X]$ such that $\varphi(X_i) = F_i$ for all i and $\det JF \in R[X]^*$, then $\frac{\partial}{\partial F_i} \circ \varphi = \varphi \circ \partial_i$.]
2. Let k be a field of characteristic zero. Let $k[Y] := k[Y_1, \ldots, Y_n] \subset k[X] := k[X_1, \ldots, X_n]$ be two polynomial rings over k. Prove that the following statement is equivalent to the Jacobian Conjecture: if every k-derivation of $k[Y]$ extends to a k-derivation of $k[X]$, then $k[Y] = k[X]$.
3. (Cancellation problem) Let $\mathbb{C}[X_1, \ldots, X_n] = A[T]$ a polynomial ring in one variable over a $\mathbb{C}$-algebra A. Is A a polynomial ring in $n-1$ variables over $\mathbb{C}$, i.e. $A = \mathbb{C}[F_1, \ldots, F_{n-1}]$ for some $F_i \in \mathbb{C}[X]$?

   Show that the Cancellation problem is equivalent to: Let D be a locally nilpotent derivation on $\mathbb{C}[X]$ having a slice s in $\mathbb{C}[X]$. Does it follow that $\mathbb{C}[X]^D = \mathbb{C}[F_1, \ldots, F_{n-1}]$ for some $F_i \in \mathbb{C}[X]$? [Use 1.3.21.]

4. Let R be a commutative $\mathbb{Q}$-algebra and $F_1, \ldots, F_n \in R[X]$ with 2.2.1. Let $D \in \mathrm{Der}_R\, R[X]$ be defined by $D(g) := \det J(F_1, \ldots, F_{n-1}, g)$. Show that $D = \det JF.\frac{\partial}{\partial F_n}$.
5. (Uniqueness of the Jacobian mate) Let k be a field of characteristic zero. Let $F_1, F_2, F_2^\star \in k[X_1, X_2]$ be such that $\det J(F_1, F_2) = \det J(F_1, F_2^\star) \in k^*$. Show that $F_2^\star = F_2 + a(F_1)$, for some $a(F_1) \in k[F_1]$. [Show that the derivations $\frac{\partial}{\partial F_2}$ and $\frac{\partial}{\partial F_2^\star}$ are equal and use 2.2.15.]
6. (Razar) Let k be a field of characteristic zero and $F \in k[X]^n$ with $\det JF \in k^*$. If $k[X] \subset K[h]$ for some $h \in k[X]$, where $K := k(F_1, \ldots, F_{n-1})$, then F is invertible. [Write $F_n = g(h) \in K[h]$ and apply $\frac{\partial}{\partial F_n}$. Deduce that $h \in k(F)$ and apply 1.1.35.]
7. Let R be a commutative ring and $R[[X]] := R[[X_1, \ldots, X_n]]$. Let $F = (F_1, \ldots, F_n) \in R[[X]]^n$ with $F(0) = 0$ and $\det JF(0) \in R^*$. Then we can define derivations $\frac{\partial}{\partial F_1}, \ldots, \frac{\partial}{\partial F_n}$ on $R[[X]]$ by the formula 2.2.2. Show that these derivations verify the analagous statements of 2.2.3, obtained by replacing $R[X]$ by $R[[X]]$.

2.3 The degree of the inverse of a polynomial automorphism

Throughout this section k denotes a field of characteristic zero and R a commutative k-algebra. For $F = (F_1, \ldots, F_n) \in R[X]^n$ we define the **degree of** F, denoted $\deg F$, as

$$\deg F := \max_i \deg F_i .$$

An interesting question in the study of polynomial automorphisms is the following: let F be a polynomial automorphism with inverse F^{-1}. What can be said about $\deg F^{-1}$?

This question was first studied for polynomial rings over a field. The remarkable result (2.3.1) is that $\deg F^{-1} \leq (\deg F)^{n-1}$. First proofs of this fact appeared in [29] and [256]. It is not difficult to extend this result to reduced k-algebras. However for non-reduced rings the above estimate no longer holds. If one analyses the example given after 2.3.4 one observes that the cause of the trouble seems to be the fact that $R^* \neq R[X]^*$, for non-reduced rings. So this naturally leads to the question: suppose that F is a polynomial automorphism over any k-algebra R such that $\det JF \in R^*$: does it follow that $\deg F^{-1} \leq (\deg F)^{n-1}$ or even weaker: does it follow that the degree of F^{-1} is bounded by a number $C(n, \deg F)$ which only depends on n and $\deg F$, i.e. does **not** depend on the k-algebra R? Now it turns out (2.3.8) that this question is equivalent to the Jacobian Conjecture.

Since for reduced rings the bound $(\deg F)^{n-1}$ works, it is natural to ask if for automorphisms with $\det JF = 1$ we can take $C(n, \deg F) = (\deg F)^{n-1}$. However we will give some easy examples in 2.3.5 which show that the answer is negative. At the end of this section we discuss some questions/conjectures concerning polynomial automorphisms over a field.

Proposition 2.3.1 *Let k be a field and $F = (F_1, \dots, F_n) \in \mathrm{Aut}_k\, k[X]$. Then* $\deg F^{-1} \leq (\deg F)^{n-1}$.

Proof: Let $(G_1, \dots, G_n)$ be the inverse of F, where $G_i \in k[Y_1, \dots, Y_n]$ for each i. By the Lefschetz principle we may assume that $k = \mathbb{C}$. Furthermore, making a linear coordinate change, we may assume that $\deg G_i = \deg_{Y_n} G_i$ for all i. Since $X_i = G_i(F_1, \dots, F_n)$ we deduce (using the notation of 1.3.32) that

$$\deg G_i = \deg_{Y_n} G_i = \rho\left(\frac{\partial}{\partial F_n}, X_i\right) - 1. \tag{2.3.2}$$

Since G_i contains Y_n ($\deg_{Y_n} G_i = \deg G_i \geq 1$) we have $\frac{\partial}{\partial F_n}(X_i) \neq 0$. So $(F_1 \dots, F_{n-1}, X_i)$ are algebraically independent over $\mathbb{C}$ by 1.2.8. Furthermore $\frac{\partial}{\partial F_n}$ is locally nilpotent on $\mathbb{C}[F] = \mathbb{C}[X]$ and $\mathbb{C}[X]^{\frac{\partial}{\partial F_n}} = \mathbb{C}[F_1, \dots, F_{n-1}]$. Whence by 2.3.2 and 1.3.32 iv) we get

$$\deg G_i \leq |\mathbb{C}(X) : \mathbb{C}(F_1 \dots, F_{n-1}, X_i)|. \tag{2.3.3}$$

Finally observe that the map $\tilde{F} = (F_1 \dots, F_{n-1}, X_i) : \mathbb{C}^n \to \mathbb{C}^n$ is dominant, (Appendix B) since $F_1 \dots, F_{n-1}, X_i$ are algebraically independent over $\mathbb{C}$. So we can apply B.2.1 and B.2.7, which completes the proof. □

Corollary 2.3.4 *Let R be a reduced ring and $F \in \mathrm{Aut}_R\, R[X]$. Then* $\deg F^{-1} \leq (\deg F)^{n-1}$.

Proof: Write again $G = (G_1, \dots, G_n)$ instead of F^{-1}. i) If R is a domain, embed R in its quotient field and apply 2.3.1.

ii) To prove the general case let $1 \leq i \leq n$ and $\alpha = (\alpha_1, \dots, \alpha_n)$ with all $\alpha_i \geq 0$ such that $|\alpha| > (\deg F)^{n-1}$. It suffices to show that $c_\alpha^{(i)} = 0$, where $c_\alpha^{(i)}$ is the coefficient of the monomial X^α in G_i. Therefore let $\wp$ be a prime ideal in R and consider the maps $\overline{F}$ and $\overline{G}$, obtained by reducing the coefficients of the F_i and G_j modulo $\wp$. Put $\overline{R} := R/\wp$. Then we deduce from i) that

$$\deg \overline{G} \leq (\deg \overline{F})^{n-1} \leq (\deg F)^{n-1}.$$

So $\overline{c_\alpha^{(i)}} = 0$, i.e. $c_\alpha^{(i)} \in \wp$. Since this holds for all prime ideals $\wp$ in R, we deduce that $c_\alpha^{(i)} \in \cap \wp = (0)$, since R is reduced. □

If R is not a reduced ring there does not exist a uniform bound depending only on n and $\deg F$. This can be seen as follows. Let $m \in \mathbb{N}$, $m \geq 2$ and let $R = k[T]/(T^m)$: put $F = (F_1, \dots, F_n)$ with $F_1 = X_1 - \overline{T} X_1^2$ and $F_i = X_i$ for all $i \geq 2$. Observe that $F_1' = 1 - 2\overline{T} X_1$ is a unit in $R[X_1]$. So by 1.1.12 and the fact that $JC(\mathbb{C}, 1)$ is true, it follows that there exists $G_1 \in R[X_1]$ such that $G_1(F(X_1)) = X_1$. Furthermore

$G = (G_1, X_2, \ldots, X_n)$ is the inverse of F. Differentiating the relation $G(F_1(X_1)) = X_1$ gives $G_1'(F_1(X_1))F_1'(X_1) = 1$, so

$$G_1'(X_1 - \overline{T}X_1^2) = 1 + 2\overline{T}X_1 + (2\overline{T}X_1)^2 + \cdots + (2\overline{T}X_1)^{m-1}.$$

Consequently $m - 1 = \deg G_1'(X_1 - \overline{T}X_1^2) \leq 2(\deg G_1 - 1)$. So

$$\deg G_1 \geq \frac{m-1}{2} + 1.$$

So $\deg G$ can be arbitrarily large, depending on the choice of R, in spite of the fact that $\deg F = 2$ (see also Exercise 3).

Looking at the above example, the cause of the trouble seems to be the strict inequality

$$R^* \neq R[X]^*.$$

On the other hand if we had assumed that $n = 1$ and $\det JF \in R^*$ instead of $\in R[X]^*$, then the bound of 2.3.1 works. So one could hope that the bound $(\deg F)^{n-1}$ holds for all $F \in \mathrm{Aut}_R\, R[X]$ satisfying the additional condition that $\det JF \in R^*$. However the example given in Chapter 2, §1 Exercise 5, which is due to Furter in [136], shows that even under this additional assumption $\deg F^{-1}$ can exceed $(\deg F)^{n-1}$. In fact it is rather easy to find counterexamples.

Example 2.3.5 Let R be a non-reduced ring and ε a non-zero element in R such that $\varepsilon^2 = 0$. Let $d \geq 2$ and consider $F = (X + Y^d, Y) \circ (X, Y + \varepsilon X^d)$. Then $\det JF = 1$ and using $\varepsilon^2 = 0$ we get $F = (X + (Y^d + dY^{d-1}\varepsilon X^d), Y + \varepsilon X^d)$. So $\deg F = 2d - 1$. Furthermore $F^{-1} = (X, Y - \varepsilon X^d) \circ (X - Y^d, Y) = (X - Y^d, Y - \varepsilon(X - Y^d)^d)$. So $\deg F^{-1} = d^2$.

In the example just given we have $\deg F = 2d - 1$, i.e. $d = \frac{1}{2}(\deg F + 1)$. So $\deg F^{-1} \leq (\frac{1}{2}(\deg F + 1))^2$. So one could still hope for a uniform bound for $\deg F^{-1}$, i.e. depending only on $\deg F$ and n. Now we will show that the existence of such a bound is equivalent to the Jacobian Conjecture. To describe the precise result we introduce the following "uniform-boundness" statements.

$UB(k, n)$ *For every* $d \in \mathbb{N}$ *there exists a positive integer* $C(n, d)$ *such that: for every k-algebra R and any* $F \in \mathrm{Aut}_R\, R[X]$ *satisfying* $\det JF = 1$ *and* $\deg F \leq d$, *the degree of* F^{-1} *is bounded by* $C(n, d)$. $\overline{UB}(k, n)$ *As in* $UB(k, n)$ *but replace "for all k-algebras R" by "for all k-algebras of the form* $k[T]/(T^l)$, $l \geq 1$*".*

Theorem 2.3.6 *Let* $n \in \mathbb{N}$. i) *If JC(R, n) is true for all k-algebras R, then UB(k, n) is true.*

ii) $\overline{UB}(k, n)$ *implies JC(k, n).*

Proof: i) Consider an n-tuple of universal polynomials $F_1^U, \ldots, F_n^U$ of degree d, i.e. each F_i^U is of the form

$$F_i^U := \sum_{|\alpha| \leq d} A_\alpha^{(i)} X^\alpha$$

where all $A_\alpha^{(i)}$ are different variables. Denote by $k[A]$ the polynomial ring $k[A_\alpha^{(i)} : |\alpha| \le d, 1 \le i \le n]$ generated by all variables $A_\alpha^{(i)}$. Consider in $k[A][X]$ the polynomial

$$P := \det(J_X F^U) - 1.$$

Let J be the ideal in $k[A]$ generated by the coefficients of all monomials X^α appearing in P and put

$$R_0 := k[A]/J.$$

The polynomials $\overline{F_1^U}, \ldots, \overline{F_n^U} \in R_0[X]$, obtained by reducing the coefficients of the F_i^U modulo J satisfy $\det J_X(\overline{F_1^U}, \ldots, \overline{F_n^U}) = 1$ in R_0. So by our hypothesis $\overline{F_U}$ has a polynomial inverse $g = (g_1, \ldots, g_n)$ in $R_0[X]^n$. Let $C(n,d)$ be the degree of g and choose $G_1, \ldots, G_n$ in $k[A][X]$ such that $\overline{G_i} = g_i$ and $\deg G_i = \deg \overline{G_i}$ for all i. So $\deg G_i \le C(n,d)$ for all i and furthermore

(2.3.7) $F_i^U(G_1, \ldots, G_n) - X_i$ and $G_i(F_1^U, \ldots, F_n^U) - X_i$ in $Jk[A][X]$, for all i.

Let R be any k-algebra, $(F_1, \ldots, F_n) \in R[X]^n$ with $\det JF = 1$ and $\deg F \le d$. Then $F_i = \sum a_\alpha^{(i)} X^\alpha$, $|\alpha| \le d$ for some $a_\alpha^{(i)} \in R$ and since $\det JF = 1$, the $a_\alpha^{(i)}$ are zeros of the ideal J. Finally, making the substitutions $A_\alpha^{(i)} \to a_\alpha^{(i)}$ in 2.3.7 we deduce that $G(a) := (G_1(a), \ldots, G_n(a))$, obtained by substitution of $a = (a_\alpha^{(i)})$ in the coefficients of $G = (G_1, \ldots, G_n)$, is the inverse of F. So $\deg F^{-1} \le C(n,d)$.

ii) Let $F \in k[X]^n$ with $\det JF = 1$. We may assume that $F(0) = 0$. Let $d = \deg F$. By the formal inverse function theorem, F has a formal inverse $G = (G_1, \ldots, G_n) \in k[[X]]^n$. Let $F_{(j)}$, resp. $G_{(j)}$, denote the homogeneous component of F, resp. G, of degree j, for all $j \ge 0$. We are going to show that $G_{(l)} = 0$ if $l > C(n,d)$. Therefore let T be a new variable and consider the clearing maps ${}^T F \in k[T][X]^n$ and ${}^T G \in k[T][[X]]^n$ (§1 in Chapter 1). Then by 1.1.22 we get

$${}^T F \circ {}^T G = X \text{ and } \det J_X({}^T F) = 1.$$

Now let $l > C(n,d)$ be fixed and assume that $G_{(l)} \ne 0$. We will derive a contradiction. Reducing the above equalities modulo T^l we get

$$\overline{{}^T F} \circ \overline{{}^T G} = X \text{ and } \det J_X(\overline{{}^T F}) = 1 \text{ in } k_l[X], \text{ where } k_l := k[T]/(T^l).$$

Observe that ${}^T G$ is a polynomial map over k_l, so the equality $\overline{{}^T F} \circ \overline{{}^T G} = X$ implies that $\overline{{}^T F}$ is invertible over k_l. Since $\det J_X(\overline{{}^T F}) = 1$ we get from $\overline{UB}(k,n)$ that $\deg_X {}^T G \le C(n,d)$. Since

$${}^T G = G_{(1)} + \overline{T} G_{(2)} + \cdots + \overline{T}^{l-1} G_{(l)}$$

and $\overline{T}^{l-1} \ne 0$ in k_l we deduce that

$$\deg_X \overline{{}^T G} = \deg G_{(l)} = l > C(n,d),$$

a contradiction. So $G_{(l)} = 0$, as desired. □

Corollary 2.3.8 *The statements $JC(\mathbb{C}, n)$, $UB(\mathbb{C}, n)$ and $\overline{U\dot{B}}(\mathbb{C}, n)$ are equivalent.*

Proof: Follows from 1.1.12 and 2.3.6. □

(2.3.9) **Final remark.** In Chapter 7, §4 and Exercise 4 below we will discuss the degree of the inverse of a so-called cubic-homogeneous polynomial map and give a counterexample to a conjecture of Drużkowski and Rusek ([96] and [109]). In Chapter 7, §4 and Exercise 5 below we discuss a question of Meisters and Olech concerning the degree of the inverse of a quadratic polynomial map, [213] and [134].

Exercises for §2.3

1. Let R be a non-reduced ring contained in a $\mathbb{Q}$-algebra. Assume that R contains a non-zero element ε with $\varepsilon^2 = 0$. Let $f(X_1) \in R[X_1]$ with $f'(X_1) \in R[X_1]^*$.
   (i) Show that there exists $g(X_1) \in R[X_1]$ with $f(g(X_1)) = g(f(X_1)) = X_1$ and $g'(X_1) \in R[X_1]^*$. [Use 1.1.12.]
   (ii) Let $c(X_1) \in R[X_1]$. Show that $F := (f(X_1), f'(X_1)^{-1}X_2 + c(X_1))$ is invertible over R with inverse $G := (g(X_1), g'(X_1)^{-1}(X_2 - c(g(X_1)))$.
   (iii) Let $m, d \geq 1$ be positive integers with $d > m + 1$. Put $f(X_1) := X_1 - \frac{\varepsilon}{m+1}X_1^{m+1}$ and $c(X_1) := X_1^d$. Show that $\deg F = d$, $\det JF = 1$ and $\deg F^{-1} = \deg F + m$.
   (iv) Let $m, d \geq 1$ be positive integers with $d > m+1$ and $n \geq 3$. Let $F_1 := f(X_1)$ be as in iii), $F_i := X_i - X_{i-1}^d$ for all $2 \leq i \leq n-1$ and $F_n := f'(X_1)^{-1}X_n - X_{n-1}^d$. Show that $\det JF = 1$, $\deg F = d$ and $\deg F^{-1} = (\deg F)^{n-1} + m$.
2. Let k be a field.
   (i) Show that for all $F \in \mathrm{Aut}_k\, k[X]$ $\deg F^{-1} = \deg F$, if $n = 2$.
   (ii) Show that there exist $F \in \mathrm{Aut}_k\, k[X]$ such that $\deg F^{-1} < (\deg F)^{n-1}$, if $n \geq 3$.
   (iii) Show that there exist $F \in \mathrm{Aut}_k\, k[X]$ such that $\deg F^{-1} = (\deg F)^{n-1}$, if $n \geq 1$.
3. (Furter) Let $2 \leq d_1 \leq d_2$ and $R = k[T]/(T^{d_2})$. Consider in $R[X, Y]$ the equation $(1 + \overline{T}X)^{d_1} = 1 + \overline{T}Y$. Show, by solving this equation with respect to X, resp. with respect to Y, that there exists $F \in \mathrm{Aut}_R\, R[X]$ such that $\deg F = d_1$ and $\deg F^{-1} = d_2$.
4. (van den Essen) Let $F : k^5 \to k^5$ be given by
$$F = (X_1 + 3X_4^2X_2 - 2X_3X_4X_5, X_2 + X_4^2X_5, X_3 + X_4^3, X_4 + X_5^3, X_5).$$
   (i) Compute F^{-1} and $\deg F^{-1}$.
   (ii) Put $H := F - X$. Show that $(JH)^2 \neq 0$ and $(JH)^3 = 0$. Deduce from i) that $\deg F^{-1} > (\deg F)^{3-1}$.
5. (Furter) Repeat the questions of the previous exercise for the map
$$F = (X_1 + 2X_2X_6 - 2X_3^2 - X_4X_5, X_2 + 2X_3X_5 - X_4X_6,$$
$$X_3 + X_5X_6, X_4 + X_5^2, X_5 + X_6^2, X_6).$$

Notes

Propositions 2.1.3 and 2.1.13 are taken from the author's paper [112]. Theorem 2.1.17 is taken from the paper [18] of Alev and Chamarie. Anick's example in Exercise 3 of §1 appeared in Martha Smith's paper [271]. Both Exercises 4 and 5 of §1 are taken from Furter's paper [136]. The automorphism in Exercise 5 was the first example of a polynomial automorphism with Jacobian determinant equal to 1 and such that the degree of its inverse exceeds the degree of the map. The derivations $\frac{\partial}{\partial F_i}$ described in §2 appeared in various papers: [289] of Wang, [229] of Nousiainen and Sweedler, [29] of Bass, Connell and Wright and for $n = 2$ in [250] of Rentschler, and several papers of Stein, [276], [277] and [278].

Proposition 2.2.4 is due to Nowicki in [230]. The kernel conjecture comes from the author's paper [104]. Theorem 2.2.16 is "classical". The equivalence of i), ii) and iii) can be found in [289] and [29] and the equivalence of i) and iii), for $k = \mathbb{C}$, goes back to Campbell in [44], 1973. Other proofs of the equivalence of i) and iii) are found in [248], [293], [234] and [3]. Exercise 6 at the end of Section 2 is Theorem 1 in the paper [248] of Razar. The first proof of 2.3.1 is due to Gabber in [29] and Rusek and Winiarski in [256]. Other proofs were given by Ploski in [244] and Yu in [299]. The proof given here appeared in the author's paper [103]. The proof of i) in Theorem 2.3.6 is similar to the one of [29] and the proof of ii) in Theorem 2.3.6 comes from the paper [76] of Derksen. The idea of relating the degree of the inverse of a polynomial automorphism to the Jacobian Conjecture is due to Bass in [26]. There he proves the equivalence of $JC(\mathbb{C}, n)$ and $UB(\mathbb{C}, n)$.

3 Invertibility criteria and inversion formulae

In §1 we give a formula for the inverse of a formal power series map F over a commutative $\mathbb{Q}$-algebra R satisfying $\det JF(0) \in R^*$. More precisely, for each integer $d \geq 1$ this formula computes the homogeneous component of degree d of the formal inverse of F. The very simple proof is an immediate consequence of Taylor's formula. Using the well-known bound for the degree of the inverse of a polynomial automorphism over a reduced ring, we obtain an invertibility criterion for a polynomial map over such a ring.

In §2 we use results from the theory of Gröbner bases to obtain an algorithm which decides if an R-algebra homomorphism between two finitely generated R-algebras is an R-isomorphism, in case R is a finitely generated algebra over a field. This algorithm also computes the inverse in case it exists.

Using this algorithm we show that an automorphism F of a polynomial ring in n variables over a reduced ring R is completely determined by its so-called face polynomials, i.e. the n^2 polynomials $F_i(X_j = 0)$, $1 \leq i, j \leq n$. In fact, in case R is a domain the proof gives an algorithm which reconstructs F from its face polynomials.

We also use the invertibility algorithm above to give an algorithm which decides if a given $(n-1)$-tuple of polynomials $F_1, \ldots, F_{n-1}$ in a polynomial ring over a field can be extended to a polynomial automorphism, i.e. decides if there exists a polynomial F_n such that $(F_1, \ldots, F_{n-1}, F_n)$ is a polynomial automorphism. In case it exists, the algorithm also computes such an F_n. Finally in §3 we give an invertibility criterion and formula for the inverse of a polynomial map in two variables over a domain, based on the resultant of two polynomials in one variable.

3.1 A formula for the formal inverse

Throughout this section R denotes a commutative $\mathbb{Q}$-algebra and $R[[X]] := R[[X_1, \ldots, X_n]]$. In §1 of Chapter 1 we showed that if $F \in R[[X]]^n$ satisfies $F(0) = 0$ and $\det JF(0) \in R^*$, then F has a uniquely defined formal inverse $G = (G_1, \ldots, G_n) \in R[[Y]]^n$ which satisfies $G(0) = 0$ and $X_i = G_i(F_1, \ldots, F_n)$ for all i. The main result of this section, 3.1.1, gives a formula for this inverse G. More precisely, if for every element g of $R[[Y]]$ and each $d \in \mathbb{N}$ we denote the homogeneous component of g of degree d by $g_{(d)}$, then we get a formula for the homogeneous component $G_{i(d)}$ of degree d of G_i. To describe this formula we use the derivations $\frac{\partial}{\partial F_j}$ on $R[[X]]$ as defined in Exercise 7, §2 of Chapter 2. They extend uniquely to $R[[Y]]$-derivations on the $2n$-variable power series ring $R[[X, Y]]$. So on $R[X, Y]$ we can form the derivation

$$D := Y_1 \frac{\partial}{\partial F_1} + \cdots + Y_n \frac{\partial}{\partial F_n}.$$

Theorem 3.1.1 *Let $F \in R[[X]]^n$ with $F(0) = 0$ and $\det JF(0) \in R^*$. Then the formal inverse G of F is given by*

$$G_{i(d)} = \frac{1}{d!} D^d (X_i)_{|X=0} \tag{3.1.2}$$

for all $d \geq 1$ and all i. Here $|X = 0$ means substitute $X_1 = 0, \ldots, X_n = 0$.

Proof: Let $1 \leq i \leq n$ be fixed. Write $G_{i(d)} = \sum_{|\alpha|=d} c_\alpha Y^\alpha$, $c_\alpha \in R$. Since $G_i = \sum_{d \geq 1} G_{i(d)}$ and $X_i = G_i(F_1, \ldots, F_n)$ we get

$$X_i = \sum_{d \geq 1} \left(\sum_{|\alpha|=d} c_\alpha F_1^{\alpha_1} \ldots F_n^{\alpha_n} \right).$$

Now choose $\alpha_1, \ldots, \alpha_n \geq 0$ with $\alpha_1 + \cdots + \alpha_n = d$. Then it follows from Exercise 1, and the fact that $F_i(0) = 0$ for all i, that

$$\frac{\partial^{\alpha_1}}{\partial F_1} \cdots \frac{\partial^{\alpha_n}}{\partial F_1} X_{i|X=0} = \alpha_1! \ldots \alpha_n! c_\alpha.$$

So

$$\begin{aligned} G_{i(d)} &= \sum_{|\alpha|=d} \frac{1}{\alpha_1! \ldots \alpha_n!} \frac{\partial^{\alpha_1}}{\partial F_1} \cdots \frac{\partial^{\alpha_n}}{\partial F_n} (X_i)_{|X=0} Y_1^{\alpha_1} \ldots Y_n^{\alpha_n} \\ &= \frac{1}{d!} \left(Y_1 \frac{\partial}{\partial F_1} + \cdots + Y_n \frac{\partial}{\partial F_n} \right)^d (X_i)_{|X=0}. \end{aligned}$$

□

(3.1.3) **An invertibility criterion**

Now we show how 3.1.1 can be used to decide if a given $F \in R[X]^n$, satisfying $F(0) = 0$, is invertible over R and how to compute the inverse of F in case it exists. As we saw before in 1.1.10, a necessary condition for invertibility is that $\det JF \in R[X]^*$. In this case the derivations $\frac{\partial}{\partial F_i}$ restrict to derivations on $R[X]$ and D gives (by restriction) a derivation on $R[X, Y]$. Furthermore we saw in 1.1.9 that F is invertible over R if and only if $\overline{F}$ is invertible over $\overline{R}$, where $\overline{R} = R/\eta$ and η is the nilradical of R. So the following result enables us to decide if a given $F \in R[X]^n$ is invertible.

Proposition 3.1.4 *Let R be a reduced ring and $F \in R[X]^n$ satisfying $F(0) = 0$ and* $\det JF \in R[X]^*$. *Put $d := \deg F$ and $N := d^{n-1}$. Then* i) *F is invertible if and only if $D^{N+1} X_i = 0$, for all i.*

ii) *If F is invertible, the inverse G of F is given by*

$$G_i = \exp D(X_i)_{|X=0} \left(= \sum_{k=0}^{N} \frac{1}{k!} D^k (X_i)_{|X=0} \right).$$

Proof: i) If $D^{N+1}X_i = 0$ for all i, then it follows from 3.1.2 that

$$G_i = \sum_{k=0}^{N} \frac{1}{k!} D^k(X_i)_{|X=0} \tag{3.1.5}$$

whence F is invertible and the inverse is given by 3.1.5.

ii) Conversely, if F is invertible then $\deg G \leq N$ by 2.3.4. So writing $X_i = G_i(F_1, \ldots, F_n)$, all monomials $F_1^{\alpha_1} \ldots F_n^{\alpha_n}$ appearing in X_i satisfy $|\alpha| \leq N$, which implies that $D^{N+1}X_i = 0$. By i) we conclude that G_i is as in 3.1.5. □

Now assume that $F \in R[X]^n$, satisfying $F(0) = 0$, is invertible. To compute the inverse of F one can proceed as follows: first compute according 3.1.4 the inverse $\overline{G}$ of $\overline{F}$ in $\overline{R}[X]^n$, where $G \in R[X]^n$. Then $\overline{G} \circ \overline{F} = X$, i.e. $G \circ F \in N(R, n)$. So, as described in section one of the previous chapter, we can write $G \circ F = \exp D$, for some locally nilpotent derivation D on $R[X]$. Consequently $F^{-1} = \exp(-D) \circ G$.

Exercises for §3.1

1. Let $g = \sum c_\alpha X^\alpha \in R[[X]]$, where $\alpha = (\alpha_1, \ldots, \alpha_n)$. Show that

$$\partial_1^{\alpha_1} \ldots \partial_n^{\alpha_n} g_{|X=0} = \alpha_1! \ldots \alpha_n! g_\alpha.$$

2. Let $R = \mathbb{Q}[T]/(T^4)$. Use 3.1.2 to compute the inverse of $F(X) = X + \overline{T}X^3$.
3. Let R and F be as in Exercise 5, §1, Chapter 2. Use 3.1.2 to compute F^{-1}.

3.2 An invertibility algorithm for morphisms between finitely generated k-algebras

In this section we give a completely different approach to the problem of deciding if a given $F \in R[X]^n$ is invertible and how to compute the inverse if it exists. The method used is based on results from the theory of Gröbner bases. Therefore we need to make some minor restrictions on the ring R; namely we assume that R is a finitely generated k-algebra, where k is any field. So this time we allow k to have positive characteristic. On the other hand, instead of considering the polynomial ring $R[X]$ and endomorphisms of it, we will consider ring homomorphisms over R between two finitely generated R-algebras and give an algorithm to decide if the given R-homomorphism is an isomorphism and, in case it is, to compute the inverse (3.2.1 and 3.2.4).

Throughout this section we have the following notation: k is any field and $0 \neq k[x] := k[x_1, \ldots, x_n]$ a finitely generated k-algebra. Let the kernel of the k-homomorphism from $k[X] \to k[x]$ sending X_i to x_i be generated by the elements $R_1(X), \ldots, R_u(X)$ in $k[X]$. Furthermore let $k[y] := k[y_1, \ldots, y_m]$, which in a similar way is k-isomorphic to $k[Y]/(S_1(Y), \ldots, S_v(Y))$ for some $S_j(Y)$ in $k[Y]$. Let $\varphi : k[y] \to k[x]$ be the k-algebra homomorphism given by $\varphi(y_j) = F_j(x_1, \ldots, x_n)$ for some $F_j \in k[X]$, all j. On the set of terms of $k[X, Y] := k[X_1, \dots X_n, Y_1, \dots, Y_m]$ we choose an admissible ordering $<$ such that $Y^\alpha < X_i$ for all $\alpha \in \bar{\mathbb{N}}^m$ and all i. Furthermore put

$$I := (Y_1 - F_1(X), \dots, Y_m - F_m(X), R_1(X), \dots, R_u(X))$$

and let B be the reduced Gröbner bases of I with respect to $<$.

Theorem 3.2.1 (van den Essen). i) *φ is a k-algebra isomorphism if and only if B is of the form*

$$B = \{X_1 - G_1(Y), \dots, X_n - G_n(Y), Q_1(Y), \dots, Q_r(Y)\} \tag{3.2.2}$$

for some $G_i, Q_j \in k[Y]$ with $Q_j \in \mathfrak{s} := (S_1(Y), \dots, S_v(Y))$ for all j.

ii) *Furthermore, if φ is an isomorphism the inverse ψ is given by $\psi(x_i) = G_i(y_1, \dots, y_m)$ for all i and the elements $Q_1, \dots, Q_r$ form a reduced Gröbner bases of $\mathfrak{s}$. If $\mathfrak{s} = (0)$ it must be understood that there are no Q_j in the above formulae.*

To prove this result we first show

Lemma 3.2.3 *If $I = (X_1 - G_1(Y), \dots, X_n - G_n(Y), Q_1(Y), \dots, Q_r(Y))$ for some $G_i, Q_j \in k[Y]$ with $Q_j \in \mathfrak{s}$ for all j, then φ is an isomorphism with inverse ψ given by $\psi(x_i) = G_i(y_1, \dots, y_m)$.*

Proof: i) First we show that ψ is well-defined. So we need to show that $R(x_1, \dots, x_n) = 0$ implies that $R(\psi(x_1), \dots, \psi(x_n)) = 0$. Since the R_i generate all relations between the x_j's, it suffices to show that $R_i(G_1(y), \dots, G_n(y)) = 0$ for all i. Therefore observe that

$$R_i(X) \in I = \sum_j k[X, Y](X_j - G_j(Y)) + \sum_l k[X, Y]Q_l(Y).$$

Making the substitutions $X_j := G_j(y)$, $Y_i := y_i$ and using that $Q_l \in \mathfrak{s}$, i.e. $Q_l(y) = 0$ for all l, we get that $R_i(G_1(y), \dots, G_n(y)) = 0$ for all i.

ii) Now we show that $\psi \circ \varphi = 1_{k[y]}$ (similarly we get $\varphi \circ \psi = 1_{k[x]}$). Since

$$Y_i - F_i(X) \in I = \sum_j k[X, Y](X_j - G_j(Y)) + \sum_l k[X, Y]Q_l(Y)$$

the substitutions $X_i := G_i(y)$ and $Y_j := y_j$ give that

$$\begin{aligned} y_i &= F_i(G_1(y), \dots, G_n(y)) = F_i(\psi(x_1), \dots, \psi(x_n)) \\ &= \psi(F_i(x)) = \psi \circ \varphi(y_i), \end{aligned}$$

i.e. $\psi \circ \varphi = 1_{k[y]}$. □

Proof of Theorem 3.2.1: First we prove i). Therefore let ψ be the inverse of φ given by $\psi(x_i) = G_i(y)$, for all i. Then $\varphi \circ \psi(x_i) = x_i$ implies that $G_i(F_1(x), \ldots, F_m(x)) = x_i$, so $X_i - G_i(F_1(X), \ldots, F_m(X)) \in (R_1(X), \ldots, R_u(X))$. Since $F_i \equiv Y_i \pmod{I}$ we deduce that $X_i - G_i(Y) \in I$ for all i. We may assume that $X_n > X_{n-1} > \cdots > X_1$. Since $X_1 - G_1(Y) \in I$, we deduce from the choice of $<$ that $X_1 \in lt(I)$, so $X_1 = t\,lt(b_1)$ for some term t and $b_1 \in B$. Consequently $t = 1$ or $t = X_1$. If $t = X_1$ then $lt(b_1) = 1$, whence $1 \in B \subset I$. This implies that $1 \in (R_1(X), \ldots, R_u(X))$, a contradiction since $k[x] \neq 0$. So $t = 1$ and hence $lt(b_1) = X_1$, i.e. $b_1 = X_1 - H_1(Y)$ for some $H_1 \in k[Y]$. Here we used the special choice of the ordering $<$. Similarly $X_2 - G_2(Y) \in I$ implies that there exists $b_2 \in B$ with $lt(b_2) = X_2$, hence again by the choice of $<$ we get $b_2 = X_2 - H_2(X_1, Y)$. However since B is a reduced Gröbner bases and $X_1 - H_1(Y)$ belongs to B, X_1 cannot appear in H_2 (otherwise it could be reduced!). So $H_2 = H_2(Y)$. Whence $X_2 - H_2(Y) \in B$. Continuing in this way we deduce that

$$\{X_1 - H_1(Y), \ldots, X_n - H_n(Y)\} \subset B, \text{ for some } H_i \in k[Y].$$

Let h be an eventual other element in B. Then $X_1, \ldots, X_n$ cannot appear in h (since otherwise it can be reduced). So $h \in k[Y]$. Consequently

$$B = \{X_1 - H_1(Y), \ldots, X_n - H_n(Y), Q_1(Y), \ldots, Q_r(Y)\}$$

for some $H_i, Q_j \in k[Y]$. Finally since $Q_i(Y) \in B \subset I$, the substitutions $Y_i := F_i(x)$ and $X_j := x_j$ give $Q_i(F_1(x), \ldots, F_m(x)) = 0$, i.e. $Q_i(\varphi(y_1), \ldots, \varphi(y_m)) = 0$, so $\varphi(Q_i(y_1, \ldots, y_m)) = 0$. Consequently $Q_i(y_1, \ldots, y_m) = 0$ since φ is injective. Whence $Q_i(Y) \in \mathfrak{s}$, which completes the proof of the first half of i). Conversely if B is of the form 3.2.2 with all $Q_j \in \mathfrak{s}$, then it follows from 3.2.3 that φ is an isomorphism.

ii) Now suppose that φ is an isomorphism. Then by i) B is of the form 3.2.2. So by 3.2.3 the inverse ψ is given by $\psi(x_i) = G_i(y_1, \ldots, y_m)$. Furthermore, since B is the reduced Gröbner bases of $I = (X_1 - G_1, \ldots, X_n - G_n, Q_1, \ldots, Q_r)$, it follows from the elimination algorithm in Appendix C that $\{Q_1, \ldots, Q_r\} = B \cap k[Y]$ is a Gröbner bases of $(X_1 - G_1, \ldots, X_n - G_n, Q_1, \ldots, Q_r) \cap k[Y] = (Q_1, \ldots, Q_r)$. The last equality follows by making the substitutions $X_i := G_i(Y)$ for all i, which completes the proof. □

Finally we will show how 3.2.1 can be extended to finitely generated R-algebras, where R in turn is a finitely generated k-algebra. Therefore let $R = k[y_1, \ldots, y_q]$ and $\varphi : R[y_{q+1}, \ldots, y_{q+m}] \to R[x_1, \ldots, x_n]$ be an R-algebra homomorphism. Then obviously φ is a k-algebra homomorphism (between finitely generated k-algebras). Now observe that if φ is a k-algebra isomorphism, then automatically φ is an R-algebra isomorphism: namely if ψ is the inverse of φ, then since $\varphi(y_i) = y_i$ for all $1 \leq i \leq q$, we get $y_i = \psi(\varphi(y_i)) = \psi(y_i)$ for all $1 \leq i \leq q$, i.e. ψ preserves all elements of R. So ψ is an R-algebra isomorphism.

Corollary 3.2.4 *Let R be a finitely generated k-algebra. Then* 3.2.1 *gives a criterion to decide if a given R-algebra homomorphism between finitely generated R-algebras is an isomorphism and to compute the inverse in case it exists.*

Face polynomials

Now we return to automorphisms of a polynomial ring $R[X]$ and show how 3.2.1 can be used to prove a remarkable property of these automorphisms over a **reduced ring** R, namely that an automorphism $F \in R[X]^n$ is completely determined by its n^2 **face polynomials**, i.e. the polynomials

$$F_i(X_j = 0), 1 \le i, j \le n.$$

Furthermore we give an algorithm which reconstructs a polynomial automorphism from its face polynomials, in case R is a domain.

Theorem 3.2.5 (McKay, Wang). *Let R be a reduced ring, $n \ge 2$ and $F = (F_1, \ldots, F_n) \in \mathrm{Aut}_R\, R[X]$. Then F is completely determined by its n^2 face polynomials.*

The proof of this result is based on the following lemma.

Lemma 3.2.6 (Li). *Let k be a field and $F \in \mathrm{Aut}_k\, k[X]$ with inverse $G = (G_1, \ldots, G_n)$. Define for every $1 \le j \le n$ the ideal*

$$\begin{aligned} I_j &:= (Y_1 - F_1(X_j = 0), \ldots, Y_n - F_n(X_j = 0)) \\ &\subset k[X_1, \ldots, X_{j-1}, X_{j+1}, \ldots, X_n, Y]. \end{aligned}$$

Then $I_j \cap k[Y] = (G_j)$.

Proof: From 3.2.1, with $\mathfrak{s} = (0)$, we get

$$(Y_1 - F_1(X), \ldots, Y_n - F_n(X)) = (X_1 - G_1(Y), \ldots, X_n - G_n(Y)).$$

So substituting $X_j = 0$ we get that I_j is equal to

$$\begin{aligned} &(X_1 - G_1(Y), \ldots, X_{j-1} - G_{j-1}(Y), -G_j(Y), \\ &\quad X_{j+1} - G_{j+1}(Y), \ldots, X_n - G_n(Y)). \end{aligned}$$

So $(G_j) \subset I_j \cap k[Y]$. Conversely, if $g(Y) \in k[Y] \cap I_j$, then

$$g(Y) = a_j(X, Y)G_j(Y) + \sum_{p \ne j} a_p(X, Y)(X_p - G_p(Y))$$

for some $a_i \in k[X, Y]$. Substitute $G_p(Y)$ for every X_p. Then $g(Y) \in (G_j(Y))$ follows. □

Proof of Theorem 3.2.5: i) We will show below (3.2.7) that, in case R is a domain, it is in fact possible to give an, algorithm which reconstructs F from its face polynomials, so obviously F is completely determined by these face polynomials.

ii) The case that R is a reduced ring follows from the "domain case" by the standard technique of reducing modulo prime ideals and using that the intersection of all prime ideals of R is the zero ideal (R is reduced). The details are left to the reader (Exercise 4).

(3.2.7) **Reconstruction algorithm in case R is a domain**

Let K be the quotient field of R and F an R-automorphism of $R[X]$. Then obviously F is a K-automorphism of $K[X]$. The inverse of this K-automorphism, which we denote by G is, because of its uniqueness, the inverse of the R-automorphism F, i.e. $G \in R[X]^n$. If we can compute G from the face polynomials of F, then by 3.2.1, applied to the K-algebra $K[X]$, we can compute the inverse of G, i.e. F, and we have reconstructed F from its face polynomials. So it remains to show how G can be constructed from the polynomials $F_i(X_j = 0)$, $1 \le i, j \le n$. To do this we use 3.2.6.

Algorithm

On the terms of $k[X, Y]$, choose an admissible order $<$ such that $Y^\alpha < X_i$ for all α, i. For every $1 \le j \le n$ compute the reduced Gröbner bases B_j of I_j. Then by the elimination algorithm of Appendix C, $B_j \cap K[Y]$ is the reduced Gröbner bases of $I_j \cap K[Y] = (G_j)$, using 3.2.6. So $B_j = \{H_j\}$ for some $H_j \in K[Y]$, whence $H_j = aG_j$ for some $a \in K^*$. If e_j denotes the j-th standard basis vector of K^n we get

$$H_j(F(e_j)) = aG_j(F(e_j)) = a,$$

since $G_j(F(e_j)) = 1$. So $G_j = H_j/H_j(F(e_j))$, i.e. G is completely determined by the face polynomials of F.

Remark 3.2.8 If R is not reduced, the statement of 3.2.5 is not true: let $a \in R$ with $0 \neq a$ and $a^2 = 0$. Then $F = (X_1 + aX_1 \dots X_n, X_2, \dots, X_n)$ is a polynomial automorphism which has the same face polynomials as the identity map.

Remark 3.2.9 In [6] Abhyankar and Li used 3.2.1 in case $\mathfrak{s} = (0)$ to attack the Jacobian Conjecture. The idea was roughly the following: if one can show that there exists a constant $D(n, d)$ such that, for all finitely generated k-algebras R, all $s \ge 1$ and all $f_1, \dots, f_s$ in $R[X]$ with $\deg f_i \le d$ for all i, there exists a minimal Gröbner bases of $(f_1, \dots, f_s)$, with respect to a fixed admissible ordering, such that the degree of its elements are bounded by $D(n, d)$, then the Jacobian Conjecture follows from 2.3.6 and 3.2.1. However it is not difficult to see that the above statement about a uniform bound $D(n, d)$ is not true: in fact take $R := k[T]/(T^{m+1})$, $m \ge 1$ and $f := Y - (X - \overline{T}X^2) \in R[X, Y]$. Then it is left to the reader to verify that, in spite of the fact that $\deg f = 2$, the degree of the elements of each minimal Gröbner bases of (f), with respect to the ordering $Y < X$, is equal to $m + 1$ (Exercise 5).

In the remainder of this section k is a field of characteristic zero. We consider the following problem.

(3.2.10) **Extendibility problem.** Let $F_1, \dots, F_{n-1} \in k[X]$. Give an algorithm to decide if this $(n-1)$-tuple is **extendible**, i.e. if there exists $F_n \in k[X]$ such that $(F_1, \dots, F_{n-1}, F_n)$ is a polynomial automorphism.

The algorithm which we describe below is based on the following easy result.

Proposition 3.2.11 *Let $F_1, \dots, F_{n-1} \in k[X]$ and let D be the derivation defined by*

(3.2.12) $$D(g) = \det J(F_1, \dots, F_{n-1}, g), \ \textit{for all } g \in k[X].$$

i) *Then $F_1, \dots, F_{n-1}$ is extendible if and only if D is a locally nilpotent derivation on $k[X]$ which has a slice s in $k[X]$ and satisfies $k[X]^D = k[F_1, \dots, F_{n-1}]$.*
ii) *Furthermore, if we are in the situation of* i) *then $(F_1, \dots, F_{n-1}, s)$ is a polynomial automorphism of $k[X]$.*

Proof: Suppose that $F_1, \dots, F_{n-1}$ is extendible. Then there exists $F_n \in k[X]$ such that $k[X] = k[F_1, \dots, F_{n-1}, F_n]$ and $\det J(F_1, \dots, F_{n-1}, F_n) = 1$. Since $D(F_i) = 0$ for all $1 \le i \le n-1$ and $D(F_n) = \det J(F_1, \dots, F_n) = 1$, we deduce that $D = \frac{\partial}{\partial F_n}$. So D is locally nilpotent on $k[F_1, \dots, F_n] = k[X]$, $k[X]^D = k[F_1, \dots, F_{n-1}]$ and F_n is a slice of D. Conversely, if D is locally nilpotent on $k[X]$ having a slice s in $k[X]$ and such that $k[X]^D = k[F_1, \dots, F_{n-1}]$, then it follows from 1.3.21 that $k[X] = k[F_1, \dots, F_{n-1}][s]$, whence $(F_1, \dots, F_{n-1}, s)$ is a polynomial automorphism of $k[X]$. □

Now we are able to give

The extendibility algorithm

Let $F_1, \dots, F_{n-1} \in k[X]$. First check if the derivation D defined in 3.2.12 is non-zero, i.e. check if $D(X_i) \neq 0$ for some i. If not, the $(n-1)$-tuple is not extendible. Otherwise check if D is locally nilpotent by the partial nilpotency criterion 1.4.17: therefore we need to estimate $|k(X) : k(F_1, \dots, F_{n-1})(X_i)|$ for each X_i with $D(X_i) \neq 0$. This can be done by B.2.7 in Appendix B. If D is not locally nilpotent, then by 3.2.11 the $(n-1)$-tuple is not extendible. If D is locally nilpotent, check if D has a slice: therefore assume that $k[X]^D = k[F_1, \dots, F_{n-1}]$ and check by the image membership algorithm 1.4.15 if $1 \in \operatorname{Im} D$. If not, the $(n-1)$-tuple is not extendible. If yes, compute $s \in k[X]$ with $D(s) = 1$. Finally use 3.2.1 to verify if $k[F_1, \dots, F_{n-1}, s] = k[X]$. If not, the $(n-1)$-tuple is not extendible (by 3.2.11). If yes, the $(n-1)$-tuple is obviously extendible. □

Exercises for §3.2

1. Let $F = (X^4 + 2(Y+Z)X^3 + (Y+Z)^2X^2 + (Y+1)X + Y^2 + ZY, X^3 + (Y+Z)X^2 + Y, X + Y + Z)$. Compute the inverse of F.
2. Let $F = (X_1 + X_5(X_2^2 - X_3^2 - X_4X_5), X_2 + X_1^2X_5, X_3 + X_1^2X_5, X_4 + 2X_1^2(X_2 - X_3), X_5)$. Compute the inverse of F.
3. Let $F \in \operatorname{Aut}_{\mathbb{Q}} \mathbb{Q}[X, Y, Z]$. The face polynomials of F are

$$\begin{aligned} F_1(X=0) &= -2Y^3 - Y^4Z, F_2(X=0) = Y + Y^2Z, F_3(X=0) = Z, \\ F_1(Y=0) &= X - X^2Z^3, F_2(Y=0) = XZ^2, F_3(Y=0) = Z, \\ F_1(Z=0) &= X - 2Y^3, F_2(Z=0) = Y, F_3(Z=0) = 0. \end{aligned}$$

Reconstruct F

4. Show that 3.2.5 holds for any reduced ring R. [First observe that the case $R = k$ is a field follows from 3.2.5 and deduce that 3.2.5 is true for any domain. To see the general case let $F, G \in \mathrm{Aut}_R\, R[X]$ having the same face polynomials. Show that the coefficients of F and G are equal by reducing modulo prime ideals.]
5. Let $m \geq 1$ and $R := k[T]/(T^{m+1})$. Let $F(X) := X - \overline{T}X^2 \in R[X]$. Let G be the inverse of F.

   (i) Show that $\deg G = m + 1$.

   (ii) Put $f := Y - (X - \overline{T}X^2) \in R[X, Y]$. So $\deg f = 2$. Show that $\{X - G(Y)\}$ is the reduced, hence minimal, Gröbner bases of $I := R[X, Y]f$ with respect to the ordering $Y < X$.

   (iii) Show that for each minimal Gröbner bases of I with respect to $<$ as in ii) the degree of the (only) Gröbner bases element has degree $m+1$. [Let U be a minimal Gröbner bases of I. Show that $U = \{v.(X - h(Y))\}$, for a unit v in R and $h(Y) \in R[Y]$. Use 3.2.1 ii) and show $h(Y) = G(Y)$.]
6. In [24] a non-constructive proof of 3.2.5 is given. However in the proof the following **incorrect** statement is used:

   "if R is a reduced ring, $a, b \in R$ such that $Ra = Rb$, then $a = ub$ for some unit u in R."

   Give a counterexample to this statement. [Take for R the ring $\mathbb{Z}[X, Y, Z]/(Z(XY - 1))$.]
7. (Minimal polynomials) Let k be a field, $n \geq 2$ and assume that $f_1, \ldots, f_n \in k[X_1, \ldots, X_{n-1}]$ contain precisely $n-1$ algebraically independent elements over k.

   (i) Show that there exists an irreducible polynomial $h \in k[X]$ such that $h(f_1, \ldots, f_n) = 0$.

   (ii) Show that h is unique up to a constant factor. Such an h is called a **minimal polynomial of** $f_1, \ldots, f_n$ **in** $k[X]$. [Assume that $f_1, \ldots, f_{n-1}$ are algebraically independent over k and let h_1 and h_2 be irreducible polynomials in $k[X]$ such that $h_i(f_1, \ldots, f_n) = 0$, for $i = 1, 2$. View both h_i in $A[X_n]$, where $A = k[X_1, \ldots, X_{n-1}]$. Put $K := Q(A)$. Consider $r := \mathrm{Res}_{X_n}(h_1, h_2)$ (= the X_n-resultant of h_1 and h_2, see the next section). Show that $r = 0$ and deduce that the h_i have a common zero in an algebraic closure of K and hence have a common factor in $K[X_n]$. Conclude that $h_1 = ch_2$ for some non-zero $c \in K$.]
8. (Płoski, Yu) Let k be a field, $n \geq 2$ and $F = (F_1, \ldots, F_n) \in \mathrm{Aut}_k\, k[X]$ with inverse $G = (G_1, \ldots, G_n)$.

   (i) Let $1 \leq i \leq n$. Show that $F_1(X_i = 0), \ldots, F_n(X_i = 0)$ contain precisely $n - 1$ algebraically independent elements over k.

   (ii) Show that G_i is a minimal polynomial of these polynomials.

9. (Płoski, Yu) Let k be a field of characteristic zero and $n \geq 2$. Then the following result is proved in [244] and [299]:

   If $f_1, \ldots, f_n \in k[X_1, \ldots, X_{n-1}]$ contain precisely $n-1$ algebraically independent polynomials and g is a minimal polynomial of the f_i, then

   $$\deg g \cdot |k(X_1, \ldots, X_{n-1}) : k(f_1, \ldots, f_{n-1})| \leq (\max_i\{\deg(f_i)\})^{n-1}. \tag{3.2.13}$$

   Deduce from 3.2.13 and Exercise 8 that, if $F = (F_1, \ldots, F_n) \in \mathrm{Aut}_k\, k[X]$ with inverse $G = (G_1, \ldots, G_n)$, then

   $$\deg G_i \leq (\max_j\{\deg F_j(X_i = 0)\})^{n-1} \leq (\deg F)^{n-1}. \tag{3.2.14}$$

10. (Libgober) Show that $f := X + X^3Y + Y^3Z + T^4$ is a coordinate. [Show that (f, Y, T) is extendible.]
11. Let k be a field of characteristic zero, $a(X, Z) \in k[X, Z]$, $b \in k^*$ and $n \in \mathbb{N}$. Show that $F_1 := X + a(X, Z)Z + bYZ^n$ is a coordinate. [Show that (F_1, Z) is extendible using that every derivation of the form $p(Z)\partial_X + q(X, Z)\partial_Y$ has a slice in $k[X, Y, Z]$ if $1 \in (p(Z), q(X, Z))$, see Chapter 9, §5.]
12. Let $m \geq 1$ and $F_2 = Y + Z(Z^{m-1}X + Y^m)$.
    (i) Show that F_2 is a coordinate.
    (ii) Let $m = 3, 4$. Find $F_1, F_3 \in k[X, Y, Z]$ such that (F_1, F_2, F_3) is a coordinate system of $k[X, Y, Z]$.

3.3 A resultant criterion and formula for the inversion of a polynomial map in two variables

Throughout this section R denotes a commutative domain and K denotes its quotient field. In the previous section we saw that if $F = (F_1, F_2) \in \mathrm{Aut}_R\, R[X_1, X_2]$, then the ideal $I = (F_1(X_1, X_2) - Y_1, F_2(X_1, X_2) - Y_2)$ contains an element of the form $X_1 - G_1(Y_1, Y_2)$. In other words, an element which does not contain X_2, i.e. a polynomial built up from $F_1(X_1, X_2) - Y_1$ and $F_2(X_1, X_2) - Y_2$ from which X_2 is eliminated. This guides us to elimination theory, whose crucial tool is the resultant. So viewing both generators of I as polynomials in, for example X_2, we are lead to consider $\mathrm{Res}_{X_2}(F_1 - Y_1, F_2 - Y_2)$. Here "$\mathrm{Res}_{X_2}$" denotes the resultant with respect to X_2. We will recall some of its properties below. For more details we refer to [190]. The main result of this section is

Theorem 3.3.1 (Adjamagbo, van den Essen). *Let $F \in R[X]^2$.* 1) *There is equivalence between*

i) $F \in \mathrm{Aut}_R\, R[X]$.
ii) *There exist c_1, c_2 in R, non-zero and $G_1, G_2 \in R[Y]$ such that*

$$\begin{aligned}\mathrm{Res}_{X_2}(F_1(X_1, X_2) - Y_1, F_2(X_1, X_2) - Y_2) &= c_1(X_1 - G_1(Y)),\\ \mathrm{Res}_{X_1}(F_1(X_1, X_2) - Y_1, F_2(X_1, X_2) - Y_2) &= c_2(X_2 - G_2(Y)).\end{aligned}$$

2) *Furthermore, if* $F \in \mathrm{Aut}_R\, R[X]$ *then* $G = (G_1, G_2)$ *is the inverse of* F.

Before we can prove this result we recall some properties of the resultant and make some preparations.

(3.3.2) **The resultant**

Let A be any commutative ring and $f = f_nT^n + \cdots + f_1T + f_0 \in A[T]$ with f_n non-zero and $g = g_mT^m + \cdots + g_1T + g_0 \in A[T]$ with g_m non-zero.

i) If $n, m \geq 1$, the **resultant** of f and g with respect to T, denoted $\mathrm{Res}_T(f, g)$, is defined as

$$\mathrm{Res}_T(f,g) := \det \begin{pmatrix} f_n & \cdots & \cdots & f_0 & & & \\ & \ddots & & & \ddots & & \\ & & f_n & \cdots & \cdots & f_0 \\ g_m & \cdots & \cdots & g_0 & & & \\ & \ddots & & & \ddots & & \\ & & g_m & \cdots & \cdots & g_0 \end{pmatrix}$$

where the above matrix consists of m rows of f's and n rows of g's.

ii) If $m = 0$, we put $\mathrm{Res}_T(f, g) := g_0^n$, if $n = 0$, we put $\mathrm{Res}_T(f, g) := f_0^m$.

Some of the main properties of the resultant which we will need are summarized in the next proposition.

Proposition 3.3.3 *Notations as above.* i) $\mathrm{Res}_T(f, g) \in A[T]f + A[T]g$.

ii) $\mathrm{Res}_T(f, g)$ *contains the terms* $f_n^m g_0^n$ *and* $(-1)^{nm} g_m^n f_0^m$.

iii) *Assume A is a domain. Then f and g have a common zero in some field extension of* $Q(A)$ *if and only if* $\mathrm{Res}_T(f, g) = 0$.

For the application we need to introduce two more variables, Y_1 and Y_2 and form $f - Y_1$ and $g - Y_2$, which we consider as polynomials in T with coefficients in $A[Y] := A[Y_1, Y_2]$. So by definition $\mathrm{Res}_T(f - Y_1, g - Y_2)$ is a well-defined element of $A[Y]$ which we denote by r. It is left to the reader to verify, Exercise 1, that

(3.3.4)
$$\deg_Y r \leq \max(m, n)$$

(3.3.5)
$$\deg_{Y_1} r = m, \deg_{Y_2} r = n, lc_{Y_1}(r) = (-1)^{m+mn}(g_m)^n, lc_{Y_2}(r) = (-1)^n(f_n)^m.$$

Proof of Theorem 3.3.1: (started) ii)→ i) From 3.3.3 i) and the hypothesis we deduce, by substituting $Y_i := F_i(X)$, for $i = 1, 2$, that $c_i(X_i - G_i(F)) = 0$. So $X = G(F)$, i.e. F is invertible with inverse G. This also gives that the last statement of 3.3.1 follows from the implication i)→ ii). So it remains to prove i)→ ii). Therefore we need

Lemma 3.3.6 *Let R be any commutative ring. Put*

$$r_i := \mathrm{Res}_{X_i}(F_1 - Y_1, F_2 - Y_2) \text{ for } i = 1, 2.$$

If $F \in \mathrm{Aut}_R\, R[X_1, X_2]$ *with inverse* $G = (G_1, G_2) \in R[Y]^2$*, then there exist* $c_1, c_2 \in R[X, Y]$ *such that* $r_1 = c_2(X_2 - G_2)$ *and* $r_2 = c_1(X_1 - G_1)$.

Proof: View $F_1(G_1, X_2) - Y_1$ and $F_2(G_1, X_2) - Y_2$ as polynomials in X_2. They have a common zero, namely $X_2 = G_2$. Then r, the X_2-resultant of these polynomials, is equal to zero by 3.3.3 i). Now observe that $r = r_{2|X_1=G_1}$, namely $\deg_{X_2} F_i(G_1, X_2) = \deg_{X_2} F_i(X_1, X_2)$, since if $a(X_1) \in R[X_1]\backslash\{0\}$, then also $a(G_1)$ is non-zero, for if $a(G_1)$ is zero then $a(G_1(F_1, F_2)) = 0$, i.e. $a(X_1) = 0$, a contradiction. So $r_{2|X_1=G_1} = 0$, whence $X_1 - G_1$ divides r_2 in $R[Y][X_1]$. So $r_2 = c_1(X_1 - G_1(Y))$ for some $c_1 \in R[Y][X_1]$. Similarly we get $r_1 = c_2(X_2 - G_2(Y))$ for some $c_2 \in R[Y][X_2]$. □

Corollary 3.3.7 *Let* $F \in \mathrm{Aut}_R\, R[X_1, X_2]$. *Then we have: if* $\deg_{X_j} F_i > 0$*, then* $lc_{X_j} F_i \in R\backslash\{0\}$*, for all* i, j.

Proof: Let $n = \deg_{X_2} F_1$, $m = \deg_{X_2} F_2$ and $G = (G_1, G_2)$ the inverse of F. It suffices to show that G has the property of 3.3.7 (for F is the inverse of G). So assume that $\deg_{Y_1} G_1 > 0$. This hypothesis implies that $lc_{Y_1}(X_1 - G_1(Y)) = -lc_{Y_1} G_1(Y)$. Looking at the highest degree coefficients with respect to the Y_1-development of both sides of the equation $r_2 = c_1(X_1 - G_1(Y))$, it follows from 3.3.5 that

$$(-1)^{m+mn}(lc_{X_2} F_2)^n = lc_{Y_1}(c_1).(-1).lc_{Y_1}(G_1). \tag{3.3.8}$$

Since the left hand side is a non-zero polynomial in $R[X_1]$ and $lc_{Y_1} G_1 \in R[Y_2]$ we deduce that $lc_{Y_1} G_1 \in R\backslash\{0\}$. Arguing in a similar way for the other $lc_{Y_j} G_i$ we find that G has the desired property 3.3.7. □

Proof of Theorem 3.3.1: (finished) We will show that c_1 is a non-zero element of R. The proof for c_2 is similar. Since R is a domain we get from $r_2 = c_1(X_1 - G_1)$ and 3.3.5 that

$$\deg_{Y_1} G_1 \le \deg_{Y_1} r_2 = \deg_{X_2} F_2 \text{ and } \deg_{Y_2} G_1 \le \deg_{Y_2} r_2 = \deg_{X_2} F_1. \tag{3.3.9}$$

Similarly we get from $r_1 = c_2(X_2 - G_2)$ and 3.3.5 that

$$\deg_{Y_2} G_2 \le \deg_{Y_2} r_1 = \deg_{X_1} F_1 \text{ and } \deg_{Y_1} G_2 \le \deg_{Y_1} r_1 = \deg_{X_1} F_2. \tag{3.3.10}$$

Exchanging the roles of F and G we get

$$\deg_{X_1} F_1 \le \deg_{Y_2} G_2, \deg_{X_2} F_1 \le \deg_{Y_2} G_1, \deg_{X_2} F_2 \le \deg_{Y_1} G_1 \tag{3.3.11}$$

and $\deg_{X_1} F_2 \le \deg_{Y_1} G_2$.

From 3.3.9, 3.3.10 and 3.3.11 we get in particular

$$\deg_{Y_1} G_1 = \deg_{Y_1} r_2 \text{ and } \deg_{Y_2} G_1 = \deg_{Y_2} r_2. \tag{3.3.12}$$

Since $r_2 = c_1(X_1 - G_1)$, we deduce that both the Y_1 and Y_2-degree of c_1 is zero, i.e. $c_1 \in R[X_1]$. If $\deg_{X_2} F_2 = 0$, then $F_2 = F_2(X_1)$ and we deduce from 3.3.2 ii) that $r_2 = (F_2(X_1) - Y_2)^n$ for some $n \geq 1$. Since $c_1 \in R[X_1]$ and c_1 divides r_2, we get that c_1 is a non-zero element of R; look at the coefficient of Y_2^n. Finally, if $\deg_{X_2} F_2 > 0$, then by 3.3.12 $\deg_{Y_1} G_1 = \deg_{Y_1} r_2 = \deg_{X_2} F_2 > 0$, so by 3.3.7 we get that both $lc_{X_2} F_2$ and $lc_{Y_1} G_1$ are non-zero elements of R. So 3.3.8 implies that $lc_{Y_1} c_1$ is a non-zero element of R. However $c_1 \in R[X_1]$ so $lc_{Y_1} c_1 = c_1$, which implies that c_1 is a non-zero element of R.

Corollary 3.3.13 (Inversion formula). *Let $F \in \mathrm{Aut}_R\, R[X_1, X_2]$. Then the inverse G is given by the formulae*

$$\begin{aligned}&\mathrm{Res}_{X_2}(F_1(0, X_2) - Y_1, F_2(0, X_2) - Y_2)/\\ &\quad \mathrm{Res}_{X_2}(F_1(0, X_2) - F_1(e_1), F_2(0, X_2) - F_2(e_1)),\end{aligned}$$

$$\begin{aligned}&\mathrm{Res}_{X_1}(F_1(X_1, 0) - Y_1, F_2(X_1, 0) - Y_2)/\\ &\quad \mathrm{Res}_{X_1}(F_1(X_1, 0) - F_1(e_2), F_2(X_1, 0) - F_2(e_2)).\end{aligned}$$

Proof: Write $r_1(X_2, Y)$ (resp. $r_2(X_1, Y)$) instead of r_1 (resp. r_2). From 3.3.7 we deduce that

$$\mathrm{Res}_{X_1}(F_1(X_1, 0) - Y_1, F_2(X_1, 0) - Y_2) = r_1(0, Y)$$

and

$$\mathrm{Res}_{X_2}(F_1(0, X_2) - Y_1, F_2(0, X_2) - Y_2) = r_2(0, Y).$$

Since $G_1(F(e_1)) = G_2(F(e_2)) = 1$ the result follows easily from 3.3.1. □

Remarks 3.3.14 Again from 3.3.13 we see that F is completely determined by its face polynomials and how to reconstruct F from them (just apply 3.3.13 two times). We refer the reader to [52] where a fast method is given to reconstruct F from its face polynomials and to decompose F as a product of elementary maps in case R is a field.

Exercises for §3.3

1. (i) Supply a proof of 3.3.4 and 3.3.5.
   (ii) Let R be a domain. Use 3.3.4 and 3.3.6 to show that if $F \in \mathrm{Aut}_R\, R[X_1, X_2]$ with inverse G, then $\deg F = \deg G$.
2. The following result on **multivariate resultants** can be found in [48]. For every commutative ring A and every n-tuple of homogeneous polynomials $P_1, \ldots, P_n$ of degrees $d_1, \ldots, d_n$ respectively in $A[X_1, \ldots, X_n]$, there exists an element $a \in A$, denoted $\mathrm{Res}(P_1, \ldots, P_n)$, which is uniquely determined by the following four properties:

   1) If $\varphi : A \to B$ is a ring homomorphism and $\tilde{\varphi}$ its natural extension as a ring homomorphism from $A[X_1, \ldots, X_n]$ to $B[X_1, \ldots, X_n]$, then

   $$\mathrm{Res}(\tilde{\varphi}(P_1), \ldots, \tilde{\varphi}(P_n)) = \varphi(\mathrm{Res}(P_1, \ldots, P_n)).$$

2) The universal resultant is irreducible, i.e. the resultant of the polynomials $P_i = \sum_{|\alpha|=d_i} U_{\alpha,i} X^\alpha$, where the $U_{\alpha,i}$ are indeterminates, is an irreducible polynomial in the ring $\mathbb{Z}[\{U_{\alpha,i}\}]$.
3) If A is a field and $\overline{A}$ an algebraic closure of A, then $\operatorname{Res}(P_1, \ldots, P_n) = 0$ if and only if the polynomials P_i have a non-trivial common zero in $\overline{A}^n$.
4) $\operatorname{Res}(X_1^{\alpha_1}, \ldots, X_n^{\alpha_n}) = 1$.

The element $\operatorname{Res}(P_1, \ldots, P_n)$ is called the **resultant** of $P_1, \ldots, P_n$ with respect to $X_1, \ldots, X_n$.

If $f_1, \ldots, f_n$ are n polynomials (not necessary homogeneous) in $n-1$ variables $X_1, \ldots, X_{n-1}$ with coefficients in A and degrees $d_1, \ldots, d_n$ respectively, then we define the **resultant** of $f_1, \ldots, f_n$ with respect to $X_1, \ldots, X_n$, denoted $\operatorname{Res}_{X_1,\ldots,X_n}(f_1, \ldots, f_n)$ as the element $\operatorname{Res}(f_1^\star, \ldots, f_n^\star)$ in A, where $f_i^\star = X_n^{d_i} f_i(X_1/X_n, \ldots, X_{n-1}/X_n)$.

(i) Show that the following is true: if $\varphi : A \to B$ is a ring homomorphism and $\tilde{\varphi}$ is the natural extension of φ from $A[X_1, \ldots, X_n]$ to $B[X_1, \ldots, X_n]$ such that $\deg \tilde{\varphi}(f_i) = \deg f_i$, for all i (hence $(\tilde{\varphi}(f_i))^\star = \tilde{\varphi}(f_i^\star)$), then

$$\operatorname{Res}_{X_1,\ldots,X_n}(\tilde{\varphi}(f_1), \ldots, \tilde{\varphi}(f_n)) = \varphi(\operatorname{Res}_{X_1,\ldots,X_n}(f_1, \ldots, f_n)).$$

(ii) Let $n \geq 3$, $F \in k[X]^n$ (k a field) and $1 \leq j \leq n$ fixed. Put $d_i := \deg F_i(X_j = 0)$. Denote by $(F_i(X_j = 0))_+$ its homogeneous component of highest degree. Suppose that $d_i > 0$ for all i and that all polynomials $(F_i(X_j = 0))_+$ with $1 \leq i \leq n$ have a non-trivial common zero in $\bar{k}^n$. Show that $R_j(Y) = 0$, where

$$R_j(Y) := \operatorname{Res}_{X_1,\ldots,\hat{X}_j,\ldots,X_n}(F_1(X_j = 0) - Y_1, \ldots, F_n(X_j = 0) - Y_n).$$

(iii) Let $n \geq 3$ and $d \geq 2$. Put $F_i = X_i + X_2^d$ if $i \neq 2$ and $F_2 = X_2$. Show that $F \in \operatorname{Aut}_k k[X]$ and that $R_1(Y) = 0$.
(iv) Also show that the Nagata map σ satisfies $R_1(Y) = 0$.

3. (Adjamagbo, van den Essen). Let k be a field and $F = (F_1, F_2) \in \operatorname{Aut}_k k[X_1, X_2]$ with inverse $G = (G_1, G_2)$. Consider the five variable polynomial ring $k[Y, V, T] := k[Y_1, Y_2, V_1, V_2, T]$. Show that

$$\begin{aligned} r &:= \operatorname{Res}_T(F_1(TV_1, TV_2) - Y_1, F_2(TV_1, TV_2) - Y_2) \\ &= a(V)(V_1G_2 - V_2G_1) \end{aligned}$$

for some non-zero element $a(V) \in k[V]$. [Show that $J := (F_1(TV) - Y_1, F_2(TV) - Y_2)$ is a prime ideal in $k[Y, V, T]$ of height 2 and deduce that $I := J \cap k[Y, V]$ is a prime ideal of $k[Y, V]$ of height 1. Use 3.2.1 i) to show that $TV_i - G_i \in J, i = 1, 2$ and deduce that $V_1G_2 - V_2G_1 \in I$. Show that this element is irreducible in $k[Y, V]$ and deduce that $I = (V_1G_2 - V_2G_1)$. Finally use that $r \in I$.]

4. Let $F \in \mathrm{Aut}_{\mathbb{Z}}\, \mathbb{Z}[X_1, X_2]$. Its face polynomials are

$$\begin{aligned} F_1(X_1 = 0) &= 3X_2^3 + 4X_2 + 36X_2^6, \\ F_2(X_1 = 0) &= X_2 + 9X_2^6, \\ F_1(X_2 = 0) &= X_1 + 4X_1^2, \\ F_2(X_2 = 0) &= X_1^2. \end{aligned}$$

Reconstruct F.

5. Show that a statement similar to 3.3.7 does not hold if $n \geq 3$. [Take the Nagata map σ.]

Notes

Formulae for the inverse of a formal power series map were described by several authors. The first one was discovered by Abhyankar and Joni in the beginning of the seventies and appeared in [2] and was later called Abhyankar's inversion formula (see also [29] and [5]). Other inversion formulae were found by Nousiainen and Sweedler in [229], Drużkowski and Rusek in [96], for the so-called cubic homogeneous polynomial maps, and Adjamagbo and the author in [16]. The use of Gröbner bases to study polynomial automorphisms was introduced by the author in 1986, [99] where Theorem 3.2.1 was proved for polynomial rings over a field. The paper was finally published in [100], 1990, since the submitted text was lost by the Journal! The general version in 3.2.1 is also due to the author. This result forms a generalization of an earlier result due to Kwieciński in [188], where the geometric case that I is a radical ideal is treated.

Face polynomials were introduced by McKay and Wang in [206]. First in dimension two, where they were called border polynomials, and later by the same authors in [208] for all dimensions. The inversion formule in Corollary 3.3.13, in case R is a field, is also due to McKay and Wang in [206]. The reconstruction algorithm is due to Kwieciński and the author in [118] and is based on the paper [193] of Li. Later in [187] Kwieciński gave a faster reconstruction algorithm. Theorem 3.3.1, for fields, was obtained by Adjamagbo and the author in [15].

Minimal polynomials (Exercise 7 in §2) were introduced by Płoski in [244] to describe the inverse of a polynomial automorphism. This concept was independently rediscovered by J. Yu in [298]. The interplay between Gröbner bases, face polynomials and inversion formulae formed the main theme in several papers. The reader is referred to the papers [299] and [300] of Yu, the papers [194] and [195] of Li and Yu and the paper [13] of Adjamagbo and Boury. Recently in [90] Drensky, Gutierrez and Yu proposed some new strategies to investigate the Nagata Conjecture using the Gröbner bases.

4 Injective morphisms

Introduction

Let k be any field. Then a well-known result from linear algebra asserts that if $\varphi : V \to W$ is an injective linear map between two finite dimensional vector spaces of the same dimension, then φ is surjective and the inverse map is again k-linear, i.e. a morphism in the category of vector spaces over k. In this chapter we will investigate to what extent this result can be generalized to the category of affine algebraic sets.

The first obstacle one meets is that algebraic sets of the same dimension need not be isomorphic. For example the inclusion $\mathbb{C}\backslash\{0\} \to \mathbb{C}$ is injective but not surjective. One way to overcome this difficulty is to restrict to endomorphisms of algebraic sets. However the map $\varphi : \mathbb{Q} \to \mathbb{Q}$ given by $\varphi(x) = x^3$ is injective but not surjective. Furthermore the map $\varphi : \mathbb{R} \to \mathbb{R}$ given by $\varphi(x) = x^3$ is bijective but its inverse is not a polynomial map. These examples lead one naturally to assume that k is an algebraically closed field. However if k is algebraically closed and char $k = p > 0$, then the Frobenius automorphism $x \to x^p$ shows that the inverse map need not be a polynomial map. Now the surprising point is that every injective endomorphism $\varphi : V \to V$ of an affine algebraic set over an algebraically closed field of characteristic zero is a polynomial automorphism! In §4 we study injective morphisms $\varphi : V \to W$ between irreducible affine algebraic sets of the same dimension. We show that if $A(W)$ is a UFD, then φ is an isomorphism if and only if the induced map $\varphi^* : A(W)^* \to A(V)^*$ on the units of the coordinate rings is bijective.

4.1 Injective endomorphisms are surjective

Throughout this section k denotes an algebraically closed field. We will show that every injective endomorphism of an affine algebraic set V over k is surjective, i.e.

Theorem 4.1.1 *Let V be an affine algebraic set over k. If $\varphi : V \to V$ is an injective endomorphism, then φ is surjective.*

Proof: Let $V \subset k^n$ be defined by the polynomials $g_1, \ldots, g_s \in k[X]$ and $\varphi : V \to V$ be given by the polynomials $f_1, \ldots, f_n \in k[X]$. i) φ is injective if and only if for all $(a, b) \in V \times V$ the equations

$$f_1(a) - f_1(b) = 0, \ldots, f_n(a) - f_n(b) = 0$$

imply that $a_1 - b_1 = 0, \ldots, a_n - b_n = 0$. In other words, if every zero $(a, b) \in k^{2n}$ of the ideal

$$I := (g_1(X), \ldots, g_s(X), g_1(Y), \ldots, g_s(Y), \\ f_1(X) - f_1(Y), \ldots, f_n(X) - f_n(Y))$$

in $k[X, Y]$ is a zero of the ideal

$$J := (X_1 - Y_1, \ldots, X_n - Y_n)$$

or equivalently (by the Nullstellensatz) if there exist $m \in \mathbb{N}, a_{ij}, b_{ij}$ and c_{ij} in $k[X, Y]$ such that

$$(4.1.2) \qquad \begin{aligned} (X_i - Y_i)^m &= \sum_j a_{ij}(f_j(X) - f_j(Y)) \\ &+ \sum_j b_{ij} g_j(X) + \sum_j c_{ij} g_j(Y). \end{aligned}$$

ii) Suppose that φ is not surjective. Then there exists a point $(c_1, \ldots, c_n) \in V$ which does not belong to $\varphi(V)$, in other words such that the system

$$f_1(X) - c_1 = 0, \ldots, f_n(X) - c_n = 0, g_1(X) = 0, \ldots, g_s(X) = 0$$

has no solution in k^n. By the Nullstellensatz this is equivalent to: there exist $a_1, \ldots, a_n$ and $b_1, \ldots, b_s$ in $k[X]$ such that

$$(4.1.3) \qquad 1 = \sum_j a_j(f_j(X) - c_j) + \sum_j b_j g_j(X).$$

iii) Finally the condition $\varphi(V) \subset V$ is equivalent to: every zero of the ideal $(g_1(X), \ldots, g_s(X))$ is a zero of the ideal

$$(g_1(f_1(X), \ldots, f_n(X)), \ldots, g_s(f_1(X), \ldots, f_n(X))).$$

By the Nullstellensatz this is equivalent to: there exist $N \in \mathbb{N}$ and $q_{i,j} \in k[X]$ such that

$$(4.1.4) \qquad g_i(f_1(X), \ldots, f_n(X))^N = \sum_j q_{i,j} g_j(X), \text{ for all } i.$$

iv) Now collect the coefficients of all polynomials appearing in 4.1.2, 4.1.3 and 4.1.4. Call this set $\{d_1, \ldots, d_t\}$ and define the subring S of k by

$$\mathbb{Z}[d_1, \ldots, d_t], \text{ if char } k = 0,$$

$$\mathbb{F}_p[d_1, \ldots, d_t], \text{ if char } k = p > 0.$$

From 4.1.5 below it follows that there exists a maximal ideal $\mathfrak{m}$ of S such that $k_0 := S/\mathfrak{m}$ is a finite field. Now passing from S to k_0 we get k_0^n and $V_0 := \{x \in k_0^n \mid \overline{g_1}(x) = \cdots = \overline{g_s}(x) = 0\}$ where $\overline{g_i}$ is obtained from g_i by reducing all its coefficients modulo $\mathfrak{m}$. By 4.1.4 the induced map $\varphi_0 : V_0 \to k_0^n$ satisfies $\varphi_0(V_0) \subset V_0$, so φ is an endomorphism of V_0. From 4.1.2 and the injectivity of φ we deduce that φ_0 is injective. From 4.1.3 we get that $(\overline{c_1}, \ldots, \overline{c_n})$ does not belong to $\varphi_0(V_0)$, i.e. φ_0 is not surjective. However φ_0 is an injective map from the finite set V_0 to itself, hence surjective, a contradiction. □

Proposition 4.1.5 *Let $A = \mathbb{Z}[x_1, \ldots, x_n]$ be a $\mathbb{Z}$ subalgebra of some $\mathbb{Q}$-algebra. Then for almost all prime numbers p there exists a maximal ideal $\mathfrak{m}_p$ containing p such that $A/\mathfrak{m}_p$ is a finite field.*

Proof: We use induction on n. If $n = 0$, i.e. $A = \mathbb{Z}$ the result is obvious. i) So let $n \geq 1$. If $x_1, \ldots, x_n$ are algebraically independent over $\mathbb{Q}$ take $\mathfrak{m} = (p, x_1, \ldots, x_n)$, for each p. So assume that there exists $0 \neq f(X_1, \ldots, X_n)$ in $\mathbb{Z}[X_1, \ldots, X_n]$ with $f(x_1, \ldots, x_n) = 0$. We may assume that X_n appears and, replacing $x_1, \ldots, x_n$ by $x_1 + x_n^{d_1}, \ldots, x_{n-1} + x_n^{d_{n-1}}$ for suitable $d_1, \ldots, d_{n-1}$ in $\mathbb{N}$, we may assume that

$$f = f_m X_n^m + f_{m-1}(X_1, \ldots, X_{n-1}) X_n^{m-1} + \cdots + f_0(X_1, \ldots, X_{n-1})$$

with $f_m \in \mathbb{Z}\backslash\{0\}$ and $f_i \in \mathbb{Z}[X_1, \ldots, X_{n-1}]$ for all $0 \leq i \leq m-1$. Put $r := f_m$. Then rx_n is integral over $B := \mathbb{Z}[x_1, \ldots, x_{n-1}]$.

ii) By the induction hypothesis we know that there exists a finite set $E_1 \subset \mathbb{Z}$ such that for each prime $p \in \mathbb{Z}\backslash E_1$ there exists a maximal ideal η_p of B such that $p \in \eta_p$ and B/η_p is finite.

iii) Let E_2 be the set of prime divisors of r and put $E := E_1 \cup E_2$. Now we claim: for each prime $p \in \mathbb{Z}\backslash E$ there exists a maximal ideal $\mathfrak{m}_p$ of A as desired. To see this let p be prime and $p \in \mathbb{Z}\backslash E$. Choose η_p as in ii) and write $\eta = \eta_p$. Since by i) rx_n is integral over B it follows that $S^{-1}B \subset S^{-1}B[x_n]$ is integral, where $S = \{1, r, r^2, \ldots\} \subset \mathbb{Z}$.

iv) Consider $S^{-1}\eta$ and observe that $S^{-1}B/S^{-1}\eta \simeq \overline{S}^{-1}(B/\eta)$. Since $(p, r) = 1$ and $p \in \eta$ it follows that $\overline{r}$ is a unit in B/η. Hence $\overline{S}^{-1}(B/\eta) = B/\eta$ which is a finite field, whence so is $S^{-1}B/S^{-1}\eta$.

v) Since by iii) $S^{-1}B \subset S^{-1}B[x_n]$ is integral there exists a maximal ideal $\mathfrak{m}$ in $S^{-1}B[x_n]$ lying over $S^{-1}\eta$. So we get the inclusion map $S^{-1}B/S^{-1}\eta \to S^{-1}B[x_n]/\mathfrak{m}$ which is also integral and hence finite. It follows from iv) that $S^{-1}B[x_n]/\mathfrak{m}$ is a finite field.

vi) Consider $\wp := \varphi^{-1}(\mathfrak{m})$, where $\varphi : B[x_n] \to S^{-1}B[x_n]$ is the canonical map. So $\wp$ is a prime ideal in $B[x_n]$ and the map $B[x_n]/\wp \to S^{-1}B[x_n]/\mathfrak{m}$ is injective, so $B[x_n]/\wp$ is finite.

vii) Finally, if $\mathfrak{m}_p$ is a maximal ideal of $A = B[x_n]$ containing $\wp$, then it follows from the surjection $A/\wp \to A/\mathfrak{m}_p$ that $A/\mathfrak{m}_p$ is a finite field. □

4.2 Injective endomorphisms of affine algebraic sets are automorphisms

Throughout this section k denotes an algebraically closed field of characteristic zero. The main result of this section is

Theorem 4.2.1 (Cynk, Rusek). *Let V be an affine algebraic set over k and $F : V \to V$ an injective endomorphism. Then F is an automorphism.*

The proof of this result, which occupies the remainder of this section is based upon 4.1.1 and the following special case of Zariski's Main Theorem (Appendix B.2)

Proposition 4.2.2 *Let $F : V \to W$ be a bijective morphism of an irreducible algebraic set V onto a normal affine algebraic set W. Then F is an isomorphism and hence V is normal.*

Proof of Theorem 4.2.1: If V is normal, the result follows from 4.2.2 and 4.1.1. So to treat the general case let $V = V_1 \cup \cdots \cup V_s$ be the decomposition of V into irreducible components. First we show that we may assume that $F(V_i) \subset V_i$ for all i: therefore observe that by 4.1.1

$$V = F(V) = F(V_1) \cup \cdots \cup F(V_s).$$

So

$$V = \bigcup_{i=1}^{s} \overline{F(V_i)}.$$

From the decomposition of V into irreducible components we deduce that there exists a permutation σ of $\{1, 2, \ldots, s\}$ such that $\overline{F(V_i)} = V_{\sigma(i)}$, for all i. Hence $F(V_i) \subset V_{\sigma(i)}$, for all i. Obviously if $\sigma^d = 1$, then $F^d(V_i) \subset V_i$ for all i. Furthermore if F^d is an automorphism, so is F. Consequently we may assume that $F(V_i) \subset V_i$, for all i.

i) For each $1 \leq i \leq s$, let $\pi_i : \tilde{V}_i \to V_i$ be a normalization of V_i and let $F_i := F_{|V_i} : V_i \to V_i$ be the restriction of F to V_i. Since F is injective, F_i is injective and consequently by 4.1.1 F_i is surjective. Furthermore by the universal property of normalization (Appendix B) there exists a morphism $\tilde{F}_i : \tilde{V}_i \to \tilde{V}_i$ such that $\pi_i \circ \tilde{F}_i = F_i \circ \pi_i$. We claim: $\tilde{F}_i : \tilde{V}_i \to \tilde{V}_i$ is an automorphism.

To see this it suffices to show that $\tilde{F}_i$ is injective (for then apply 4.1.1 and 4.2.2, using that $\tilde{V}_i$ is normal). First we show that $\tilde{F}_i$ is dominant: namely if $\overline{\tilde{F}_i(\tilde{V}_i)}$ is not equal to $\tilde{V}_i$ then the irreducibility of $\tilde{V}_i$ implies that

$$\dim \overline{\tilde{F}_i(\tilde{V}_i)} < \dim \tilde{V}_i. \tag{4.2.3}$$

However, from $\pi_i \circ \tilde{F}_i = F_i \circ \pi_i$ and the surjectivity of π_i and F_i we get

$$\pi_i(\tilde{F}_i(\tilde{V}_i)) = V_i. \tag{4.2.4}$$

So by 4.2.3 and 4.2.4 we get

$$\dim V_i = \dim \pi_i(\tilde{F}_i(\tilde{V}_i)) \leq \dim \overline{\tilde{F}_i(\tilde{V}_i)} < \dim \tilde{V}_i = \dim V_i,$$

a contradiction. So $\tilde{F}_i$ is dominant. Furthermore $\tilde{F}_i$ has finite fibers, since $\pi_i \circ \tilde{F}_i = F_i \circ \pi_i$ and both F_i and π_i have finite fibers. Finally to show that $\tilde{F}_i$ is injective, it then follows from B.2.2 that it suffices to show that $d(\tilde{F}_i) = 1$. Therefore observe that $d(\pi_i) = 1$ since π_i is birational and that $d(F_i) = 1$ (apply B.2.1 and use that F_i is injective). So

$$d(\tilde{F}_i) = d(\pi_i)d(\tilde{F}_i) = d(\pi_i \circ \tilde{F}_i) = d(F_i \circ \pi_i) = d(F_i)d(\pi_i) = 1$$

as desired, so $\tilde{F}_i$ is injective.

ii) From i) and 4.2.2 it follows that $\tilde{F}_i$ is an automorphism, so in particular $\tilde{F}_i^{-1}$: $\tilde{V}_i \to \tilde{V}_i$ is a morphism and more generally $(\tilde{F}_i)^{-q} : \tilde{V}_i \to \tilde{V}_i$ is a morphism for all $q \geq 1$. Consequently

$$(F^{-q})_{|V_i} \circ \pi_i = \pi_i \circ (\tilde{F}_i)^{-q} : \tilde{V}_i \to V_i$$

is a morphism, all $q \geq 1$, for all i. Since V is affine, we have $V \subset k^n$ for some $n \geq 1$. So if we define

$$M_V(n) := \{G : V \to k^n \mid G_{|V_i} \circ \pi_i \text{ is a morphism, for all } i\},$$

then we see that F^{-q} belongs to $M_V(n)$ for all $q \geq 1$. Obviously $M_V(n)$ is a left $A(V)$-module and we show below in 4.2.5 that $M_V(n)$ is a Noetherian $A(V)$-module. So the $A(V)$-submodule $\sum_{i=1}^{\infty} A(V)F^{-i}$ of $M_V(n)$ is finitely generated. Consequently there exists $\ell \in \mathbb{N}$ such that

$$F^{-(\ell+1)} = \sum_{j=1}^{\ell} r_j F^{-j}, \text{ for some } r_j \in A(V).$$

So

$$F^{-1} = \sum_{j=1}^{\ell} r_j F^{\ell-j}.$$

which implies that F^{-1} is a morphism of V, as desired.

Lemma 4.2.5 *Notations as above. Put*

$$M_V(1) = \{f : V \to k \mid f_{|V_i} \circ \pi_i \in A(V_i), \text{ for all } i\}.$$

Then i) *$M_V(1)$ is a Noetherian $A(V)$-module.* ii) *$M_V(n) \simeq M_V(1)^n$ as $A(V)$-modules. So $M_V(n)$ is a Noetherian $A(V)$-module.*

Proof: It is well-known that $A(\tilde{V}_i)$ is a Noetherian $A(V_i)$-module. Consequently $A(\tilde{V}_i)$ is a Noetherian $A(V)$-module, since $A(V) \to A(V_i)$ is surjective. Now observe that the map

$$M_V(1) \ni f \to (f_{|V_1} \circ \pi_1, \ldots, f_{|V_s} \circ \pi_s) \in \oplus_{i=1}^{s} A(\tilde{V}_i)$$

is an injective $A(V)$-module homomorphism, which implies that $M_V(1)$ is a Noetherian $A(V)$-module. Finally observe that the map $G \to (G_1, \ldots, G_n)$ from $M_V(n)$ to $M_V(1)^n$ is an isomorphism of $A(V)$-modules, where $G_i = p_i \circ G : V \to k$ and $p_i : k^n \to k$ the projection on the i-th component. This completes the proof. □

4.3 A short proof of Theorem 4.2.1 in case $V = k^n$ and an application to the Jacobian Conjecture

Throughout this section, k denotes an algebraically closed field of characteristic zero. We will consider the case $V = k^n$ and give a short proof of 4.2.1. Furthermore we deduce that the Jacobian Conjecture is true for all $F : k^n \to k^n$ satisfying $\det JF \in k^*$ and $\deg F \leq 2$.

Proof of Theorem 4.2.1: in case $V = k^n$ By 4.1.1, F is surjective, hence dominant, and so by B.2.1 $k(X_1, \ldots, X_n) = k(F_1, \ldots, F_n)$, since F is injective. So there exist polynomials $g_i, h_i \in k[Y]$ such that

$$X_i = g_i(F_1, \ldots, F_n)/h_i(F_1, \ldots, F_n) \text{ and } \gcd(g_i, h_i) = 1 \text{ for all } i.$$

Now let $1 \leq i \leq n$ be fixed and suppose that $h_i \notin k^*$. Then there exists y in k^n such that $h_i(y) = 0$. Since F is surjective we have that $y = F(x)$ for some x in k^n. From $h_i(F(x))x_i = g_i(F(x))$ we deduce that $g_i(y) = 0$. So $V(h_i) \subset V(g_i)$, whence $g_i^m \in (h_i)$ for some $m \in \mathbb{N}$ (Nullstellensatz). So $\gcd(h_i, g_i) \neq 1$, contradiction. So $h_i \in k^*$. Consequently $X_i \in k[F_1, \ldots, F_n]$. This holds for all i, whence $k[X] = k[F]$ as desired.

So by 4.2.1 the Jacobian Conjecture can be reformulated as

Jacobian Conjecture. If $F : k^n \to k^n$ is a polynomial map with $\det JF \in k^*$, then F is injective.

Proposition 4.3.1 (Wang). *If* $\deg F \leq 2$ *and* $\det JF \in k^*$, *then* F *is invertible.*

Proof: It suffices to prove that F is injective. So suppose that $F(a) = F(b)$ for some $a, b \in k^n, a \neq b$. We first show that we can assume that $b = 0$. To see this we define $G(X) := F(X + a) - F(a)$. Then $\deg G \leq 2$, $G(0) = 0$ and putting $c := b - a$ we have $c \neq 0$ and $G(c) = 0$. Observe $JG(X) = JF(X + a)$, so $\det JG \in k^*$. Now write $G = G_{(1)} + G_{(2)}$, its decomposition in homogeneous components. Consider $G(tc) = tG_{(1)}(c) + t^2 G_{(2)}(c)$. Differentiating gives

$$G_{(1)}(c) + 2tG_{(2)}(c) = \frac{d}{dt}G(tc) = JG(tc)c \neq 0 \text{ for all } t \in k$$

since $c \neq 0$ and $\det JG \in k^*$. Substituting $t = \frac{1}{2}$ gives $G(c) \neq 0$, a contradiction with $G(c) = 0$. So F is injective. □.

Remark 4.3.2 In Chapter 10 we will show that if $F : k^2 \to k^2$ satisfies $\det JF \in k^*$ and is injective on a line L (i.e. $F_{|L} : L \to k^2$ is injective), then $F : k^2 \to k^2$ is injective, hence invertible.

Exercises for §4.3

1. Let $n, m \in \mathbb{N}$, $n \le m$. Consider the following statement

   $GJC(n, m)$: If $F : k^n \to k^m$ is a polynomial map such that JF has a left inverse, i.e. $A.JF = I_n$ for some $A \in M_{n,m}(k[X_1, \ldots, X_n])$, then F is injective.

   Show that $GJC(n, m)$ is true for all F with $\deg F \le 2$.
2. Consider $F : k \to k^2$ defined by $F(t) = (t^2(1-t), t(1-t))$.
   (i) Show that F is a counterexample to $GJC(1, 2)$.
   (ii) Let $n, m \in \mathbb{N}$, $n \le m$. Use the map F defined above to construct a counterexample to $GJC(n, m)$.

4.4 Injective morphisms between irreducible affine varieties of the same dimension

Throughout this section k denotes an algebraically closed field of characteristic zero, V and W are irreducible affine varieties over k and $f : V \to W$ a morphism. As already observed in the introduction of this chapter, the inclusion map $i : k^* \to k$ shows that an injective morphism between two irreducible affine varieties of the same dimension need not be an isomorphism. The reason is that their coordinate rings do not have the same units. Obviously if $f : V \to W$ is an isomorphism, then f^* induces an isomorphism between $A(W)^*$ and $A(V)^*$, their groups of units. However even this necessary condition on the units is not sufficient to guarantee that f is an isomorphism, as can be seen from the next example.

Example 4.4.1 (A bijective morphism between irreducible affine varieties of dimension one having the same units, but which is not an isomorphism)

Let $V := \mathbb{A}^1$, $W := V(X^2 - Y^3) \subset \mathbb{A}^2$ and $f : V \to W$ be the morphism defined by $f(x) = (x^3, x^2)$, for all $x \in \mathbb{A}^1$. Obviously f is bijective. However f is not an isomorphism since the induced map $f^* : k[\overline{X}, \overline{Y}] \to k[X]$ given by

$$f^*(\overline{X}) = X^3 \quad \text{and} \quad f^*(\overline{Y}) = X^2$$

is not surjective ($X \notin \operatorname{Im} f^*$). On the other hand, since $X = f^*(\overline{X}/\overline{Y})$ we get that $f^*(k(\overline{X}, \overline{Y})) = k(X)$, i.e. f is birational.

However if we assume that $A(W)$ is a UFD we get

Theorem 4.4.2 (Adjamagbo). *Let $f : V \to W$ be an injective morphism between irreducible affine varieties of the same dimension. If $A(W)$ is a UFD then there is* *equivalence between* i) f *is an isomorphism and* ii) $f^* : A(W)^* \to A(V)^*$ *is an isomorphism of groups.*

Proof: The implication i)$\to$ii) is obvious. Let us assume ii). Since f is injective, $\# f^{-1}(w) \leq 1$ for all $w \in W$, so f has finite fibers. So f is dominant by 5.4.1 [146]. It follows from B.2.1 that f^* is birational. So by Zariski's Main Theorem (Appendix B.2) f is an open immersion. In particular $f^* : A(f(V))_{f(v)} \to A(V)_v$ is a ring isomorphism for all v in V. Since $f(V) \subset W$ is open, we have that $A(f(V))_{f(v)} = A(W)_{f(v)}$, so we deduce that $f^* : A(W)_{f(v)} \to A(V)_v$ is flat, for all $v \in V$. Then it follows from Appendix D.2 that $f^* : A(W) \to A(V)$ is flat. By our hypothesis on the units of both coordinate rings and the hypothesis that $A(W)$ is a UFD, we get that the hypothesis of D.1.7 are satisfied. So $f^*(A(W)) = A(V)$, i.e. f^* is surjective. Also f^* is injective since f is dominant, as observed above. So f^* is an isomorphism, hence so is f. □

Notes

The proof of Theorem 4.1.1 is taken from Kang's paper [176]. Theorem 4.2.1 in its full generality is due to Cynk and Rusek in [66] and the proof we give here closely follows their arguments. The short proof of Theorem 4.2.1 in case $V = k^n$ is taken from Rusek's nice survey paper [255]. The simple proof of Proposition 4.3.1 comes from [29]. This result was first obtained by Wang in [289]. Finally, Theorem 4.4.2 is due to Adjamagbo in [12]. He gave a more general result in the language of schemes in [11].

5 The tame automorphism group of a polynomial ring

Throughout this chapter R denotes a commutative ring, however in almost all results R will be a domain, and $R[X] := R[X_1, \ldots, X_n]$ the polynomial ring in n variables over R. We will consider the following subgroups of $\mathrm{Aut}_R\, R[X]$: $\mathrm{Aff}(R, n) :=$ the **affine subgroup** of $\mathrm{Aut}_R\, R[X]$ consisting of all R-automorphisms F such that $\deg F_i = 1$ for all i. $J(R, n) :=$ the **"de Jonquières"** subgroup of $\mathrm{Aut}_R\, R[X]$ consisting of all R-automorphisms F of the form

$$F = (a_1X_1 + f_1(X_2, \ldots, X_n), a_2X_2 + f_2(X_3, \ldots, X_n), \ldots, a_nX_n + f_n)$$

where each a_i belongs to R^* and $f_i \in R[X_{i+1}, \ldots, X_n]$ for all $1 \leq i \leq n-1$ and $f_n \in R$. $E(R, n) :=$ the subgroup of $\mathrm{Aut}_R\, R[X]$ generated by the **elementary** automorphisms, i.e. the automorphisms of the form

$$F = (X_1, \ldots, X_{i-1}, X_i + a, X_{i+1}, \ldots, X_n)$$

for some $a \in R[X_1, \ldots, \hat{X}_i, \ldots, X_n]$ and $1 \leq i \leq n$.

$T(R, n)$:= the **tame subgroup** of $\mathrm{Aut}_R\, R[X]$ is the subgroup generated by $\mathrm{Aff}(R, n)$ and $E(R, n)$.

It is not difficult to verify (Exercise 1) that every element of $J(R, n)$ is a product of an element of $\mathrm{Aff}(R, n)$ and a finite number of elementary automorphisms. So $J(R, n) \subset T(R, n)$. Also, conjugating the de Jonquierès automorphisms with suitable permutation maps, one readily verifies that all elementary automorphisms belong to the subgroup of $\mathrm{Aut}_R\, R[X]$ generated by $\mathrm{Aff}(R, n)$ and $J(R, n)$. So we get

$$T(R, n) = \langle \mathrm{Aff}(R, n), J(R, n) \rangle.$$

In §1 we consider the case $n = 2$ and assume that R is an **integral domain**. We will show that $T(R, 2)$ is the amalgamated free product of $\mathrm{Aff}(R, 2)$ and $J(R, 2)$ over their intersection. Furthermore we describe an algorithm which decides if a polynomial endomorphism of $R[X, Y]$ is tame. Using this algorithm we will show that if R is not a field then $T(R, 2) \neq \mathrm{Aut}_R\, R[X, Y]$. However if R is a field then it turns out that we have equality, i.e. every automorphism in dimension two over a field is tame! This is the famous Jung-van der Kulk theorem (5.1.11).

In §2 we consider the group $T(R, n)$ in case $n \geq 3$. We will prove a remarkable result, due to Harm Derksen, asserting that, in case R is a field, $T(R, n)$ is generated by $\mathrm{Aff}(R, n)$ and only one non-linear automorphism, namely

$$\varepsilon = (X_1 + X_2^2, X_2, X_3, \ldots, X_n).$$

In §3 we discuss the relationship between tame automorphisms and embedding problems of affine varieties in affine n-space. The first one of this kind is the famous Abhyankar-Moh-Suzuki theorem which asserts that every embedding of the line in the plane is rectifiable (5.3.5). The proof of this result will be given in §4. One easily deduces from it a proof of the Jung-van der Kulk theorem.

Embeddings of the line in $\mathbb{A}^n$ turn out to be rectifiable for all $n \geq 4$ and in fact they can be rectified by tame automorphisms. This result was originally proved independently by Craighero and Jelonek, but we will present it as a corollary of a much more general result due to Srinivas (5.3.7). The case $n = 3$ remains open.

At the end of §3 we consider embeddings of $\mathbb{A}^{n-1}$ in $\mathbb{A}^n$ and briefly discuss the Abhyankar-Sathaye conjecture and describe a family of candidate counterexamples due to Choudary and Dimca.

5.1 The tame automorphism group of $R[X, Y]$

Throughout this section R denotes an integral domain. First we show that $T(R, 2)$ is the amalgamated free product of $\text{Aff}(R, 2)$ and $J(R, 2)$ over their intersection. Furthermore we describe an algorithm which decides if a given endomorphism F of $R[X, Y]$ is tame. This algorithm can also be used to decompose F as a product of elementary and linear automorphisms, in case F is tame. We deduce that in case R is not a field, then $T(R, 2)$ is a proper subgroup of $\text{Aut}_R\, R[X, Y]$. On the other hand, if R is a field, we will show that $\text{Aut}_R\, R[X, Y] = T(R, 2)$ and hence it follows that $\text{Aut}_R\, R[X, Y]$ is the amalgamated free product of $\text{Aff}(R, 2)$ and $J(R, n)$ over their intersection. This is the so-called **Jung-van der Kulk theorem**, also known as the **Automorphism theorem**.

To prove the results announced above we need the following lemmas.

Lemma 5.1.1 *Let G be a group generated by two subgroups H and K. Then every element g of G can be written as*

$$g = h_0 k_1 h_1 \dots k_\ell h_\ell$$

for some $\ell \geq 1$, where $h_i \in H \backslash K$ for all $1 \leq i \leq \ell - 1$ and $k_i \in K \backslash H$ for all $1 \leq i \leq \ell$ and $h_0 \in H$.

Proof: If $g \in G$ we have $g = h_0 k_1 h_1 \dots k_\ell h_\ell$ for some $\ell \geq 1$, $h_i \in H$ and $k_i \in K$. If the additional condition on the h_i and k_i is not satisfied, then we can obtain an expression for g of the same shape but with ℓ replaced by $\ell - 1$ as follows: if for example $h_{i_0} \in K$ for some $1 \leq i_0 \leq \ell - 1$ then

$$g = h_0 k_1 \dots h_{i_0 - 1}(k_{i_0 - 1} h_{i_0} k_{i_0 + 1}) h_{i_0 + 1} \dots k_\ell h_\ell$$

and $k_{i_0-1} h_{i_0} k_{i_0+1} \in K$. After a finite number of such reductions we obtain the desired expression for g. □

To formulate 5.1.2 below and the announced algorithm, we need the following notations: let $F = (F_1, F_2) \in R[X, Y]^2$. Then

$$\begin{aligned} \text{bideg}\, F &:= (\deg F_1, \deg F_2), \\ \text{tdeg}\, F &:= \deg F_1 + \deg F_2. \end{aligned}$$

Now let $F \in T(R,2)$. Then applying 5.1.1 to $G := T(R,2)$, $H := \mathrm{Aff}(R,2)$, $K := J(R,2)$ and $g := F$ we can write

$$F = \lambda_0\tau_1\lambda_1 \dots \tau_\ell\lambda_\ell$$

with $\lambda_i \in \mathrm{Aff}(R,2)\backslash J(R,2)$ for all $1 \le i \le \ell - 1$ and $\tau_i \in J(R,2)\backslash \mathrm{Aff}(R,2)$ for all $1 \le i \le \ell$. Write

$$\lambda_i = (a_iX + b_iY + c_i, d_iX + e_iY + f_i).$$

Lemma 5.1.2 *For all* $1 \le i \le \ell$ *we have*

$$\mathrm{bideg}(\tau_i\lambda_i \dots \tau_\ell\lambda_\ell) = (\Pi_{j=i}^{\ell} \deg\tau_j, \Pi_{j=i+1}^{\ell} \deg\tau_j)$$

where the second product is by definition 1 *if* $i = \ell$.

Proof: By (decreasing) induction on i. The case $i = \ell$ is obvious. So let us assume that the statement is true for some $2 \le i \le \ell$ and consider $\mathrm{bideg}(\tau_{i-1}\lambda_{i-1} \dots \tau_\ell\lambda_\ell)$. First observe that, since $\lambda_{i-1} \notin J(R,2)$, we have that $d_{i-1} \ne 0$. Consequently

$$\mathrm{bideg}(\lambda_{i-1}\tau_i\lambda_i \dots \tau_\ell\lambda_\ell) = (p_i, \Pi_{j=i}^{\ell} \deg\tau_j) \text{ where } p_i \le \Pi_{j=i}^{\ell} \deg\tau_j.$$

Finally, since $\tau_{i-1} \notin \mathrm{Aff}(R,2)$ we have that $\deg\tau_{i-1} \ge 2$. So

$$\mathrm{bideg}(\tau_{i-1}\lambda_{i-1}\tau_i\lambda_i \dots \tau_\ell\lambda_\ell) = (\Pi_{j=i-1}^{\ell} \deg\tau_j, \Pi_{j=i}^{\ell} \deg\tau_j)$$

which completes the proof. □

Corollary 5.1.3 $T(R,2)$ *is the amalgamated free product of* $\mathrm{Aff}(R,2)$ *and* $J(R,2)$ *over their intersection, i.e.* $T(R,2)$ *is generated by these two groups and if* $\tau_i \in J(R,2)\backslash \mathrm{Aff}(R,2)$ *and* $\lambda_i \in \mathrm{Aff}(R,2)\backslash J(R,2)$ *then* $\tau_1\lambda_1 \dots \tau_n\lambda_n\tau_{n+1}$ *does not belong to* $\mathrm{Aff}(R,2)$.

Proof: As already observed above, $T(R,2)$ is generated by $\mathrm{Aff}(R,2)$ and $J(R,2)$. So suppose that

$$\tau_1\lambda_1 \dots \tau_n\lambda_n\tau_{n+1} = \lambda \in \mathrm{Aff}(R,2)$$

with $\tau_i \in J(R,2)\backslash \mathrm{Aff}(R,2)$ and $\lambda_i \in \mathrm{Aff}(R,2)\backslash J(R,2)$ for all i. Then

$$\tau_1\lambda_1 \dots \tau_n\lambda_n\tau_{n+1}\lambda^{-1} = (X, Y).$$

So $\mathrm{bideg}(\tau_1\lambda_1 \dots \tau_{n+1}\lambda^{-1}) = (1,1)$. On the other hand it follows from 5.1.2 that

$$\mathrm{bideg}(\tau_1\lambda_1 \dots \tau_{n+1}\lambda^{-1}) = (\Pi_{j=1}^{n} \deg\tau_j, \Pi_{j=2}^{n} \deg\tau_j).$$

So this bidegree is unequal to $(1,1)$, since $\deg\tau_1 \ge 2$, a contradiction. □

(5.1.4) **Warning.** Let J_0 be the subgroup of $J(R,2)$ consisting of the elements $(X + g(Y), Y)$ with $g(Y) \in R[Y]$. One easily verifies that $T(R,2)$ is generated by

$\mathrm{Aff}(R,2)$ and J_0. However $T(R,2)$ is not the amalgamated free product of $\mathrm{Aff}(R,2)$ and J_0 over their intersection: for example let $\tau_1 = (X - Y^2, Y)$, $\lambda_1 = (X, Y+1)$ and $\tau_2 = (X + (Y+1)^2, Y)$. Then $\tau_1\lambda_1\tau_2 = \lambda_1 \in \mathrm{Aff}(R,2)$, but none of $\tau_1, \lambda_1, \tau_2$ belong to $\mathrm{Aff}(R,2) \cap J_0$.

Remark 5.1.5 If R is not a domain, then $T(R,2)$ is not the amalgamated free product of $\mathrm{Aff}(R,2)$ and $J(R,2)$ over their intersection (Exercise 3).

Another consequence of 5.1.2 is

Corollary 5.1.6 *Let* $F = (F_1, F_2) \in T(R,2)$ *with* bideg $F = (d_1, d_2)$. *Let* h_i *denote the homogeneous component of* F_i *of degree* d_i. *Then*

a) $d_1 | d_2$ *or* $d_2 | d_1$.
b) *If* $\deg F > 1$, *then we have*
   i) *if* $d_1 < d_2$, *then* $h_2 = c h_1^{d_2/d_1}$, *for some* $c \in R$.
   ii) *if* $d_2 < d_1$, *then* $h_1 = c h_2^{d_1/d_2}$, *for some* $c \in R$.
   iii) *if* $d_1 = d_2$, *then there exists* $\lambda \in \mathrm{Aff}(R,2)$ *such that* $\deg F_1' > \deg F_2'$, *where* $(F_1', F_2') := \lambda \circ F$.

Proof: By 5.1.2 we have $F = \lambda_0\tau_1\lambda_1 \dots \tau_\ell\lambda_\ell$ with

$$\mathrm{bideg}(\tau_1\lambda_1 \dots \tau_\ell\lambda_\ell) = (\Pi_{j=1}^{\ell} \deg \tau_j, \Pi_{j=2}^{\ell} \deg \tau_j)$$

which gives a). Furthermore b) follows by checking that we are in the case i), ii) or iii) according to $a_0 = 0$, $d_0 = 0$ or $a_0 d_0 \neq 0$ where $\lambda_0 = (a_0X + b_0Y + c_0, d_0X + e_0Y + f_0)$ (Exercise 9). □

Remark 5.1.7 Notation as in 5.1.6. If $d_1 = d_2$ then in general there does not exist $c \in R$ such that $h_1 = ch_2$ or $h_2 = ch_1$ (Exercise 4).

Remark 5.1.8 If R is not a domain, the statement a) of 5.1.6 is false (Exercise 3).

Observe that if F is as in i) (resp. ii)) of 5.1.6, then $\mathrm{tdeg}\, \tau^{-1} \circ F < \mathrm{tdeg}\, F$, where $\tau = (X, Y + cX^{d_2/d_1})$ (resp. $\tau = (X + cY^{d_1/d_2}, Y)$). Consequently we get:

Algorithm to decide if $F = (F_1, F_2) \in R[X,Y]^2$ belongs to $T(R,2)$.

Input: $F = (F_1, F_2) \in R[X,Y]^2$.

1. Let $(d_1, d_2) = \mathrm{bideg}(F_1, F_2)$.
2. If $d_1 = d_2 = 1$, go to 7.
3. If $d_1 \neq d_2$, go to 5.
4. If there exists $\lambda \in \mathrm{Aff}(R,2)$ with $\mathrm{tdeg}\, \lambda F < \mathrm{tdeg}\, F$, replace F by λF and go to 1, else stop: $F \notin T(R,2)$.
5. If $d_2 < d_1$, replace F by (F_2, F_1).
6. If $d_1 | d_2$ and there exists $c \in R$ with $h_2 = c h_1^{d_2/d_1}$, replace F by $(X, Y - cX^{d_2/d_1}) \circ F$ and go to 1, else stop: $F \notin T(R,2)$.
7. If $\det JF \in R^*$, stop: $F \in T(R,2)$, else stop: $F \notin T(R,2)$.

Now we will use this algorithm, or 5.1.6, to address the following question.

Question: *Under which conditions on R is every R-automorphism of $R[X,Y]$ tame, i.e. under which conditions on R do we have the equality $T(R,2) = \operatorname{Aut}_R R[X,Y]$?*

Proposition 5.1.9 *If R is not a field, then $T(R,2) \neq \operatorname{Aut}_R R[X,Y]$. More precisely, let $0 \neq z \in R$ be a non-unit. Consider*

$$F = (X - 2Y(zX + Y^2) - z(zX + Y^2)^2, Y + z(zX + Y^2)).$$

Then $F \in \operatorname{Aut}_R R[X,Y] \backslash T(R,2)$.

Proof: One readily verifies that $F = \exp D$, where D is the locally nilpotent derivation $D = (zX + Y^2)(-2Y\partial_X + z\partial_Y)$ on $R[X,Y]$. So $F \in \operatorname{Aut}_R R[X,Y]$. If $F \in T(R,2)$ then by 5.1.6 ii) we get that $-zY^4 = c(zY^2)^2$ for some $c \in R$. Hence $-z = cz^2$, so dividing by $-z$ we get that z is a unit in R, contradiction. □

Remark 5.1.10 The automorphism F described in 5.1.9 is the so-called **Nagata automorphism**, and was introduced by Nagata in [224]. It is interesting to remark that Nagata defined his automorphism as the composition $F = {\sigma_1}^{-1}\sigma_2\sigma_1$, where $\sigma_1, \sigma_2 \in \operatorname{Aut}_K K[X,Y]$ ($K :=$ the quotient field of R) are defined by

$$\sigma_1(X,Y) := (X + z^{-1}Y^2, Y), \quad \sigma_2(X,Y) := (X, Y + z^2X).$$

Now the remarkable point is that the necessary condition "R is a field" is also sufficient to have the equality $\operatorname{Aut}_R R[X,Y] = T(R,2)$, i.e.

Theorem 5.1.11 (Jung, van der Kulk). *If k is a field, then $\operatorname{Aut}_k k[X,Y] = T(k,2)$. More precisely, $\operatorname{Aut}_k k[X,Y]$ is the amalgamated free product of $\operatorname{Aff}(k,2)$ and $J(k,2)$ over their intersection.*

The proof we will give below is based on the following strong version of Rentschler's theorem (see 1.3.48). Therefore our proof of 5.1.11 only works for the case char $k = 0$. For the general case we refer to the papers [185] of van der Kulk, [198] of Makar-Limanov and [87] of Dicks. So from now on k denotes a field of characteristic zero.

Theorem 5.1.12 (Rentschler). *Let $0 \neq D$ be a locally nilpotent derivation on $k[X_1, X_2]$. Then there exist $h \in T(k,2)$ and $f(X_2) \in k[X_2]$ such that $h^{-1}Dh = f(X_2)\partial_{X_1}$.*

Before we prove this result let us show how it implies the Jung-van der Kulk theorem.

Proof of Theorem 5.1.11: Write $A := k[X_1, X_2]$. By 5.1.3 we only need to show that $T(k,2) = \operatorname{Aut}_k A$. Therefore let $F = (F_1, F_2) \in \operatorname{Aut}_k A$. Then $\frac{\partial}{\partial F_1}$ is locally nilpotent, so by 5.1.12 there exist $h \in T(k,2)$ and $f(X_2) \in k[X_2]$ such that $h^{-1}\frac{\partial}{\partial F_1}h = f(X_2)\partial_{X_1}$. Let $g \in A$. Then $h^{-1}\frac{\partial}{\partial F_1}h(g) = 0$ if and only if $h(g) \in \ker \frac{\partial}{\partial F_1} = k[F_2]$, i.e. $\ker h^{-1}\frac{\partial}{\partial F_1}h = k[h^{-1}(F_2)]$. On the other hand $\ker f(X_2)\partial_{X_1} = k[X_2]$, so

$k[X_2] = k[h^{-1}(F_2)]$ which implies that $h^{-1}(F_2) = cX_2 + d$, for some $c \in k^*$ and $d \in k$. So $F_2 = ch(X_2) + d$.

Furthermore $(h^{-1}\frac{\partial}{\partial F_1}h)(h^{-1}(F_1)) = 1$, so $f(X_2)\partial_{X_1}(h^{-1}(F_1)) = 1$ which implies that $\partial_{X_1}(h^{-1}(F_1)) \in k^*$, so $h^{-1}(F_1) = c'X_1 + d'(X_2)$ for some $c' \in k^*$ and $d'(X_2) \in k[X_2]$. Consequently $F_1 = c'h(X_1) + d'(h(X_2))$. Summarizing

$$(F_1, F_2) = (c'h(X_1) + d'(h(X_2)), ch(X_2) + d).$$

Since $h \in T(k, 2)$ one deduces easily that $F \in T(k, 2)$ (Exercise 7), which completes the proof. □

Before we give the proof of Rentschler's theorem, we discuss some generalities concerning derivations and gradings. The starting point is the following result.

Proposition 5.1.13 *Let $R = \oplus_{m \in \mathbb{Z}} R_m$ be a graded ring and D a non-zero derivation on R. Suppose that D can be written as a finite sum of derivations of the form*

$$D = D_p + D_{p+1} + \cdots + D_d$$

such that $D_n R_m \subset R_{n+m}$, for all $n, m \in \mathbb{Z}$. If D is locally nilpotent, then so is D_d.

Proof: It suffices to verify that for every $m \in \mathbb{Z}$ each element of R_m is annihilated by some power of D_d. So let $g \in R_m$. Since D is locally nilpotent, there exists N such that $D^N g = 0$. The component of $D^N g$ which belongs to R_{m+Nd} is equal to $D_d^N g$. Since $D^N g = 0$ it follows that $D_d^N g = 0$, as desired. □

From now on we restrict to the case $R := k[X] = k[X_1, \ldots, X_n]$. To each non-zero vector w in $\mathbb{Z}^n$ we associate the so-called **w-grading on** R by defining for each $d \in \mathbb{Z}$ $R_d(w)$ as the k-vector space generated by all monomials X^a with $\langle a, w \rangle = d$. Here $\langle\ ,\ \rangle$ denotes the standard inner product on $\mathbb{R}^n$. If there is no confusion possible, we simply write R_d instead of $R_d(w)$. In order to apply the previous proposition we show

Proposition 5.1.14 *Let $0 \neq w \in \mathbb{Z}^n$ and consider on R the w-grading. Let D be a non-zero derivation on R. Then D can be written uniquely as a finite sum of derivations $\sum D_p$ such that $D_p R_d \subset R_{p+d}$ for all $p, d \in \mathbb{Z}$.*

Proof: The uniqueness follows readily from the definitions, so it remains to prove the existence. Therefore let $T := cX^a\partial_i$ be a term appearing in D. Put $s := a - e_i$, where e_i denotes the i-th standard basis vector of $\mathbb{R}^n$. Then $T(X^m) \in kX^{m+s}$ for all m. We call s the **strength of** T and put

$$\operatorname{supp} D := \{s \in \mathbb{Z}^n \mid \text{D contains a term of strength } s\}.$$

Finally denote by $D(s)$ the sum of all terms in D of strength s and put

$$D_p := \sum_{\langle s, w \rangle = p} D(s).$$

Obviously $D = \sum D_p$, since every non-zero term appearing in D has some strength, say s_0 and hence belongs to D_{p_0} where $p_0 = \langle s_0, w\rangle$. So it remains to show that $D_p R_d \subset R_{p+d}$ for all p, d or equivalently that $D(s)X^m \in R_{p+d}$ for all $X^m \in R_d$ and all $s \in \mathbb{Z}^n$ with $\langle s, w\rangle = p$. To see this first observe that $D(s)X^m \in kX^{m+s}$, so we need to show that $\langle m+s, w\rangle = p+d$. But this follows readily since $\langle m+s, w\rangle = \langle m, w\rangle + \langle s, w\rangle = d + p$, using that $X^m \in R_d$. □

The decomposition of D described in the previous proposition is called the **w-homogeneous** decomposition of D. If p is maximal with $D_p \neq 0$, p is called the **w-degree of** D, denoted $w\deg D$. In case $w = (1, 1, \ldots, 1)$ p is simply called the **degree of** D and is denoted by $\deg D$.

The following result plays a crucial role in the proof of Rentschler's theorem.

Proposition 5.1.15 *Let $R = \oplus_{d\in\mathbb{Z}} R_d$ be any grading on $R = k[X]$, so it need not even be a w-grading, and let $D = \sum D_p$ be a finite decomposition satisfying $D_p R_d \subset R_{p+d}$ for all $p, d \in \mathbb{Z}$. Let p be maximal with $D_p \neq 0$. If $D_p = D(s)$ for some $s = (s_1, \ldots, s_n) \in \mathbb{Z}^n$, where all $s_i \geq 0$, then D is not locally nilpotent.*

Proof: Let $s \in \mathbb{Z}^n$. By Exercise 8 it follows that $D(s) = X^s \sum c_j X_j \partial_j$ for some $c_j \in k$. Since $D(s) = D_p \neq 0$ there exists j with $c_j \neq 0$. Now observe that if all $s_i \geq 0$, then $D(s)(X_j) = c_j X^s X_j$ implies that $c_j X^s$ is a non-zero eigenvalue of $D(s)$ in $k[X]$. So by 1.3.32 $D(s)$ is not locally nilpotent, i.e. D_p is not locally nilpotent. Then apply 5.1.13. □

Now we restrict to the case $n = 2$ and write $R = k[X, Y]$ instead of $k[X_1, X_2]$. Let D be a non-zero derivation on R. An element of supp D will be denoted by (s, t). So $s, t \geq -1$ and for example $(s, -1) \in \operatorname{supp} D$ means that D contains a term of the form $cX^s\partial_Y$ with $c \in k^*$.

Corollary 5.1.16 *If D is locally nilpotent, then either*

i) $D = f(Y)\partial_X$, *for some* $f(Y) \in k[Y]$ *or*
ii) $D = f(X)\partial_Y$, *for some* $f(X) \in k[X]$ *or*
iii) *there exist* $s_0, t_0 \geq 0$ *such that* $(s_0, -1)$ *and* $(-1, t_0)$ *belong to* supp D *and furthermore* supp D *is contained in the triangle with vertices* $(s_0, -1)$, $(-1, -1)$, $(-1, t_0)$.

Proof: Let $p_0 := \deg D$. So $p_0 \geq -1$. If $p_0 = -1$ then $D = a\partial_X + b\partial_Y$ for some $a, b \in k$, not both zero and we are done. So let $p_0 \geq 0$ and let ℓ denote the line $x + y = p_0$.

Case 1: ℓ contains a point of the form $(-1, t_0)$. By definition of ℓ all points of supp D are on or below this line. Now turn ℓ around the point $(-1, t_0)$ (clockwise) until one hits another point of supp D. If ℓ contains more than one point of supp D we do not have to turn ℓ, i.e. we turn over an angle of degree zero. In case one does not meet any other point of supp D, then obviously supp $D = \{(-1, t_0)\}$ and consequently $D = cY^{t_0}\partial_X$ for some $c \in k^*$ and we are in case i). If one only meets points of the form $(-1, t)$ then all these points satisfy $t \leq t_0$ and $D = f(Y)\partial_X$, with $\deg_Y f(Y) \leq t_0$. So we are also in case i). Hence we may assume that we meet at least one point $(s, t) \in \operatorname{supp} D$ with $s \geq 0$. From all these points we choose the point with second coordinate minimal. Call this point (s', t'). We will show that $t' = -1$, which puts us in case iii). Therefore suppose that $t' \geq 0$. We will arrive at a contradiction. We also know that $s' \geq 0$, so both $s', t' \geq 0$. Now turning ℓ over a very small angle around (s', t') (clockwise) we find that there exists a line ℓ': $w_1x + w_2y = d$ with $w_1, w_2, d \in \mathbb{Z}$ such that (s', t') is the only point of $\operatorname{supp} D$ on this line and furthermore that all other points of $\operatorname{supp} D$ are below ℓ'. Putting $w := (w_1, w_2)$ and $p := w\deg D$, it follows that $D_p = D((s', t'))$. Applying 5.1.15 shows that D is not locally nilpotent, a contradiction. So $t' = -1$.

Case 2: ℓ contains a point of the form $(s', -1)$. This case is similar to case 1.

Case 3: ℓ contains neither a point of the form $(s, -1)$ nor a point of the form $(-1, t)$. Let $(s, t) \in \ell$ with t minimal. Since $s, t \geq 0$ we can again find a line ℓ' such that (s, t) is the only point of $\operatorname{supp} D$ on ℓ' and all other points of $\operatorname{supp} D$ are below ℓ'. Then arguing as above we get a contradiction, which completes the proof. □

Proof of Rentschler's theorem

i) We use 5.1.16. If $D = f(X)\partial_Y$ take $h = (Y, X)$. Then $h^{-1}Dh = f(Y)\partial_X$. So we may assume that we have s_0, t_0 as in 5.1.16. We write $s_0(D), t_0(D)$ instead of s_0, respectively t_0 if necessary. Let ℓ' be the line passing through both $(s_0, -1)$ and $(-1, t_0)$. So ℓ' is given by the equation $(t_0 + 1)x + (s_0 + 1)y = p$, where $p = s_0t_0 - 1$. Let $w := (t_0 + 1, s_0 + 1)$. So by 5.1.16 $w\deg D = p$ and by 5.1.13 D_p is locally nilpotent. Write $D_p = gD_1$ where $D_1 = a\partial_X + b\partial_Y$ satisfies $\gcd(a, b) = 1$. By 1.3.34 D_1 is locally nilpotent and $D_1(g) = 0$. Since D_p is w-homogeneous it follows that g and D_1 are also w-homogeneous.

ii) Since D_1 is locally nilpotent with $\gcd(a, b) = 1$ it follows from 1.3.46 that D_1 has a slice in $k[X, Y]$, so in particular $a(0) \neq 0$ or $b(0) \neq 0$. We may assume that $a(0) \neq 0$. So D_1 contains a term of the form $c\partial_X$ with $c \in k^*$. Since $(s_0, -1) \in \operatorname{supp} D_p$, i.e. D_p contains a term of the form $c'X^{s_0}\partial_Y$ and since $D_p = gD_1$ it follows that D_1 also contains a term of the form $dX^r\partial_Y$ with $r \geq 0$ and $d \in k^*$. But then it follows that $D_1 = c\partial_X + dX^r\partial_Y$, namely D_1 is w-homogeneous whence $\operatorname{supp} D_1$ is on a line, obviously the line passing through $(-1, 0)$ and $(r, -1)$. However this line does not contain any other point with integer coordinates ≥ -1!

iii) Since $D_1(g) = 0$ we get that $g \in \ker D_1 = k[Y - \frac{d}{(r+1)c}X^{r+1}]$. The homogeneity of g implies that $g = a(Y - \frac{d}{(r+1)c}X^{r+1})^N$ for some $a \in k^*$ and $N \in \mathbb{N}$. So

$$D_p = a\left(Y - \frac{d}{(r+1)c}X^{r+1}\right)^N (c\partial_X + dX^r\partial_Y).$$

iv) Finally let h be the automorphism given by $h(X) = X$ and $h(Y) = Y - \frac{d}{(r+1)c} X^{r+1}$.

Then one easily verifies that $h^{-1} D_p h = aY^N c\partial_X$. Also one easily shows that $h(R_d) \subset R_d$ and hence $h(R_d) = R_d$ for all $d \in \mathbb{Z}$. Consequently $(h^{-1}Dh)_m = h^{-1} D_m h$ for all $m \in \mathbb{Z}$. In particular the w-degree of $h^{-1}Dh$ is equal to the w-degree of D which equals p. Since $h^{-1}D_p h = acY^N \partial_X$ there is no contribution for a point of the form $(s, -1)$ in $\operatorname{supp} h^{-1}Dh$ coming from $(h^{-1}Dh)_p$. Consequently $s_0(h^{-1}Dh) < s_0(D)$. Then the theorem follows by induction on $s_0(D) + t_0(D)$.

Remark 5.1.17 Another proof of the equality $\operatorname{Aut}_k k[X, Y] = T(k, 2)$ is given in §3 of this chapter, see 5.3.6. This proof is an immediate consequence of the Abhyankar-Moh theorem.

As a consequence of the above proof we get some information about a coordinate in $k[X, Y]$. First if $0 \neq f = \sum f_{ij} X^i Y^j \in k[X, Y]$ we put

$$\operatorname{Supp} f := \{(i, j) \in \mathbb{Z}^2 \mid f_{ij} \neq 0\}.$$

The **Newton polygon of** f, denoted $N(f)$, is the convex hull of $\operatorname{Supp} f \cup \{(0, 0)\}$. Finally we say that f **has** $r \geq 1$ **points at infinity** if f_+, the highest degree homogeneous part of f according to the (1,1)-grading has r distinct (linear) prime factors in $\bar{k}[X, Y]$.

Proposition 5.1.18 *If f is a coordinate in $k[X, Y]$ then*

i) *$N(f)$ is a triangle with vertices $(0, 0)$, $(n, 0)$, $(0, m)$, where n, m are integers ≥ 0.*

ii) *f has one point at infinity.*

Proof: i) Since f is a coordinate there exists $g \in k[X, Y]$ such that $k[X, Y] = k[f, g]$ and $\det J(f, g) = 1$. So $D := \frac{\partial}{\partial g}$ is locally nilpotent on $k[X, Y]$ and $D = f_Y \partial_X - f_X \partial_Y$. Using this description of D one easily verifies that

$$(i, j) \in \operatorname{Supp} f \text{ if and only if } (i-1, j-1) \in \operatorname{Supp} D.$$

Using 5.1.16 we obtain that $N(f)$ is a triangle.

ii) By i) we know that $N(f)$ is a triangle. If $n > m$ then $f_+ = c_1 X^n$ for some $c_1 \in k^*$ and if $n < m$ obviously $f_+ = c_2 Y^m$ with $c_2 \in k^*$. Finally if $n = m$, then according to the (1,1)-grading we get $D_p = f_{+Y} \partial_X - f_{+X} \partial_Y$, where p is the (1,1)-degree of D. From the proof of Rentschler's theorem we know that

$$D_p = a\left(Y - \frac{d}{(r+1)c} X^{r+1}\right)^N (c\partial_X + dX^r \partial_Y)$$

for some $a, c, d \in k^*$, $N \geq 0$ and $r \geq 0$. Furthermore $Y - \frac{d}{(r+1)c}X^{r+1}$ is (1,1)-homogeneous. So $r = 0$ and consequently

$$D_p = a\left(Y - \frac{d}{c}X\right)^N (c\partial_X + d\partial_Y).$$

Since, as observed above, $D_p = f_{+Y}\partial_X - f_{+X}\partial_Y$ we obtain

$$f_+ = ac\frac{1}{N+1}\left(Y - \frac{d}{c}X\right)^{N+1}$$

which concludes the proof. □

Exercises for §5.1

1. Show that every element of $J(R, n)$ is of the form $\lambda E_1 \ldots E_n$ where $\lambda \in \mathrm{Aff}(R, n)$ and E_i is an elementary automorphism.
2. Show that each elementary automorphism belongs to $\langle \mathrm{Aff}(R, n), J(R, n)\rangle$.
3. Let R be a commutative ring which is not a domain, say $ab = 0$ where a and b are non-zero elements in R.
   (i) Let $\tau := (X + aY^2, Y)$ and $\lambda := (X, Y + bX)$. Show that both τ and λ do not belong to $\mathrm{Aff}(R, 2) \cap J(R, 2)$ and that $\tau\lambda\tau^{-1} = \lambda$. So $T(R, 2)$ is not the amalgamated free product of $\mathrm{Aff}(R, 2)$ and $J(R, 2)$ over their intersection.
   (ii) Let $F = (X, Y + aX^3) \circ (X + bY^2, Y)$. Show that $F = (X + bY^2, Y + aX^3)$, so $(d_1, d_2) := \mathrm{bideg}\, F = (2, 3)$ and d_1 does not divide d_2.
4. Suppose that R is a domain which is not a field and $z \in R$ is such that both z and $1-z$ are non-units in R. Put $F = \lambda\tau$, where $\lambda = ((1-z)X + zY, -zX + (1+z)Y)$ and $\tau = (X + Y^2, Y)$.
   (i) Show that $\lambda \in \mathrm{Aff}(R, 2)$, $F \in T(R, 2)$ and $\mathrm{bideg}\, F = (2, 2)$.
   (ii) Let h_1, h_2 be as in 5.1.6. Show that there does not exist $c \in R$ such that $h_1 = ch_2$ or $h_2 = ch_1$.
5. Let z be a unit in R and F the Nagata automorphism defined in 5.1.9. Show that $F \in T(R, 2)$.
6. Let R be a ring containing the elements a, b. Let $f(T) \in R[T]$ and define $F := (X + bf(aX + bY), Y - af(aX + bY))$.
   (i) Show that $F \in \mathrm{Aut}_R\, R[X, Y]$. [Show that $D := f(aX + bY)(b\partial_X - a\partial_Y)$ is a locally nilpotent derivation on $R[X, Y]$ and that $F = \exp D$.]
   (ii) Now assume that R is a domain such that $Ra + Rb$ is not a principal ideal. Let $\deg f(T) \geq 2$. Show that $F \notin T(R, 2)$.
7. Complete the proof of 5.1.11.
8. Let $s \in \mathbb{Z}^n$. Show that $D(s) = X^s \sum_{i=1}^n c_i X_i \partial_i$ for some $c_i \in k$.
9. Complete the proof of 5.1.6.

5.2 The tame automorphism group in dimension ≥ 3

Throughout this section R denotes a commutative $\mathbb{Q}$-algebra, $R[X]$ the polynomial ring in n variables over R and $n \geq 3$. The aim is to study $T(R,n)$. It turns out that there are several differences with the case $n = 2$ studied in §1. One such difference is that although $T(R,n)$ is generated by $\text{Aff}(R,n)$ and $J(R,n)$, $T(R,n)$ is not the amalgamated free product over their intersection (5.2.2).

However, the most remarkable result (which does not hold in case $n = 2$, Excercise 1) is Theorem 5.2.1 which, in case R is a field, asserts that $T(R,n)$ is generated by $\text{Aff}(R,n)$ and only one non-linear automorphism, namely $\varepsilon = (X_1 + X_2^2, X_2, \ldots, X_n)$. This result was observed by Harm Derksen in 1994 during the Curaçao conference and remained unpublished. By a simple trick, due to Stefan Maubach, Derksen's result can be generalised to the case that R is a local ring (see 5.2.1 for the exact statement). Finally we show that for any $\mathbb{Q}$-algebra R, $E(R,n)$ is equal to its commutator subgroup (5.2.3). At the end of this section we discuss some open problems concerning tame automorphisms.

In the sequel we will use the following notations: the elementary automorphism

$$(X_1, \ldots, X_{i-1}, X_i + a, X_{i+1}, \ldots, X_n),$$

where $a \in R[X_1, \ldots, \hat{X}_i, \ldots, X_n]$ will be denoted by $E_i(a)$. Furthermore, if $c \in R^*$ we denote the element $(X_1, \ldots, X_{i-1}, cX_i, X_{i+1}, \ldots, X_n)$ of $\text{Aff}(R,n)$ by $D_i(c)$, and by $P_{i,j}$ we denote the element of $\text{Aff}(R,n)$ interchanging X_j and X_i. If G is any group and g_1, g_2 are elements of G, we denote the **commutator** $g_1 g_2 g_1{}^{-1} g_2{}^{-1}$ by $[g_1, g_2]$ and the **commutator subgroup**, i.e. the subgroup of G generated by all such commutators, by $[G,G]$. Finally we say that R contains sufficiently many units if R^* generates R as a $\mathbb{Q}$-vectorspace. For example this condition is satisfied if R is a local ring (Exercise 6) and hence if R is a field.

Theorem 5.2.1 (Derksen). *Let $n \geq 3$. Then $T(R,n)$ is generated by* $\text{Aff}(R,n)$ *and the elementary automorphisms $E_1(cX_2^2)$, with $c \in R$. Furthermore, if R has sufficiently many units, then $T(R,n) = \langle \text{Aff}(R,n), \varepsilon \rangle$, where $\varepsilon = (X_1 + X_2^2, X_2, \ldots, X_n)$.*

Once having this description of $T(R,n)$, it is not difficult to show that $T(R,n)$ is not the amalgamated free product of $\text{Aff}(R,n)$ and $J(R,n)$ over their intersection. To see this we give the following example which is due to Jacques Alev.

Example 5.2.2 Let $n \geq 3$. Define $\lambda = P_{2,3}$ and $\tau = \lambda \circ \varepsilon \circ \lambda$. Then $\tau = (X_1 + X_3^2, X_2, X_3, \ldots, X_n)$. So both τ and ε belong to $J(R,n) \setminus \text{Aff}(R,n)$. However

$$\tau^{-1} \lambda \varepsilon = \lambda^{-1} \in \text{Aff}(R,n).$$

Consequently $T(R,n)$ is not the amalgamated free product of $\text{Aff}(R,n)$ and $J(R,n)$ over their intersection.

Arguing as in the proof of 5.2.1, we will show

Proposition 5.2.3 *Let $n \geq 3$. Then $[E(R,n), E(R,n)] = E(R,n)$.*

A crucial ingredient in the proof of both 5.2.1 and 5.2.3 is the following elementary lemma, which proof is left to the reader (Exercise 2).

Lemma 5.2.4 i) *$D_1(c^{-1})D_2(c^{-1})\varepsilon D_2(c)D_1(c) = E_1(cX_2^2)$, for all $c \in R^*$.*
ii) *$P_{i,2}E_1(aX_2^d)P_{i,2} = E_1(aX_i^d)$, for all $i \geq 2$, all $d \in \mathbb{N}$ and all $a \in R$.*
iii) *Let $2 \leq i \leq n$ and*

$$L_i := (X_1, \ldots, X_{i-1}, c_2X_2 + \cdots + c_nX_n, X_{i+1}, \ldots, X_n)$$

with $c_j \in R$ for all j and $c_i \in R^*$. Then for all $a \in R$, $d \in \mathbb{N}$ we have

$$E_1(a(c_2X_2 + \cdots + c_nX_n)^d) = L_i^{-1}E_1(aX_2^d)L_i.$$

iv) *$E_1(X_2X_3) = E_1(-(\frac{1}{2}X_2 - \frac{1}{2}X_3)^2)E_1((\frac{1}{2}X_2 + \frac{1}{2}X_3)^2)$.*
v) *$E_1(aX_2^d) = [E_1(X_2X_3), E_3(aX_2{}^{d-1})]$, for all $a \in R$ and all $d \geq 1$.*
vi) *$E_i(aX_j) = [E_i(aX_k), E_k(X_j)]$, for all $a \in R$ and all triples (i,j,k) with $k \neq i, j$ and $i \neq j$.*

Proof of Theorem 5.2.1: Let H be the subgroup of $T(R,n)$ generated by $\mathrm{Aff}(R,n)$ and the elements $E_1(cX_2^2)$, $c \in R$. We show that $H = T(R,n)$. Since $T(R,n) = \langle \mathrm{Aff}(R,n), E(R,n)\rangle$ it is sufficient to prove that $E_i(g) \in H$ for all $g \in R[X_1, \ldots, \hat{X}_i, \ldots, X_n]$ and all $1 \leq i \leq n$. Furthermore using permutation matrices, it suffices to prove that $E_1(g) \in H$ for all $g \in R[X]$ not containing X_1. Writing g as the sum of its monomials and using that $E_1(g+h) = E_1(g) + E_1(h)$ for all g and h not containing X_1, it suffices to show that $E_1(m) \in H$ for each monomial not containing X_1. The case $m = X_2X_3$ follows from 5.2.4 iii) and iv). Then by induction on d we deduce from 5.2.4 v) that $E_1(aX_2^d) \in H$ for all $d \geq 2$ and all $a \in R$. Consequently by 5.2.4 iii) we get

$$(5.2.5) \qquad E_1(a(X_2 + c_3X_3 + \cdots + c_nX_n)^d) \in H, \text{ for all } a, c_i \in R, \text{ all } d \geq 2.$$

Since by Exercise 7 every monomial of degree d in $\mathbb{Q}[X_2, \ldots, X_n]$ is a finite $\mathbb{Q}$-linear combination of elements of the form $(X_2 + q_3X_3 + \cdots + q_nX_n)^d$, $q_i \in \mathbb{Q}$, it follows from (5.2.5) that $E_1(m) \in H$ for all monomials m in $R[X_2, \ldots, X_n]$, which completes the proof of the first part of the theorem. So to see the second part we only need to show that $E_1(cX_2^2)$ belongs to $\langle \mathrm{Aff}(R,n), \varepsilon\rangle$ in case R has sufficiently many units. First if $c \in R^*$ the result follows from 5.2.4 i). So let $c \in R$ be arbitrary. Then by the units hypothesis on R, c can be written as a sum of units, say $c = c_1 + \cdots + c_t$, $c_i \in R^*$. Finally, since $E_1(cX_2^2) = E_1(c_1X_2^2) \circ \cdots \circ E_1(c_tX_2^2)$, the desired result follows.

Proof of Proposition 5.2.3: Arguing as in the proof of 5.2.1 it suffices to prove that

$$E_1(a(\langle q, X'\rangle)^d) \in [E(R,n), E(R,n)]$$

for all $a \in R$, all $d \geq 2$ and all $0 \neq q \in \mathbb{Q}^{n-1}$. So let $q = (q_2, \ldots, q_n) \in \mathbb{Q}^{n-1}$, say with $q_i \neq 0$. Then by 5.2.4 iii) we get

$$(5.2.6) \qquad E_1(a(\langle q, X'\rangle)^d) = L_i^{-1} E_1(aq_i^d X_i^d) L_i$$

where $L_i := (X_1, \ldots, X_{i-1}, \langle \tilde{q}, X'\rangle, X_{i+1}, \ldots, X_n)$ and $\tilde{q} := q_i^{-1} q$. Since

$$L_i = E_i(\tilde{q}_2 X_2) \ldots E_i(\tilde{q}_{i-1} X_{i-1}) E_i(\tilde{q}_{i+1} X_{i+1}) \ldots E_i(\tilde{q}_n X_n)$$

it follows from 5.2.4 vi) that

$$(5.2.7) \qquad L_i \in [E(R,n), E(R,n)].$$

Finally observe that $E_1(aq_i^d X_i^d) \in [E(R,n), E(R,n)]$ by 5.2.4 v), which together with 5.2.7 and 5.2.6 implies that

$$E_1(a(\langle q, X'\rangle)^d) \in [E(R,n), E(R,n)].$$

This completes the proof of 5.2.3. □

To conclude this section we briefly discuss some open problems in the case that $R = k$ is a **field**. If $n \geq 3$ we do not know an algorithm to decide if a given k-endomorphism of $k[X]$ is tame. Even more concretely: consider the $k[Z]$-endomorphism F of $k[X,Y,Z]$ described in 5.1.9. Then $F \in \mathrm{Aut}_{k[Z]} k[X,Y,Z] \backslash T(k[Z],2)$. Now we view F as a k-endomorphism of $k[X,Y,Z]$ and denote this endomorphism by σ, i.e.

$$\sigma = (X - 2Y(ZX + Y^2) - Z(ZX + Y^2)^2, Y + Z(ZX + Y^2), Z).$$

Then $\sigma \in \mathrm{Aut}_k k[X,Y,Z]$. However we do not know if $\sigma \in T(k,3)$. As already remarked before it was conjectured by Nagata in 1972 that σ is not tame, i.e.

Conjecture (Nagata) $\sigma \notin T(k,3)$.

This conjecture is still open. We refer the reader to [17] were it is shown that σ does not admit a certain kind of decomposition, which gives some support for Nagata's conjecture. On the other hand it was observed by Martha Smith in [271] that if we extend σ to $\tilde{\sigma} : k[X,Y,Z,W] \to k[X,Y,Z,W]$ by putting $\tilde{\sigma}(W) = W$, then $\tilde{\sigma} \in T(k,4)$. (see Chapter 6, §1 and Chapter 7, §3 for more details).

Apart from the Nagata automorphism given above, there are many other candidate non-tame automorphisms. One of the simplest is the so-called **Anick-automorphism** given by

$$\tau = (X + Z(XW + YZ), Y - W(XW + YZ), Z, W)$$

which is obviously of degree 3. Quadratic automorphisms are usually better behaved. For example the Jacobian Conjecture is known to be true for quadratic maps, see Chapter 4, §3. This has led Rusek in [255] to the following conjecture.

Conjecture (Rusek) Let $F \in \mathrm{Aut}_k\, k[X]$ and $\deg F \leq 2$. Then $F \in T(k,n)$.

In his thesis, [157] Hubbers considers quadratic homogeneous polynomial maps in dimension 5 and quadratic maps in dimension 4. Amongst the various types of quadratic maps he describes there is one family which does not satisfy the generalized dependence problem given in Chapter 7, §1. More precisely, let k be any field of characteristic $\neq 2$ and let $Q = (Q_1, \ldots, Q_4)$ be given by

$$\begin{aligned}
Q_1 &= g_3X_1 - f_3X_2,\\
Q_2 &= \tfrac{(12g_3^2+2f_3g_2)}{2f_3}X_1 - 3g_3X_2 + f_3X_3 + \tfrac{1}{2}X_1^2,\\
Q_3 &= g_1X_1 + g_2X_2 + g_3X_3 + \tfrac{f_3}{2d_5}X_4 + X_1X_2,\\
Q_4 &= \tfrac{2d_5g_3^4}{f_3^3}X_1 + \tfrac{(-8d_5f_3g_3^3+2d_5f_3^3g_1)}{f_3^3}X_2 + g_3X_4 + d_5X_2^2,
\end{aligned}$$

where $d_5, f_3, g_1, g_2, g_3 \in k$; then JQ is nilpotent and $Q_1, \ldots, Q_4$ are linearly independent over k. In particular $F := X + Q$ has an invertible Jacobian matrix and hence by Wang's Proposition 4.3.1 F is invertible over k. In fact it is left to the reader to verify that F satisfies Rusek's conjecture, i.e. F is tame (Exercise 5).

On the other hand if k is not a field, a quadratic automorphism need not be tame: it was shown in Exercise 6 of §1 that if R is a ring in which $Ra + Rb$ is not a principal ideal, then

$$(X + b(aX + bY)^2, Y - a(aX + bY)^2) \in \mathrm{Aut}_R\, R[X,Y] \backslash T(R,2).$$

Exercises for §5.2

1. Show that if $n = 2$ the statement analogous to 5.2.1 is false.
2. Give a proof of 5.2.4.
3. Let $n \geq 3$ and assume that R has sufficiently many units. Show that every $F \in T(R,n)$ can be written as a finite product $\lambda_1\varepsilon\lambda_2\varepsilon \ldots \varepsilon\lambda_n$, for some $\lambda_i \in \mathrm{Aff}(R,n)$.
4. Let $F \in \mathrm{Aut}_k\, k[X_1, \ldots, X_5]$ given by
$$F = (X_1 - X_2X_4, X_2 + X_3X_4 - X_1X_5, X_3 - X_2X_5, X_4, X_5).$$
Show that F is a product of three elementary maps.
5. Show that the map $F = X + Q$ described above is tame. [Compose F from the right with elementary automorphisms.]
6. Let R be a local ring. Show that R has sufficiently many units.
7. Let $d \geq 1$. Show that the $\mathbb{Q}$-vectorspace generated by all monomials of degree d in $\mathbb{Q}[X_1, \ldots, X_n]$ is generated by the polynomials of the form $(X_1 + q_2X_2 + \cdots + q_nX_n)^d$, where $q_i \in \mathbb{Q}$. [The case $n = 2$, use a Vandermonde determinant. For the general case, consider $(X_1 + q_2(X_2 + q_3X_3 + \cdots + q_nX_n))^d$ and use induction on n and d.]

5.3 Embeddings of affine algebraic varieties and tame automorphisms

Throughout this section k denotes a field of characteristic zero. We discuss the relationship between tame automorphisms and the problem of embeddings of affine algebraic varieties. More precisely we investigate when an embedding of some affine space $\mathbb{A}^k$ in some affine space $\mathbb{A}^n$ is **rectifiable** (5.3.3). The first case we discuss is the famous Abhyankar-Moh-Suzuki theorem, which asserts that every embedding of a line in $\mathbb{A}^2$ is rectifiable. In more algebraic terms this result can be formulated as follows: given polynomials $f(T), g(T) \in k[T]$ such that $k[f(T), g(T)] = k[T]$, then either $\deg f(T)$ divides $\deg g(T)$ or $\deg g(T)$ divides $\deg f(T)$. It is interesting to remark that this result was used by Segre in [263] to "prove" the Jacobian Conjecture. Therefore sometimes the above result is called **Segre's lemma**. Later Canals and Lluis found an error in Segre's proof and published a correction, [47]. In the fundamental paper [9] the authors remark that also in the corrected version of Canals and Lluis there is an error. The proof given by Abhyankar and Moh is rather complicated and based upon earlier results in [7] and [8]. In the meantime several new proofs of the Abhyankar-Moh-Suzuki theorem have been given. In §4 we will give an elementary proof devised by Andrzej Nowicki. We will use the Segre lemma to give a new proof of the automorphism theorem, i.e. we deduce that every k-automorphism of $k[X, Y]$ is tame.

It was conjectured by Abhyankar in [4] that for each $n \geq 3$ there exists lines in $\mathbb{A}^n$ which are not rectifiable. However, as we will see below, for all $n \geq 4$ every embedding of the line in $\mathbb{A}^n$ is rectifiable (5.3.10). This result was first obtained by Craighero in [65a] and Jelonek in [161]. A more general result was obtained by Srinivas in [275]. His main result asserts the following: let A be a finitely generated k-algebra. Then there exists a positive integer $n(A)$ such that if $f, g : k[X_1, \ldots, X_n] \to A$ are two surjections with $n > n(A)$, then there exists $F \in T(k, n)$ such that $f = g \circ F$. In case $A = k[T]$ it turns out that $n(A) = 3$ and we recover the Craighero-Jelonek result. Also one deduces immediately that if V and W are closed algebraic subsets of $\mathbb{A}^n$ with $n > n(A(V))$, then any isomorphism $\varphi : V \to W$ can be extended to a polynomial automorphism of $\mathbb{A}^n$.

At the end of this section we consider embeddings of $\mathbb{A}^{n-1}$ in $\mathbb{A}^n$ and briefly discuss the Abhyankar-Sathaye conjecture and describe some candidate counterexamples to this conjecture due to Choudary and Dimca.

Definition 5.3.1 Let $V \subset \mathbb{A}^n$ and $W \subset \mathbb{A}^m$ be closed algebraic subsets and $\varphi : V \to W$ a regular map. Then φ is called an **embedding** if

1) $\varphi(V)$ is closed in the Zariski topology of W and
2) $\varphi : V \to \varphi(V)$ is an isomorphism, i.e. $\varphi^{-1} : \varphi(V) \to V$ exists and is a regular mapping.

Example 5.3.2 i) Let $V \subset \mathbb{A}^n$ be a closed algebraic set. Then the inclusion map $i : V \to \mathbb{A}^n$ is an embedding. Furthermore, if $F \in \mathrm{Aut}_k\, \mathbb{A}^n$ then $F \circ i : V \to \mathbb{A}^n$ is also an embedding.

ii) Let $1 \leq k < n$. Then the map $j : \mathbb{A}^k \to \mathbb{A}^n$ defined by

$$j(x_1, \ldots, x_k) = (x_1, \ldots, x_k, 0, \ldots, 0)$$

is an embedding, called the **standard embedding**. Furthermore, if $F \in \mathrm{Aut}_k \mathbb{A}^n$ then $F \circ j : \mathbb{A}^k \to \mathbb{A}^n$ is also an embedding.

Definition 5.3.3 i) Let $V \subset \mathbb{A}^n$ be a closed algebraic subset and $\varphi : V \to \mathbb{A}^n$ an embedding. Then φ is called an **exotic embedding** if φ is not of the form $F \circ i$ as in 5.3.2 above.

ii) Let $1 \leq k < n$. An embedding $\varphi : \mathbb{A}^k \to \mathbb{A}^n$ is called **rectifiable** if it is of the form $F \circ j$, with $F \in \mathrm{Aut}_k \mathbb{A}^n$ and j as in 5.3.2.

Of course, if an embedding $\varphi : \mathbb{A}^k \to \mathbb{A}^n$ is rectifiable, say $\varphi = F \circ j$, then composing φ with the automorphism F^{-1} we get $F^{-1} \circ \varphi = j$, the standard linear embedding of $\mathbb{A}^k$ in $\mathbb{A}^n$, which explains the name "rectifiable".

We will investigate under which conditions embeddings are not exotic and for which k and n every embedding of $\mathbb{A}^k$ in $\mathbb{A}^n$ is rectifiable. The first case we investigate is $k = 1, n = 2$, i.e. the embedding of the line in the plane.

Theorem 5.3.4 (Abhyankar, Moh). *Let $f(T), g(T) \in k[T]$ with $n = \deg f(T)$, $m = \deg g(T)$ such that $k[f(T), g(T)] = k[T]$. Then n divides m or m divides n.*

The proof of this result will be postponed until the next section. However we deduce immediately.

Theorem 5.3.5 (Abhyankar, Moh, Suzuki). *Every embedding of a line in the affine plane is rectifiable.*

Proof: Let $\varphi = (f, g) : k \to k^2$ be an embedding, where $f(T), g(T) \in k[T]$. We use induction on $\mathrm{tdeg}\, \varphi := \deg f + \deg g$. The case $\mathrm{tdeg}\, \varphi = 1$ is easy, so let $\mathrm{tdeg}\, \varphi \geq 2$. Since φ is an embedding, the induced map $\varphi^* : k[X, Y]/I(\varphi(k)) \to k[T]$ given by $\overline{X} \to f(T), \overline{Y} \to g(T)$, is an isomorphism. In particular $k[T] = k[f(T), g(T)]$, so by 5.3.4 we may assume that $n = \deg f(T)$ divides $m = \deg g(T)$, i.e. $m = dn$ for some $d \in \mathbb{N}$. So for suitable $c \in k^*$ the elementary automorphism $E := (X, Y - cX^d)$ satisfies $\mathrm{tdeg}\, E \circ \varphi < \mathrm{tdeg}\, \varphi$. Then the desired result follows from the induction hypothesis. □

Before we continue to investigate embeddings in $\mathbb{A}^n$, let us first show how 5.3.4 can be used to give another proof of the automorphism theorem, i.e.

Corollary 5.3.6 $\mathrm{Aut}_k\, k[X, Y] = T(k, 2)$.

Proof: Let $F = (F_1, F_2) \in \mathrm{Aut}_k\, k[X, Y]$. We will show that $F \in T(k, 2)$ by induction on $\mathrm{tdeg}\, F := \deg F_1 + \deg F_2$. We may assume that $\deg F_1$ and $\deg F_2 > 1$ (Exercise 3). Making a linear coordinate change we may assume that $\deg F_i = \deg_X F_i(X, 0)$ for $i = 1, 2$ and that $d_1 := \deg F_1 \leq d_2 := \deg F_2$. Since $k[F_1, F_2] = k[X, Y]$ we get $k[F_1(X, 0), F_2(X, 0)] = k[X]$, whence by 5.3.4 we get that d_1 divides d_2. So by Exercise 9 it follows that, for suitable $c \in k^*$, $\mathrm{tdeg}\, E \circ F < \mathrm{tdeg}\, F$, where

$E = (X, Y - cX^d)$ and $d := d_2/d_1$. Then the result follows from the induction hypothesis. □

Now let us return to the study of embeddings in n-space. The natural next question to ask is: is every embedding of $\mathbb{A}^1$ in $\mathbb{A}^n$ is rectifiable if $n \geq 3$?

It turns out that the answer is yes if $n \geq 4$. In fact this follows from a general result due to Srinivas which we will discuss now. To describe his result, let A be a finitely generated k-algebra. We put

$$m(A) := \text{ the Krull dimension of the symmetric algebra of } \Omega^1_{A/k} \text{ over } A.$$

Here $\Omega^1_{A/k}$ denotes the A-module of Kähler differentials, [203], Chapter 9, §25 and

$$n(A) := \sup\{m(A), 2\dim A + 1\}.$$

Theorem 5.3.7 (Srinivas). *Let A be a finitely generated k-algebra. Suppose that $f : k[X_1, \ldots, X_n] \to A$ and $g : k[X_1, \ldots, X_n] \to A$ are two surjections of k-algebras with $n > n(A)$. Then there exists $F \in T(k, n)$ such that $f = g \circ F$.*

The proof is based on the following two lemmas:

Lemma 5.3.8 *Let f, g be as in 5.3.7. Put $f_i := f(X_i)$ and $g_i := g(X_i)$ for all i. Then after a linear coordinate change in $k[X]$ we may assume that any subset of $\{f_1, \ldots, f_n, g_1, \ldots, g_n\}$ of cardinality $\geq n(A)$ generates A as a k-algebra.*

Proof: We refer to [275] for a proof of this result. □

To formulate the next lemma we introduce some notation: if $f : k[X_1, \ldots, X_n] \to A$ and $g : k[X_1, \ldots, X_n] \to A$ are two surjections of k-algebras such that $f = g \circ F$ for some $F \in T(k, n)$, then we write $f \sim g$. Finally, if $f(X_i) = f_i$ for all i, we also write $(f_1, \ldots, f_n)$ instead of f.

Lemma 5.3.9 *Let $H : k[X_1, \ldots, X_n] \to A$ be a surjection of k-algebras with $H(X_i) = h_i \in A$. Suppose that for some t $h_1, \ldots, h_t$ already generate A over k. For arbitrary $a_1, \ldots, a_{n-t}$ in A let $\tilde{H} : k[X_1, \ldots, X_n] \to A$ be defined by*

$$\tilde{H}(X_i) = h_i, \text{ if } 1 \leq i \leq t \text{ and } \tilde{H}(X_i) = a_{i-t}, \text{ if } t < i \leq n.$$

Then $\tilde{H}$ is also a surjection and $H \sim \tilde{H}$.

Proof: Since $h_1, \ldots, h_t$ generate A over k, we get $h_{t+i} = P_{t+i}(h_1, \ldots, h_t)$ and $a_i = Q_i(h_1, \ldots, h_t)$, for some polynomials P_{t+i}, Q_i in $k[X_1, \ldots, X_t]$. Let $F \in T(k, n)$ be defined by

$$F_i = X_i, \text{ all } 1 \leq i \leq t, \; F_{i+t} = X_{i+t} - P_{i+t} + Q_i, \text{ all } 1 \leq i \leq n - t.$$

Then one easily verifies that $\tilde{H} = H \circ F$. □

Proof of Theorem 5.3.7: Let $n > n(A)$. So by 5.3.8 $f_1, \ldots, f_{n-1}$ generate A, whence by 5.3.9

$$(f_1, \ldots, f_n) \sim (f_1, \ldots, f_{n-1}, g_1) \sim (g_1, f_1, \ldots, f_{n-1}).$$

Again by 5.3.8 $(g_1, f_1, \ldots, f_{n-2})$ generate A and hence by 5.3.9

$$(g_1, f_1, \ldots, f_{n-2}, f_{n-1}) \sim (g_1, f_1, \ldots, f_{n-2}, g_2) \sim (g_1, g_2, f_1, \ldots, f_{n-2}).$$

Continuing in this way we finally arrive at $(g_1, \ldots, g_n)$. So $(f_1, \ldots, f_n) \sim (g_1, \ldots, g_n)$ as desired.

Corollary 5.3.10 *If $n \geq 2k+2$, then every embedding of $\mathbb{A}^k$ in $\mathbb{A}^n$ is rectifiable. In particular, if $n \geq 4$ every embedding of a line in $\mathbb{A}^n$ is rectifiable.*

Proof: Let $\varphi = (\varphi_1, \ldots, \varphi_n) : \mathbb{A}^k \to \mathbb{A}^n$ be an embedding and $j : \mathbb{A}^k \to \mathbb{A}^n$ the standard embedding. Then the induced maps $\varphi^*, j^* : k[X] \to k[X_1, \ldots, X_k]$ are surjections of k-algebras. Since obviously $n(k[X_1, \ldots, X_k]) = 2k+1$, we deduce from 5.3.7 that there exists $F \in T(k, n)$ such that $\varphi^* \circ F = j^*$, if $n \geq 2k+2$, which implies the desired result. □

Taking into account 5.3.5 we see that, concerning the question if an embedding of a line in $\mathbb{A}^n$ is rectifiable, only the case $n = 3$ remains open. In fact we do not know the answer. Therefore we formulate

Problem 5.3.11 Is every embedding of a line in $\mathbb{A}^3$ rectifiable?

Another consequence of 5.3.7 is

Proposition 5.3.12 (Kaliman). *Let V, W be closed algebraic subsets of $\mathbb{A}^n$. Assume that $n > n(A(V))$. Then any isomorphism $\varphi : V \to W$ can be extended to an automorphism F of $\mathbb{A}^n$.*

Proof: Write $A(V) := k[X]/I$ and $A(W) := k[X]/J$. So we have surjections $p_V : k[X] \to A(V)$ and $p_W : k[X] \to A(W)$. Let $\varphi : V \to W$ be an isomorphism. Then the induced map $\varphi^* : A(W) \to A(V)$ is an isomorphism of k-algebras, so $\varphi^* \circ p_W : k[X] \to A(V)$ is a surjection. So by 5.3.7 there exists $F \in T(k, n)$ such that $p_V \circ F = \varphi^* \circ p_W$, which implies the desired result. □

To conclude this section we consider embeddings of $\mathbb{A}^{n-1}$ in $\mathbb{A}^n$ and assume that k is **algebraically closed**. Observe that if $\varphi : \mathbb{A}^{n-1} \to \mathbb{A}^n$ is an embedding, then $\varphi(\mathbb{A}^{n-1}) \simeq \mathbb{A}^{n-1}$. So $\varphi(\mathbb{A}^{n-1}) = V(f)$, for some irreducible polynomial f in $k[X]$.

Lemma 5.3.13 *The following statements are equivalent*

i) *Every embedding of $\mathbb{A}^{n-1}$ in $\mathbb{A}^n$ is rectifiable.*
ii) *If f is an irreducible polynomial in $k[X]$ such that $V(f) \simeq \mathbb{A}^{n-1}$, then f is a coordinate.*

Proof: i) → ii). Let $\varphi : \mathbb{A}^{n-1} \to V(f)$ be an isomorphism and $i : V(f) \to \mathbb{A}^n$ the inclusion map. Then $i \circ \varphi$ is an embedding of $\mathbb{A}^{n-1}$ in $\mathbb{A}^n$, so there exists $F \in \operatorname{Aut}_k \mathbb{A}^n$ such that $i \circ \varphi = F \circ j$, where $j : \mathbb{A}^{n-1} \to \mathbb{A}^n$ is the standard embedding. So $V(f) = F \circ j(\mathbb{A}^{n-1}) = V(G_n)$, where G_n is the n-th component of the inverse of F. So $f = \lambda G_n$, for some $\lambda \in k^*$, whence f is a coordinate.

ii) → i). Let $\varphi : \mathbb{A}^{n-1} \to \mathbb{A}^n$ be an embedding. Then $\varphi(\mathbb{A}^{n-1}) = V(f)$ and since $\mathbb{A}^{n-1} \simeq \varphi(\mathbb{A}^{n-1})$ it follows from the hypothesis that f is a coordinate, say f is the n-th component of the automorphism, F, i.e. $f = F_n$. The restriction of F to $V(f)$ induces an isomorphism $F_{|V(f)} : V(f) \simeq V(X_n)$. So $h := F_{|V(f)} \circ \varphi : \mathbb{A}^{n-1} \simeq V(X_n) \subset \mathbb{A}^n$. One readily verifies that there exists an automorphism ψ of $\mathbb{A}^n$ such that $\psi \circ h = j$, in other words $(\psi \circ F) \circ \varphi = j$, as desired. □

From 5.3.5 we get that if a polynomial $f \in k[X, Y]$ satisfies $k[X, Y]/(f) \simeq k[T]$, then f is a coordinate. Since conversely a coordinate f satisfies $k[X, Y]/(f) \simeq k[T]$ (Exercise 6) we get

Corollary 5.3.14 *Let $f \in k[X, Y]$, then f is a coordinate if and only if $k[X, Y]/(f) \simeq k[T]$.*

In particular we see that if $k[X, Y]/(f_1) \simeq k[X, Y]/(f_2) \simeq k[T]$ then there exists $\varphi \in \operatorname{Aut}_k k[X, Y]$ such that $\varphi(f_1) = f_2$. In general if f_1 and f_2 are irreducible curves such that $k[X, Y]/(f_1) \simeq k[X, Y]/(f_2)$ there need not exist an $\varphi \in \operatorname{Aut}_k k[X, Y]$ such that $\varphi(f_1) = f_2$. For examples of this phenomenon we refer the reader to Appendix F, Example 5.6.

Remark 5.3.15 If $n \geq 3$, then the condition $k[X]/(F_1, \ldots, F_{n-1}) \simeq k[T]$ in general does not imply that $(F_1, \ldots, F_{n-1})$ is extendible (Exercise 10).

Now let $n \geq 3$. If f is a coordinate in $k[X]$ then $k[X]/(f) \simeq k[T_1, \ldots, T_{n-1}]$, see Exercise 6. However we do not know if the converse is true, i.e. we have

Abhyankar-Sathaye Conjecture. Let $n \geq 3$. If $f \in k[X]$ is such that $k[X]/(f) \simeq k[T_1, \ldots, T_{n-1}]$, then f is a coordinate.

So by 5.3.13 this conjecture asserts that every embedding of $\mathbb{A}^{n-1}$ in $\mathbb{A}^n$ is rectifiable. The following special case was proved by Sathaye in [260] and by Russell in [257].

Theorem 5.3.16 *Let k be a field and $f \in k[X, Y, Z]$ such that $k[X, Y, Z]/(f) \simeq k[T_1, T_2]$. If furthermore f is linear in Z, then f is a coordinate.*

For some generalizations of this result we refer to [38], where k is replaced by a discrete valuation ring. It is left to the reader to verify (Exercise 8) that the Abhyankar-Sathaye Conjecture implies

Sathaye Conjecture Let $n \geq 3$. If $f^{-1}(0) \simeq \mathbb{A}^{n-1}$, then $f^{-1}(c) \simeq \mathbb{A}^{n-1}$, for all $c \in k$.

It seems unlikely that this conjecture and hence the Abhyankar-Sathaye Conjecture is true; in fact there are several candidate counterexamples.

Theorem 5.3.17 (Choudary, Dimca). *Let m, d, a be positive integers such that $m > 1, 0 < a < d - 1$ and $\gcd(a, d) = \gcd(a, d-1) = 1$. Let $f_{d,a} \in$*

$\mathbb{C}[X_0, X_1, \ldots, X_{2m-1}]$ be defined by

$$f_{d,a} := X_0^a X_1^{d-a} + X_1 X_2^{d-1} + \cdots + X_{2m-3} X_{2m-2}^{d-1} + X_{2m-2} + X_{2m-1}^d.$$

Then

i) *$V(f_{d,a})$ is a smooth hypersurface diffeomorphic to $\mathbb{C}^{2m-1}$.*
ii) *All fibers $f_{d,a}^{-1}(t)$ are smooth hypersurfaces diffeomorphic to $\mathbb{C}^{2m-1}$.*
iii) *For all $a > 1$ the fibers $f_{d,a}^{-1}(t)$ are not contractible for all $t \neq 0$. So they are not algebraically isomorphic to $\mathbb{C}^{2m-1}$.*

Question 5.3.18 *Does there exist a triple (m, d, a) as above with $a > 1$ such that $V(f_{d,a}) \simeq \mathbb{C}^{2m-1}$ as algebraic varieties?*

Such a triple would give a counterexample to the Sathaye Conjecture and so would give an exotic embedding of $\mathbb{C}^{n-1}$ in $\mathbb{C}^n$! On the other hand it was shown by Edo in [97] that all $f_{d,1}$ are coordinates (see Exercise 12 at the end of §5 in Chapter 9).

Finally we mention a recent result relevant to the Jacobian Conjecture,

Theorem 5.3.19 (Jelonek). *The n-cross K_n given by the equation $x_1.x_2 \ldots x_n = 0$ in $\mathbb{C}^n$ does not admit exotic embeddings.*

For a proof of this non-trivial result we refer to [166]. In fact the existence of an exotic embedding of K_n in $\mathbb{C}^n$ would have given a counterexample to the Jacobian Conjecture as was observed by Jelonek in [162]. On the other hand the non-existence of exotic embeddings of K_n is equivalent with Conjecture 12 of [208]. So Jelonek's result gives an affirmative answer to the **Complementary Conjecture** of McKay and Wang. The two dimensional case of this conjecture was proved by Mckay, Wang and Moh in [209] and Yu in [303].

Exercises for §5.3

1. Show that the examples given in 5.3.2 are indeed embeddings.
2. Let k be a field of characteristic $p > 0$ and $a \in \mathbb{N}$ with $\gcd(a, p) = 1$. Let $f(T) = T^{p^2}$ and $g(T) = T^{ap} + T$. Show that $k[f(T), g(T)] = k[T]$. However $f(T)$ and $g(T)$ do not verify the conclusion of 5.3.4.
3. Let $F = (F_1, F_2) \in \mathrm{Aut}_k\, k[X, Y]$ with $\deg F_1 = 1$. Show directly that $F \in T(k, 2)$.
4. Let $\varphi : k \to k^3$ be the embedding given by $\varphi(t) = (t + t^5, t^4, t^3)$. Show that φ is rectifiable. [Use the fact that for all $a(X, Z) \in k[X, Z], b \in k^*$ and $n \in \mathbb{N}$ the polynomial $X + a(X, Z)Z + bYZ^n$ is a coordinate, see Chapter 3, §2, Exercise 11.]
5. Let $\varphi : k \to k^2$ be given by

$$\varphi(t) = (2t^2 + t^3 + 4t^4 + 2t^6 + 8t^7 + 8t^9 + 8t^{12}, t + t^3 + 2t^6).$$

Show that φ is rectifiable.

6. Let $f \in k[X_1, \ldots, X_n]$ be a coordinate. Show that $k[X]/(f) \simeq k[T_1, \ldots, T_{n-1}]$, a polynomial ring in $n-1$ variables.
7. (Kaliman) We give two curves $V \subset k^3$ and $W \subset k^3$ and an isomorphism $g : V \to W$ which cannot be extended to an automorphism of k^3. This shows that the hypothesis $n > n(A(V))$ in 5.3.12 cannot be improved.

   Consider $\rho_1, \rho_2 : k \to k^3$ given by

   $$\rho_1(t) = (t^7, t^{11}, t^{13}) \text{ and } \rho_2(t) = (t^7 + t^{14}, t^{11}, t^{13}).$$

   Let $V := \rho_1(k)$ and $W := \rho_2(k)$.

   (i) Show that $G : k^3 \to k^3$ given by $G(x, y, z) = (x + x^2, y, z)$ induces an isomorphism $g := G_{|V}$ between V and W.
   (ii) Suppose that there exists an automorphism $F : k^3 \to k^3$ such that $\varphi := F_{|V} : V \to W$ is an isomorphism. Show that φ induces an isomorphism $\tilde{\varphi} : k \to k$ such that $\rho_2 \circ \tilde{\varphi} = \varphi \circ \rho_1$.
   (iii) Deduce that $\tilde{\varphi}(t) = ct$ for some $c \in k^*$. [Use $\varphi(0) = 0$, since an isomorphism maps singular points to singular points.]
   (iv) Let $\gamma : k^3 \to k^3$ be given by $\gamma(x, y, z) = (c^{-7}x, c^{-11}y, c^{-13}z)$. Put $H := F \circ \gamma$. Show that $H_{|V} = G_{|V}$.
   (v) Deduce that H is of the form

   $$H = (X + X^2 + p_1(X, Y, Z), Y + p_2(X, Y, Z), Z + p_3(X, Y, Z))$$

   where all p_i are polynomials vanishing on V.
   (vi) Deduce that $\det JH$ is of the form $1 + 2X + h(X, Y, Z)$, where the polynomial h does not contain the monomial X with non-zero coefficient. So $\det JH \notin k^*$, a contradiction since H is an automorphism.

8. Show that the Abhyankar-Sathaye Conjecture implies the Sathaye Conjecture. [If f is a coordinate and $c \in k$, then $f - c$ is a coordinate.]
9. Let $F_1, F_2 \in k[X, Y]$ with $\det J(F_1, F_2) \in k^*$ and assume that both $\deg F_i > 1$. Let h_i denote the homogeneous component of F_i of highest degree.

   (i) Prove that $\det J(h_1, h_2) = 0$.
   (ii) Deduce from i) and Exercise 10, §2, Chapter 1 that there exists a homogeneous polynomial h such that $h_1 = c_1 h^{d_1}$ and $h_2 = c_2 h^{d_2}$ for some $c_i \in k^*$ and some positive integers d_i.
   (iii) Assume that $\deg F_1$ divides $\deg F_2$. Deduce from ii) that there exists an elementary map of the form $E = (X, Y - cX^d)$ such that $\operatorname{tdeg} E \circ F < \operatorname{tdeg} F$, where $d := d_2/d_1$.

10. Let $n \geq 3$. Show that the condition $k[X]/(F_1, \ldots, F_{n-1}) \simeq k[T]$ in general does not imply that $(F_1, \ldots, F_{n-1})$ is extendible in the sense of Chapter 3. [Consider $F_1 = X_1 + X_1^2 X_2$, $F_2 = X_2 - X_1 X_2^2$, $F_i = X_i$ if $i \geq 3$.]

11. Let k be an algebraically closed field of characteristic zero and $f, g \in k[X, Y]$ with $\det J(f, g) \in k^*$.
    (i) Show that $k[f] \cap gk[X, Y] = (0)$, i.e. that $k[f]$ can be considered as a subring of $k[X, Y]/gk[X, Y]$.
    (ii) Show that if this ring extension is finite and g is irreducible in $k[X, Y]$, then (f, g) is invertible over k. [Use Corollary 1.3.33 and 5.3.14.]

5.4 The Abhyankar-Moh theorem

Throughout this section k denotes a field of characteristic zero and f and g are fixed non-constant polynomials in the one variable polynomial ring $k[t]$. On $k[t]$ we have the usual degree function deg which extends to $k(t)$ by defining $\deg \frac{p}{q} = \deg p - \deg q$ for all $p, q \in k[t]$ with $q \neq 0$. The main result is

Theorem 5.4.1 (Abhyankar-Moh). *If* $k[f, g] = k[t]$, *then* $\deg f \mid \deg g$ *or* $\deg g \mid \deg f$.

The remainder of this section is devoted to the proof of this result. The proof we will give here closely follows the proof given by Nowicki in [232].

To start first observe that t is integral over $A := k[g]$, namely write $g = \sum_{i=0}^{n} g_i t^i$, $g_i \in k$ and $g_n \neq 0$. So $n \geq 1$ ($g \notin k$!) and multiplying the equation $g - g = 0$ by g_n^{-1} we get

$$t^n + g_n^{-1} g_{n-1} t^{n-1} + \cdots + g_n^{-1} g_1 t + g_n^{-1}(g_0 - g) = 0,$$

i.e. t is integral over A. So $A \subset k[t]$ is finite, hence an integral extension. In particular f is integral over A. So $A[f]$ is a finite A-module. More precisely,

Proposition 5.4.2 *$A[f]$ is a free A-module with A-basis* $1, f, f^2, \ldots, f^{N-1}$, *where* $N := |K(f) : K|$ *and K is the quotient field of A.*

Proof: Let $P(X)$ be the minimal polynomial of f over K. Since f is integral over A the coefficients of $P(X)$ are also integral over A (see [190]). However they obviously belong to K and A is integrally closed. So $P(X) \in A[X]$ which implies $f^N \in \sum_{i=0}^{N-1} Af^i$. From here the result follows easily. □

Observe that if $N = 1$ then $A[f] = A$, so $f \in k[g]$ and hence $\deg g \mid \deg f$. Therefore we assume from now on: $N \geq 2$.

Now let $K[f]_0 := K$, $A[f]_0 := A$ and define for each $1 \leq n \leq N - 1$

$$\begin{aligned} K[f]_n &:= \sum_{i=0}^{n} Kf^i, \\ A[f]_n &:= \sum_{i=0}^{n} Af^i, \\ K_n &:= f^n + K[f]_{n-1} \subset K[f]_n \quad \text{and} \\ A_n &:= f^n + A[f]_{n-1} \subset A[f]_n. \end{aligned}$$

Using 5.4.2 one readily verifies (Exercise 1) that

(5.4.3) $$K_n \cap A[f] = A_n.$$

A sequence $h_0, h_1, \ldots, h_{N-1}$ of elements in $K(f) \subset k(t)$ is called an **α-sequence** (resp. **β-sequence**) if $h_n \in K_n$ (resp. $h_n \in A_n$) for all $1 \leq n \leq N-1$, $h_0 = 1$ and $\deg h_0, \deg h_1, \ldots, \deg h_{N-1}$ are pairwise incongruent modulo $\deg g$.

One readily verifies (Exercise 2) that for all $0 \leq n \leq N-1$ we have

(5.4.4) If $(h_0, h_1, \ldots, h_{N-1})$ is an α-sequence, then $K[f]_n = \sum_{i=0}^{N-1} K h_i$.

(5.4.5) If $(h_0, h_1, \ldots, h_{N-1})$ is a β-sequence, then $A[f]_n = \sum_{i=0}^{N-1} A h_i$.

The usefulness of the "incongruent modulo $\deg g$" property is illustrated by

Lemma 5.4.6 *Let $p_1, \ldots, p_m \in k(t)\backslash\{0\}$, with degrees incongruent modulo* $\deg g$. *If $0 \neq w \in \sum K p_i$ (resp. $\sum A p_i$), then* $\deg w = \deg g^s p_j$ *for some j and $s \in \mathbb{Z}$ (resp. $s \geq 0$).*

Proof: Let $w = \sum a_j p_j$ for some $a_j \in K = k(g)$. So $\deg a_j \equiv 0 \pmod{\deg g}$ for all j and hence the hypothesis implies that the numbers $\deg a_j p_j$ are pairwise distinct modulo $\deg g$ and hence pairwise distinct. So $\deg w = \deg a_j p_j$ for some j. Since $\deg a_j = s \deg g = \deg g^s$, for some integer s, the result for K follows. In case $a_j \in A$ for all j, obviously $s \geq 0$. □

Corollary 5.4.7 *If $(h_0, h_1, \ldots, h_{N-1})$ is a β-system and $w \in A[f]_n$, where $0 \leq n \leq N-1$, then* $\deg w = \deg g^s h_i$ *for some $s \geq 0$ and $0 \leq i \leq n$.*

However we do not know, yet, that there exists a β-system.

Theorem 5.4.8 *There exists a β-system.*

We prove this below. So let us first assume this result and give a

Proof of Theorem 5.4.1: i) Put $d_f := \deg f$, $d_g := \deg g$ and $d := \gcd(d_f, d_g)$. So $d_f = ad$ and $d_g = bd$ where $\gcd(a, b) = 1$ and there exist integers m, n such that $1 = ma + nb$ and $0 \leq m < b$. Hence $d = md_f + nd_g$, i.e. $d \equiv \deg f^m \pmod{d_g}$.

ii) Now let $(h_0, h_1, \ldots, h_{N-1})$ be a β-system. So by 5.4.2 and 5.4.5 $k[f, g] = A[f]_{N-1} = \sum_{i=0}^{N-1} A h_i$. Since $t^d \in k[t] = k[f, g]$ we get $t^d \in \sum A h_i = A[f]_{N-1}$. So by 5.4.7 $d = \deg t^d = \deg g^s h_r$ for some $s \geq 0$ and $0 \leq r \leq N-1$. Also since $f^m \in A[f]_m$ we get $\deg f^m = \deg g^{s_1} h_{r_1}$, for some $s_1 \geq 0$ and $0 \leq r_1 \leq m$.

iii) By i) and the congruences for d in i) and ii) we get $\deg g^s h_r \equiv \deg g^{s_1} h_{r_1} \pmod{d_g}$, whence $\deg h_r \equiv \deg h_{r_1} \pmod{d_g}$. So $r = r_1 \leq m < b$, which implies $h_r \in A[f]_{b-1}$.

iv) The degrees of the polynomials $1, f, \ldots, f^{b-1}$ are pairwise incongruent modulo d_g, namely if $\deg f^i \equiv \deg f^j \pmod{d_g}$ with $0 \le i, j \le b-1$, then $(i-j)d_f = ud_g$ for some integer u. So, using $d_f = da, d_g = db$ and $\gcd(a, b) = 1$ we get $b|i-j$, i.e. $i = j$.

v) From iii), iv) and 5.4.6 we get that $\deg h_r = \deg g^q f^i$ for some $q \ge 0$ and $0 \le i \le b-1$. So by ii) $d = sd_g + \deg h_r = (s+q)d_g + id_f$ with $s, q, i \ge 0$. Consequently $d \ge \min(d_f, d_g) \ge \gcd(d_f, d_g) = d$. Hence $\gcd(d_f, d_g) = \min(d_f, d_g)$, i.e. $d_f | d_g$ or $d_g | d_f$.

The remainder of this section is used to prove 5.4.8. First we show that there exists an α-system.

Let $S \subset k[t]\backslash\{0\}$ be such that the degrees of all elements of S are different. Then for each $0 \ne h \in k[t]$ such that $\deg h \in \deg S$, there exists a unique element $s \in S$ and $a \in k$ such that $\deg(h - as) < \deg h$. This enables us to define the **reduction operator** $R_S : k[t] \to k[t]$: $R_S(0) = 0$ and if $h \ne 0$ put $R_S(h) = h$ if $\deg h \notin \deg S$ and $R_S(h) := R_S(h - as)$ if $\deg h \in \deg S$. It is left to the reader to verify (Exercise 3) that

$$\deg R_S(h) \notin \deg S. \tag{5.4.9}$$

$$R_S(h) = 0 \leftrightarrow h = a_1 s_1 + \cdots + a_p s_p, \text{ for some } a_i \in k, s_i \in S. \tag{5.4.10}$$

Proposition 5.4.11 *There exists an α-system.*

Proof: i) Put $h_0 := 1$. For each $1 \le n \le N-1$ we are going to construct polynomials $h_1, \ldots, h_n$ such that $h_i \in A[f]_i \backslash A[f]_{i-1}$, for all $i \le i \le n$ and $\deg h_0, \deg h_1, \ldots, \deg h_n$ are pairwise incongruent modulo $\deg g$. First define $h_1 := R_S(f)$, where $S := \{g^m | m \ge 0\}$. Then $h_1 \in A[f]_1 \backslash A[f]_0 = A$ (for otherwise $f \in A$, contradicting $N > 1$). By 5.4.9 $\deg h_1 \notin \deg S$, so $\deg h_1$ is incongruent $\deg h_0$ modulo $\deg g$. Now let $0 \le n \le N-1$ and assume that $h_0, h_1, \ldots, h_n$ are already constructed. We are going to construct h_{n+1}. Let S_0 be the set of all elements $h_r g^i$ with $0 \le r \le n$ and $i \ge 0$. Then all these elements are non-zero and have pairwise different degrees. So we can define $y_0 := R_{S_0}(f^{n+1})$. Obviously $y_0 \in f^{n+1} + A[f]_n$ (and $y_0 \ne 0$ because $n+1 \le N-1$ and the f^i with $0 \le i \le N-1$ are linearly independent over K). Also by 5.4.9, $\deg y_0 \notin S_0$. So if we put $S_1 := S_0 \cup \{y_0\}$ we can define $y_1 = R_{S_1}(gf^{n+1})$. Then $y_1 \in (g + a_0)f^{n+1} + A[f]_n$, for some $a_0 \in k$. Again $y_1 \ne 0$ and $\deg y_1 \notin \{\deg y_0\} \cup \deg S_0$. Continuing this process we can construct an infinite sequence $y_0, y_1, \ldots$ of non-zero polynomials, where $y_i = R_{S_i}(g^i f^{n+1})$ and $S_i := S_0 \cup \{y_0, y_1, \ldots, y_{i-1}\}$. One readily verifies that $y_i \in (g^i + kg^{i-1} + \cdots + k)f^{n+1} + A[f]_n$, where $y_i \in A[f]_{n+1} \backslash A[f]_n$.

ii) Now we show: there exists j such that $\deg y_j$ is incongruent $\deg h_0, \ldots, \deg h_n$ $(\mathrm{mod} \deg g)$. Assume that this is not true. Then there exists $0 \le s \le n$ and an infinite set U of j's such that $\deg y_j \equiv \deg h_s \pmod{\deg g}$ for all $j \in U$, say $\deg y_j + p_j \deg g = \deg h_s$, with $p_j \in \mathbb{Z}$. If $p_j \le 0$ then $\deg y_j = \deg g^{-p_j} h_s$, i.e.

$\deg y_j \in \deg S_0$ (since $-p_s \geq 0$), a contradiction. So $p_j \geq 0$. However this implies that $\deg y_j < \deg h_s$ for all j. But all $\deg y_j$ are different and U is infinite. This is a contradiction. So there exists j such that $\deg y_j$ is incongruent to $\deg h_0, \dots, \deg h_n$ (mod $\deg g$). Then define $h_{n+1} := y_j$ and we have constructed h_{n+1} as desired.

iii) Finally let $h_1, \dots, h_{N-1}$ be as above. Since for each $1 \leq n \leq N-1$, $h_n \in A[f]_n \backslash A[f]_{n-1}$, the coefficient of f^n in h_n is non-zero. Dividing h_n by this coefficient we get $h'_n \in K_n$. Then $(h'_0, h'_1, \dots, h'_{N-1})$ is an α-system. □

Let $(h_0, h_1, \dots, h_{N-1})$ be an α-system. For each $0 \leq n \leq N-1$ let G_n be the ideal in $\mathbb{Z}$ generated by the elements $\deg w$, where $0 \neq w \in K[f]_n$. By 5.4.6 it follows that $G_n = (d_0, d_1, \dots, d_n)$, where $d_i := \deg h_i$ for all $1 \leq i \leq n$ and $d_0 := \deg g$. So

$$G_0 \subset G_1 \subset \cdots \subset G_{N-1}.$$

We consider the subsequence

$$G_0 \subsetneqq G_{c(1)} \subsetneqq \cdots \subsetneqq G_{c(p)}$$

where $G_{c(n)} = G_{c(n)+1} = \cdots = G_{c(n+1)-1}$ and $c(p+1) = N$. Observe that $c(1) = 1$, since we saw in the proof of 5.4.11 that the element h_1 constructed there belongs to $A[f]_1$ but $\deg h_1 \notin G_0$. Furthermore, if we define

$$D_0 := (d_0) \text{ and } D_j := (d_0, d_1, \dots, d_{c(j)}), \text{ for } 1 \leq j \leq p$$

we have $d_{c(s)} \notin D_{s-1}$. So if e_s denotes the smallest non-zero natural number n such that $nd_{c(s)} \in D_{s-1}$, then $e_s \geq 2$ for all s.

Proposition 5.4.12 *If $1 \leq s \leq p$, then $e_s c(s) = c(s+1)$.*

To prove this proposition we introduce for each $1 \leq s \leq p$

$$B_s := \{h_n h^j_{c(s)} \mid 0 \leq n < c(s),\ 0 \leq j < e_s\}.$$

Lemma 5.4.13 i) *The degrees of the elements of B_s are incongruent modulo $d_0 := \deg g$.*

ii) *The set B_s is a K-basis of $K[f]_{e_s c(s)-1}$ (hence $e_s c(s) \leq N$).*

Proof: i) Suppose $\deg h_n h^j_{c(s)} \equiv \deg h_m h^i_{c(s)}$ (mod d_0). If $i = j$ then $\deg h_n \equiv \deg h_m$ (mod d_0), so $n = m$ by the definition of α-system. So $i \neq j$, say $i > j$. Then $(i-j)d_{c(s)} \in (d_0) \subset D_{s-1}$. But $0 < i - j < e_s$, contradicting the minimal choice of e_s. So $i = j$ as desired.

ii) The second statement is left to the reader (Exercise 4). □

Corollary 5.4.14 i) *If $1 \leq s \leq p$, then $e_s c(s) \leq c(s+1)$.*

ii) *Let $0 \leq s \leq p$ and $B_0 := \{1\}$. If $w \in D_s$ then $w \equiv \deg b$ (mod d_0), for some $b \in B_s$.*

Proof: i) The case $s = p$ follows from 5.4.13 ii) since $c(p+1) = N$. So let $s < p$. Then $d_{c(s+1)} \notin D_s$. We will show: if $m < e_s c(s)$, then $d_m \in D_s$ (this implies that

$c(s+1) \geq e_s c(s)$). So let $m < e_s c(s)$. Then $h_m \in K[f]_{e_s c(s)-1}$. Hence by 5.4.6 and 5.4.13 i), ii) we get

$$d_m = \deg h_m = \deg g^i b = i d_0 + \deg b, \text{ for some } i \in \mathbb{Z}, b \in B_s.$$

One easily verifies that $\deg b \in D_s$, whence $d_m \in D_s$, as desired.

ii) To prove ii) we use induction on s. The case $s = 0$ is obvious. So let $s \geq 0$ and $w \in D_{s+1}$. Since $D_{s+1} = D_s + (d_{c(s+1)})$ we have $w = w' + m d_{c(s+1)}$, where $w' \in D_s$ and $m \in \mathbb{Z}$. Since $e_{s+1} d_{c(s+1)} \in D_s$ we may assume that $0 \leq m < e_{s+1}$. By the induction hypothesis $w' \equiv \deg b' \pmod{d_0}$ for some $b' \in B_s$. So by 5.4.13 ii), 5.4.14 and 5.4.4 we get

$$b' \in B_s \subset K[f]_{e_s c(s)-1} \subset K[f]_{c(s+1)-1} = \sum_{i=0}^{c(s+1)-1} K h_i.$$

Then by 5.4.6 $\deg b' \equiv \deg h_r \pmod{d_0}$, for some $0 \leq r < c(s+1)$. So we get

$$w = w' + m d_{c(s+1)} \equiv \deg b' + \deg h^m_{c(s+1)} \equiv \deg h_r + \deg h^m_{c(s+1)} = \deg b$$

$\pmod{d_0}$ where $b = h_r h^m_{c(s+1)} \in B_{s+1}$, as desired. □

Proof of Proposition 5.4.12: Put $n := e_s c(s)$. So by Lemma 5.4.13 ii) $n \leq N$. First assume that $n = N$. Suppose $s < p$. Then $N = e_s c(s) \leq c(s+1) \leq c(p) = N - 1$, a contradiction. So $n = N$ implies that $s = p$, whence $e_p c(p) = N = c(p+1)$.

Now assume $n < N$. It suffices to show that $d_n \notin D_s$, for then $n \geq c(s+1)$, i.e. $e_s c(s) \geq c(s+1)$, so $e_s c(s) = c(s+1)$ by 5.4.14 i). So suppose $d_n \in D_s$. Then by 5.4.14 ii) $d_n \equiv \deg b \pmod{d_0}$ for some $b \in B_s$. Since by 5.4.13 ii) and 5.4.4 $B_s \in K[f]_{e_s c(s)-1} = \sum_{i=0}^{n-1} K h_i$ it follows from 5.4.6 that $\deg b \equiv \deg h_j \pmod{d_0}$ for some $0 \leq j < n$. Hence $\deg h_n = d_n \equiv \deg h_j \pmod{d_0}$, contradicting that $h_0, h_1, \ldots, h_n$ have pairwise incongruent degrees modulo d_0. □

Corollary 5.4.15 *If* $1 \leq s \leq p$, *then* $c(s+1) = e_1.e_2 \ldots e_s$.

Proof: By induction on s. □

Let us give one more application of 5.4.12 which will play a crucial role later. Since by 5.4.13, B_s is a K-basis of $K[f]_{e_s c(s)-1} = K[f]_{c(s+1)-1}$ (for $e_s c(s) = c(s+1)$) every element $w \in K[f]_{c(s+1)-1}$ can be written uniquely as $a_1 b_1 + \cdots + a_r b_r$, where $B_s = \{b_1, \ldots, b_r\}$ and each $a_i \in K$. Put $h := h_{c(s)}$ and $e := e_s$. Since each b_i is of the form $h_n h^j$ for some $0 \leq n < c(s)$ and $0 \leq j < e$ we can, after collecting powers of h, write w of the form

$$w_1 h^{e-1} + w_2 h^{e-2} + \cdots + w_{e-1} h + w_e$$

where each w_i belongs to $K h_0 + \cdots + K h_{c(s)-1} = K[f]_{c(s)-1}$ (by 5.4.4). The elements w_i are uniquely determined, since the $h_n h^j$ are linearly independent over

K. So for example $w_1h^{e-1} = \sum a_ib_i$, where the sum runs over all i with b_i of the form h_nh^{e-1}, i.e. which have a factor h^{e-1}. Summarizing,

Lemma 5.4.16 *Let $e := e_s$ and $h := h_{c(s)}$. For each $w \in K[f]_{c(s+1)-1}$ there exist uniquely determined elements $w_1, \ldots, w_e$ in $K[f]_{c(s)-1}$ such that*

$$w = w_1h^{e-1} + w_2h^{e-2} + \cdots + w_{e-1}h + w_e.$$

In particular if we write w in the K-basis $B_s = \{B_1, \ldots, b_r\}$, i.e. $w = \sum a_ib_i$ with $a_i \in K$, then w_1h^{e-1} is the sum of all a_jb_j where b_j is of the form h_nh^{e-1} with $0 \leq n \leq c(s) - 1$.

We will use 5.4.12 to improve our α-system, i.e. we will show

Theorem 5.4.17 *There exists an α-system $H = (h_0, h_1, \ldots, h_{N-1})$ with $h_{c(s)} \in A_{c(s)}$ for all s.*

Before we prove this result, we show how it gives

Proof of Theorem 5.4.8: Let H be as in 5.4.17 and let H' be the set of elements $h_{c(1)}^{j_1} \ldots h_{c(p)}^{j_p}$ with $0 \leq j_s < e_s$ for all $1 \leq s \leq p$. Observe that

$$h_{c(1)}^{j_1} \cdots h_{c(p)}^{j_p} \in A_{j_1c(1)+\cdots+j_pc(p)}.$$

So to show that H' can be arranged in a sequence $(1, h'_1, \ldots, h'_{N-1})$ which is a β-system, it suffices to prove that

$$\text{(5.4.18)} \quad \begin{aligned} J &:= \{j_1c(1) + \cdots + j_pc(p) \mid 0 \leq j_s < e_s, 1 \leq s \leq p\} = E \\ &:= \{0, 1, 2, \ldots, N-1\}. \end{aligned}$$

(5.4.19) The degrees of the elements of H' are incongruent modulo d_0.

To see 5.4.18 observe that by induction on s and 5.4.12

$$j_1c(1) + \cdots + j_sc(s) < c(s+1)$$

for all $1 \leq s \leq p$ where $1 \leq j_i \leq e_i$ for each $1 \leq i \leq s$. So in particular for $s = p$, using $c(p+1) = N$, this implies that $J \subset E$. Furthermore if $j_1c(1)+\cdots+j_pc(p) = i_1c(1) + \cdots + i_pc(p)$, where $0 \leq i_1, j_1 < e_1, \ldots, 0 \leq i_p, j_p < e_p$, and using $c(1) = 1, c(2) = e_1, \ldots, c(p) = e_1e_2 \ldots e_{p-1}$ (5.4.15), we deduce that $|i_1 - j_1|$ is divisible by e_1, whence $i_1 = j_1$ since $0 \leq i_1, j_1 < e_1$. Consequently $|i_2 - j_2|$ is divisible by e_2, whence $i_2 = j_2$ etc. We get $i_s = j_s$ for all $1 \leq s \leq p$, so all elements of J are different. Hence J contains $e_1.e_2 \ldots e_p = N$ elements (by 5.4.15 and $c(p+1) = N$). This together with $J \subset E$ proves 5.4.18. To see 5.4.19 let

$$a = h_{c(1)}^{i_1} \cdots h_{c(p)}^{i_p} \text{ and } b = h_{c(1)}^{j_1} \cdots h_{c(p)}^{j_p}$$

with $0 \le i_s, j_s < e_s$ for all $1 \le s < p$. If $\deg a = \deg b \pmod{d_0}$ then

$$(i_1 - j_1)d_{c(1)} + \cdots + (i_p - j_p)d_{c(p)} \equiv 0 \pmod{d_0}$$

whence $|i_p - j_p|d_{c(p)} \in D_{p-1}$. So $i_p = j_p$ by the minimal choice of e_p. In a similar way we get $i_{p-1} = j_{p-1}$ etc. So $a = b$.

To prove Theorem 5.4.17 we use the following

Lemma 5.4.20 *Let $R \subset S$ be commutative rings containing the rationals and $P = X^n + p_{n-1}X^{n-1} + \cdots + p_0 \in S[X]$ with $n \ge 1$. Let $e \ge 1$ and write*

$$P^e = X^{ne} + s_{ne-1}X^{ne-1} + \cdots + s_1X + s_0.$$

If $s_{ne-1}, s_{ne-2}, \ldots, s_{ne-n} \in R$, then $P \in R[X]$.

Proof: The coefficient of $X^{(e-1)n-n-1}$ in P^e is equal to $ep_{n-1} = s_{ne-1} \in R$. So $p_{n-1} \in R$. Now suppose that we know that $p_{n-1}, \ldots, p_{t+1} \in R$. Consider the coefficient of $X^{ne-(n-t)}$ in p^e. This coefficient is a polynomial in the p_i. However it cannot contain a p_i with $i < t$ since the corresponding X-monomial has at most degree $(e-1)n + i = ne - (n-i) < ne - (n-t)$ because $i < t$. Furthermore, the only monomial in the p_i's which contains p_t is ep_t. So we get $s_{(e-1)n+t} = ep_t + g$, where g is a polynomial in $p_{n-1}, \ldots, p_{t+1}$ with integer coefficients. So $g \in R$. Since $s_{(e-1)n+t} \in R$ we get $ep_t \in R$, whence $p_t \in R$. □

Corollary 5.4.21 *If $H = (1, h_1, \ldots, h_{N-1})$ is an α-system such that $h_{c(s)}^{e_s} - h_{c(s+1)} \in K[f]_{c(s)e_s - c(s)-1}$ for all $1 \le s \le p$, where $h_{c(p+1)} = 0$, then $h_{c(s)} \in A_{c(s)}$ for all $1 \le s \le p$.*

Proof: Induction on $p - s$. First let $s = p$. Put $n := c(p)$ and $e := e_p$. Then $en = e_pc(p) = c(p+1) = N$. Furthermore $h_{c(p)} = P(f)$ with

$$P = X^n + p_{n-1}X^{n-1} + \cdots + p_0 \in K[X]$$

and

$$h_{c(p+1)} = 0 = f^N + \sum_{i=0}^{N-1} a_i f^i \text{ for some } a_i \in A.$$

So if we put $T := X^N + \sum_{i=0}^{N-1} a_i X^i$, then using $ne = N$ we get $P^e - T = \sum_{i=0}^{N-1} s_i X^i$ for some $s_i \in K$. Since by hypothesis $(P^e - T)(f) = P(f)^e - T(f) \in K[f]_{ne-n-1}$ the linear independence of $1, f, \ldots, f^{N-1}$ over K implies that $s_{ne-1}, s_{ne-2}, \ldots, s_{ne-n} = 0$. Hence $P(X) \in A[X]$ by 5.4.20. So $h_{c(p)} = P(f) \in A[f] \cap K_{c(p)} = A_{c(p)}$ (by 5.4.3).

Now let $1 \le s \le p-1$ and assume $h_{c(s+1)} \in A_{c(s+1)}$. Then again using 5.4.20 and a similar argument as above we get $h_{c(s)} \in A_{c(s)}$. □

To prove the existence of an α-system as in 5.4.21 we study the expressions $h_{c(s)}^{e_s} - h_{c(s+1)}$.

Proposition 5.4.22 *Let $H = (1, h_1, \ldots, h_{N-1})$ be an α-system. Put $h_N := 0$. Let $1 \le s \le p$, $e := e_s$ and $h := h_{c(s)}$. Then*

i) *There exist uniquely determined elements* $w_1, \dots, w_e \in K[f]_{c(s)-1}$ *such that*

(5.4.23) $$h^e - h_{c(s+1)} = w_1 h^{e-1} + \cdots + w_{e-1}h + w_e.$$

ii) $\deg w_1 < \deg h$.

Proof: i) Let $v := h^e - h_{c(s+1)}$. By 5.4.16 we need to show that $v \in K[f]_{c(s+1)-1}$. The case $s = p$ is obvious since $v \in K[f] = K[f]_{N-1} = K[f]_{c(p+1)-1}$. So let $1 \le s < p$. Then $h^e \in f^{c(s)e} + K[f]_{c(s)e-1}$ and $h_{c(s+1)} \in f^{c(s+1)} + K[f]_{c(s+1)-1}$. Since $c(s+1) = c(s)e$ it follows that $v \in K[f]_{c(s+1)-1}$.

ii) Write $v = \sum a_i b_i$ on the K-basis B_s of $K[f]_{c(s+1)-1}$. Then by i) and 5.4.16 $w_1 h^{e-1} = \sum a_j b_j$ where each b_j is of the form $b_j = h_n h^{e-1}$ for some $0 \le n < c(s)$. For such b_j we have $\deg b_j \notin D_{s-1}$, namely if $\deg b_j = \deg h_n + (e-1)d_{c(s)} \in D_{s-1}$, then $(e-1)d_{c(s)} \in D_{s-1}$ ($\deg h_n = d_n \in D_{s-1}$, because $n \le c(s) - 1$), which contradicts the minimal choice of e. Now we claim

(5.4.24) $$\text{if } \deg b_j \notin D_{s-1}, \text{ then } \deg a_j b_j < e \deg h.$$

Let us assume this for a moment. Then indeed $\deg w_1 h^{e-1} < e \deg h$, i.e. $\deg w_1 + (e-1)\deg h < e \deg h$, so $\deg w_1 < \deg h$.

iii) Now we prove 5.4.24. First observe that by 5.4.13 and 5.4.6 $\deg v = \deg a_q b_q$ for some q and $\deg v > \deg a_i b_i$ for all $i \ne q$. Since $\deg a_q \equiv 0 \pmod{d_0}$ and $b_q = h_n h^j$ for some $j \ge 0$ and some $0 \le n \le c(s) - 1$ we have $\deg b_q = d_n + j d_{c(s)} \in D_s$, so $\deg v \in D_s$. Moreover $\deg h^e = e d_{c(s)} \in D_s$ and $\deg h_{c(s+1)} = d_{c(s+1)} \notin D_s$. Since $v = h^e - h_{c(s+1)}$ this implies that $\deg h_{c(s+1)} < \deg h^e$, whence $\deg v = e \deg h$. So

(5.4.25) $$\deg a_q b_q = e \deg h \text{ and } \deg a_i b_i < e \deg h, \text{ if } i \ne q.$$

Now suppose that $\deg b_j \notin D_{s-1}$, then $j \ne q$, for $\deg a_q b_q = e \deg h = e_s d_{c(s)} \in D_{s-1}$, whence $\deg b_q = \deg a_q b_q - \deg a_q \in D_{s-1}$. So by 5.4.25, $\deg a_j b_j < e \deg h$, which proves 5.4.24. □

The uniquely determined element w_1 in 5.4.22 we denote by $w(H, s)$. One easily verifies that if $w(H, s) = 0$, then $h^e_{c(s)} - h_{c(s+1)} \in K[f]_{c(s)e_s - c(s)-1}$, so by 5.4.21 $h_{c(s)} \in A_{c(s)}$. Consequently, if we can find an α-system H such that $w(H, s) = 0$ for all $1 \le s \le p$, then by 5.4.21 $h_{c(s)} \in A_{c(s)}$ for all s and as we have seen before this implies the existence of a β-system which in turn proves Theorem 5.4.1.

So let $H = (1, h_1, \dots, h_{n-1})$ be an α-system and $1 \le s \le p$. Then, in the light of 5.4.23 it is natural to introduce the sequence $\overline{H} := (1, \bar{h}_1, \dots, \bar{h}_{N-1})$ where $\bar{h}_n := h_n$ for all $n \ne c(s)$ and $\bar{h}_{c(s)} := h_{c(s)} - \frac{1}{e_s} w(H, s)$.

Proposition 5.4.26 $\overline{H}$ *is an* α*-system.*

Proof: Since $w(H,s) \in K[f]_{c(s)-1}$ and $h_{c(s)} \in f^{c(s)} + K[f]_{c(s)-1}$ we get $\bar{h}_{c(s)} \in A_{c(s)}$. Finally by 5.4.22 $\deg w(H,s) < \deg h_{c(s)}$, whence $\deg \bar{h}_{c(s)} = \deg h_{c(s)}$. □

Lemma 5.4.27 *Let $1 \le s \le p$, $h := h_{c(s)}$, $e := e_s$, $\bar{h} := \bar{h}_{c(s)}$ and $u := a_2 h^{e-1} + a_3 h^{e-3} + \cdots + a_e$, where $a_i \in K[f]_{c(s)-1}$. Then there exist uniquely determined elements $b_2, \ldots, b_e \in K[f]_{c(s)-1}$ such that $u = b_2\bar{h}^{e-2} + \cdots + b_{e-1}\bar{h} + b_e$.*

Proof: One easily verifies that $u \in K[f]_{(e-1)c(s)-1} \subset K[f]_{c(s+1)-1}$ (since $e_s c(s) = c(s+1)$). So by 5.4.15 there exist uniquely determined elements $b_1, \ldots, b_e \in K[f]_{c(s)-1}$ such that $u = b_1\bar{h}^{e-1} + b_2\bar{h}^{e-2} + \cdots + b_e$. It remains to show that $b_1 = 0$. So let $b_1 \ne 0$. Put $K[f]_{-1} := 0$. Then $b_1 \in K[f]_n \backslash K[f]_{n-1}$ for some $0 \le n < c(s)$. Then $b_1\bar{h}^{e-1} \in K[f]_{n+(e-1)c(s)} \backslash V$, where $V := K[f]_{n+(e-1)c(s)-1}$. Putting $\bar{u} := b_2\bar{h}^{e-2} + b_3\bar{h}^{e-3} + \cdots + b_e$ one verifies $\bar{u} \in K[f]_{(e-1)c(s)-1} \subset V$. By definition of $\bar{u}$, $b_1\bar{h}^{e-1} = u - \bar{u}$. Since $b_1\bar{h}^{e-1} \notin V$ and $u - \bar{u} \in V$ we get a contradiction. So $b_1 = 0$. □

Proposition 5.4.28 *Let H be an α-system, $1 \le s \le p$ and $\bar{w} := w(\bar{H}, s)$. Let $\bar{H}$ be as in 5.4.27 and assume $w \ne 0$.*

i) *If $w \in K[f]_0 = K$, then $\bar{w} = 0$.*
ii) *If $w \in K[f]_j \backslash K[f]_{j-1}$, where $1 \le j < c(s)$, then $\bar{w} = 0$ or $\bar{w} \in K[f]_i \backslash K[f]_{i-1}$ for some $0 \le i < j$.*

Proof: Let $e := e_s$, $h := h_{c(s)}$ and $\bar{h} := \bar{h}_{c(s)} (= h - \frac{1}{e}w)$. Then by 5.4.23 $h^e - h_{c(s+1)} = wh^{e-1} + \sum_{i=2}^{e} w_i h^{e-i}$ for some $w_i \in K[f]_{c(s)-1}$. Hence $\bar{h}^e - \bar{h}_{c(s)} = (h^e - \bar{h}_{c(s+1)}) - wh^{e-1} + R = \sum_{i=2}^{e} w_i h^{e-i} + R$, where $R = \sum_{r=0}^{e-2} \binom{e}{r} h^r (\frac{-1}{e} w)^{e-r}$. So by Lemma 5.4.27 there exist $b_2, \ldots, b_e \in K[f]_{c(s)-1}$ such that

$$\bar{h}^e - \bar{h}_{c(s)} = \sum_{i=2}^{e} b_i \bar{h}^{e-i} + R. \tag{5.4.29}$$

Now look at R. Let $w \in K[f]_j$ for some $0 \le j < c(s)$. Then for each $0 \le r \le e-2$, $h^r w^{e-r} \in K[f]_{rc(s)+j(e-r)}$. Since $rc(s) + j(e-m) = rc(s) + j(e-r-2) + j + j \le rc(s) + c(s)(e-r-2) + c(s) - 1 + j = (e-1)c(s) + j - 1$ we get

$$R \in K[f]_{(e-1)c(s)+j-1}. \tag{5.4.30}$$

Since $(e-1)c(s)+j-1 = ec(s)+j-c(s)-1 = c(s+1)-j-c(s)-1 \le c(s+1)-1$ we get $R \in K[f]_{c(s+1)-1}$ and so by 5.4.16 there exists uniquely determined elements $z_1, \ldots, z_e \in K[f]_{c(s)-1}$ such that

$$R = z_1\bar{h}^{e-1} + z_2\bar{h}^{e-2} + \cdots + z_e. \tag{5.4.31}$$

From 5.4.29 and 5.4.31 we get $\bar{w} = z_1$. Since

$$z_2\bar{h}^{e-2} + z_3\bar{h}^{e-3} + \cdots + z_e \in K[f]_{(e-1)c(s)-1}$$

it follows from 5.4.30 that

$$(5.4.32) \qquad \bar{w}\bar{h}^{e-1} = z_1\bar{h}^{e-1} \in K[f]_{(e-1)c(s)+j-1}.$$

Finally we prove i) and ii). Assume $\bar{w} \neq 0$. Then $\bar{w} \in K[f]_i \backslash K[f]_{i-1}$ for some $0 \leq i < c(s)$ (where $K[f]_{-1} := 0$). Consequently

$$(5.4.33) \qquad \bar{w}\bar{h}^{e-1} \in K[f]_{(e-1)c(s)+i} \backslash K[f]_{(e-1)c(s)+i-1}.$$

So if $j = 0$, i.e. $w \in K[f]_0 = K$ then 5.4.31 contradicts 5.4.33, so $\bar{w} = 0$, which proves i). Now let $j \geq 1$. Then 5.4.32 and 5.4.33 implies that $(e-1)c(s) + i - 1 < (e-1)c(s) + j - 1$, i.e. $i < j$, which proves ii). □

Corollary 5.4.34 *There exists an α-system H such that $w(H, s) = 0$ for all $1 \leq s \leq p$.*

Proof: Let H be an α-system. Apply 5.4.28 several times to the case $s = p$. We get an α-system H' with $w(H', p) = 0$. Then apply 5.4.27 with $s = p - 1$ to H' and we arrive at an α-system H'' with both $w(H'', p-1) = 0$ and $w(H'', p) = 0$ etc. □

Exercises for §5.4

1. Prove 5.4.3 [Use 5.4.2.]
2. Prove 5.4.4 and 5.4.5.
3. Prove 5.4.9 and 5.4.10.
4. Prove 5.4.13 ii) using the following steps.
   (i) Show that $B_s \subset K[f]_{e_s c(s)-1}$.
   (ii) Show that the elements of B_s are linearly independent over K. [Use 5.4.6 and i).]
   (iii) Deduce that B_s is a K-basis of $K[f]_{e_s c(s)-1}$.
5. (i) Prove the Abhyankar-Moh theorem, using only 5.4.2, when one of the numbers $\deg f$ or $\deg g$ is prime.
   (ii) Show that if char $k = 2$ then $k[t^4, t^6 + t] = k[t]$.
   (iii) Give similar examples for any characteristic $p > 0$. [Consider $k[t^{p^2}, t^{p(p+1)} + t]$.]

Notes

The Jung-van der Kulk theorem was first proved by Jung in [168], 1942 for fields of characteristic zero. In 1953 van der Kulk in [185] extended Jung's result to fields with arbitrary characteristic. Since then several proofs appeared: Gutwirth [148], 1961, Shafarevich [268], 1966, Rentschler [250], Makar-Limanov [198], Nagata [224], 1972, Abhyankar and Moh [9], 1975, Dicks [87], 1983, McKay and Wang [207], 1988. The stronger statement that $\text{Aut}_k\, k[X, Y]$ is the amalgamated free product of $\text{Aff}(k, 2)$ and $J(k, 2)$ over their intersection was deduced in [185], [174], [224], [87],

[19], and stated in [268] without proof. This result is also called the Shafarevich-Nagata-Kambayashi theorem. The proof of the Jung-van der Kulk theorem we gave is due to Rentschler in [250]. Proposition 5.1.18 i) was generalized to the n-variable case by Hadas in [151]. A more general study of Newton polytopes of constants of locally nilpotent derivations was given by Hadas and Makar-Limanov in [152]. In particular a new proof of Rentschler's Theorem is obtained. See also the earlier paper [79] of Derksen, Hadas and Makar-Limanov. The proof of the amalgamated free product structure given in Corollary 5.1.3 and the algorithm for tameness are taken from Furter's paper [135]. Theorem 5.2.1 is due to Derksen. He explained the idea of the proof of this nice result to me in more or less five lines during the Curaçao conference in July 1994. Rusek's conjecture concerning tameness of quadratic automorphisms is taken from [255] (Conjecture 11.2). There he also conjectures (Conjecture 11.3) that, if for linear forms $\ell_1, \ldots, \ell_n$ the matrix $J(\ell_1^2, \ldots, \ell_n^2)$ is nilpotent, then $\ell_1^2, \ldots, \ell_n^2$ are linearly dependent over k (See also Chapter 7, §1). Then he claims that Conjecture 11.3 implies Conjecture 11.2. This claim is still open! Some results concerning Conjecture 11.2 can be found in Furter's thesis [134]. Theorem 5.3.4 and Theorem 5.3.5 were proved by Abhyankar and Moh in [9] and Theorem 5.3.5 was also obtained independently by Suzuki in [283]. Corollary 5.3.10 was proved independently by Craighero and Jelonek in [65a] resp. [161] under the stronger assumption that $n \geq 3k + 1$. The proof we give in §4 is due to Nowicki in [232] following ideas from Richman, who gave a proof of the Abhyankar-Moh theorem in [251], 1986. §4 was written by Nowicki.

Originally Theorem 5.4.1 appeared in the fifties with an incorrect proof in Segre's paper [264]. After the proof given by Abhyankar and Moh other proofs appeared in the years 1977–1982. They were published by Miyanishi in [215] (see also [216]), Ganong [139] and Rudolph [254] who used results from knot theory. Yet another proof was given by Kang in [175] who noticed that there is a gap in Richman's proof and corrected it.

It is interesting to observe that there do exist holomorphic embeddings of the complex line in $\mathbb{C}^2$, which cannot be rectified by any holomorphic automorphism of $\mathbb{C}^2$. We refer to the paper [126] of Forstneric, Globevnik and Rosay. This result was used by Derksen and Kutzschebauch in [80] to show that the holomorphic linearization conjecture is false. Theorem 5.3.7 and its proof are taken from the paper [275] of Srinivas. His proof was inspired by (unpublished) ideas of Nori. Theorem 5.3.12 was obtained by Kaliman in [169] and appeared almost simultaneously with Srinivas' paper. Theorem 5.3.17 was taken from the paper [53] of Choudary and Dimca which is concerned with the study of **exotic algebraic structures on $\mathbb{C}^n$** (Chapter 9, §4).

6 Stabilization Methods

In K-theory the "stabilization philosophy" that a problem posed in dimension n may become simpler if looked at in dimension $n + m$, is applied successfully in many places. For example for suitable rings R a projective R-module M which is not free may become free after stabilization, i.e. after replacing M by $M \oplus R^m$ for suitable m, or a matrix $A \in Gl_n(R)$ which is not a product of elementary matrices may become a product of such matrices after stabilization, i.e. after replacing A by $\begin{pmatrix} A & 0 \\ 0 & I_m \end{pmatrix} \in Gl_{n+m}(R)$, where I_m denotes the $m \times m$ identity matrix. The idea of using stabilization methods in the context of polynomial maps was first introduced by H. Bass, E. Connell and D. Wright in [29] and Yagzhev in [297]. Among other things they showed that it suffices to prove the Jacobian Conjecture for all polynomial maps of degree ≤ 3 in any dimension. For the precise formulation see 6.3.1. A little later this result was improved by Drużkowski in [93] (see 6.3.2). Both results are now considered as classical. More recently J. Yu in his thesis, [299], pursues the stabilization ideas and shows that the Jacobian Conjecture is true for a special class of polynomial maps (see 6.3.4 and [301]). To explain the stabilization idea, we start in §1 with some instructive examples. Then in §2 we lay the foundations for the main results of this chapter which can be found in §3.

Finally in §4 we discuss a theory of pairing between cubic homogeneous and cubic linear polynomial mappings. These results are due to Gorni and Zampieri in [145].

6.1 The stabilization principle: some instructive examples

As already mentioned in the beginning of this chapter, the idea behind the stabilization philosophy is that a problem posed in dimension n may become simpler if looked at in dimension $n + m$, for some $m \geq 1$.

To demonstrate this principle we consider two questions posed in the previous chapters, namely: "is the Nagata automorphism σ tame?" and "is every embedding of the line in $\mathbb{A}^3$ rectifiable?" (5.3.11).

(6.1.1) **The Nagata automorphism**

Let k be a field and $\sigma : k^3 \to k^3$ be given by

$$\sigma(x, y, z) = (x - 2y(zx + y^2) - z(zx + y^2)^2, y + z(zx + y^2), z).$$

As already observed in Chapter 5 we do not know if $\sigma \in T(k, 3)$. Now let us extend σ to a map $\tilde{\sigma} : k^4 \to k^4$ by defining

$$\tilde{\sigma}(x, y, z, t) = (\sigma(x, y, z), t).$$

Question 6.1.2 *Is $\tilde{\sigma} : k^4 \to k^4$ tame?*

The answer to this question is yes! This can be seen by an elegant argument due to Martha Smith [271], which gives a more general result: let R be a commutative ring, $R[X] := R[X_1, \ldots, X_n]$ and D a locally nilpotent R-derivation on $R[X]$. Let $a \in \ker(D, R[X])$. Extend D to a derivation $\tilde{D}$ on $R[X, X_{n+1}]$ by defining $\tilde{D}(X_{n+1}) = 0$. Finally define the elementary automorphism $\tau \in \mathrm{Aut}_R\, R[X, X_{n+1}]$ by $\tau(X_i) = X_i$ for all $1 \le i \le n$ and $\tau(X_{n+1}) = X_{n+1} + a$. Then

Lemma 6.1.3 $\exp(a\tilde{D}) = \tau^{-1} \exp(-X_{n+1}\tilde{D})\tau \exp(X_{n+1}\tilde{D})$.

Proof: Left to the reader (Exercise 1). □

Now we are able to prove

Proposition 6.1.4 (M. Smith). *If D is a triangular derivation on $R[X]$ and $a \in \ker(D, R[X])$, then* $(\exp(aD), X_{n+1}) \in T(R, n+1)$.

Proof: Observe that $(\exp(aD), X_{n+1}) = \exp(a\tilde{D})$. So by 6.1.3 it suffices to prove that $\exp(X_{n+1}\tilde{D})$ is tame, since τ is tame. However $X_{n+1}\tilde{D}$ is a triangular derivation on $R[X, X_{n+1}]$. So $\exp(X_{n+1}\tilde{D})$ is tame by Exercise 2. □

Corollary 6.1.5 *$\tilde{\sigma}$ is tame.*

Proof: One easily verifies that $\tilde{\sigma} = \exp(a\tilde{D})$, where $a := ZX + Y^2$ and $D := -2Y\partial_X + Z\partial_Y$, which is a triangular derivation on $k[X, Y, Z]$ satisfying $D(a) = 0$. □

Remark 6.1.6 The same argument also shows that if R is a ring in which $0 \neq z \in R$ is not a unit, then the non-tame map $F \in \mathrm{Aut}_R\, R[X, Y]$ defined in 5.1.9 becomes tame when it is extended to the map $(F, X_3) \in \mathrm{Aut}_R\, R[X, Y, X_3]$.

Remark 6.1.7 In the next chapter 6.1.4 will be used to show that members of a large class of polynomial automorphisms are stably tame (6.1.8).

This leads us to formulate

(6.1.8) **Stable tameness conjecture.** Let k be a field and $F \in \mathrm{Aut}_k\, k[X_1, \ldots, X_n]$. Then F is **stably tame**, i.e. there exist $m \in \mathbb{N}$ and new variables $X_{n+1}, \ldots, X_{n+m}$ such that the extended map $(F, X_{n+1}, \ldots, X_{n+m})$ is tame, i.e. belongs to $T(k, n+m)$.

To conclude this section we give another example of the stabilization principle.

(6.1.9) **Embeddings and stabilization**

Although we do not know if any embedding is rectifiable, we will show below that any embedding $\varphi : V \to \mathbb{A}^n$ is **stably rectifiable**, i.e., there exists $m \ge 1$ such that the extended map $(\varphi, 0) : V \to \mathbb{A}^{n+m}$ defined by $(\varphi, 0)(v) := (\varphi(v), 0)$ is rectifiable. In fact this result follows easily from the following proposition.

Proposition 6.1.10 *Let V be an algebraic set and $\varphi_1 : V \to \mathbb{A}^n$ and $\varphi_2 : V \to \mathbb{A}^m$ embeddings. Let $(\varphi_1, 0) : V \to \mathbb{A}^{n+m}$ and $(0, \varphi_2) : V \to \mathbb{A}^{n+m}$ be the extensions of φ_1 respectively φ_2. Then there exists $F \in \operatorname{Aut} \mathbb{A}^{n+m}$ such that $F \circ (\varphi_1, 0) = (0, \varphi_2)$.*

Proof: i) Since $\varphi_1 : V \to \varphi_1(V) \subset \mathbb{A}^n$ and $\varphi_2 : V \to \varphi_2(V) \subset \mathbb{A}^m$ are isomorphisms we get an isomorphism $\varphi_1 \circ \varphi_2^{-1} : \varphi_2(V) \to \varphi_1(V)$. So there exists a regular map $\mu_1 : \mathbb{A}^m \to \mathbb{A}^n$ extending $\varphi_1 \circ \varphi_2^{-1}$, hence $\mu_1 \circ \varphi_2 = \varphi_1$. Similarly there exists a regular map $\mu_2 : \mathbb{A}^n \to \mathbb{A}^m$ such that $\mu_2 \circ \varphi_1 = \varphi_2$.

ii) Now let $\beta_1, \beta_2 \in \operatorname{Aut} \mathbb{A}^{n+m}$ be defined by

$$\beta_1(x, y) = (x, y + \mu_2(x)),\ \beta_2(x, y) = (x + \mu_1(y), y)$$

and put

$$F := \beta_2^{-1} \circ \beta_1.$$

Then one easily verifies that F is as desired. □

Corollary 6.1.11 *Let $V \subseteq \mathbb{A}^d$ be an algebraic set and $\varphi : V \to \mathbb{A}^n$ an embedding. Then $(\varphi, 0) : V \to \mathbb{A}^{n+d}$ is rectifiable.*

Proof: Take $\varphi_1 = \varphi$ and $\varphi_2 = i$, the inclusion map of V into $\mathbb{A}^d$. Then by Proposition 6.1.10 there exists $F \in \operatorname{Aut} \mathbb{A}^{n+d}$ such that $F \circ (\varphi_1, 0) = (0, \varphi_2)$, i.e., $F(\varphi_1(v), 0) = (0, v)$ for all $v \in V$. Finally let $s \in \operatorname{Aut} \mathbb{A}^{n+d}$ be defined by $s(x, y) = (y, x)$ for all $(x, y) \in \mathbb{A}^n \times \mathbb{A}^d$. Then we get that $(s \circ F)(\varphi(v), 0) = (v, 0)$ for all $v \in V$, i.e., $(\varphi, 0)$ is rectifiable. □

Exercises for §6.1

1. Prove 6.1.3. [Check that both sides of the equality coincide on each X_i.]
2. Let D be a triangular derivation on $R[X_1, \ldots, X_n]$. Show that $\exp D$ is tame.
3. Let $\varphi : \mathbb{A}^1 \to \mathbb{A}^3$ be given by $\varphi(t) = (t + t^6, t^5, t^4)$.

   i) Find $p \in k[X_1, X_2, X_3]$ such that $p(\varphi(t)) = t$.
   ii) Let $\tilde{\varphi} := (\varphi, 0) : \mathbb{A}^1 \to \mathbb{A}^4$. Find $F \in \operatorname{Aut}_k \mathbb{A}^4$ such that $F(\tilde{\varphi}(t)) = (t, 0, 0, 0)$, for all $t \in \mathbb{A}^1$.

4. Let $F \in \operatorname{Aut}_k k[X, Y, Z, W]$ be given by

   $$F := (X + W(ZX + WY), Y - Z(ZX + WY), Z + W^3, W).$$

   Let T be a new variable. Write (F, T) as a product of elementary automorphisms in $\operatorname{Aut}_k k[X, Y, Z, W, T]$.

6.2 Stable equivalence

Throughout this section R denotes a commutative ring, $R[X] := R[X_1, \ldots, X_n]$ and $\operatorname{End}(R, n)$ the set of all polynomial mappings $F = (F_1, \ldots, F_n) \in R[X]^n$, with the usual componentwise addition and multiplication, which is the composition of maps, i.e.

$$F \circ G := (F_1(G), \dots, F_n(G))$$

where $F_i(G) := F_i(G_1, \dots, G_n)$ for all $G = (G_1, \dots, G_n) \in R[X]^n$. We will use the notation of Chapter 5. For each $m \geq 1$ we define the **stabilization map** from $\mathrm{End}(R, n)$ to $\mathrm{End}(R, n+m)$ by $F \to F^{[m]}$, where

$$F^{[m]} := (F_1, \dots, F_n, Y_1, \dots, Y_m)$$

and $Y_1, \dots, Y_m$ are m new variables. Observe that

(6.2.1) $\quad JF$ is invertible if and only if $JF^{[m]}$ is invertible.

Furthermore it is left to the reader to verify (Exercise 1)

(6.2.2) $\quad F$ is invertible over R if and only if $F^{[m]}$ is invertible over R.

From 6.2.1 and 6.2.2 we deduce that, to investigate the Jacobian Conjecture we may replace F by $F^{[m]}$.

Definition 6.2.3 Let $F, F' \in \mathrm{End}(R, n)$. Then we say that F and F' are **elementarily equivalent** or just **equivalent** resp. **tamely equivalent** if there exist $E_1, E_2 \in E(R, n)$ resp. in $T(R, n)$ such that $E_1 \circ F \circ E_2 = F'$. Notation: $F \sim F'$ (resp. $F \sim_T F'$).

Furthermore, if $m \geq 1$ and $G \in \mathrm{End}(R, n+m)$ we say that F is **stably equivalent** resp. **stably tamely equivalent** to G if $F^{[m]}$ is equivalent resp. tamely equivalent to G. Notation: $F \sim^m G$ resp. $F \sim^m_T G$.

Example 6.2.4 Let $F \in \mathrm{End}(R, n)$, $m \geq 1$ and $a_1, \dots, a_m \in R[X]$. Then F is stably equivalent to $(F, Y_1 + a_1(X), \dots, Y_m + a_m(X))$.

Proof: Let $E_i := (X, Y_1, \dots, Y_{i-1}, Y_i + a_i(X), Y_{i+1}, \dots, Y_m)$, for all $1 \leq i \leq m$. Put $E := E_1 \circ E_2 \circ \cdots \circ E_m$. Then $E \in E(R, n+m)$ and $F^{[m]} \circ E = (F, Y_1 + a_1(X), \dots, Y_m + a_m(X))$. □

As a first example of stable equivalence and the use of 6.2.4 we give

Lemma 6.2.5 *Let* $F = (F_1, \dots, F_n) \in \mathrm{End}(R, n)$. *Then for some* $m \geq 1$, F *is stably equivalent to some* $G \in \mathrm{End}(R, n+m)$ *with* $\deg G \leq 3$.

Proof: Let M be a monomial of degree $d \geq 4$ appearing in some F_i. We may assume that $i = 1$. Write $M = PQ$ with $\deg P = 2 \leq d - 2$, since $d \geq 4$. Then by 6.2.4

$$F^{[2]} \sim (F, Y_1 + P(X), Y_2 + Q(X)).$$

Now define

$$E_1 := (X_1 - Y_1 Y_2, X_2, \dots, X_n, Y_1, Y_2).$$

Then

$$\begin{aligned} F^{[2]} &\sim E_1 \circ (F, Y_1 + P(X), Y_2 + Q(X)) \\ &= (F_1', F_2, \ldots, F_n, Y_1 + P(X), Y_2 + Q(X)) \end{aligned}$$

where

$$\begin{aligned} F_1' &= F_1 - (Y_1 + P)(Y_2 + Q) \\ &= F_1 - PQ - Y_1 Y_2 - Y_1 Q - Y_2 P. \end{aligned}$$

So we see that the monomial M does not appear any more in F_1'. Furthermore, the terms $Y_1 Y_2$, $Y_1 Q$, $Y_2 P$ are all of degree $\leq d-1$. So after a finite number of steps we have eliminated all terms of degree d, introducing only new terms of degree $\leq d-1$. So we arrive at a map of degree $\leq d-1$. Then the lemma follows by induction on d. □

Remark 6.2.6 In the proof of 6.2.5 the "right E_2" in 6.2.3 is given by an application of 6.2.4 and is used to "make space" and to prepare the "simplification" performed by the "left E_1" in 6.2.3. This procedure will return in several of the stabilization results below.

In order to state the next result we introduce

$$\mathrm{End}^1(R, n) := \{F \in \mathrm{End}(R, n) | F(0) = 0 \text{ and } JF(0) = I_n\}.$$

Remark 6.2.7 The proof of 6.2.5 shows that if $F \in \mathrm{End}^1(R, n)$, then $G \in \mathrm{End}^1(R, n+m)$.

Theorem 6.2.8 *Let $F \in \mathrm{End}^1(R, n)$ with $JF \in Gl_n(R[X])$. Then there exists $m \geq 1$ such that F is stably equivalent to some $G \in \mathrm{End}(R, n+m)$ of the form $G = X^{[m]} + H$ where H satisfies*

i) $\deg H \leq 3$.
ii) *JH is nilpotent.*

Proof: (started) By 6.2.5 and 6.2.7 we may assume that $F = X + F_{(2)} + F_{(3)}$, where $F_{(i)}$ is either zero or homogeneous of degree i. Let ${}^T F$ be the clearing map (1.1.20). Write $F(T)$ instead of ${}^T F$. Then $F(T) \in \mathrm{End}^1(R[T], n)$ and by 1.1.22 $JF(T) \in Gl_n(R[T][X])$ since $JF \in Gl_n(R[X])$. Introduce n new variables $Y_1, \ldots, Y_n$ and put

$$\begin{aligned} E_1(T) &:= (X + TY, Y) \in E(R[T], 2n), \\ E_2(T) &:= (X, Y - TF_{(3)}) \in E(R[T], 2n) \quad \text{and} \\ G(T) &:= E_1(T) \circ F(T)^{[n]} \circ E_2(T) \in \mathrm{End}(R[T], 2n). \end{aligned}$$

Then

$$\begin{aligned} G(T) &= (X + TY, Y) \circ (F(T), Y - TF_{(3)}) \\ &= (F(T) + TY - T^2F_{(3)}, Y - TF_{(3)}) \\ &= (X, Y) + T(Y + F_{(2)}, -F_{(3)}). \end{aligned}$$

So if we put $H := (Y + F_{(2)}, -F_{(3)})$ we get $G(T) = (X, Y) + TH$. Since $JF(T)$ is invertible over $R[T]$, it follows that

(6.2.9) $\qquad JG(T) = I + JH.T$ is invertible.

From this we will deduce below that

(6.2.10) $\qquad JH$ is nilpotent.

Finally setting $T = 1$ we get that $E_1(1), E_2(1) \in E(R, 2n)$ and $G(1) = E_1(1) \circ F^{[n]} \circ E_2(1) = (X, Y) + H$ with JH nilpotent, as desired. So it remains to prove 6.2.10. Therefore we use the following

Lemma 6.2.11 *Let $A = A_0 \oplus A_1 \oplus \cdots$ be a graded ring (A need not be commutative). Let $a \in A_d$, for some $d \geq 1$. Then $1 + a$ is invertible in A if and only if a is nilpotent.*

Proof: If a is nilpotent, say $a^N = 0$, then $\sum_{i=0}^{N-1}(-a)^i$ is the inverse of $1 + a$.

Conversely, let b be the inverse of $1 + a$. We may assume that $a \neq 0$. So $b = 1 + b_{i_1} + \cdots + b_{i_m}$, for some $m \geq 1$, where all b_{i_j} are non-zero and homogeneous of degree c_{i_j}. Comparing degrees in the equality

$$(1 + a)(1 + b_{i_1} + \cdots + b_{i_m}) = 1$$

we get that $b_{i_1} + a = 0$, so $b_{i_1} = -a$. By induction on j it follows that $b_{i_j} = (-a)^j$, for all $1 \leq j \leq m$. In particular $b_{i_m} = (-a)^m$. Since the highest degree term appearing in $(1 + a)b$ equals ab_{i_m} we deduce that $a^{m+1} = 0$, so a is nilpotent. □

Proof of Theorem 6.2.8: (finished) Put $A := M_{2n}(R[X, Y][T])$. For each $d \geq 0$ let A_d be the subset of A consisting of all matrices of the form $M.T^d$, where $M \in M_{2n}(R[X, Y])$. Then obviously $A = \oplus_{d \geq 0} A_d$ is a graded ring. Now observe that $JG(T) = I + JH.T$ is an invertible element of A by 6.2.9 and that the element $JH.T$ is homogeneous of degree 1. So it follows from 6.2.11 that $JH.T$ is nilpotent and hence so is JH, which proves 6.2.10 and completes the proof of 6.2.8.

The proof of 6.2.8 immediately gives

Corollary 6.2.12 *If $F = X + F_{(2)} + F_{(3)}$ and JF is invertible, then JH is nilpotent where $H = (Y + F_{(2)}, -F_{(3)})$.*

In fact, introducing $(d - 1)n$ new variables $X^{(2)} := (X_1^{(2)}, \ldots, X_n^{(2)}), \ldots, X^{(d)} := (X_1^{(d)}, \ldots, X_n^{(d)})$, we can generalize 6.2.12 to

Proposition 6.2.13 *Let $F = X + F_{(2)} + \cdots + F_{(d+1)} \in \mathrm{End}(R, n)$ with $\deg F \leq d + 1$. Then*

i) $F^{[(d-1)n]}$ *is equivalent to* $X+H$ *where*

$$\begin{aligned} H &:= (F_{(2)}-X^{(d)}, F_{(d+1)}, F_{(d)}-X^{(2)},\\ &\quad F_{(d-1)}-X^{(3)},\ldots,F_{(3)}-X^{(d-1)}). \end{aligned}$$

ii) JF *is invertible if and only if* JH *is nilpotent.*

Proof: Write again $F(T)$ instead of ${}^{T}F$. So $F(T)\in \operatorname{End}(R[T],n)$. Put

$$\begin{aligned} G_{(1)} &:= (X-T^{d-1}X^{(2)}, X^{(2)}) \quad \text{and}\\ H_{(1)} &:= (X, X^{(2)}+TF_{(d+1)}). \end{aligned}$$

Then

$$\begin{aligned} K &:= G_{(1)}\circ F(T)^{[n]}\circ H_{(1)}\\ &= G_{(1)}\circ (F(T), X^{(2)}+TF_{(d+1)})\\ &= (F(T)-T^{d-1}X^{(2)}-T^{d}F_{(d+1)}, X^{(2)}+TF_{(d+1)})\\ &= (X+F_{(2)}T+\cdots+F_{(d)}T^{d-1}-T^{d-1}X^{(2)},\\ &\quad X^{(2)}+TF_{(d+1)})\\ &= (X+F_{(2)}T+\cdots+F_{(d-1)}T^{d-2}+(F_{(d)}-X^{(2)})T^{d-1},\\ &\quad X^{(2)}+TF_{(d+1)}) \end{aligned}$$

Next take

$$\begin{aligned} G_{(2)} &:= (X-T^{d-2}X^{(3)}, X^{(2)}, X^{(3)}) \quad \text{and}\\ H_{(2)} &:= (X, X^{(2)}, X^{(3)}+T(F_{(d)}-X^{(2)})). \end{aligned}$$

Then

$$\begin{aligned} G_{(2)}\circ K^{[n]}\circ H_{(2)} &= G_{(2)}\circ (K, X^{(3)}+T(F_{(d)}-X^{(2)}))\\ &= (X+F_{(2)}T+\cdots+(F_{(d)}-X^{(2)})T^{d-1}-T^{d-2}X^{(3)}-T^{d-1}\\ &\quad (F_{(d)}-X^{(2)}), X^{(2)}+TF_{(d+1)}, X^{(3)}+T(F_{(d)}-X^{(2)}))\\ &= (X+F_{(2)}T+\cdots+F_{(d-2)}T^{d-3}+(F_{(d-1)}-X^{(3)})T^{d-2}, X^{(2)}\\ &\quad +TF_{(d+1)}, X^{(3)}+T(F_{(d)}-X^{(2)})). \end{aligned}$$

More generally define

$$\begin{aligned} G_{(j)} &:= (X-T^{d-j}X^{(j+1)}, X^{(2)},\ldots,X^{(j+1)}),\\ H_{(j)} &:= (X, X^{(2)},\ldots,X^{(j)}, X^{(j+1)}+T(F_{(d-(j-2))}-X^{(j)})) \end{aligned}$$

for all $2\le j\le d-1$.

Then repeating the arguments above, we finally end up with

$$
\begin{aligned}
&F(T)^{[(d-1)n]} \\
&\sim (X + T(F_{(2)} - X^{(d)}), X^{(2)} + TF_{(d+1)}, X^{(3)} \\
&+ T(F_{(d)} - X^{(2)}), \ldots, X^{(d)} + T(F_{(3)} - X^{(d-1)})) \\
&= (X, X^{(2)}, \ldots, X^{(d)}) + TH.
\end{aligned}
\tag{6.2.14}
$$

So setting $T = 1$ in 6.2.14 we get the conclusion i) in 6.2.13.

Now we prove ii). Therefore first observe that by 1.1.22

$$JF \in Gl_n(R[X]) \text{ if and only if } JF(T) \in Gl_n(R[T][X]).$$

Since $JF(T)$ is invertible if and only if $JF(T)^{[dn]}$ is (by 6.2.1), we deduce from (6.2.14) that JF is invertible if and only if $I + TJH$ is invertible. Then the desired result follows from 6.2.11, since TJH is a homogeneous element of degree 1 in the T-grading, as in the proof of 6.2.8. □

Exercises for §6.2

1. Let $F \in \text{End}(R, n)$ and $m \geq 1, m \in \mathbb{N}$. Show

   (i) F is invertible if and only if $F^{[m]}$ is invertible.

   (ii) $F : R^n \to R^n$ is injective if and only if $F^{[m]}$ is injective.

2. Let $F \in \text{End}(R, n)$. Show that there exists an m and $G \in \text{End}(R, n + m)$ such that $F^{[m]}$ is equivalent to G which satisfies: $\deg G \leq 3$ and each variable in G appears linearly, i.e. G has no monomials of the form X_i^3 or $X_i^2 X_j$.
3. Let $F \in \text{End}(R, n)$. Show that F is stably equivalent to a polynomial map of the form $(AX, Y_1 + M_1(X), \ldots, Y_m + M_m(X))$ where $X = (X_1, \ldots, X_n)$, $A \in M_n(R)$ and each $M_i(X)$ is a monomial in $X_1, \ldots, X_n$.
4. Give an example of a ring A and a non-zero element $a \in A$ such that $1 + a$ is invertible in A but a is not nilpotent.
5. Let k be a field of characteristic $p > 0$. Give a counterexample to the following statement: if $F = (F_1, \ldots, F_n) \in \text{End}(k, n)$ such that $\deg F_i < p$ for all i and $\det JF = 1$, then F is invertible. [Use 6.2.5.]
6. (i) Let $H = (H_1, \ldots, H_n) \in \text{End}(R, n)$ with $H_n \in R$. Show that if JH is nilpotent, then $J_{X_1,\ldots,X_{n-1}}(H_1, \ldots, H_{n-1})$ is nilpotent.

   (ii) Let $F := (X_1 + X_2^2, X_2 + (X_1 + X_2^2)^2) \in \text{End}(\mathbb{C}, 2)$. Apply 6.2.13 to F and use i) to construct $H = (H_1, \ldots, H_4) \in \text{End}(\mathbb{C}, 4)$ such that JH is nilpotent and $H_1, \ldots, H_4$ are linearly independent over $\mathbb{C}$. (See also Chapter 8, Example 8.4.4).
7. Show that if F is stably equivalent with G then $d(F) = d(G)$, where $d(F)$ denotes the geometric degree of F (Appendix B.2).

6.3 Applications to the Jacobian conjecture

We begin this section by describing two "classical" reduction theorems. The first one, due to Bass, Connell and Wright in [29] and Yagzhev in [297] asserts that to prove the Jacobian conjecture it suffices to prove it for all polynomial maps of the form $F = X + H$ where $H = (H_1, \ldots, H_n)$ and each H_i is either zero or homogeneous of degree 3. We call such maps **cubic homogeneous polynomial maps** or also **Yagzhev maps**. If furthermore each H_i is a third power of a linear form, we call such an F a **cubic linear polynomial map** or a **Drużkowski map**. The second reduction theorem, due to Drużkowski in [93], asserts that to prove the Jacobian Conjecture it suffices to prove the conjecture for all Drużkowski maps! As another application of the stabilization idea it was shown by Jie-Tai Yu in [301] that it suffices to prove the Jacobian Conjecture for polynomial maps having positive real coefficients. On the other hand he showed that the conjecture is true for so-called negative polynomial maps: more precisely he showed that these maps are stably tame. This result was recently improved upon by Chau and Nga in [49]. We will prove this last version in 6.3.4.

Throughout this section R denotes a commutative $\mathbb{Q}$-algebra and $R[X] := R[X_1, \ldots, X_n]$.

Theorem 6.3.1 (Bass, Connell, Wright, Yagzhev). *It suffices to prove the Jacobian Conjecture for all $n \geq 2$ and all cubic homogeneous polynomial maps $F \in \mathrm{End}^1(R, n)$.*

Proof: i) Replacing F by $F - F(0)$ we may assume that $F(0) = 0$ and replacing F by $JF(0)^{-1} \circ F$ we may assume that $F \in \mathrm{End}^1(R, n)$. Then it follows from 6.2.8 that it suffices to prove the Jacobian Conjecture for all $n \geq 2$ and all $F \in \mathrm{End}(R, n)$ of the form $F = X + H$ with $\deg H \leq 3$ and JH nilpotent (observe that H may have linear terms, as can be seen from the proof of 6.2.8).

ii) Write $H = H_{(1)} + H_{(2)} + H_{(3)}$, where $H_{(i)}$ is homogeneous of degree i. Define $\tilde{F}(T) := (X + \tilde{H}(T), T) \in \mathrm{End}(R, n+1)$, where

$$\tilde{H}(T) := T^2 H_{(1)} + T H_{(2)} + H_{(3)} \quad (= T^3 H(X/T)).$$

Now observe that if $\tilde{F}(T) \in \mathrm{Aut}_R\, R[X, T]$, then it follows easily that F belongs to $\mathrm{Aut}_R\, R[X]$, i.e. F is invertible over R. So to prove the theorem it suffices to show that $J(\tilde{H}(T), 0)$ is nilpotent, i.e. that $J_X \tilde{H}(T)$ is nilpotent. However this follows readily, since $\tilde{H}(T) = T^3 H(X/T)$ implies that $J_X \tilde{H}(T) = T^3 JH(X/T)\, T^{-1} = T^2 JH(X/T)$ and the fact that JH is nilpotent. □

Theorem 6.3.2 (Drużkowski). *It suffices to prove the Jacobian Conjecture for all $n \geq 2$ and all Drużkowski maps.*

Proof: By 6.3.1 we may assume that $F = (X_1 + H_1, \ldots, X_n + H_n)$, where each H_i is either zero or cubic homogeneous. Using the algebraic identities

$$\begin{aligned} ab^2 &= \frac{1}{6}((a+b)^3 + (a-b)^3 - 2a^3), \\ abc &= \frac{1}{24}((a+b+c)^3 + (a-b-c)^3 - (a+b-c)^3 - (a-b+c)^3) \end{aligned}$$

we can write every third degree monomial appearing in H_j as an R-linear combination of cubic powers of linear forms in the X_i, i.e.

$$H_j = \sum_{i=1}^{r_j} c_{j,i} l_{j,i}^3, \text{ for some } r_j \in \mathbb{N}, \quad c_{j,i} \in R$$

and

$$l_{j,i} \in \mathbb{Z}X_1 + \cdots + \mathbb{Z}X_n.$$

So we have $r := r_1 + \cdots + r_n$ of such cubic-linear powers $l_{j,i}^3$. Introducing r new variables $Y_i^{(j)}$, $1 \le j \le n$, $1 \le i \le r_j$ it follows from 6.2.4 that $F^{[r]}$ is equivalent to

$$F' := (F, Y_1^{(1)} + l_{1,1}^3, \ldots, Y_{r_1}^{(1)} + l_{1,r_1}^3, \ldots, Y_1^{(n)} + l_{n,1}^3, \ldots, Y_{r_n}^{(n)} + l_{n,r_n}^3).$$

Now we are going to subtract $\sum_{i=1}^{r_1} c_{1,i} l_{1,i}^3$ from F_1, by multiplying F' from the left by the elementary map

$$E := \left(X_1 - \sum_{i=1}^{r_1} c_{1,i} Y_i^{(1)}, X_2, \ldots, X_n, Y_1^{(1)}, \ldots, Y_{r_n}^{(n)} \right).$$

Then the first component of the new F becomes $X_1 - \sum c_{1,i} Y_i^{(1)}$. Applying a similar operation to the second component of F until the n-th component of F we obtain that $F^{[r]}$ is tamely equivalent to

$$\left(X_1 - \sum_{i=1}^{r_1} c_{1,i} Y_i^{(1)}, \ldots, X_n - \sum_{i=1}^{r_n} c_{n,i} Y_i^{(n)}, Y_1^{(1)} + l_{1,1}^3, \ldots, Y_{r_n}^{(n)} + l_{n,r_n}^3 \right).$$

Finally performing the linear coordinate change

$$X_j' := X_j - \sum_{i=1}^{r_j} c_{j,i} Y_i^{(j)}, \text{ for all } 1 \le j \le n$$

we get that $F^{[r]}$ is tamely equivalent to

$$(X, Y_1^{(1)} + \tilde{l}_{1,1}^3, \ldots, Y_{r_n}^{(n)} + \tilde{l}_{n,r_n}^3).$$

where $\tilde{l}_{j,i}$ is obtained from $l_{j,i}$ by replacing each X_j by $X_j + \sum_{i=1}^{r_j} c_{j,i} Y_i^{(j)}$. □

Now we will describe the two theorems of Yu announced in the introduction. These theorems concern polynomial mappings with real coefficients. To describe these results we need the following definition.

Definition 6.3.3 Let $F = X + F_{(2)} + \cdots + F_{(d)} : \mathbb{R}^n \to \mathbb{R}^n$ be a polynomial map. Then F is called **positive** (resp. **negative**) if all non-zero coefficients appearing in the $F_{(i)}$ are positive (resp. negative).

The main results in the remainder of this section are

Theorem 6.3.4 (Yu, Chau, Nga). *Let $F = X + F_{(2)} + \cdots + F_{(d)} : \mathbb{R}^n \to \mathbb{R}^n$ be a negative polynomial map with* $\det JF = 1$. *Then there exists a permutation matrix P such that $P^{-1} \circ F \circ P$ is of triangular form. So F is tame, hence invertible.*

Theorem 6.3.5 (Yu). *If F is injective for all $n \geq 2$ and all positive polynomial maps $F : \mathbb{R}^n \to \mathbb{R}^n$ with* $\det JF = 1$, *then the Jacobian Conjecture is true, i.e. $JC(\mathbb{C}, n)$ is true for all $n \geq 1$ and hence $JC(R, n)$ is true for any commutative ring R contained in a $\mathbb{Q}$-algebra (by* 1.1.12*).*

Before we can prove these results we need several preliminaries.

Definition 6.3.6 Let R be a commutative ring and $A \in M_n(R)$. Then a **principal minor** of A is the determinant of a submatrix obtained by deleting $(n-k)$ rows and columns of A with the same indices.

Lemma 6.3.7 *Let $A = (a_{i,j}) \in M_n(\mathbb{R})$ with $a_{i,j} \geq 0$ for all pairs (i, j). If A is nilpotent then every principal minor of A is zero.*

Proof: Let A_* be the matrix obtained from A by replacing the entries in the n-th column and row by zeroes. Denote the entries of A_* by $a_{*i,j}$. Consider the universal matrix $A_U := (A_{i,j})$ (Appendix A). Write

$$A_U^n = (p_{i,j}(A_{1,1}, \ldots, A_{n,n})).$$

So each $p_{i,j}$ is a polynomial in $A_{1,1}, \ldots, A_{n,n}$ with coefficients in the non-negative integers. Now we claim that A_* is nilpotent: namely the nilpotency of A implies that $A^n = 0$, so $p_{i,j}(a_{1,1}, \ldots, a_{n,n}) = 0$ for all i, j. Obviously

$$0 \leq p_{i,j}(a_{*1,1}, \ldots, a_{*n,n}) \leq p_{i,j}(a_{1,1}, \ldots, a_{n,n}) = 0$$

whence $p_{i,j}(a_{*1,1}, \ldots, a_{*n,n}) = 0$, i.e. $A_*^n = 0$. Consequently the submatrix obtained from A by deleting the n-th row and column is nilpotent and in particular has determinant zero. The result follows by induction on n. □

Corollary 6.3.8 *Let $M \in M_n(\mathbb{R}[X_1, \ldots, X_n])$ be a nilpotent matrix such that each $M_{i,j}$ is a polynomial whose non-zero coefficients are positive. Then all principal minors of M are zero.*

Proof: Let $y = (y_1, \ldots, y_n) \in \mathbb{R}^n$ with $y_i \geq 0$ for all i. Since M is nilpotent, so is $M(y)$. Then it follows from 6.3.7 that all principal minors of $M(y)$ are zero for all

$y = (y_1, \ldots, y_n) \in \mathbb{R}^n$ with $y_i \geq 0$ for all i. So each principal minor of M is an element of $\mathbb{R}[X_1, \ldots, X_n]$ which vanishes on the set of $y = (y_1, \ldots, y_n) \in \mathbb{R}^n$ with $y_i \geq 0$ for all i. Consequently each principal minor of M is zero. □

Proposition 6.3.9 *Let R be an integral domain and $A \in M_n(R)$ such that every principal minor of A is zero. Then there exists a permutation matrix P such that $P^{-1}AP$ is an upper triangular matrix with zero diagonal.*

Proof: By induction on n. The case $n = 1$ is obvious, so let $n \geq 2$. Replacing R by its quotient field we may assume that R is a field.

i) Since $\det A = 0$ we have that one column of A is a linear combination of the others and hence, replacing A by $P^{-1}AP$, where P is some suitable permutation matrix, we may assume that the n-th column of A depends linearly on the first $n - 1$ columns. Denote by A_0 the $(n-1) \times (n-1)$ submatrix $(A_{i,j})_{1 \leq i,j \leq n-1}$ of A. This matrix satisfies the induction hypothesis, so there exists an $(n-1) \times (n-1)$ permutation matrix Q such that

(6.3.10) $\quad Q^{-1}A_0Q$ is upper triangular with zero diagonal.

We claim

(6.3.11) $\quad$ The $(n-1)$-*th* row of $\tilde{A} := \tilde{Q}^{-1}A\tilde{Q}$ is zero

where $\tilde{Q} := \begin{pmatrix} Q & 0 \\ 0 & 1 \end{pmatrix}$.

Namely $\tilde{A}_{n-1,i} = 0$ for all $1 \leq i \leq n-1$ by 6.3.10. Consequently we get $\tilde{A}_{n-1,n} = 0$ if we also show that the n-th column of $\tilde{A}$ depends linearly on the first $n - 1$ columns of $\tilde{A}$.

To see this recall that Ae_n (= the n-th column of A) $= \sum_{i=1}^{n-1} \lambda_i Ae_i$ for some $\lambda_i \in R$, hence

$$\tilde{Q}^{-1}A\tilde{Q}(\tilde{Q}^{-1}e_n) = \sum_{i=1}^{n-1} \lambda_i \tilde{Q}^{-1}A\tilde{Q}(\tilde{Q}^{-1}e_i).$$

So, since $\tilde{Q}^{-1}e_n = e_n$ we get that

$$\tilde{A}e_n = \sum_{i=1}^{n-1} \lambda_i \tilde{A}(Q^{-1}e_i).$$

Since Q^{-1} permutes the vectors $(e_1, \ldots, e_{n-1})$ it follows that the n-th column of $\tilde{A}$ is a linear combination of the first $n - 1$ columns of $\tilde{A}$. So, as observed above, this proves 6.3.11.

ii) Consequently we may assume that the $(n-1)$-th row of A is zero and hence, replacing A by $P^{-1}AP$ for some suitable permutation matrix P, we may assume that the n-th row of A is zero. Now repeat the argument in i) above, i.e. consider again A_0, Q and $\tilde{Q}$ as above. Then $\tilde{A} := \tilde{Q}^{-1}A\tilde{Q}$ has the desired form. □

Corollary 6.3.12 *Let R be a domain of characteristic zero and $F = X + H \in \mathrm{End}^1(R, n)$ such that every principal minor of JH is zero. Then there exists a permutation matrix P such that*

$$P^{-1} \circ F \circ P = (X_1 + g_1(X_2, \ldots, X_n), X_2 + g_2(X_3, \ldots, X_n), \ldots, X_{n-1} + g_{n-1}(X_n), X_n).$$

In particular F is tame.

Proof: Since R is a domain, hence so is $R[X]$. So applying 6.3.9 there exists a permutation matrix P such that $P^{-1}.JH.P$ is upper triangular with zero diagonal. Put $G := P^{-1} \circ F \circ P = X + P^{-1} \circ H \circ P$. Then $JG = I + P^{-1}JH(PX)P$ is upper triangular and every entry on the main diagonal of JG is equal to 1. Using that R has characteristic zero it follows readily that G is as desired. □

Proof of Theorem 6.3.4:

i) Let $N := -(F_{(2)} + \cdots + F_{(d)})$. It suffices to prove that JN is nilpotent, namely then by 6.3.8 every principal minor of JN is zero, so by 6.3.12 $P^{-1} \circ F \circ P$ is of triangular form for some permutation matrix P, hence F is tame.

ii) Now we show that JN is nilpotent. Therefore we put for each $x \in \mathbb{R}^n$

$$r(x) := \max\{|\lambda| \mid \lambda \in \mathbb{C} \text{ eigenvalue of } JN(x)\}.$$

So r defines a continuous function on $\mathbb{R}^n$ with values ≥ 0 and $r(0) = 0$. We need to show that $r(x) = 0$ for all $x \in \mathbb{R}^n$.

iii) Let $x \geq 0$, i.e. $x \in \mathbb{R}^n$ with $x_i \geq 0$ for all i. Then all entries of $JN(x)$ are ≥ 0. So by the Perron-Frobenius theorem $r(x)$ is an eigenvalue of $JN(x)$.

Claim: $r(x) < 1$.

Namely, suppose that $r(x_0) \geq 1$ for some $x_0 \geq 0$. Since r is continuous on the connected cone $\{x \geq 0\}$ and $r(0) = 0$ there exists a point $x_* \geq 0$ such that $r(x_*) = 1$. So 1 is an eigenvalue of $JN(x_*)$ and hence 0 is an eigenvalue of $I - JN(x_*) = JF(x_*)$, contradicting the fact that $JF(x_*)$ is invertible. So $r(x) < 1$ for all $x \geq 0$.

iv) From iii) it follows that if $x \geq 0$ the absolute values of all eigenvalues of $JN(x)$ are bounded by 1. Consequently the coefficients of the characteristic polynomial of $JN(x)$, which are the elementary symmetric polynomials in the eigenvalues of $JN(x)$, are bounded. On the other hand these coefficients are polynomials in the x_i. So these polynomials are bounded on $\{x \geq 0\}$. But the only polynomials with this property are the constants. So $r(x) = r(0)$ for all $x \in \mathbb{R}^n$. Finally since $r(0) = 0$ we get that $r(x) = 0$ for all $x \in \mathbb{R}^n$, which completes the proof.

Now we are going to prove 6.3.5. Again we need some preliminaries

Lemma 6.3.13 *Assume that for all $n \geq 2$ all Keller maps $F : \mathbb{R}^n \to \mathbb{R}^n$ are injective. Then the Jacobian Conjecture is true.*

Proof: Let $F : \mathbb{C}^n \to \mathbb{C}^n$ be a Keller map. By 4.2.1 it suffices to prove that F is injective. Now put

$$\tilde{F} := (ReF_1, ImF_1, \ldots, ReF_n, ImF_n) : \mathbb{R}^{2n} \to \mathbb{R}^{2n}.$$

It is well-known that $\det J\tilde{F} = |\det JF|^2$ ([41], Theorem 3, page 39). So $\det J\tilde{F} \in \mathbb{R}^*$ and the hypothesis implies that $\tilde{F}$ is injective. Consequently F is injective. □

Corollary 6.3.14 *If F is injective for all $n \geq 2$ and all polynomial maps $F : \mathbb{R}^n \to \mathbb{R}^n$ of the form $F = -X + H$, where H is homogeneous of degree 4 and JH nilpotent, then all Keller maps $F : \mathbb{R}^n \to \mathbb{R}^n$ are injective.*

Proof: From the proof of 6.3.1, replacing "invertibility" by "injective" it follows that it suffices to prove that all polynomial maps of the form $X + N_{(3)} : \mathbb{R}^n \to \mathbb{R}^n$, where $N_{(3)}$ is homogeneous of degree 3 and $JN_{(3)}$ is nilpotent, are injective. So let $G := X + N_{(3)}$ be such a map. Define

$$G(T) := (X, T) + T(N_{(3)}, 0) : \mathbb{R}^{n+1} \to \mathbb{R}^{n+1}.$$

Obviously if $G(T)$ is injective, then so is G. Furthermore $J(TN_{(3)}, 0) = (TJN_{(3)}, 0)$ is nilpotent since $JN_{(3)}$ is. Finally replacing $G(T)$ by $-G(T) = -(X, T) + H$, where $H := (-TN_{(3)}, 0)$, the result follows. □

Proof of Theorem 6.3.5: Let $F = -X + H : \mathbb{R}^n \to \mathbb{R}^n$ with H homogeneous of degree 4 and JH nilpotent. By 6.3.13 and 6.3.14 it suffices to prove that F is injective. Therefore we show that F is stably tamely equivalent to a positive polynomial map, which obviously concludes the proof of 6.3.5.

If every non-zero monomial appearing in H has a negative coefficient, then replace F by $-F = (-X) \circ F$ and we are done. Otherwise let M be a monomial appearing in some H_i with positive coefficient. We may assume that $i = 1$. Write $M = PQ$ where P and Q are monomials of degree 2 with positive coefficients. Then arguing as in the proof of 6.2.5 we get that

$$F^{[2]} \sim (F_1', F_2, \ldots, F_n, Y_1 + P, Y_2 + Q)$$

where

$$F_1' = (F_1 - M) - Y_1Y_2 - Y_1Q - Y_2P.$$

So M does not appear in F_1'. Continuing in this way we obtain that, after a finite number of steps, all monomials in H with positive coefficients are replaced by monomials of degree 2 or 3, all having negative coefficients. So we get for some $m \geq 1$

$$F^{[m]} \sim (-X_1 - H_1', \ldots, -X_n - H_n', Y_1 + P_1, \ldots, Y_m + P_m)$$

where all H_i' and P_j have positive coefficients. Furthermore $2 \leq \deg H_i' \leq 4$ for all i and $\deg P_j = 2$ for all j. Finally replacing $F^{[m]}$ by $E \circ F^{[m]}$, where

$$E = (-X_1, \ldots, -X_n, Y_1, \ldots, Y_m)$$

we get that F is stably tamely equivalent to a positive polynomial map, which concludes the proof. □

Exercises for §6.3

1. (i) Let $F(X_1) = X_1 + X_1^5$. Find $m \geq 1$ and $G \in \text{End}(R, m+1)$ such that $\deg G \leq 3$ and $F^{[m]} \sim G$.
   (ii) Compute the Yagzhev map deduced from G according to the proof of 6.3.1.
2. Let $F = (X_1 + X_4 d, X_2 - X_3 d, X_3 + X_4^3, X_4)$, where $d := (X_3 X_1 + X_4 X_2)$. Compute the Drużkowski map associated to F according to the proof of 6.3.2.

6.4 Gorni-Zampieri pairing

In the previous section we showed that it suffices to prove the Jacobian Conjecture for cubic homogeneous polynomial maps and even for cubic-linear maps. In this section we show that there is a notion of "pairing" between cubic homogeneous and cubic-linear polynomial maps over a field k of characteristic $\neq 2, 3$, i.e. to a cubic homogeneous polynomial map $f : k^n \to k^n$ we associate a cubic-linear map $F : k^N \to k^N$ with $N > n$ and conversely to a cubic-linear map with $\text{rk}\, A = n < N$ (see below for the definition of A) we associate a cubic homogeneous polynomial map $f : k^n \to k^n$. The point is that if f and F are paired, then several properties for one of these mappings implies the same property for the other. Before we give the definition of pairing let us fix some notations and make some simple remarks concerning cubic-linear mappings.

In this section k^m denotes $M_{m,1}(k)$. Furthermore, if $x, y \in k^m$ then $x * y$ denotes the vector $(x_1 y_1, \ldots, x_m y_m)^t$ in k^m. So if $F : k^N \to k^N$ is a cubic-linear mapping, then for each $1 \leq i \leq N$ there exist $a_{i,j} \in k$ such that

$$F_i(y) = y_i + \left(\sum a_{i,j} y_j\right)^3, \text{ for all } y \in k^N.$$

In other words, if we put $A := (a_{i,j}) \in M_N(k)$, then

$$F(y) = y + (Ay)^{*3}, \text{ for all } y \in k^N.$$

Sometimes we will write F_A to denote the dependence on A.

Definition 6.4.1 Let $f : k^n \to k^n$ be a cubic homogeneous map and $F_A : k^N \to k^N$ a cubic-linear map with $N > n$. We say that f **and** F_A **are paired** through the matrices $B \in M_{n,N}(k)$ and $C \in M_{N,n}(k)$ if

i) $f(x) = BF(Cx)$, for all $x \in k^n$.
ii) $BC = I_n$.
iii) $\ker B = \ker A$.

Observe that ii) implies that $\text{rk}\, B = \text{rk}\, C = n$. Consequently $\text{rk}\, A = n$ (by iii)). Since $n < N$ the matrix A does not have full rank. So $\det A = 0$.

Theorem 6.4.2 (Gorni, Zampieri). *Let $f : k^n \to k^n$ be a cubic homogeneous polynomial map. Then there exist $N > n$ and a cubic-linear map $F_A = X + (AX)^{*3}$*

with $A \in M_N(k)$ *such that* f *and* F_A *are paired through some matrices* $B \in M_{n,N}(k)$ *and* $C \in M_{N,n}(k)$.

Proof: i) Using the algebraic identities for ab^2 and abc described in the proof of 6.3.2, we can write every third degree monomial appearing in the components of f as a k-linear combination of the cubic powers of linear forms in x. Write all linear forms which appear in this way in some order in a column and then write this column as a matrix product D_0x, where $D_0 \in M_{N_0,n}(k)$ and N_0 denotes the number of linear forms. Then writing each $f_i - x_i$ as a linear combination of the third powers of these linear forms, we can write $f(x) = x + B_0(D_0x)^{*3}$ for all $x \in k^n$, where $B_0 \in M_{n,N_0}(k)$. By adding null columns to B_0 at the right end of B_0 and an equal number of null rows below D_0, we may assume that the number of columns of B_0 is $> n$.

The rank of B_0 need not be n. However one can arrange that rk $B_0 = n$ by adding a suitable number of columns to the right side of B_0 and the same number of zero rows below D_0. For example one can add an $n \times n$ identity matrix at the right end of B_0 and an $n \times n$ zero matrix to the bottom of D_0. In a similar way we can arrange that D_0 has rank n too. Denote by B resp. D the resulting matrices. So rk $B =$ rk $D = n$. Let N be the number of columns of B. So we have $N > n$ and

$$f(x) = x + B(Dx)^{*3}, \text{ for all } x \in k^n.$$

Since rk $B = n$ there exists $C \in M_{N,n}(k)$ with $BC = I_n$.

ii) Now we need to define $A \in M_N(k)$ such that $f(x) = BF_A(Cx)$ for all $x \in k^n$ and ker $A =$ ker B. Therefore let M be a matrix whose columns form a basis of ker B. So $M \in M_{N,N-n}(k)$, since rk $B = n$, and A must satisfy $AM = 0$. Furthermore since $f(x) = x + B(Dx)^{*3}$ the condition $f(x) = BF_A(Cx)$ for all $x \in k^n$ gives that $x + B(Dx)^{*3} = B(Cx + (ACx)^{*3})$, i.e. $B(Dx)^{*3} = B(ACx)^{*3}$, for all $x \in k^n$. Therefore it is natural to impose the following condition on A:

$$AC = D.$$

Together with $AM = 0$ this shows that A must satisfy

$$A(C|M) = (D|0)$$

where $(C|M)$ is the matrix obtained by joining the two blocks of columns of C and of M. The matrix $(D|0)$ is defined similarly. Now observe that $(C|M) \in Gl_N(k)$: namely Im C and Im $M =$ ker B are complementary, i.e. Im $C \cap$ ker $B = (0)$, since $BC = I_n$, and dim Im $C +$ dim Im $M =$ rk $C +$ dim ker $B = n + (N -$ rk $B) = n + (N - n) = N$. Consequently we can define

$$A := (D|0)(C|M)^{-1}.$$

iii) Finally we show that A, B and C are as desired. Namely since $(C|M)$ is invertible we get

$$\text{rk } A = \text{rk } D = n = \text{rk } B.$$

So $\dim \ker A = N - n = \dim \ker B$, whence $\ker A = \ker B$ (since $\ker B \subset \ker A$ because $AM = 0$). Furthermore since $AC = D$ and $BC = I_n$ we get

$$BF_A(Cx) = B(Cx + (ACx)^{*3}) = x + B(Dx)^{*3} = f(x)$$

for all $x \in k^n$, as desired. □

Conversely, let $F_A : k^N \to k^N$ be a cubic-linear mapping. Put $n := \text{rk}\, A$. We assume that $n < N$. In case $\det JF_A \in k^*$ we have that $\det A = 0$ by Exercise 3, whence $n = \text{rk}\, A < N$. Now it is easy to find a cubic linear mapping $f : k^n \to k^n$ which is paired to F_A, as follows. Choose $B \in M_{n,N}(k)$ with $\ker B = \ker A$. So $\text{rk}\, B = N - (N - n) = n$. Hence B has a right inverse $C \in M_{N,n}(k)$, i.e. $BC = I_n$. Consequently, for any such C the map

(6.4.3) $$f(x) := BF_A(Cx), \text{ for all } x \in k^n$$

is paired with F_A. Furthermore we will show below, 6.4.5, that f does not depend on the choice of the right inverse C. The proof is based on the following useful lemma, whose proof is left to the reader (Exercise 4).

Lemma 6.4.4 i) *If* $y \in k^N$ *and* $y_0 \in \ker A$, *then* $F_A(y + y_0) = F_A(y) + y_0$.
ii) *If* A, B, C *satisfy* ii) *and* iii) *of* 6.4.1, *then* $CBy - y \in \ker A = \ker B$, *for all* $y \in k^N$.

Proposition 6.4.5 *Let* $B \in M_{n,N}(k)$ *with* $\ker B = \ker A$. *Then the mapping* f *defined in* 6.4.3 *does not depend on the choice of the right inverse* C. *Furthermore* $f(By) = BF_A(y)$ *for all* $y \in k^N$.

Proof: i) Let C and $\tilde{C}$ be two right inverses of B. Then $Cx - \tilde{C}x \in \ker A = \ker B$ for all $x \in k^n$ since

$$B(Cx - \tilde{C}x) = BCx - B\tilde{C}x = x - x = 0.$$

Consequently

$$\begin{aligned} BF_A(\tilde{C}x) &= B(F(\tilde{C}x) + (Cx - \tilde{C}x)) \\ &= B(F(\tilde{C}x + Cx - \tilde{C}x)) \quad \text{by 6.4.4 i)} \\ &= B(F(Cx)). \end{aligned}$$

So indeed f defined in 6.4.3 does not depend on the choice of C.
ii) Finally

$$\begin{aligned} f(By) &= BF_A(CBy) \\ &= B(F_A(CBy) + (y - CBy)) \quad \text{(by 6.4.4 ii))} \\ &= B(F_A(CBy + y - CBy)) \quad \text{(by 6.4.4 i))} \\ &= BF_A(y), \quad \text{for all } y \in k^N \end{aligned}$$

as desired. □

The interest of the concept of pairing comes from the fact that if f and F_A are paired, several properties for one of the two mappings imply the same property for the other. In the remainder of this section we discuss various such properties.

Proposition 6.4.6 *Let $f : k^n \to k^n$ and $F : k^N \to k^N$ be paired through B and C, then*

i) $\det Jf(x) = \det JF(Cx)$, *for all* $x \in k^n$.
ii) $\det JF(y) = \det Jf(By)$, *for all* $y \in k^n$.

In particular $\det Jf(x) \in k^*$ *for all* $x \in k^n$ *if and only if* $\det JF(y) \in k^*$ *for all* $y \in k^N$.

Proof: Let $F = F_A$. In point ii) of the proof of 6.4.2 we saw that $k^N = \operatorname{Im} C \oplus \ker A$. Choose a basis of k^N of the form $(Ce_1, \ldots, Ce_n, h_1, \ldots, h_{N-n})$, where $(e_1, \ldots, e_n)$ is the canonical basis of k^n and the h_i form a basis of $\ker A$. With respect to this basis of k^N the mappings represented by the matrices $JF(y)$, $y \in k^N$, B and C have the following form

$$JF(y) = \begin{pmatrix} R(y) & S(y) \\ 0 & I_n \end{pmatrix}, C = \begin{pmatrix} I_n \\ 0 \end{pmatrix} \text{ and } B = (\, I_n \quad 0\,)$$

for matrices $R(y)$ and $S(y)$ of suitable dimensions.

i) Now let $x \in k^n$ and put $y := Cx$. Then

$$\begin{aligned} \det Jf(x) &= \det(BJF(Cx)C) \\ &= \det R(Cx) \\ &= \det JF(Cx). \end{aligned}$$

ii) Finally let $y \in k^N$. Then

$$\begin{aligned} \det JF(y) &= \det JF(CBy + y - CBy) \\ &= \det JF(CBy) \quad \text{(by 6.4.4 i))} \\ &= \det Jf(By) \quad \text{(by i)).} \end{aligned}$$

□

Proposition 6.4.7 *Let $f : k^n \to k^n$ and $F := F_A : k^N \to k^N$ be paired through B and C.*

i) *If F is injective, so is f.*
ii) *If F is surjective, so is f.*
iii) *If F is bijective, so is f. Furthermore $f^{-1}(z) = BF^{-1}(Cz)$, for all $z \in k^n$. In particular, if F^{-1} is a polynomial map, so is f^{-1} and $\deg f^{-1} \leq \deg F^{-1}$.*

Proof: i) Assume that F is injective and suppose that $f(x_1) = f(x_2)$ for some x_1, x_2 in k^n. Then $BF(Cx_1) = BF(Cx_2)$, whence $y_0 := F(Cx_1) - F(Cx_2) \in \ker B = \ker A$. So by 6.4.4 i) $F(Cx_1 - y_0) = F(Cx_2)$. The injectivity of F implies that $Cx_1 - y_0 = Cx_2$, so $y_0 \in \operatorname{Im} C$. Hence $y_0 \in \ker B \cap \operatorname{Im} C = 0$. So $C(x_1 - x_2) = 0$. Multiplying by B, using that $BC = I_n$, gives $x_1 - x_2 = 0$, i.e. $x_1 = x_2$.

ii) Suppose that F is surjective. Let $z \in k^n$. We need to show that $z \in \operatorname{Im} f$. Therefore observe that since F is surjective there exists $y \in k^N$ with $F(y) = Cz$, whence $BF(y) = BCz = z$. Since $f(By) = BF(y)$ (by 6.4.5) this implies that $z = f(By) \in \operatorname{Im} f$.

iii) From ii) we get that $f^{-1}(z) = By$ and $y = F^{-1}(Cz)$, so $f^{-1}(z) = BF^{-1}(Cz)$ for all $z \in k^n$, which completes the proof. □

Proposition 6.4.8 *Let $f : k^n \to k^n$ and $F = F_A : k^N \to k^N$ be paired through B and C.*

i) *If f is injective, then so is F.*
ii) *If f is surjective, then so is F.*
iii) *If f is bijective, then so is F. Furthermore in this case*

$$F^{-1}(w) = w - (ACf^{-1}(Bw))^{*3}, \text{ for all } w \in k^N.$$

In particular, if f^{-1} is a polynomial map so is F^{-1} and $\deg F^{-1} \le 3 \deg f^{-1}$.

Proof: i) Assume that f is injective. Suppose that $F(y_1) = F(y_2)$ for some $y_1, y_2 \in k^N$. Then $F(CBy_1 + y_1 - CBy_1) = F(CBy_2 + y_2 - CBy_2)$ and by 6.4.4 i) and 6.4.4 ii) we get

$$F(CBy_1) + y_1 - CBy_1 = F(CBy_2) + y_2 - CBy_2. \tag{6.4.9}$$

Since $y_i - CBy_i \in \ker B$ (by 6.4.4 i)) we get $BF(CBy_1) = BF(CBy_2)$, i.e. $f(By_1) = f(By_2)$. So $By_1 = By_2$ by the injectivity of f. Then (6.4.9) gives $y_1 = y_2$, as desired.

ii) Suppose that f is surjective and let $w \in k^N$, so $Bw = f(x)$ for some $x \in k^n$ by the surjectivity of f. We will show that $w = F(w - (ACx)^{*3})$ which implies that F is surjective. Therefore consider

$$\begin{aligned} F(w - (ACx)^{*3}) &= F((w - CBw) + CBw - (ACx)^{*3}) \\ &= (w - CBw) + F(CBw - (ACx)^{*3}) \\ &\quad \text{(by 6.4.4 i), ii))} \\ &= (w - Cf(x)) + F((Cf(x) - F(Cx)) \\ &\quad +(F(Cx) - (ACx)^{*3})). \end{aligned}$$

Observe that $B(Cf(x) - F(Cx)) = f(x) - BF(Cx) = f(x) - f(x) = 0$, i.e. $Cf(x) - F(Cx) \in \ker B$. Furthermore $F(Cx) - (ACx)^{*3} = Cx$. So we get by 6.4.4 i) again

$$F(w - (ACx)^{*3}) = w - Cf(x) + Cf(x) - F(Cx) + F(Cx) = w$$

as desired.

iii) Finally suppose that f is bijective. Since $x = f^{-1}(Bw)$ and $F(w-(ACx)^{*3}) = w$ we get

$$F^{-1}(w) = w - (ACf^{-1}(Bw))^{*3}.$$

□

The concept of pairing was introduced by Gorni and Zampieri in [145] in order to find out whether certain dilatations of a particular Drużkowski mapping are global analytic linearizable (see 6.4.10 below and Chapter 8, §3 and §4). In order to state the results obtained in [145] we start with a definition.

Definition 6.4.10 A polynomial map $F : \mathbb{C}^n \to \mathbb{C}^n$ with $F(0) = 0$ is called **linearizable** (resp. **global analytic linearizable**) if there exists a polynomial automorphism (resp. a complex analytic automorphism) $h : \mathbb{C}^n \to \mathbb{C}^n$ such that $h^{-1} \circ F \circ h$ is equal to the linear part of F, i.e. the linear map $x \to JF(0)x$.

It is a classical result, due to Poincaré and Siegel ([21], Chapter VIII, Theorem 3.1), that under some weak conditions on the eigenvalues of $JF(0)$ (non-resonance) such an h exists **locally** in a neighborhood of $0 \in \mathbb{C}^n$ and is unique if we assume that $h(0) = 0$ and $Jh(0) = I_n$. Furthermore it follows from Exercise 1 of §3 in Chapter 8 that, for each $s \in \mathbb{C}^*$ with $|s| \neq 1$ and each cubic homogeneous map f, the eigenvalues of $J(sf)(0)(= sI_n)$ satisfy the non-resonance condition. So in a neighborhood of $0 \in \mathbb{C}^n$ there exists a uniquely defined analytic map h_s such that $h_s(0) = 0$, $Jh_s(0) = I_n$ and which satisfies the equation

$$sf \circ h_s = h_s \circ sI_n.$$

Suppose now that f is paired with F_A through some matrices B and C. Since F_A is obviously cubic homogeneous, there also exists a uniquely defined analytic map H_s defined in a neighborhood of $0 \in \mathbb{C}^N$ such that

$$sF_A \circ H_s = H_s \circ sI_N$$

and which satisfies the normalization properties $H_s(0) = 0$ and $JH_s(0) = I_N$.

Theorem 6.4.11 (Gorni, Zampieri). *Let $s \in \mathbb{C}^*$ with $|s| \neq 1$ and f, F_A, h_s and H_s as above.*

i) *If H_s is globally analytic on $\mathbb{C}^N$, then h_s is globally analytic on $\mathbb{C}^n$. More precisely $h_s(x) = BH_s(Cx)$ for all $x \in \mathbb{C}^n$ and $h_s(By) = BH_s(y)$ for all $y \in \mathbb{C}^N$. In particular if H_s is a polynomial map, so is h_s and $\deg h_s \leq \deg H_s$.*
ii) *Suppose that H_s is globally analytic (then so is h_s by i)). If H_s is injective (resp. surjective), hence so is h_s. If H_s is bijective, so is h_s and $h_s^{-1}(z) = BH_s^{-1}(Cz)$ for all $z \in \mathbb{C}^n$. In particular if H_s^{-1} is a polynomial map, then so is h_s^{-1}.*

To prove this result we need one lemma. As already observed above, the existence of H_s (locally) follows from a result of Poincaré and Siegel. However, in the case of a map of the form sF_A, a new elegant proof of the local existence of H_s is given in [145]. Moreover it is also shown that if $|s| > 1$, then H_s is globally defined on $\mathbb{C}^N$. A closer look at the proof also reveals that if H_s is analytic in a neighborhood U of 0 and $y_0 \in \ker A$, then for each point $y \in U$ we have $y + y_0 \in U$ and $H_s(y + y_0) = H_s(y) + y_0$. Summarizing we have the following useful result (we refer to [145], Proposition 4.1 and Proposition 5.1 for a proof).

Lemma 6.4.12 i) *Let H_s be analytic in a neighborhood dom H_s of $0 \in \mathbb{C}^N$ and $y_0 \in \ker A$, then $y + y_0 \in$ dom H_s and $H_s(y + y_0) = H_s(y) + y_0$, for all $y \in$ dom H_s.*

ii) *If $|s| > 1$, then H_s is globally defined on $\mathbb{C}^N$ and analytic.*

Proof of Theorem 6.4.11: Write F instead of F_A.

i) Put $p(x) := BH_s(Cx)$ for all $x \in \mathbb{C}^n$. One readily verifies that $p(0) = 0$ and $Jp(0) = I_n$. Furthermore

$$\begin{aligned}
sf(p(x)) &= sBF(Cp(x)) \\
&= sBF(CBH_s(Cx)) \\
&= sB(F(CBH_s(Cx)) + H_s(Cx) \\
&\quad -CBH_s(Cx)) \quad \text{(by 6.4.4 ii))} \\
&= sBF(CBH_s(Cx) + H_s(Cx) \\
&\quad -CBH_s(Cx)) \quad \text{(by 6.4.4 i))} \\
&= sBF(H_s(Cx)) \\
&= BH_s(sCx) \\
&= p(sx), \quad \text{for all } x \in \mathbb{C}^n.
\end{aligned}$$

So p is an analytic map on $\mathbb{C}^n$, which has the same properties as h_s, so $h_s = p$ by the uniqueness of h_s, i.e. $h_s(x) = BH_s(Cx)$. In particular h_s is globally analytic on $\mathbb{C}^n$. Furthermore

$$H_s(y) = H_s(CBy + y - CBy) = H_s(CBy) + y - CBy \text{ (by 6.4.12).}$$

Multiplying by B gives

$$BH_s(y) = BH_s(CBy) = h_s(By), \text{ for all } y \in \mathbb{C}^N.$$

which proves i).

ii) Now we assume that H_s, and hence h_s, are globally analytic on $\mathbb{C}^N$ resp. $\mathbb{C}^n$. Suppose that H_s is injective and let $h_s(x_1) = h_s(x_2)$ for some x_1, x_2 in $\mathbb{C}^n$. Then $BH_s(Cx_1) = BH_s(Cx_2)$. So

$$y_0 := H_s(Cx_1) - H_s(Cx_2) \in \ker B = \ker A$$

whence

$$H_s(Cx_1) = H_s(Cx_2) + y_0 = H_s(Cx_2 + y_0) \text{ (by 6.4.12).}$$

The injectivity of H_s implies that $Cx_1 = Cx_2 + y_0$, so $y_0 \in \ker B \cap \operatorname{Im} C = 0$, i.e. $C(x_1 - x_2) = 0$. Multiplying by B gives $x_1 - x_2 = 0$, i.e. $x_1 = x_2$. So h_s is injective.

iii) Now suppose that H_s is surjective. Let $z \in \mathbb{C}^n$. Then there exists $w \in \mathbb{C}^N$ such that $Cz = H_s(w)$. So $z = BCz = BH_s(w) = h_s(Bw)$ (by i)), i.e. $z \in \operatorname{Im} h_s$. So h_s is surjective.

iv) Consequently, if H_s is bijective, so is h_s and from iii) we get

$$h_s^{-1}(z) = Bw = BH_s^{-1}(Cz), \text{ since } Cz = H_s(w)$$

which completes the proof of 6.4.11. □

To give the "complementary" result of 6.4.11, i.e. properties for h_s imply the same properties for H_s, we need

Lemma 6.4.13 *If h_s is globally defined and analytic on $\mathbb{C}^n$, then H_s is also globally defined and analytic on $\mathbb{C}^N$. Moreover, if h_s is a polynomial map, so is H_s and* $\deg H_s \le 3 \deg h_s$.

Proof: i) In a neighborhood of $0 \in \mathbb{C}^N$ we define the analytic map $Q(y) := H_s(y) - Ch_s(By)$. The argument given above in i) of the proof of 6.4.11 shows that $h_s(By) = BH_s(y)$ for all small y (replace everywhere "for all" by "for small"). Consequently

$$BQ(y) = BH_s(y) - BCh_s(By) = h_s(By) - h_s(By) = 0$$

so $Q(y) \in \ker B = \ker A$. Then by the definition of $Q(y)$ and 6.4.4 i) we get

$$sF_A(H_s(y)) = sF_A(Ch_s(By) + Q(y)) = sF_A(Ch_s(By)) + sQ(y).$$

On the other hand, using the definitions of H_s, Q and h_s we get

$$\begin{aligned} sF_A(H_s(y)) &= H_s(sy) \\ &= Q(sy) + Ch_s(sBy) \\ &= Q(sy) + Csf(h_s(By)) \\ &= Q(sy) + sCBF_A(Ch_s(By)). \end{aligned}$$

Combining the equalities above we deduce

$$Q(sy) - sQ(y) = sG(y) \tag{6.4.14}$$

where $G(y) := (I_N - CB)F_A(Ch_s(By))$. Let $\sum G_m(y)$ be the Taylor expansion of G at the origin. Let $\sum q_m(y)$ be the Taylor expansion of Q at the origin $0 \in \mathbb{C}^N$. Then 6.4.14 gives

$$(s^{m-1} - 1)q_m(y) = G_m(y), \text{ for all } m \ge 1 \text{ and small } y. \tag{6.4.15}$$

ii) Suppose now that h_s is defined globally on $\mathbb{C}^n$ and $0 < |s| < 1$. Then G is defined globally on $\mathbb{C}^N$ and by 6.4.14 we can extend Q analytically to the whole of $\mathbb{C}^N$. So Q is globally defined. So by the definition of Q it follows that H_s is also globally defined. If $|s| > 1$ then H_s is also globally defined by 6.4.12 i).

iii) Finally, if h_s is a polynomial mapping, then from the definition of G it follows that $G_m = 0$ if $m > 3 \deg h_s$. Then the same holds for q_m, by 6.4.15, and the result follows from the definition of Q. □

Theorem 6.4.16 (Gorni, Zampieri). *Let $s \in \mathbb{C}^*$ with $|s| \neq 1$ and f, F_A, h_s and H_s as above. Suppose that h_s is globally defined (so the same holds for H_s by* 6.4.13).

i) *If h_s is injective (resp. surjective), so is H_s.*

ii) *If h_s is bijective, so is H_s. Furthermore the inverse is given by*

$$H_s^{-1}(w) = w - H_s(Ch_s^{-1}(Bw)) + Ch_s^{-1}(Bw), \textit{ for all } w \in \mathbb{C}^N$$

In particular, if h_s^{-1} is a polynomial map so is H_s^{-1} and $\deg H_s^{-1} \leq \deg H_s \cdot \deg h_s^{-1}$.

Proof: i) Suppose that h_s is injective and assume that $H_s(y_1) = H_s(y_2)$ for some $y_i \in \mathbb{C}^N$. Then $H_s(CBy_1 + (y_1 - CBy_1)) = H_s(CBy_2 + (y_2 - CBy_2))$. Since $y_i - CBy_i \in \ker B = \ker A$ we deduce from 6.4.12 i)

$$H_s(CBy_1) + (y_1 - CBy_1) = H_s(CBy_2) + (y_2 - CBy_2) \tag{6.4.17}$$

whence $BH_s(CBy_1) = BH_s(CBy_2)$. So by 6.4.11 i) $h_s(By_1) = h_s(By_2)$. Since h_s is injective we deduce that $By_1 = By_2$. Now, using 6.4.17 we get $y_1 = y_2$, whence H_s is injective.

ii) Suppose now that h_s is surjective. Let $w \in \mathbb{C}^N$. Then $Bw \in \mathbb{C}^n$, so there exists $x \in \mathbb{C}^n$ with $h_s(x) = Bw$. By 6.4.11 i) this gives $BH_s(Cx) = Bw$, so $w - H_s(Cx) \in \ker B = \ker A$. Consequently

$$w = H_s(Cx) + (w - H_s(Cx)) = H_s(Cx + w - H_s(Cx)) \in \operatorname{Im} H_s,$$

i.e. H_s is surjective.

iii) Suppose that h_s is bijective. Let $w \in \mathbb{C}^N$. Then using the formulae in ii) we see that $H_s^{-1}(w) = Cx + w - H_s(Cx)$, where $h_s(x) = Bw$, i.e. $x = h_s^{-1}(Bw)$. Substituting this formula for x in H_s^{-1} gives the desired formula. □

Remark 6.4.18 To conclude this section we describe the question which led Gorni and Zampieri to the theory developed in this section.

Let $F : \mathbb{C}^{15} \to \mathbb{C}^{15}$ be the Drużkowski mapping given by

$$\begin{pmatrix} X_1 \\ X_2 \\ X_3 + (-2X_4 - X_5 + X_6 + X_7 + X_8 - X_{11} - X_{14})^3 \\ X_4 + (-X_3 - X_5 + \frac{1}{2}X_7 + \frac{1}{2}X_{10} - \frac{1}{2}X_{12} - \frac{1}{2}X_{13})^3 \\ X_5 + (X_3 - 2X_4 + X_8 - X_9 - X_{10} - X_{11} + X_{15})^3 \\ X_6 + (X_1 + X_3 - 2X_4 + X_8 - X_9 - X_{10} - X_{11} + X_{15})^3 \\ X_7 + (X_2 + X_3 - 2X_4 + X_8 - X_9 - X_{10} - X_{11} + X_{15})^3 \\ X_8 + (X_1 - X_3 - X_5 + \frac{1}{2}X_7 + \frac{1}{2}X_{10} - \frac{1}{2}X_{12} - \frac{1}{2}X_{13})^3 \\ X_9 + (X_1 - 2X_4 - X_5 + X_6 + X_7 + X_8 - X_{11} - X_{14})^3 \\ X_{10} + (X_2 - 2X_4 - X_5 + X_6 + X_7 + X_8 - X_{11} - X_{14})^3 \\ X_{11} + (X_1 + X_3 + X_5 - \frac{1}{2}X_7 - \frac{1}{2}X_{10} + \frac{1}{2}X_{12} + \frac{1}{2}X_{13})^3 \\ X_{12} + (X_2 - X_3 + 2X_4 - X_8 + X_9 + X_{10} + X_{11} - X_{15})^3 \\ X_{13} + (X_2 + 2X_4 + X_5 - X_6 - X_7 - X_8 + X_{11} + X_{14})^3 \\ X_{14} + (X_1 + X_2 + X_3 - 2X_4 + X_8 - X_9 - X_{10} - X_{11} + X_{15})^3 \\ X_{15} + (X_1 + X_2 - 2X_4 - X_5 + X_6 + X_7 + X_8 - X_{11} - X_{14})^3 \end{pmatrix}$$

Question: *Is sF globally analytically linearizable for all $s \in \mathbb{C}^*$, $|s| \neq 1$?*

The solution given by the authors in [145] is the following: let $f : \mathbb{C}^5 \to \mathbb{C}^5$ be given by

$$\begin{aligned} f \quad = \quad & (X_1, X_2, X_3 + 3(X_1^2X_2 + X_1X_2^2 + 2X_1X_2X_4 - 2X_1^2X_5), \\ & X_4 - 3(X_1^2X_2 + X_1X_2^2 + 2X_1X_2X_3 \\ & +2X_1^2X_5), X_5 - 3(X_2^2X_3 + X_2^2X_4)). \end{aligned}$$

Then they show that f is paired with F and that both h_s and its inverse are polynomial mappings of degree 7. Furthermore it follows from 6.4.13 that H_s is a polynomial mapping of degree ≤ 21 and from 6.4.16 that $\deg H_s^{-1} \leq 21.7 = 147$. In particular sF is linearizable for all $|s| \neq 1$, $s \in \mathbb{C}^*$.

Exercises for §6.4

1. Let $f = (X + Y^3 + XZ^2 + Z^3, Y + XYZ, Z + X^3)$. Find a matrix $A \in M_N(\mathbb{Q})$ with $N > 3$ such that F_A is paired with f.
2. Let $F = (X + (X + Y + Z)^3, Y + (X + Z)^3, Z + (2X + Y + 2Z)^3)$. Find a cubic homogeneous polynomial map $f : \mathbb{Q}^2 \to \mathbb{Q}^2$ which is paired with F. Is F invertible?
3. Let $F = X + (AX)^{*3} \in \mathrm{End}_k\, k[X_1, \ldots, X_n]$.
   (i) Show that $JF = I_n + 3(\mathrm{diag}\, AX)^2 A$, where $\mathrm{diag}\, AX$ is the diagonal matrix $(n \times n)$ which i-th diagonal element is equal to the i-th component of the column AX.
   (ii) Deduce from i) and 6.2.11 that $\det JF \in k^*$ implies that $\det A = 0$.
4. Give a proof of 6.4.4.

Notes

Both Proposition 6.1.4 and Lemma 6.1.3 are due to M. Smith in [271]. The results of §2 come from the "classical" paper [29] of Bass, Connell and Wright. It is interesting to compare Exercise 3, §2 with a "result" of Lang and Maslamani in [189], which asserts that if a Keller map $F : k^n \to k^n$ is of the form $F = (X_1+M_1, \ldots, X_n+M_n)$, where each M_i is either a monomial or zero, then F is invertible. However the proof given in [189] contains several serious mistakes which make the proof invalid.

The proof of Drużkowski's theorem given in §3 can be found in [94]. The proof of Theorem 6.3.5 is taken from the paper [301] of Yu. The proof of Proposition 6.3.9 can be found in Drużkowski's paper [95]. All results in §4 come from the paper [145] of Gorni and Zampieri.

7 Polynomial maps with nilpotent Jacobian

In the previous chapter we showed that every polynomial map whose Jacobian matrix is invertible is stably equivalent to a polynomial map of the form $X + H$, where JH is nilpotent. For polynomial maps over a field of characteristic zero, such maps with the additional assumption that H is cubic homogeneous were first studied by Wright in dimension 3, [294] and Hubbers in dimension 4, [156]. In both cases it turned out that the rows of the Jacobian matrix are linearly dependent over the coefficient field. This gave rise to the formulation of a (still open) dependence problem, 7.1.5. This problem led Hubbers and the author to the discovery of a large class of polynomial maps (over an arbitrary commutative ring A) having nilpotent Jacobian matrix. This class will be defined and studied in §2 and is denoted by $H(n, A)$. So $H(n, A)$ is a subset of $N(n, A)$, the set of all polynomial maps $H \in A[X]^n$ with JH nilpotent. If A is a UFD of characteristic zero we will show that $H(2, A) = N(2, A)$. However if $n \geq 3$ the class $N(n, A)$ is essentially larger than $H(n, A)$. Furthermore we will show that all maps of the form $X + H$ with $H \in H(n, A)$ are polynomial automorphisms. In fact we show that every such map is a finite product of exponentials of a special kind of locally nilpotent derivations, the so-called nice derivations, 7.3.18, and we deduce that all these maps are stably tame. At the end of §3 we give a shorter proof of this stable tameness result due to Hubbers and Wright in [158].

In §4 we consider a special class of polynomial maps such that their Jacobian matrix is strongly nilpotent. We show that after a linear coordinate change, these maps have an upper triangular Jacobian matrix with zeros on the main diagonal.

7.1 Hubbers' theorem and a dependence problem

Throughout this section k denotes a field of characteristic zero. In the previous chapter it was shown that every polynomial map with invertible Jacobian matrix is stably equivalent to a polynomial map of the form $X+H$ where the Jacobian matrix of H is nilpotent. The study of such maps was initiated by work of Wright in [294]. See also the Notes at the end of this chapter. For cubic homogeneous maps in dimension 3 over k he showed.

Proposition 7.1.1 (Wright). *Let $H \in k[X_1, X_2, X_3]^3$ be cubic homogeneous such that JH is nilpotent. Then there exists $T \in Gl_3(k)$ such that*

$$T^{-1}HT = (h_1(X_2, X_3), h_2(X_3), 0).$$

In 1994, Engelbert Hubbers in his master's thesis extended the method of Wright to the cubic homogeneous case in dimension 4 using a computer search. He completely classified all cubic homogeneous polynomial maps in dimension 4 having nilpotent Jacobian matrix. His main result in [156] is

Theorem 7.1.2 (Hubbers). *Let $H \in k[X_1, X_2, X_3, X_4]^4$ be cubic homogeneous such that JH is nilpotent. Then there exists $T \in Gl_4(k)$ such that $T^{-1}HT$ is of one of the following forms*

$$\text{i)} \begin{pmatrix} 0 \\ 0 \\ 0 \\ a_4 X_1^3 + b_4 X_1^2 X_2 + c_4 X_1^2 X_3 + e_4 X_1 X_2^2 + f_4 X_1 X_2 X_3 \\ \quad + h_4 X_1 X_3^2 + k_4 X_2^3 + l_4 X_2^2 X_3 + n_4 X_2 X_3^2 + q_4 X_3^3 \end{pmatrix}$$

$$\text{ii)} \begin{pmatrix} 0 \\ \frac{1}{3} X_1^3 + h_2 X_1 X_3^2 + q_2 X_3^3 \\ 0 \\ X_1^2 X_3 + h_4 X_1 X_3^2 + q_4 x_3^3 \end{pmatrix}$$

$$\text{iii)} \begin{pmatrix} 0 \\ \frac{1}{3} X_1^3 + c_1 X_1^2 X_4 - 3 c_1 X_1 X_2 X_3 + \frac{1}{48} \frac{(16 q_4 c_1^2 - r_4^2)}{c_1^2} X_1 X_3^2 \\ \quad + \frac{1}{2} r_4 X_1 X_3 X_4 - \frac{3}{4} r_4 X_2 X_3^2 + \frac{1}{12} \frac{r_4 q_4}{c_1} X_3^3 + \frac{1}{16} \frac{r_4^2}{c_1} X_3^2 X_4 \\ 0 \\ X_1^2 X_3 - \frac{1}{4} \frac{r_4}{c_1} X_1 X_3^2 + 3 c_1 X_1 X_3 X_4 - 9 c_1 X_2 X_3^2 \\ \quad + q_4 X_3^3 + \frac{3}{4} r_4 X_3^2 X_4 \end{pmatrix}$$

$$\text{iv)} \begin{pmatrix} 0 \\ \frac{1}{3} X_1^3 \\ X_1^2 X_2 + e_3 X_1 X_2^2 + k_3 X_2^3 \\ e_4 X_1 X_2^2 + k_4 X_2^3 \end{pmatrix}$$

$$\text{v)} \begin{pmatrix} 0 \\ \frac{1}{3} X_1^3 - i_3 X_1 X_2 X_4 + j_2 X_1 X_4^2 - s_3 X_2 X_4^2 - i_3^2 X_3 X_4^2 + t_2 X_4^3 \\ X_1^2 X_2 + 2 \frac{s_3}{i_3} X_1 X_2 X_4 + i_3 X_1 X_3 X_4 + j_3 X_1 X_4^2 + \frac{s_3^2}{i_3^2} X_2 X_4^2 \\ \quad + s_3 X_3 X_4^2 + t_3 X_4^3 \\ 0 \end{pmatrix}$$

$$\text{vi)} \begin{pmatrix} 0 \\ \frac{1}{3} X_1^3 + j_2 X_1 X_4^2 + t_2 X_4^3 \\ X_1^2 X_2 + e_3 X_1 X_2^2 + g_3 X_1 X_2 X_4 + j_3 X_1 X_4^2 \\ \quad + k_3 X_2^3 + m_3 X_2^2 X_4 + p_3 X_2 X_4^2 + t_3 X_4^3 \\ 0 \end{pmatrix}$$

$$\text{vii)} \begin{pmatrix} 0 \\ \frac{1}{3} X_1^3 \\ X_1^2 X_2 + e_3 X_1 X_2^2 + k_3 X_2^3 \\ X_1^2 X_3 + e_4 X_1 X_2^2 + f_4 X_1 X_2 X_3 + h_4 X_1 X_3^2 + k_4 X_2^3 \\ \quad + l_4 X_2^2 X_3 + n_4 X_2 X_3^2 + q_4 X_3^3 \end{pmatrix}$$

$$\text{viii)} \begin{pmatrix} 0 \\ \frac{1}{3}X_1^3 \\ X_1^2X_2 + e_3X_1X_2^2 - g_4X_1X_2X_3 + k_3X_2^3 - m_4X_2^2X_3 - g_4^2X_2^2X_4 \\ X_1^2X_3 + e_4X_1X_2^2 + 2\frac{m_4}{g_4}X_1X_2X_3 + g_4X_1X_2X_4 \\ \quad +k_4X_2^3 + \frac{m_4^2}{g_4^2}X_2^2X_3 + m_4X_2^2X_4 \end{pmatrix}$$

An immediate consequence of this result is

Corollary 7.1.3 (Hubbers). *The Jacobian Conjecture is true for all cubic homogeneous polynomial maps in dimension* 4 *(and hence in any dimension* ≤ 4*).*

Proof: We only need to show the case $n = 4$. Let $F = X + H$ be a Keller map. By 6.2.11 JH is nilpotent. So by Hubbers' theorem there exists $T \in Gl_4(k)$ such that $T^{-1}HT$ is of one of the eight forms described above. Since $T^{-1}FT = X + T^{-1}HT$ it suffices to prove that $X + H$ is invertible for each of the eight forms above. This can be checked easily by the invertibility criterion given in Chapter 3. □

Another consequence of Hubbers' theorem is

Corollary 7.1.4 *If* $H \in k[X_1, X_2, X_3, X_4]^4$ *is a cubic homogeneous polynomial map such that* JH *is nilpotent, then* H_1, H_2, H_3, H_4 *are linearly dependent over* k.

This leads us to formulate

Problem 7.1.5 (Dependence problem). Let $d \in \mathbb{N}, d \geq 1$ and $H = (H_1, \ldots, H_n) \in k[X_1, \ldots, X_n]^n$ homogeneous of degree d such that JH is nilpotent. Does it follow that $H_1, \ldots, H_n$ are linearly dependent over k?

It is not difficult to show that the problem of dependence of the H_i's is equivalent to the dependence of the rows of the corresponding Jacobian matrix JH (Exercise 1).

Let us assume for a moment that 7.1.5 has an affirmative answer and let $H \in k[X]^n$ be homogeneous of some degree d with JH nilpotent. Then, replacing H by $T^{-1}HT$ for suitable $T \in Gl_n(k)$ we may assume that $H_n = 0$. Consequently $J_{X_1,\ldots,X_{n-1}}(H_1, \ldots, H_{n-1})$ is nilpotent. So viewing $H_1, \ldots, H_{n-1}$ as polynomials in $k(X_n)[X_1, \ldots, X_{n-1}]$ we have an "induction-like" situation. However the polynomials H_i, $1 \leq i \leq n-1$ need not be homogeneous, considered as polynomials in $X_1, \ldots, X_{n-1}$ with coefficients in $k(X_n)$. To overcome this point it is therefore natural to formulate.

Problem 7.1.6 (Generalized Dependence Problem). Let $H \in k[X]^n$ be such that JH is nilpotent. Are the rows of JH dependent over k?

The answer to this question turned out to be yes if $n \leq 2$ and no if $n \geq 3$. More precisely we show

Theorem 7.1.7 i) *If* JH *is nilpotent and* rk $JH \leq 1$, *then the rows of* JH *are linearly dependent over* k *(here* rk *is the rank as an element of* $M_n(k(X))$*).*

ii) *Let* $r \geq 2$. *Then for any dimension* $n \geq r + 1$ *there exists a polynomial map* $H : k^n \to k^n$ *such that* JH *is nilpotent,* rk $JH = r$ *and the rows of* JH *are linearly independent over* k.

To prove the first part of this theorem we need one preparation.

Lemma 7.1.8 *Let $f \in k[X]\backslash k$ and denote by r the dimension of the k-vector space generated by all partial derivatives $\partial_i f$. Then $r \geq 1$ and there exists $U \in Gl_n(k)$ such that $g := f(UX)$ depends only on the first r X_i's and $\partial_1 g, \ldots, \partial_r g$ are linearly independent over $k[g]$.*

Proof: Since $f \notin k$ we get $r \geq 1$. By Exercise 2 below there exists $U \in Gl_n(k)$ such that $\partial_1 g, \ldots, \partial_r g$ are linearly independent over k and $\partial_j g = 0$ if $j \geq r+1$, where $g := f(UX)$. Suppose now that $\sum_{i=1}^r a_i(g)\partial_i g = 0$, where each $a_i(g) \in k[g]$ and $a_i(g) \neq 0$ for some i. Dividing by the greatest common divisor we may assume that $\gcd(a_1(T), \ldots, a_r(T)) = 1$ in $k[T]$. Consequently $a_i(0) \neq 0$ for some i, otherwise T divides all $a_i(T)$. Finally write $a_i = a_i(0) - T\tilde{a}_i$, with $\tilde{a}_i \in k[T]$. Then we deduce that $\sum_{i=1}^r a_i(0)\partial_i(g) = g\sum_{i=1}^r \tilde{a}_i(g)\partial_i(g)$. So the left hand side is zero (since otherwise its X-degree is strictly smaller than the X-degree of g, which in turn is smaller than or equal to the X-degree of the right hand side), a contradiction since $\partial_1(g), \ldots, \partial_r(g)$ are linearly independent over k and some $a_i(0)$ is non-zero. □

Proof of Theorem 7.1.7: First we prove i). The case of rank zero is obvious, so let us assume that $\operatorname{rk} JH = 1$. Let K denote the quotient field of $k[H] := k[H_1, \ldots, H_n]$ and let A be the integral closure of $k[H]$ in K. Then A is Noetherian (by [22], Proposition 5.17), $\dim A = 1$ (by 1.2.9) and hence A is a Dedekind ring by [22], Theorem 9.3. Furthermore A is contained in $k[X]$: namely if $a \in A$ then a is certainly integral over $k[X]$ and since K is contained in $k(X)$ it follows that $a \in k[X]$ because $k[X]$ is a UFD and hence integrally closed. So A is a Dedekind ring contained in $k[X]$, whence by Zaks' theorem, 1.2.26 we deduce that $A = k[f]$ for some $f \in k[X]$. In particular every H_i is a polynomial in f, say $H_i = h_i(f)$ for some $h_i \in k[T]$. In particular $f \notin k$ since $\operatorname{rk} JH = 1$. So by 7.1.8 there exist $U \in Gl_n(k)$ and $r \geq 1$ such that $g := f(UX)$ satisfies $\partial_1 g, \ldots, \partial_r(g)$ are linearly independent over $k[g]$ and g depends only on the first r X_i's. Put $Q := U^{-1}HU$. Then

$$Q = U^{-1}(h_1(g), \ldots, h_n(g))^t = (q_1(g), \ldots, q_n(g))^t$$

for some $q_i(g) \in k[g]$. Observe that

$$JQ = U^{-1}JH(UX)U$$

is also nilpotent. We will show that the rows of JQ are linearly dependent over k and hence so are the rows of JH. To see this, observe that since JQ is nilpotent $\operatorname{Tr} JQ = 0$, so

$$\sum_{i=1}^r q_i'(g)\partial_i(g) = 0$$

using that g depends only on the first r X_i's. Since $\partial_1(g), \ldots, \partial_r(g)$ are linearly independent over $k[g]$ we deduce that $q_i'(g) = 0$, hence $q_i(g) \in k$ for all $1 \leq i \leq r$. So the first r (≥ 1) rows of JQ are zero and hence the rows of JQ are linearly dependent over k. Finally the statement ii) follows immediately from

Proposition 7.1.9 *Let $r \geq 2$, $n \geq r+1$ and $a \in k[X_1]$ with $\deg a = r$. Put $f := X_2 - a$, $H_1 := f$, $H_i := X_{i+1} + \frac{(-1)^i}{(i-1)!} a^{(i-1)} f^{i-1}$, then for all $2 \leq i \leq r$*

$$H_{r+1} := \frac{(-1)^{r+1}}{r!} a^{(r)} f^r \text{ and } H_j := f^{j-1} \text{ if } r+1 < j \leq n.$$

Let $H := (H_1, \ldots, H_n)$. Then JH is nilpotent with rk $JH = r$ *and the rows of JH are linearly independent over k.*

Proof: Since the last $n - (r+1)$ columns of JH are zero, it suffices (for proving the first two statements) to show that $J :=$ the upper left hand $r+1 \times r+1$ matrix consisting of the elements $(JH)_{i,j}$ with $1 \leq i, j \leq r+1$, is nilpotent of rank r. To see this let e_i denote the i-th standard basis vector of k^{r+1}. Then $Je_{i+1} = e_i$ for all $2 \leq i \leq r$. Furthermore it is left to the reader to verify that $J^2 e_2 = 0$ (Exercise 3), which implies that $J^{r+1} = 0$. So J is nilpotent. Consequently $\det J = 0$, whence rk $J \leq r$. Obviously rk $J \geq r$ since $Je_{i+1} = e_i$ for all $2 \leq i \leq r$ and $Je_2 = e_1 + \sum_{i \geq 2} c_i e_i$ for some $c_i \in k[X]$. So rk $J = r$. Finally we show that the H_i are linearly independent over k. Therefore let $\sum \lambda_i H_i = 0$ for some $\lambda_i \in k$. Then $\lambda_2 = \cdots = \lambda_r = 0$ since for all $2 \leq i \leq r$ X_{i+1} only appears in H_i. So $\lambda_1 H_1 + \sum_{j \geq r+1} \lambda_j H_j = 0$. Then looking at the X_2-degrees of the H_j's we conclude that $\lambda_1 = 0$ and $\lambda_j = 0$ for all $j \geq r+1$, as desired. □

Remark 7.1.10 At the end of §4, Chapter 8 we describe a method to construct nilpotent Jacobian matrices whose rows are linearly independent over k (see in particular 8.4.4).

Exercises for §7.1

1. Let $\bar{k}$ be an algebraic closure of k. Assume that $H_1, \ldots, H_n$ in $k[X]$ have a common zero in $\bar{k}^n$.

   (i) Show that the H_i are linearly dependent over k if and only if the rows of the Jacobian matrix JH are linearly dependent over k.

   (ii) Give a counterexample to the equivalence of i) in case the H_i have no common zero in $\bar{k}^n$.

2. (i) Let V be a vector space over k and $v_1, \ldots, v_n$ vectors in V which span an $r \geq 1$ dimensional subspace of V. Show that there exists $T \in Gl_n(k)$ such that $(w_1, \ldots, w_r) := (v_1, \ldots, v_n)T$ satisfy: $w_1, \ldots, w_n$ are linearly independent over k and $w_{r+1} = \cdots = w_n = 0$.

   (ii) Prove 7.1.8. [Apply i) to $V := k[X]$ and $(v_1, \ldots, v_n) := Jf_{|TX} . T (= J(f(TX)))$.]

3. Complete the proof of 7.1.9 by showing that $J^2 e_2 = 0$.
4. Let k be an algebraically closed field. Let $H = (H_1, \ldots, H_n) \in k[X]^n$. Show that if JH is nilpotent, then for each i a power of $\partial_i H_i$ is contained in the ideal generated by all partial derivatives of H_i with respect to X_j, $j \neq i$. [Use the Nullstellensatz.]

5. (i) Give an example of a cubic homogeneous polynomial map H such that JH is nilpotent but $J(H \circ H)$ is not nilpotent. [Take $H = (X_4 d, -X_3 d, X_4^3, 0)$, where $d = X_3 X_1 + X_4 X_2$.]
   (ii) Give an example of a polynomial map H with $H(0) = 0$ and such that JH is nilpotent but $H^k \neq 0$ for all $k \geq 1$. Compare this with 7.2.16 in the next section. [Use 7.1.9.]

7.2 The class $H(n, A)$

Throughout this section A denotes an arbitrary commutative ring. By $N(n, A)$ we denote the set of all $H \in A[X]^n$ whose Jacobian matrix is nilpotent. In this section $A[X]^n$ denotes $M_{n,1}(A[X])$. We will introduce a subset of $A[X]^n$, denoted by $H(n, A)$, and show that it is a subset of $N(n, A)$. Furthermore we show that for each H in $H(n, A)$ the map $F := X + H$ is a polynomial automorphism of $A[X]$ and that $H^n := H \circ \cdots \circ H = 0$ if $H(0) = 0$.

If $n = 2$ and A a UFD we will show that $N(2, A) = H(2, A)$. However if A is a domain which is not a UFD, then we need not have equality. Moreover if $n \geq 3$, then for any domain A containing $\mathbb{Q}$ we have a strict inclusion of $H(n, A)$ in $N(n, A)$ (7.2.28).

Now we come to the main definition of this section.

Definition 7.2.1 $H(1, A) = A$. Let $n \geq 2$, then $H \in A[X]^n$ belongs to $H(n, A)$ if and only if there exist $T \in M_n(A)$, $c \in A^n$ and $h \in H(n-1, A[X_n])$ such that

$$H = \operatorname{Adj} T \begin{pmatrix} h \\ 0 \end{pmatrix}_{|TX} + c \tag{7.2.2}$$

where $\operatorname{Adj} T$ denotes the adjoint matrix of T, i.e. the matrix satisfying $\operatorname{Adj} T . T = T . \operatorname{Adj} T = \det T . I_n$ and $_{|TX}$ the "evaluation at the vector TX". The element in the right-hand side of 7.2.2 will be denoted by (h, T, c) or (h, T) in case $c = 0$.

Example 7.2.3 Let $H \in A[X_1, X_2]^2$. Then it is easy to verify (Exercise 1) that $H \in H(2, A)$ if and only if there exist $a_i, c_i \in A$ and $h(X_2) \in A[X_2]$ such that

$$\begin{aligned} H_1 &= a_2 h(a_1 X_1 + a_2 X_2) + c_1, \\ H_2 &= -a_1 h(a_1 X_1 + a_2 X_2) + c_2. \end{aligned}$$

Remark 7.2.4 In 7.2.25 below we will show that in case A is a $\mathbb{Q}$-algebra and a UFD, every element in $N(2, A)$ is as in 7.2.3, i.e. $H(2, A) = N(2, A)$. On the other hand if A is a domain, but not an UFD then $H(2, A)$ can be strictly smaller than $N(2, A)$ (see 7.2.24 below).

Let us first show that $H(n, A)$ is indeed a subset of $N(n, A)$. This result is based on the following simple observation: let $H = (h, T, c) \in H(n, A)$. Then

$$JH = \operatorname{Adj} T \begin{pmatrix} J_{n-1} h_{|TX} & \star \\ 0 & 0 \end{pmatrix} T \tag{7.2.5}$$

where $J_{n-1} h$ denotes the Jacobian matrix of h with respect to the first $n-1$ X_i's.

Proposition 7.2.6 *For all A and all $n \geq 1$ $H(n, A) \subset N(n, A)$.*

Proof: Induction on n. The case $n = 1$ is obvious. So let $n \geq 2$ and $H = (h, T, c)$ be in $H(n, A)$. By the induction hypothesis $J_{n-1}h$ is nilpotent. Hence so is $J_{n-1}h_{|TX}$. Then it follows readily from 7.2.5 that JH is nilpotent too. □

In the sequel we will frequently use some elementary properties of the elements of $H(n, A)$. These properties are collected in 7.2.10 below.

Let $H = (h, T, c) \in H(n, A)$, $a \in A$, $d \in A^n$ and $S \in M_n(A)$. Then it is easy to verify (Exercise 2) that

(7.2.7) $$aH + d = (ah, T, ac + d).$$

(7.2.8) $$\text{Adj}\, S.H_{|SX} = (h, TS, \text{Adj}\, S.c).$$

(7.2.9) If $f : A \to B$ is a ring homomorphism, then $f(H) = (f(h), f(T), f(c))$, where $f(H)$ etc.is obtained by applying f to the coefficients of H.

Corollary 7.2.10 *If $H \in H(n, A)$, $a \in A$, $d \in A^n$ and $S \in M_n(A)$, then*

i) $aH + d \in H(n, A)$.
ii) $\text{Adj}\, S.H_{|SX} \in H(n, A)$.
iii) *If $f : A \to B$ is a ring homomorphism, then $f(H) \in H(n, B)$.*

Proof: i) and iii) follow by induction on n from 7.2.7 resp. 7.2.9 and ii) follows directly from 7.2.8. □

The proofs of the results to be given below are all by induction on n and use the method of **universal coefficients** (Appendix A). In our situation this means that if $H = (h, T, c)$ we replace the matrix $T = (t_{i,j}) \in M_n(A)$ by the matrix $T_U = (T_{i,j})$ of universal coefficients, i.e. the $T_{i,j}$ are n^2 new variables and consider $H_U := (h, T_U, c)$. Furthermore put $R := A[T_U]$, the polynomial ring in the n^2 variables $T_{i,j}$ with coefficients in A and let $d := \det(T_U)$. So $d \in R$.

Lemma 7.2.11 *d is not a zero-divisor in R.*

Proof: We use induction on n. The case $n = 1$ is obvious, so let $n \geq 2$. Write $d(n-1) := \det(T_{i,j})_{1 \leq i,j \leq n-1}$. Put

$$A_\star := A[T_{n,i}, T_{i,n};\ 1 \leq i \leq n-1]$$

and

$$B := A_\star[T_{i,j};\ 1 \leq i, j \leq n-1].$$

So $B = A[T_{i,j};\ (i, j) \neq (n, n)]$. Develop d with respect to the n-th column. We get

$$d = d(n-1)T_{n,n} + b$$

for some $b \in B$. In particular b does not contain any $T_{n,n}$. Suppose that d is a zero-divisor in R. Then $dg = 0$ for some non-zero g in R. Expand g in powers of $T_{n,n}$, i.e.

$$g = g_m T_{n,n}^m + \cdots + g_1 T_{n,n} + g_0$$

with all g_i in B and g_m non-zero. Looking at the coefficient of highest degree in $T_{n,n}$ in the equation $dg = 0$ we get $d(n-1)g_m = 0$. Since by the induction hypothesis, applied to $A_\star$, $d(n-1)$ is not a zero-divisor in $A_\star[T_{i,j}; 1 \leq i, j \leq n-1] = B$ we get $g_m = 0$, a contradiction. So d is not a zero-divisor in R. □

So R is a subring of $A_U := R[d^{-1}]$ and T_U is invertible over A_U. Now we are able to prove the main result of this section.

Theorem 7.2.12 (van den Essen, Hubbers). *Let $H \in H(n, A)$. Put $F := X + H$. Then*

i) $\det JF = 1$ *and*
ii) *F is invertible over A.*

Proof: Let $H = (h, T, c)$, where $T = (t_{i,j})$ for some $t_{i,j}$ in A. Put $H_U := (h, T_U, c)$ and $F_U := X + H_U$.

i) It suffices to show that $\det JF_U = 1$, for then specializing $T_{i,j} \to t_{i,j}$ we get $\det JF = 1$. Therefore observe that, since T_U is invertible over A_U we have

(7.2.13) $$H_U = T_U{}^{-1} \begin{pmatrix} dh \\ 0 \end{pmatrix}_{|T_U X} + c, \text{ in } A_U[X]^n.$$

So putting $X' := (X_1, \ldots, X_{n-1})$ we get

(7.2.14) $$\det JF_U = \det(J_{n-1}(X' + dh))_{|T_U X}.$$

By 7.2.10 i) and iii) applied to the inclusion map from A to R we get that $dh \in H(n-1, R[X_n])$. So the induction hypothesis implies that $\det J_{n-1}(X' + dh) = 1$, whence by 7.2.14 $\det JF_U = 1$.

ii) It suffices to prove that F_U is invertible over R $(= A[T_U])$, for then specializing $T_{i,j} \to t_{i,j}$ we get that F is invertible over A. Since by i) $\det JF_U = 1$ it follows from 1.1.8 that it suffices to prove that F_U is invertible over A_U. Finally observe

(7.2.15) $$F_U = T_U{}^{-1} \left(X + \begin{pmatrix} dh \\ 0 \end{pmatrix} \right)_{|T_U X}.$$

Since $dh \in H(n-1, R[X_n])$, the induction hypothesis implies that $X' + dh$ is invertible over $R[X_n]$. Using this, it follows from 7.2.15 that F_U is invertible over A_U, as desired. □

Now we will show that the elements of $H(n, A)$ have a nice "convergence" property.

Theorem 7.2.16 *Let $H \in H(n, A)$. Then $H^n := H \circ \cdots \circ H \in A^n$. In particular if $H(0) = 0$, then $H^n = 0$.*

To prove this result we use the following lemma.

Lemma 7.2.17 *Let $H = \begin{pmatrix} H' \\ a \end{pmatrix}$ where $H' \in A[X]^{n-1}$ and $a \in A$. Put $H'_a := H'(X_n = a)$ and let $H'_{a,i}$ be its i-th component, i.e. $H'_{a,i} := H'_i(X_n = a)$, for all $1 \le i \le n-1$. Then for all $p \ge 2$ we have*

$$H^p = \begin{pmatrix} {H'_a}^{p-1}(H') \\ a \end{pmatrix}.$$

Proof: We use induction on p. First observe that

$$H_i(H) = H'_{a,i}(H') \text{ for all } 1 \le i \le n-1, \tag{7.2.18}$$

which proves the case $p = 2$. So let $p \ge 3$. The induction hypothesis gives

$$H^p = H^{p-1} \circ H = \begin{pmatrix} {H'_a}^{p-2}(H') \\ a \end{pmatrix} \circ H.$$

So by 7.2.18 we get

$$H^p = \begin{pmatrix} {H'_a}^{p-2}(H'_a(H')) \\ a \end{pmatrix} = \begin{pmatrix} {H'_a}^{p-1}(H') \\ a \end{pmatrix}.$$

□

Corollary 7.2.19 *If $H = (H', a)^t$ with $a \in A$ and $H' \in H(n-1, A[X_n])$, then $H^n \in A^n$.*

Proof: Induction on n. The case $n = 2$ is obvious. So let $n \ge 3$. With the notation of 7.2.17 we write $H' = H'_a + (X_n - a)H''$. Then by 7.2.17

$$H^n = \begin{pmatrix} {H'_a}^{n-1}(H') \\ a \end{pmatrix}. \tag{7.2.20}$$

Furthermore by 7.2.10 iii), applied to the substitution homomorphism $f : A[X_n] \to A$ sending X_n to a, we deduce that $H'_a \in H(n-1, A)$. So the induction hypothesis implies that ${H'_a}^{n-1} \in A^{n-1}$, whence by 7.2.20 $H^n \in A^n$. □

Proof of Theorem 7.2.16: Let $H = (h, T, c)$, where $T = (t_{i,j})$ for some $t_{i,j}$ in A and put $H_U := (h, T_U, c)$. It suffices to prove that ${H_U}^n \in {A_U}^n$, for then

${H_U}^n \in {A_U}^n \cap R[X]^n = R^n$, so making the substitutions $T_{i,j} \to t_{i,j}$ we deduce that $H^n \in A^n$. Now consider ${H_U}^n$. By 7.2.13 we get

$$H_U = {T_U}^{-1}\begin{pmatrix} g \\ b \end{pmatrix}_{|T_U X} \quad \text{with} \quad \begin{pmatrix} g \\ b \end{pmatrix} = \begin{pmatrix} dh \\ 0 \end{pmatrix} + T_U c.$$

So

$$(7.2.21) \qquad {H_U}^n = T_U^{-1}\begin{pmatrix} g \\ b \end{pmatrix}^n_{|T_U X}.$$

Finally observe that $b \in A_U$ and $g \in H(n-1, A_U[X_n])$ (7.2.10). So by 7.2.19 applied to $\begin{pmatrix} g \\ b \end{pmatrix}$ and the ring A_U we get $\begin{pmatrix} g \\ b \end{pmatrix}^n \in {A_U}^n$, so by 7.2.21 we get ${H_U}^n \in {A_U}^n$ as desired. □

In the remainder of this section we assume that A is a **domain**. One easily verifies that the set $N(n, A)$ is saturated with respect to the set $A \backslash \{0\}$, i.e. if $H \in A[X]^n$ is such that aH belongs to $N(n, A)$ for some non-zero a in A, then also H belongs to $N(n, A)$. This leads to the question: is $H(n, A)$ saturated with respect to the set $A \backslash \{0\}$? As we will see below (7.2.24) the answer is no in general. This naturally leads to the following definition.

Definition 7.2.22 $\overline{H}(1, A) = A$. For $n \geq 2$, $H \in A[X]^n$ belongs to $\overline{H}(n, A)$ if and only if there exist $0 \neq a \in A$, $T \in M_n(A)$, $c \in A^n$ and $h \in \overline{H}(n-1, A[X_n])$ such that $aH = (h, T, c)$.

It follows readily that $\overline{H}(n, A)$ is saturated with respect to $A \backslash \{0\}$ and furthermore $\overline{H}(n, A)$ is a subset of $N(n, A)$: namely if $aH = (h, T, c)$ with $h \in \overline{H}(n-1, A[X_n])$, then by induction on n we get that $J_{n-1}h$ is nilpotent, whence arguing as in the proof of 7.2.6 we get that $J(aH)$ is nilpotent and hence JH is nilpotent.

In fact also the proofs of 7.2.12 and 7.2.16 can be easily extended to the set $\overline{H}(n, A)$. The details are left to the reader as Exercise 3. So we have

Theorem 7.2.23

i) $\overline{H}(n, A) \subset N(n, A)$, *for all* $n \geq 1$.
ii) *Let* $H \in \overline{H}(n, A)$ *and* $F := X + H$. *Then* $\det JF = 1$ *and* F *is invertible over A.*
iii) *Let* $H \in \overline{H}(n, A)$. *Then* $H^n \in A^n$ *for all* $n \geq 1$.

To conclude this section we consider the inclusions

$$H(n, A) \subset \overline{H}(n, A) \subset N(n, A).$$

First we show that in case $n = 2$ the first inclusion can be strict.

Example 7.2.24 Let $A = \mathbb{Z}[X, Y, Z]/(X^2 + YZ)$, $H_1 = xX_1 + yX_2$, $H_2 = zX_1 - xX_2$ in $A[X_1, X_2]$ where $x = \overline{X}$, $y = \overline{Y}$ and $z = \overline{Z}$. Then

i) $H \notin H(2, A)$.
ii) $yH \in H(2, A)$, i.e. $H \in \overline{H}(2, A)$.

Proof: i) Suppose that $H \in H(2, A)$. Then by 7.2.3 there exist $a_1, a_2 \in A$ and $h \in A[T]$ with $h(0) = 0$ such that

$$\begin{aligned} H_1 &= a_2 h(a_1 X_1 + a_2 X_2), \\ H_2 &= -a_1 h(a_1 X_1 + a_2 X_2). \end{aligned}$$

Since both H_i have degree one it follows that $h = bT$ for some $b \in A \backslash \{0\}$. So $x = ba_1 a_2$ and $y = ba_2^2$. Let $A_i, B \in \mathbb{Z}[X, Y, Z]$ such that $a_i = \overline{A}_i$ and $b = \overline{B}$. Then multiplying x by a_2 and y by a_1 we get $a_2 x = a_1 y$, i.e. $A_2 X - A_1 Y = C(X^2 + YZ)$, for some C in $\mathbb{Z}[X, Y, Z]$. Consequently $X(A_2 - CX) = Y(A_1 + CZ)$. So $A_2 - CX = DY$ for some D in $\mathbb{Z}[X, Y, Z]$ and hence $A_1 + CZ = DX$. Summarizing

$$A_1 = DX - CZ \text{ and } A_2 = CX + DY.$$

Consequently the equation $x = ba_1 a_2$, i.e. $X - BA_1 A_2 \in (X^2 + YZ)$ implies that $X \in (X, Y, Z)^2$, a contradiction. So $H \notin H(2, A)$. Finally, since $yH = \begin{pmatrix} yxX_1 + y^2 X_2 \\ yzX_1 - xyX_2 \end{pmatrix}$ and $yz = -x^2$ we see that we can take $a_1 = x$, $a_2 = y$ and $h(T) = T$ to get the desired form of 7.2.3. □

The "accident" in this example is caused by the fact that the ring $\mathbb{Z}[x, y, z]$ subject to the relation $x^2 + yz = 0$ is not a UFD. In fact we have

Theorem 7.2.25 *Let A be a UFD of characteristic zero. Then $H(2, A) = N(2, A)$.*

Proof: i) First we assume that $A = k$ is a field. Let $H \in N(2, A)$. Replacing H by $H - H(0)$ we may assume that $H(0) = 0$. Then by 7.1.7 and Exercise 1 of §1 H_1 and H_2 are linearly dependent over k, so there exists $T \in Gl_2(k)$ such that the second component of $T^{-1}H$ is zero. So

(7.2.26) $$T^{-1} H_{|TX} = \begin{pmatrix} h \\ 0 \end{pmatrix}, \text{ for some } h \in k[X_1, X_2].$$

Since JH is nilpotent, the Jacobian matrix of the left hand side of 7.2.26 is also nilpotent, hence so is the Jacobian of the right hand side of 7.2.26, which implies that $h \in k[X_2]$ and hence that $H \in H(2, k)$.

ii) Now let A be a UFD and let $H \in N(2, A)$. Then $H \in N(2, K)$, where K is the quotient field of A. So by i) and 7.2.3 there exist $h(T) \in K[T]$ with $h(0) = 0$ and $b_i, d_i \in K$ such that

$$\begin{aligned} H_1 &= b_2 h(b_1 X_1 + b_2 X_2) + d_1, \\ H_2 &= -b_1 h(b_1 X_1 + b_2 X_2) + d_2. \end{aligned}$$

Clearing denominators we get: there exist $0 \neq a \in A$, $f(T) \in A[T]$ with $f(0) = 0$ and a_i, c_i in A such that

$$\text{(7.2.27)} \qquad \begin{aligned} aH_1 &= a_2 f(a_1X_1 + a_2X_2) + c_1 \text{ and } aH_2 \\ &= -a_1 f(a_1X_1 + a_2X_2) + c_2. \end{aligned}$$

Substituting $X_1 = X_2 = 0$ in 7.2.27 we obtain $c_i = ac_i'$ for some c_i' in A. So replacing H_i by $H_i - c_i'$ we may assume that $c_1 = c_2 = 0$.

iii) Now we show that we may assume that $\gcd(a_1, a_2) = 1$. Therefore let $a_i = a_i'd$, where $d = \gcd(a_1, a_2)$. Then $\gcd(a_1', a_2') = 1$ and

$$a_i f(a_1X_1 + a_2X_2) = a_i'\tilde{f}(a_1'X_1 + a_2'X_2)$$

where $\tilde{f}(T) = df(dT)$. So we may assume that $\gcd(a_1, a_2) = 1$.

iv) Write $f = \sum_{i=1}^N f_iT^i$ with f_i in A. Using 7.2.27 we see that we may assume that $\gcd(a, f_1, \ldots, f_N) = 1$.

Claim: *a is a unit in A and hence we are done.*

Suppose that p is a prime factor of a. Then 7.2.27 implies that p divides $f(a_1X_1 + a_2X_2)$ since $\gcd(a_1, a_2) = 1$. So in particular p divides both $f(a_1X_1)$ and $f(a_2X_2)$, so p divides $f_ia_1^i$ and $f_ia_2^i$ for all $i \geq 1$. Hence p divides each f_i, contradicting $\gcd(a, f_1, \ldots, f_N) = 1$. So a is a unit in A. □

In case $n \geq 3$ the situation is completely different. More precisely we have

Theorem 7.2.28 *If $n \geq 3$, then $\overline{H}(n, A) \neq N(n, A)$.*

To prove this result we need the following lemma.

Lemma 7.2.29 *Let A be a domain, $n \geq 1$ and $H \in \overline{H}(n, A)$ with $H(0) = 0$. Then the H_i are linearly dependent over A.*

Proof: If $H = 0$ we are done. So let H be non-zero. By definition $aH = (h, T, c)$ for some $h \in \overline{H}(n-1, A[X_n])$, $0 \neq a \in A$, $T \in M_n(A)$ and $c \in A^n$. Observe that T is non-zero: otherwise $aH = c$ which implies $aH = 0$ (use $H(0) = 0$), whence $H = 0$ since A is a domain, a contradiction. Multiplying the equation $aH = (h, T, c)$ from the left by T we get

$$\text{(7.2.30)} \qquad aTH = \det T \begin{pmatrix} h \\ 0 \end{pmatrix}_{|TX} + Tc.$$

i) If $\det T = 0$ it follows that $aTH = Tc$, whence $TH = 0$ ($H(0) = 0$). Since T is non-zero, some row of T is non-zero, which implies the linear dependence of the H_i over A.

ii) If $\det T \neq 0$ the last row of T is certainly non-zero. Then equating the n-th components of the vectors in 7.2.30 gives $a(TH)_n = (Tc)_n$. Again using $H(0) = 0$ gives the desired result. □

Proof of Theorem 7.2.28: Let $n \geq 3$ and put $a := X_1^2$, $f := X_2 - X_1^2$, $H_1 := f$, $H_2 := X_3 + 2X_1 f$, $H_3 := -f^2$ and $H_j := f^{j-1}$ if $3 < j \leq n$. Then it is shown in 7.1.9 that $H \in N(n, A)$ and that the H_i are linearly independent over the quotient field of A, hence over A. So by 7.2.29 $H \notin \overline{H}(n, A)$, as desired.

Exercises for §7.2

1. Prove the statement of 7.2.3.
2. Prove the statements 7.2.7, 7.2.8 and 7.2.9. [Use induction on n.]
3. Prove 7.2.23. [Localize.]
4. Let k be a field of characteristic zero. Show that $H(n, k) = \overline{H}(n, k)$ for all $n \leq 4$. [Use 7.2.29 and 7.2.25.]
5. Let A be a commutative ring and $H \in H(2, A)$. Put $F := X + H$. Show that $F = \exp D_1 \circ \exp D_2$, where D_1 and D_2 are locally nilpotent derivations on $A[X_1, X_2]$. [Use 7.2.3 and consider $D_2 := h(a_1X_1 + a_2X_2)(a_2\partial_1 - a_1\partial_2)$.]
6. Show that all eight forms of 7.1.2 are a product of at most three automorphisms of the form $\exp D$, where D is locally nilpotent on $k[X_1, X_2, X_3, X_4]$. [Use Exercise 5 and 7.2.25.]
7. Show that the eight forms of 7.1.2 are stably tame. [Use Exercise 6 and 6.1.4.]
8. Let A be a commutative ring, η its nilradical and $\overline{A} := A/\eta$. Assume that $\overline{A}$ is a UFD of characteristic zero. Show that each $H = (H_1, H_2)$ of $N(2, A)$ is of the form

$$\begin{aligned} H_1 &= a_2h(a_1X_1 + a_2X_2) + c_1 + g_1(X_1, X_2), \\ H_2 &= -a_1h(a_1X_1 + a_2X_2) + c_2 + g_2(X_1, X_2), \end{aligned}$$

where $h(X_2) \in A[X_2]$, $a_i, c_i \in A$ and $g_i \in \eta A[X_1, X_2]$, for all $i = 1, 2$.
9. (Peretz) Let k be a field of characteristic zero, $p, q \in k[X, Y]$ and $u, v \in k[T]$ such that $(\deg u - 1)\deg p + 1 > \deg q$ and $(\deg v - 1)\deg q + 1 > \deg p$. Let $F := (X + u(p), Y + v(q))$. Show that $F = X + H$ with $H \in H(2, k)$ if $\det JF = 1$. [Using $\det JF = 1$ and the degree hypothesis deduce that $\det J(p, q) = 0$. From this and Exercise 10, Chapter 1, §2 deduce that $F = (X + R(g(X, Y)), Y + S(g(X, Y)))$, for some $R, S \in k[T]$ and $g(X, Y) \in k[X, Y]$. Finally, using $\det JF = 1$ again, show that $F = X + H$ with JH nilpotent.]

7.3 $H(n, A)$, $D(n, A)$ and stable tameness

In this section we show that all polynomial maps of the form $F = X + H$ with H in $H(n, A)$ satisfy the exponential conjecture. More precisely we show that each such map F is a product of n automorphisms of the form $\exp D$, where D is a so-called nice derivation on $A[X]$ (7.3.12). In particular D satisfies $D^2(X_i) = 0$ for all i. We deduce that F is stably tame.

To prove these results we introduce a set $D(n, A)$ for each $n \geq 2$ as follows.

Definition 7.3.1 Let $n \geq 2$. Then $D(n, A)$ is the set of all $(2n-1)$-tuples $(T, c) := (T_2, \ldots, T_n, c_1, \ldots, c_n)$ where $T_n \in M_n(A)$, $T_i \in M_i(A[X_{i+1}, \ldots, X_n])$ for all $2 \leq i \leq n-1$, $c_n \in A^n (= M_{n,1}(A))$ and $c_i \in M_{i,1}(A[X_{i+1}, \ldots, X_n])$ for all $1 \leq i \leq n-1$.

If $n \geq 3$ we get a natural map $\pi : D(n, A) \to D(n-1, A[X_n])$ defined by

$$\pi((T_2, \ldots, T_n, c_1, \ldots, c_n)) = (T_2, \ldots, T_{n-1}, c_1, \ldots, c_{n-1}).$$

Instead of $\pi((T, c))$ we will write (T', c'). To each element (T, c) of $D(n, A)$ we associate $n-1$ polynomial maps as follows.

Definition 7.3.2 Let $n \geq 2$ and $0 \leq p \leq n-2$. Then $E_{n,p} : D(n, A) \to A[X]^n$ is given by

i) $E_{n,0}((T, c)) := \mathrm{Adj}(T_n) \begin{pmatrix} c_{n-1} \\ 0 \end{pmatrix}_{|T_n X}$, for all $(T, c) \in D(n, A)$.

ii) If $n \geq 3$ and $1 \leq p \leq n-2$, then inductively (with respect to n)

$$E_{n,p}((T, c)) := \mathrm{Adj}(T_n) \begin{pmatrix} E_{n-1,p-1}((T', c')) \\ 0 \end{pmatrix}_{|T_n X}.$$

Instead of $E_{n,p}((T, c))$ we simply write $E_{n,p}(T, c)$. Furthermore the polynomial map $X + E_{n,p}(T, c)$ will be denoted by $F_{n,p}(T, c)$. One readily verifies by induction on n that

(7.3.3) $\qquad aE_{n,p}(T, c) = E_{n,p}(T, ac)$, for all $a \in A$ and $(T, c) \in D(n, A)$.

Now we will show that each polynomial map $F = X + H$ with $H \in H(n, A)$ can be written as a product of the maps $F_{n,p}(T, c)$.

Theorem 7.3.4 i) *Let $n \geq 2$ and $H \in A[X]^n$. Then $H \in H(n, A)$ if and only if there exists $(T, c) \in D(n, A)$ such that*

$$H = \sum_{p=0}^{n-2} E_{n,p}(T, c) + c_n.$$

ii) *Let $H \in H(n, A)$ and $(T, c) \in D(n, A)$ as in* i). *Put $F = X + H$. Then $F = (X + c_n) \circ F_{n,0}(T, c) \circ F_{n,1}(T, c) \circ \cdots \circ F_{n,n-2}(T, c)$.*

Proof: The "if" part of i) is left to the reader (Exercise 1). To prove the "only if" part of i) and statement ii) we use induction on n and argue as in the proof of 7.2.12, i.e. we write $H = (h, T_n, c_n)$ where $T_n = (t_{i,j}) \in M_n(A)$, $c_n \in A^n$ and put $H_U := (h, T_U, c_n)$ and $F_U := X + H_U$. Since T_U is invertible over A_U we get

(7.3.5) $$F_U = {T_U}^{-1} \left(X + \begin{pmatrix} dh \\ 0 \end{pmatrix} \right)_{|T_U X} + c_n, \text{ where } d := \det T_U.$$

By the induction hypothesis we have

(7.3.6) $$h = \sum_{p=0}^{n-3} E_{n-1,p}(T',c') + c'_{n-1}, \text{ for some } (T',c') \in D(n-1, A[X_n]).$$

Put $(T,c) := (T', T_U, c', c_n)$. Observe that $(T,c) \in D(n, A_U)$ and $(T',c') = \pi((T,c))$. So our notation is consistent with the one introduced after Definition 7.3.1. From 7.3.6 and the definition of the $E_{i,j}(T,c)$ we deduce

(7.3.7)
$$\begin{aligned} H_U &= \sum_{p=0}^{n-3} \operatorname{Adj} T_U \begin{pmatrix} E_{n-1,p}(T',c') \\ 0 \end{pmatrix}_{|T_U X} \\ &\quad + \operatorname{Adj} T_U \begin{pmatrix} c'_{n-1} \\ 0 \end{pmatrix}_{|T_U X} + c_n \\ &= \sum_{p=1}^{n-2} E_{n-1,p}(T,c) + E_{n,0}(T,c) + c_n \end{aligned}$$

which implies i). Furthermore using 7.3.3 we get from 7.3.6 that

(7.3.8) $$dh = \sum_{p=0}^{n-3} E_{n-1,p}(T',dc') + dc'_{n-1}.$$

Consequently the induction hypothesis applied to dh gives

(7.3.9) $$X' + dh = (X' + dc'_{n-1}) \circ \Pi_{p=1}^{n-2}(X' + E_{n-1,p-1}(T',dc')).$$

So by 7.3.5

(7.3.10)
$$\begin{aligned} F_U = {T_U}^{-1} &\begin{pmatrix} X' + dc'_{n-1} \\ X_n \end{pmatrix}_{|T_U X} \circ \Pi_{p=1}^{n-2} {T_U}^{-1} \\ &\begin{pmatrix} X' + E_{n-1,p-1}(T',dc') \\ X_n \end{pmatrix}_{|T_U X} + c_n. \end{aligned}$$

Finally observe that

$$\begin{aligned} {T_U}^{-1} &\begin{pmatrix} X' + E_{n-1,p-1}(T',dc') \\ X_n \end{pmatrix}_{|T_U X} \\ &= X + d{T_U}^{-1} \begin{pmatrix} E_{n-1,p-1}(T',c') \\ 0 \end{pmatrix}_{|T_U X} \\ &= X + \operatorname{Adj} T_U \begin{pmatrix} E_{n-1,p-1}(T',c') \\ 0 \end{pmatrix}_{|T_U X}. \end{aligned}$$

Substituting this formula in 7.3.10 and a similar formula for the first factor of F_U in 7.3.10 we get the desired result in ii) by specializing $T_{i,j} \to t_{i,j}$ for all i, j. □

In the remainder of this section we will show that each of the polynomial maps $F_{n,p}$ is of the form $\exp D$ for some locally nilpotent derivation D and from the special structure of D we deduce that all these maps are stably tame.

In order to prove these results we introduce a special class of locally nilpotent derivations.

(7.3.11) **Nice derivations**

Let $B := A[x_1, \ldots, x_n]$ be a finitely generated A-algebra and Δ a subset of $\mathrm{Der}_A B$. By B^Δ we denote the set of all $b \in B$ such that $d(b) = 0$ for all $d \in \Delta$.

Definition 7.3.12 Let Δ be a finite subset of $\mathrm{Der}_A B$ and $\tau \in \mathrm{Der}_A B$.

i) We say that τ **is derived from** Δ **in at most one step** if τ is of the form

$$\tau = \sum_{d\in\Delta} b_d d, \quad \text{where } b_d \in B^\Delta, \quad \text{for all } d \in \Delta.$$

ii) Let $m \geq 2$. We say that τ **is derived from** Δ **in at most** m **steps** if there exists a sequence of finite subsets of $\mathrm{Der}_A B$

$$\Delta_0 := \Delta, \Delta_1, \ldots, \Delta_m$$

such that $\tau \in \Delta_m$ and for each $1 \leq i \leq m$ all elements of Δ_i are derived from Δ_{i-1} in at most one step.

iii) If furthermore the elements of Δ satisfy the condition that $d_1 d_2(x_i) = 0$ for all $d_1, d_2 \in \Delta$ and all i, then τ is called **nice of order** $\leq m$, with respect to $x_1, \ldots, x_n$ and Δ.

Example 7.3.13 (The Winkelmann derivation). Let $\tau = (1 + X_4X_2 - X_5X_3)\partial_1 + X_5\partial_2 + X_4\partial_3$ be a derivation on $B := A[X_1, X_2, X_3, X_4, X_5]$ and $\Delta := \{\partial_1, \partial_2, \partial_3\}$. Then τ is nice of order ≤ 2 with respect to $X_1, \ldots, X_5$ and Δ. To see this put

$$\Delta_1 := \{\partial_1, X_5\partial_2 + X_4\partial_3\} \text{ and } \Delta_2 := \{\tau\}.$$

Since $X_4, X_5 \in B^\Delta$, it follows that the two elements of Δ_1 are derived from Δ in at most one step. Since $1 + X_4X_2 - X_5X_3 \in B^{\Delta_1}$, it follows that τ is derived from Δ_1 in at most one step. Finally, since obviously $d_1d_2(X_i) = 0$ for all d_1, d_2 in Δ and all i, we conclude that τ is nice of order ≤ 2 with respect to $X_1, \ldots, X_5$ and Δ.

Proposition 7.3.14 *Notation as in* 7.3.12. *Assume that* $d_1d_2(x_i) = 0$ *for all* d_1, $d_2 \in \Delta$ *and all* i. *Then for all* $0 \leq k \leq m$, $d_1d_2(x_i) = 0$ *for all* $d_1, d_2 \in \Delta_k$. *In particular all elements of* Δ_k *commute pairwise and* $d^2(x_i) = 0$ *for every nice derivation.*

Proof: We use induction on k to prove the first statement. The case $k = 0$ is obvious since $\Delta_0 = \Delta$. Now let $k \geq 1$. Then

$$d_1 = \sum_{d \in \Delta_{k-1}} b_d d \text{ and } d_2 = \sum_{d' \in \Delta_{k-1}} b'_{d'} d' \text{ with } b_d, b_{d'} \in B^{\Delta_{k-1}},$$

and

$$d_1 d_2(x_i) = \sum_{d,d'} b_d d(b'_{d'}) d'(x_i) + \sum_{d,d'} b_d b'_{d'} d d'(x_i). \tag{7.3.15}$$

Now observe that $d(b'_{d'}) = 0$ since $b'_{d'} \in B^{\Delta_{k-1}}$ and $d \in \Delta_{k-1}$. The induction hypothesis gives that $dd'(x_i) = 0$ for all $d, d' \in \Delta_{k-1}$ and all i. So 7.3.15 implies that $d_1 d_2(x_i) = 0$. It is left to the reader to complete the proof (Exercise 2). □

Remark 7.3.16 One may wonder if a converse of 7.3.14 holds. In particular one can ask if all $\mathbb{C}$-derivations D on $\mathbb{C}[X_1, \ldots, X_n]$ such that $D^2 X_i = 0$ for all i are nice of some order $\leq m$ with respect to $X_1, \ldots, X_n$ and $\Delta = \{\partial_1, \ldots, \partial_n\}$. A negative answer to this question was given by Harm Derksen in dimension 8 based on a result of Z. Wang in [290]. For more details we refer to Exercise 6 at the end of this section.

To prove the result announced in the introduction of this section, namely that every map $X + H$ with $H \in H(n, A)$ is a finite product of exponential maps it suffices, according to 7.3.4 ii), to show that each map $F_{n,p}(T, c)$ is of the form $\exp D$ for some locally nilpotent derivation D on $A[X]$.

In order to prove this we need to extend some notions defined for the polynomial algebras $A[X_1, \ldots, X_n]$ to the finitely generated A-algebras $A[x_1, \ldots, x_n]$. This makes it possible to use induction on n. More precisely let $\varphi : A[X_1, \ldots, X_n] \to B$ $(= A[x_1, \ldots, x_n])$ be the A-algebra homomorphism defined by $\varphi(X_i) = x_i$ for all i. For each $p, q \geq 1$ consider the natural extension

$$\varphi : M_{p,q}(A[X]) \to M_{p,q}(B)$$

obtained by applying φ to the entries of the elements of $M_{p,q}(A[X])$. Then for each $(T, c) \in D(n, A)$ we define

$$E_{n,p}(T, c)(x) := \varphi(E_{n,p}(T, c)) \in B^n.$$

To each column $b := (b_1, \ldots, b_n)^t \in B^n$ and each column $d := (d_1, \ldots, d_n)^t \in (\mathrm{Der}_A B)^n$, we associate the following A-derivation of B

$$D(b, d) := \sum b_i d_i (= b^t d).$$

Proposition 7.3.17 *Let $d_1, \ldots, d_n$ be A-derivations of B and assume that there exists an element a in A such that $d_i(x_j) = a\delta_{i,j}$ for all i, j. Put $d := (d_1, \ldots, d_n)^t$ and let $(T, c) \in D(n, A)$. Let $n \geq 2$ and $0 \leq p \leq n-2$. Then $D := D(E_{n,p}(T, c)(x), d)$ is nice with respect to $x_1, \ldots, x_n$ and $\Delta_0 := \{d_1, \ldots, d_n\}$.*

Before we prove this result we deduce the main result of this section,

Theorem 7.3.18 (van den Essen, Hubbers). *Let $F = X + H$ with $H \in H(n, A)$. Then there exists $(T, c) \in D(n, A)$ such that*

$$F = \exp D(c_n, \partial)\Pi_{p=0}^{n-2}\exp D(E_{n,p}(T, c), \partial)$$

where $\partial := (\partial_1, \ldots, \partial_n)^t$. Furthermore each derivation D appearing in this expression is nice and satisfies $D^2(X_i) = 0$ for all i.

Proof: Observe that $\exp D(c_n, \partial) = X + c_n$ and $\exp D(E_{n,p}(T, c), \partial) = X + E_{n,p}(T, c)$ for all $0 \le p \le n-2$, since $D := D(E_{n,p}(T, c), \partial)$ is nice by 7.3.17 and hence by 7.3.14 satisfies $D^2(X_i) = 0$ for all i. Then apply 7.3.4 ii). □

The proof of 7.3.17 is based on 7.3.19 below. To formulate this lemma we need some notation: let $(T, c) \in D(n, A)$ (fixed), put

$$\begin{aligned}
x &:= (x_1, \ldots, x_n)^t, \\
y &:= T_n x, \\
y' &:= (y_1, \ldots, y_{n-1})^t, \\
d &:= (d_1, \ldots, d_n)^t, \\
\delta &:= (\operatorname{Adj} T_n)^t d, \\
\delta' &:= (\delta_1, \ldots, \delta_{n-1})^t \text{ and finally} \\
(T'(y_n), c'(y_n)) &:= (T'(X_n = y_n), c'(X_n = y_n)) \in D(n-1, A[y_n]).
\end{aligned}$$

Lemma 7.3.19 *Let $n \ge 3$ and $1 \le p \le n-2$. Then*

i) $D(E_{n,p}(T, c)(x), d) = D(E_{n-1,p-1}(T'(y_n), c'(y_n))(y'), \delta')$.
ii) *If there exists $a \in A$ such that $d_i(x_j) = a\delta_{i,j}$ for all i, j, then*

$$\delta_i(y_j) = a \det T_n \delta_{i,j} \text{ for all } i, j.$$

Proof: i) $D(E_{n,p}(T, c)(x), d) = (E_{n,p}(T, c)(x))^t d$

$$\begin{aligned}
&= ((E_{n-1,p-1}(T', c')_{|T_n x})^t, 0)(\operatorname{Adj} T_n)^t d \\
&= (E_{n-1,p-1}(T'(X_n = y_n), c'(X_n = y_n))(y'))^t \delta' \\
&= D(E_{n-1,p-1}(T'(y_n), c'(y_n))(y'), \delta').
\end{aligned}$$

ii) Let $1 \le i, j \le n$. Denote the i-th column of $\operatorname{Adj} T_n$ by $(p_1, \ldots, p_n)^t$ and the j-th row of T_n by $(q_1, \ldots, q_n)$. Then $\delta_i = \sum p_s d_s$ and $y_j = \sum q_s x_s$. So by the hypothesis on the d_i's we get

$$\delta_i(y_j) = a \sum p_s q_s = a(T_n . \operatorname{Adj} T_n)_{j,i} = a \det T_n \delta_{i,j}.$$

□

Proof of Proposition 7.3.17: i) The hypothesis on the d_i implies that $d_i d_j(x_k) = 0$ for all $d_i, d_j \in \Delta_0$ and all k.

ii) First we consider the case $p = 0$. Since $E_{n,0}(T, c) = \operatorname{Adj} T_n \begin{pmatrix} c_{n-1} \\ 0 \end{pmatrix}_{|T_n X}$ we get

$$D = c_{n-1}(y_n)^t \delta' = \sum_{i=1}^{n-1} c_{n-1,i}(y_n)\delta_i \tag{7.3.20}$$

where $c_{n-1,i}(y_n) \in A[y_n]$ for all $1 \le i \le n-1$.

Put $\Delta_1 := \{\delta_1, \ldots, \delta_{n-1}\}$ and observe that $\Delta_1 \subset Der_A B$ and that each element of Δ_1 is derived from Δ_0 in at most one step. Furthermore since $\delta_i(y_n) = 0$ for all $1 \le i \le n-1$ (by 7.3.19 ii)) we get that $c_{n-1,i}(y_n) \in B^{\Delta_1}$ for all $1 \le i \le n-1$. So 7.3.20 implies that D is derived from Δ_1 in at most one step. Consequently D is derived from Δ_0 in at most two steps. So, using i), we deduce that D is nice with respect to $x_1, \ldots, x_n$ and Δ_0.

iii) Now we prove the proposition by induction on n. If $n = 2$ we are in case ii). So let $n \ge 3$. By ii) we may assume that $p \ge 1$. Then 7.3.19 i) gives

$$D = D(E_{n-1,p-1}(T'(y_n), c'(y_n))(y'), \delta')$$

and $(T'(y_n), c'(y_n)) \in D(n-1, A[y_n])$. By 7.3.19 ii) we can apply the induction hypothesis to the ring $A' := A[y_n]$ and the $n-1$-tuple of A'-derivations $\delta_1, \ldots, \delta_{n-1}$ on the A'-algebra $B' := A'[y_1, \ldots, y_{n-1}]$. So the A'-derivation D on B' is nice with respect to $\Delta'_0 := \{\delta_1, \ldots, \delta_{n-1}\}$ and $y_1, \ldots, y_{n-1}$. So there exists a sequence

$$\Delta'_0, \Delta'_1, \ldots, \Delta'_m$$

of finite subsets of $\operatorname{Der}_{A'} B'$ such that $D \in \Delta'_m$ and each element of Δ'_i is derived from Δ'_{i-1} in at most one step, for all $1 \le i \le m$. Now observe that $\Delta'_0 \subset \operatorname{Der}_A B$ and that $B' \subset B$ since obviously $y_i \in B$ for all i. Consequently, if D' is an A'-derivation of B' derived from Δ'_0 in at most one step, then $D' \in \operatorname{Der}_A B$. Hence $\Delta'_1 \subset \operatorname{Der}_A B$. Arguing in a similar way we conclude by induction on i that $\Delta'_i \subset \operatorname{Der}_A B$ for all $0 \le i \le m$. Since, as observed in ii) above, all elements of Δ'_0 ($= \Delta_1$ in the notation of ii)) are derived from Δ_0 in at most one step, we deduce that D is derived from Δ_0 in at most $m+1$ steps. Finally define $\Delta_i := \Delta'_{i-1}$ for all $1 \le i \le m+1$. Then, using i) again, we get that D is nice with respect to $x_1, \ldots, x_n$ and Δ_0.

Now we will show that all maps of the form $X + H$ with $H \in H(n, A)$ are stably tame (6.1.8).

Theorem 7.3.21 *Let $F = X + H$ with $H \in H(n, A)$. Then F is stably tame.*

Proof: Follows directly from 7.3.18 and 7.3.22 below. □

Lemma 7.3.22 *Let $m \ge 1$ and τB a nice derivation on $A[X]$ of order $\le m$ with respect to $X_1, \ldots, X_n$ and $\Delta_0 := \{\partial_1, \ldots, \partial_n\}$. Then $\exp(g\tau)$ is stably tame for all $g \in A[X]^\tau$.*

Proof: We use induction on m. If $m = 1$ then $\tau = \sum_{d \in \Delta_0} b_d d$ with

$$b_d \in A[X]^{\Delta_0} = \cap_i \ker \partial_i = A.$$

Hence $\tau(X_i) \in A$ for all i and consequently τ is of triangular form. So by 6.1.4 $\exp(g\tau)$ is stably tame. Now let $m > 1$. So $\tau = \sum_{d \in \Delta_{m-1}} b_d d$, with $b_d \in A[X]^{\Delta_{m-1}}$ for all d. Introduce a new variable, say t. Extend all elements of $\Delta := \{\tau\} \cup \bigcup \Delta_i$ to derivations of $A[X][t]$ by putting $d(t) = 0$, for all $d \in \Delta$. Then by 6.1.3 there exists an elementary automorphism E of $A[X][t]$ such that

$$(\exp(a\tau), t) = E^{-1} \circ \exp(-t\tau) \circ E \circ \exp(t\tau).$$

So it suffices to show that $\exp(t\tau)$ is stably tame. Since $t\tau = \sum_{d \in \Delta_{m-1}} t b_d d$ and all elements $t b_d d$, $d \in \Delta_{m-1}$ commute pairwise (by 7.3.14) we get

$$\exp(t\tau) = \Pi_{d \in \Delta_{m-1}} \exp(t b_d d).$$

So it suffices to prove that each map $\exp(t b_d d)$ with $d \in \Delta_{m-1}$ is stably tame. However this follows from the induction hypothesis applied to the ring $A[t]$, since clearly $t b_d \in \ker d$ and $d \in \Delta_{m-1}$. □

To conclude this section we give another, more direct proof of 7.3.21 due to Hubbers and Wright in [158]. Their proof is based on the following lemma.

Lemma 7.3.23 *Let $H \in H(n, A)$, $d \in A$ and $a \in A^n$. Then $H(dX + a) \in H(n, A)$.*

Proof: We use induction on n. The case $n = 1$ is obvious, so let $n \geq 2$ and $H = (h, T, c)$. Then

$$H(dX + a) = \operatorname{Adj} T \begin{pmatrix} h(TdX + Ta) \\ 0 \end{pmatrix} + c.$$

Put $\tilde{a} := Ta$. Since $TdX = dTX$ we get

$$H(dX + a) = \operatorname{Adj} T \begin{pmatrix} h(dX + \tilde{a}) \\ 0 \end{pmatrix}_{|TX} + c = (h(dX + \tilde{a}), T, c).$$

So it suffices to show that $h(dX + \tilde{a}) \in H(n - 1, A[X_n])$. Therefore observe that since $h \in H(n-1, A[X_n])$ it follows from 7.2.10, applied to the ring homomorphism $A[X_n] \to A[X_n]$ sending X_n to $dX_n + \tilde{a}_n$, that

$$\tilde{h} := h(X_n = dX_n + \tilde{a}_n) \in H(n - 1, A[X_n]).$$

Applying the induction hypothesis to $\tilde{h}$ gives that $h(dX + \tilde{a}) = \tilde{h}(dX_1 + \tilde{a}_1, \ldots, dX_{n-1} + \tilde{a}_{n-1}) \in H(n - 1, A[X_n])$ as desired. □

Now we are able to prove

Theorem 7.3.24 *Let $F = X + H$ with $H \in H(n, A)$. Then there exist tame automorphisms R and S of $A[X_1, \ldots, X_n, Y_1, \ldots, Y_{n-1}]$ such that $R \circ F^{[n-1]} \circ S$ is of the form $(X, Y + H')$ where $H' \in H(n-1, A[X_1, \ldots, X_n])$ (with respect to $Y_1, \ldots, Y_{n-1}$).*

Proof: Let $H = (h, T, c), d := \det T$ and $TX := (L_1, \ldots, L_n)^t$. Put

$$R_1 := \begin{pmatrix} X - c \\ Y \end{pmatrix}, \quad R_2 := \begin{pmatrix} X - \operatorname{Adj} T \begin{pmatrix} Y \\ 0 \end{pmatrix} \\ Y \end{pmatrix},$$

$$S_1 := \begin{pmatrix} X \\ Y + h(X = TX) \end{pmatrix}, \quad S_2 := \begin{pmatrix} X + \operatorname{Adj} T \begin{pmatrix} Y \\ 0 \end{pmatrix} \\ Y \end{pmatrix},$$

$$R := R_2 \circ R_1 \text{ and } S := S_1 \circ S_2.$$

Then

$$R \circ F^{[n-1]} \circ S = \begin{pmatrix} X \\ Y + H' \end{pmatrix}$$

where $H' := h\left(X = TX + d\begin{pmatrix} Y \\ 0 \end{pmatrix}\right)$. So it suffices to prove that $H' \in H(n-1, A[X_1, \ldots, X_n])$ with respect to the variables $Y_1, \ldots, Y_{n-1}$. Therefore observe that $h \in H(n-1, A[X_n])$, whence $h(X_1 = Y_1, \ldots, X_{n-1} = Y_{n-1}) \in H(n-1, A[X_n])$ with respect to the variables $Y_1, \ldots, Y_{n-1}$. Now apply 7.2.10 iii) to the ring homomorphism $f : A[X_n] \to A[X_1, \ldots, X_n]$ sending X_n to L_n. So $\hat{h} := h(X_1 = Y_1, \ldots, X_{n-1} = Y_{n-1}, X_n = L_n) \in H(n-1, A[X_1, \ldots, X_n])$ with respect to the variables $Y_1, \ldots, Y_{n-1}$. Finally apply 7.3.23 to $\hat{h}(Y_1, \ldots, Y_{n-1})$ and the ring $A[X_1, \ldots, X_n]$. This gives that

$$\hat{h}(dY_1 + L_1, \ldots, dY_{n-1} + L_{n-1}) \in H(n-1, A[X_1, \ldots, X_n])$$

with respect to $Y_1, \ldots, Y_{n-1}$. However this last element is equal to

$$h(X_1 = dY_1 + L_1, \ldots, X_{n-1} = dY_{n-1} + L_{n-1}, X_n = L_n) = H'$$

as desired. □

Corollary 7.3.25 *Let F be as in 7.3.24. Then $F^{[\frac{n(n-1)}{2}]}$ is tame.*

Proof: Follows readily from 7.3.24 using that $(n-1) + (n-2) + \cdots + 2 + 1 = \frac{1}{2}(n-1)n$. □

Remark 7.3.26 The result in 7.3.25 is certainly not optimal. It is shown in [158] that $F^{[n-1]}$ is tame.

Exercises for §7.3

1. Prove the "if" part of 7.3.4.
2. Complete the proof of 7.3.14.

3. Let k be a field of characteristic zero. Let $H \in k[X_1, \ldots, X_4]^4$ be defined by

$$\begin{aligned} H_1 &= X_2X_4^2 + X_3^2X_4 + 2X_2X_3X_4 + X_1X_3X_4 + X_3^3 + X_2X_3^2 + X_1X_3^2, \\ H_2 &= X_3X_4^2 + X_3^2X_4 + X_2X_3X_4 + X_3^3 - X_2X_3^2 - X_1X_3^2, \\ H_3 &= X_4^3, \\ H_4 &= 0. \end{aligned}$$

(i) Show that $H \in H(4, k)$.
(ii) Find $(T, c) \in D(4, k)$ such that $H = \sum_{p=0}^{2} E_{4,p}(T, c) + c_4$. [To prove i) show that H_1, H_2, H_3 are linearly dependent over $k[X_4]$. Use this dependence relation to find $T_3 \in Sl_3(k[X_4])$ such that $T_3h = (f, 0)^t$, where $h = (H_1, H_2, H_3)^t$ and $f \in k[X_1, \ldots, X_4]^2$. Finally use 7.2.25.]

4. Let $D = a_1(X_2, X_3)\partial_1 + a_2(X_3)\partial_2 + a_3\partial_3$ be a triangular derivation on $A[X_1, X_2, X_3]$. Assume that $D^2(X_i) = 0$ for all i. Show that D is nice with respect to X_1, X_2, X_3 and $\Delta := \{\partial_1, \partial_2, \partial_3\}$.
5. Let k be a field of characteristic zero. Show that every locally nilpotent derivation D on $k[X_1, X_2]$ is equivalent to a nice derivation, i.e. there exists $\varphi \in \mathrm{Aut}_k\, k[X_1, X_2]$ such that $\varphi^{-1}D\varphi$ is nice with respect to X_1, X_2 and $\{\partial_1, \partial_2\}$. [Use 1.3.48]
6. (Derksen) Consider the 8-variable polynomial ring $\mathbb{C}^{[8]} := \mathbb{C}[P, X, Y, Z, A, B, C, Q]$ and on it the following three locally nilpotent derivations:

$$\begin{aligned} D_A &:= X\partial_P + C\partial_Y - B\partial_Z + Q\partial_A, \\ D_B &:= Y\partial_P + A\partial_Z - C\partial_X + Q\partial_B, \\ D_C &:= Z\partial_P + B\partial_X - A\partial_Y + Q\partial_C. \end{aligned}$$

(i) Let $R := AX + BY + CZ - PQ$. Show that Q and R are constants of D_A, D_B and D_C.
(ii) Put $D := R^2D_A + RQ^2D_B + Q^4D_C$. Show that $D^2P = D^2X = D^2Y = D^2Z = D^2A = D^2B = D^2C = DQ = 0$ and that D is homogeneous of degree 3.
(iii) Put $L := \mathbb{C}(X, Y, Z, A, B, C, Q)$ and let f be a linear form in $\mathbb{C}^{[8]}$ with $D(f) = 0$. Show that $f \in \mathbb{C}Q$.
(iv) It is proved by Z. Wang in [290] that any homogeneous derivation which is nice of some order $\leq m$ contains at least two (over $\mathbb{C}$) linearly independent linear forms in its kernel. Deduce that D is not nice. [Ad iii) Observe that $D_A(f)$, $D_B(f)$ and $D_C(f)$ belong to L and that R is algebraically independent over L. So $D_A(f) = D_B(f) = D_C(f) = 0$.]

7.4 Strongly nilpotent Jacobian matrices

To conclude this chapter we consider a special class of nilpotent Jacobian matrices, the so-called strongly nilpotent ones. This notion was introduced by Meisters and Olech in [213] in order to simplify expressions for the inverse of a quadratic homogeneous polynomial map. They studied the question whether for quadratic homogeneous polynomial maps $X + Q : \mathbb{R}^n \to \mathbb{R}^n$ the nilpotency of JQ implies the so-called strong nilpotency of JQ, i.e. the property that $JQ(v_1) \dots JQ(v_n) = 0$ for all vectors $v_i \in \mathbb{R}^n$. They showed that the answer to this question is yes if $n \leq 4$ and presented counterexamples, due to Vasyunin, in case $n \geq 5$.

In [113] Hubbers and the author generalized the notion of strong nilpotency to polynomial maps $X + H : k^n \to k^n$, where k is a field of characteristic zero and H an arbitrary polynomial map, which need not be homogeneous. We showed that strong nilpotency of JH is equivalent to so-called **linear triangulability**, i.e. the existence of an invertible matrix T over k such that $J(T^{-1}HT)$ is upper triangular with zeros on the main diagonal 7.4.4. At the end of this section we discuss some problems and conjectures concerning quadratic and cubic homogeneous polynomial automorphisms.

Throughout this section k denotes an arbitrary field of characteristic zero and $H = (H_1, \dots, H_n) \in k[X]^n$. To define the notion of strong nilpotency we introduce n sets of n variables

$$Y_{(1)} = (Y_{(1),1}, \dots, Y_{(1),n}), \dots, Y_{(n)} = (Y_{(n),1}, \dots, Y_{(n),n}).$$

So for each i, $JH(Y_{(i)})$ belongs to the ring of $n \times n$ matrices with entries in the n^2 variable polynomial ring $k[Y_{(i),j};\ 1 \leq i, j \leq n]$.

Definition 7.4.1 The Jacobian matrix JH is called **strongly nilpotent** if and only if $JH(Y_{(1)}) \dots JH(Y_{(n)})$ is the zero matrix.

Example 7.4.2 If JH is upper triangular with zeros on the main diagonal, then one readily verifies that JH is strongly nilpotent. The main result of this section, 7.4.4 below, asserts that up to a linear coordinate change the converse is true!

Remark 7.4.3 It is left to the reader to verify (Exercise 1) that the notion of strong nilpotency is equivalent to

$$JH(v_1) \dots JH(v_n) = 0$$

for all vectors $v_1, \dots, v_n$ in k^n. So for $k = \mathbb{R}$ and H quadratic homogeneous we obtain the strong nilpotency introduced by Meisters and Olech.

Theorem 7.4.4 (van den Essen, Hubbers). *There is equivalence between* i) JH *is strongly nilpotent.*

ii) *There exists* $T \in Gl_n(k)$ *such that* $J(T^{-1}HT)$ *is upper triangular with zeros on the main diagonal.*

The proof of this result is based on the following two lemmas.

Lemma 7.4.5 *Let* $JH = \sum_{|\alpha| \leq d} A_\alpha X^\alpha$ *where* $d = \max_i(\deg H_i - 1)$ *and* $A_\alpha \in M_n(k)$ *for all* α. *Then* JH *is strongly nilpotent if and only if* $A_{\alpha_{(1)}} \dots A_{\alpha_{(n)}} = 0$, *for all multi-indices* $\alpha_{(i)}$ *with* $|\alpha_{(i)}| \leq d$.

Proof: By 7.4.1 we obtain

$$\left(\sum_{|\alpha(1)|\le d} A_{\alpha(1)} Y_{(1)}^{\alpha(1)}\right) \cdots \left(\sum_{|\alpha(n)|\le d} A_{\alpha(n)} Y_{(n)}^{\alpha(n)}\right) = 0.$$

The result then follows by looking at the coefficients of the monomials $Y_{(1)}^{\alpha(1)} \cdots Y_{(n)}^{\alpha(n)}$. □

Lemma 7.4.6 *Let V be a finite dimensional k-vector space and $l_1, \ldots, l_p$ k-linear maps from V to V. Let $r \in \mathbb{N}$, $r \ge 1$. If $l_{i_1} \circ \cdots \circ l_{i_r} = 0$ for each r-tuple $l_{i_1}, \ldots, l_{i_r}$ with $1 \le i_1, \ldots, i_r \le p$, then there exists a basis (v) of V such that $Mat(l_i, (v)) = D_i$, where D_i is an upper triangular matrix with zeros on the main diagonal.*

Proof: Let $d := \dim V$. We use induction on d. First let $d = 1$. Then the hypothesis implies that $l_i^r = 0$ for each i. So $l_i = 0$ for each i, and we are done. So let $d > 1$, and assume that the assertion is proved for all $(d-1)$-dimensional vector spaces. Now we also use induction on r. If $r = 1$ then each l_i is zero. So let $r \ge 2$. Then for each $(r-1)$-tuple $l_{i_2}, \ldots, l_{i_r}$, with $1 \le i_2, \ldots, i_r \le p$ we have

$$\text{(7.4.7)} \qquad l_1 l_{i_2} \ldots l_{i_r} = 0, \ldots, l_p l_{i_2} \ldots l_{i_r} = 0.$$

If $l_{i_2} \ldots l_{i_r} = 0$ for each such $(r-1)$-tuple we are done by the induction hypothesis on r. So we may assume that for some $(r-1)$-tuple $l_{i_2}, \ldots, l_{i_r}$ there exists $0 \ne v \in V$ with $v_1 := l_{i_2} \ldots l_{i_r} v \ne 0$. From 7.4.7 we deduce that $l_i v_1 = 0$ for all i. Then consider $\overline{V} := V/kv_1$. Since $l_i v_1 = 0$ for all i, we get induced k-linear maps $\overline{l_i} : \overline{V} \to \overline{V}$. Since $\dim \overline{V} = d - 1$, the induction hypothesis implies that there exist $v_2, \ldots, v_d$ in V such that $(\overline{v_2}, \ldots, \overline{v_d})$ is a k-basis of $\overline{V}$ and $Mat(\overline{l_i}, \overline{v_2}, \ldots, \overline{v_d}))$ is of upper triangular form with zeros on the main diagonal. Then $(v) := (v_1, v_2, \ldots, v_d)$ is as desired. □

Corollary 7.4.8 *Let $A_1, \ldots, A_p \in M_n(k)$. Let $r \in \mathbb{N}, r \ge 1$. If $A_{i_1} \ldots A_{i_r} = 0$ for each r-tuple $A_{i_1}, \ldots, A_{i_r}$ with $1 \le i_1, \ldots, i_r \le p$, then there exists $T \in Gl_n(k)$ such that $T^{-1} A_i T = D_i$, where D_i is an upper triangular matrix with zeros on the main diagonal.*

Now we are able to give

Proof of Theorem 7.4.4: ii) → i) follows directly from 7.4.2 and the fact that

$$J(T^{-1}HT) = T^{-1} JH(TX)T.$$

So it remains to prove i) → ii). Let JH be strongly nilpotent. Write

$$JH = \sum_{|\alpha|\le d} A_\alpha X^\alpha.$$

Then by 7.4.5 $A_{\alpha(1)} \ldots A_{\alpha(n)} = 0$ for all n-tuples with $|\alpha(i)| \leq d$, where $\deg H = d + 1$. So by 7.4.8 there exists $T \in Gl_n(k)$ such that $T^{-1}A_\alpha T = D_\alpha$ for all α with $|\alpha| \leq d$, where D_α is an upper triangular matrix with zeros on the main diagonal. Consequently so is $T^{-1}JH(X)T$ $(= \sum T^{-1}A_\alpha T X^\alpha)$, and hence so is $J(T^{-1}HT) = T^{-1}JH(TX)T$, which is obtained by replacing X by TX in $T^{-1}JH(X)T$. □

Now we address the question: under what conditions is a nilpotent Jacobian matrix strongly nilpotent? In other words, when is "nilpotent" equivalent to "strongly nilpotent"?

As already described in the introduction of this section the following result was obtained by Meisters and Olech. We refer to [213] for the elementary but tedious proof.

Proposition 7.4.9 (Meisters, Olech). *Let $n \leq 4$ and $H : k^n \to k^n$ a quadratic homogeneous polynomial map. Then JH is strongly nilpotent if and only if JH is nilpotent.*

However if $n \geq 5$ the statement of 7.4.9 is no longer true, as can be seen from the following example.

Example 7.4.10 (Vasyunin). Let $n \geq 5$, $H = (H_1, \ldots, H_n)$, where $H_i = 0$ for all $i > 5$ and

$$H_1 = 0,\ H_2 = X_1X_3,\ H_3 = X_1X_4 + \frac{1}{2}X_2^2,$$
$$H_4 = X_1X_5 - X_2X_3,\ H_5 = \frac{1}{2}X_3^2.$$

Then one easily verifies that JH is nilpotent. However $JH(e_1)JH(e_2)$ is the matrix with second column equal to e_2, third column $-e_3$ and all other columns zero. So this matrix is clearly not nilpotent, and hence JH is not strongly nilpotent.

For cubic homogeneous maps the situation is the following.

Proposition 7.4.11 (Wright). i) *If $n \leq 3$ and $H : k^n \to k^n$ is cubic homogeneous, then JH is strongly nilpotent if and only if JH is nilpotent.*

ii) *Let $n \geq 4$. Put $d := X_3X_1 + X_4X_2$. Then*

$$H = (X_4d, -X_3d, 0, \ldots, 0)$$

satisfies: JH is nilpotent, however JH is not strongly nilpotent.

Proof: i) follows from 7.1.1. Since $(H_1, H_2) \in H(2, k[X_3, X_4, \ldots, X_n])$ it follows that JH is nilpotent. Furthermore one easily verifies that the first column of $JH(e_4)JH(e_3)$ equals $-e_1$, so this matrix is not nilpotent and hence JH is not strongly nilpotent. □

Finally we leave it to the reader to prove (Exercise 3).

Proposition 7.4.12 i) *If $n \leq 2$ and $H : k^n \to k^n$ is any polynomial map, then JH is strongly nilpotent if and only if JH is nilpotent.*

ii) *If* $n \geq 3$, *then* $H = (d_1, -X_3 d_1, 0, \ldots, 0)$, *where* $d_1 := X_3 X_1 + X_2$, *satisfies: JH is nilpotent but JH is not strongly nilpotent.*

To conclude this section we mention several conjectures and problems concerning (strongly) nilpotent Jacobian matrices.

To describe the first conjecture, let us recall that for all polynomial automorphisms F of k^n the degree of the inverse is bounded by $(\deg F)^{n-1}$. Suppose now that $F = X + H$ is a cubic homogeneous polynomial automorphism. Then JH is nilpotent since $\det JF = 1$. Let e denote the **nilpotence index of** JH, i.e. $JH^e = 0$ and $JH^{e-1} \neq 0$.

Conjecture 7.4.13 (Drużkowski, Rusek). $\deg F^{-1} \leq (\deg F)^{e-1}$.

Using Hubbers' theorem it is not difficult to verify that this conjecture is true for all $n \leq 4$. However if $n \geq 5$ there exist counterexamples $F = X + H$ with JH strongly nilpotent and in fact of triangular form.

Example 7.4.14 (van den Essen). Let $n \geq 5$ and

$$H = (3X_4^2 X_2 - 2X_3 X_4 X_5, X_4^2 X_5, X_4^3, X_5^3, 0, \ldots, 0).$$

Then $F = X + H$ is a counterexample to 7.4.13.

Proof: One easily verifies that $(JH)^3 = 0$ and $(JH)^2 \neq 0$. So $e = 3$, hence $(\deg F)^{e-1} = 9$. However, for example, by 3.2.1 one readily verifies that $\deg F^{-1} = 13$. □

Also for quadratic homogeneous automorphisms a statement similar to 7.4.13 is false.

Example 7.4.15 (Furter). Let $n \geq 6$ and H given by

$$H = (2X_2 X_6 - 2X_3^2 - X_4 X_5, 2X_3 X_5 - X_4 X_6, X_5 X_6, X_5^2, X_6^2, 0, \ldots, 0).$$

Then $(JH)^3 = 0$ and $(JH)^2 \neq 0$. However $F = X + H$ satisfies $\deg F^{-1} = 6 > 2^{3-1} = 4$.

Finally we return to the dependence problem which formed the starting point of this chapter. From 7.4.4 it is clear that if JH is strongly nilpotent, then the rows of JH are linearly dependent over k. However for quadratic homogeneous maps we do not know the answer to the dependence problem, i.e. to

Question 7.4.16 *Let H be quadratic homogeneous and JH nilpotent. Are the rows of JH linearly dependent over k?*

It would be interesting to know the answer to this question, since it would shed some light on the following problem (see also Exercise 5).

Problem 7.4.17 (Meisters, Olech). Let H be quadratic homogeneous. Does $(JH)^3 = 0$ imply that JH is strongly nilpotent?

In fact in [134] Furter showed that for all $n \leq 6$ the answer to this problem is yes, if the answer to 7.4.16 is yes.

Also it would be interesting to know the answer to 7.4.16 in connection with Rusek's conjecture (Chapter 5, §2), which asserts that every quadratic homogeneous automorphism is tame; in fact it is claimed by the author in [255] that a positive answer to the dependence problem for quadratic linear automorphisms, i.e. maps of the form $(X_1 + l_1^2, \ldots, X_n + l_n^2)$, where each l_i is a linear form, would imply the truth of Rusek's conjecture. However the unpublished "proof" contained an error, so this implication is still open. Nevertheless for low dimensional cases one can show that a positive answer to 7.4.16 does imply the truth of Rusek's conjecture for those cases.

Finally we like to mention a very recent result of Cheng in [50]. We use the notation as in Definition 6.4.1.

Theorem 7.4.18 (Cheng). *Let $F_A: k^n \to k^n$ be a Drużkowski mapping with $\det JF_A = 1$. If $\operatorname{rk} A \leq 2$ or $n - \operatorname{rk} A \leq 2$ then F_A is linearly triangularizable.*

In other words, $J(F_A - X)$ is strongly nilpotent by 7.4.4.

Remark 7.4.19 Theorem 7.4.18 is an improvement of an earlier result of Drużkowski in [95], who showed that under the same hypothesis F_A is tame. Combining this result with Hubbers' Theorem and the Gorni-Zampieri pairing one easily deduces that the Jacobian Conjecture holds for all Drużkowski mappings in case $n \leq 7$.

Exercises for §7.4

1. Show that JH is strongly nilpotent if and only if $JH(v_1)\ldots JH(v_n) = 0$ for all vectors $v_i \in k^n$. [Observe that k is infinite since k has characteristic zero and use that a polynomial $p(Y_1, \ldots, Y_m) \in k[Y_1, \ldots, Y_m]$ is zero if $p(y) = 0$ for all $y \in k^m$.]
2. Let $H \in k[X]^n$ be such that JH is nilpotent. Define the **strong nilpotency index of** JH, denoted $i_{SN}(JH)$, to be the supremum of all positive integers e with $JH(v_1)\ldots JH(v_e) = 0$ for all vectors $v_i \in k^n$. Show that $i_{SN}(JH) = \infty$ or $i_{SN}(JH) \leq n$. See also Notes at the end of this chapter. [Use 7.4.4.]
3. Give a proof of 7.4.12. [Use 7.2.25.]
4. Verify the statements of 7.4.15.
5. Let $H \in k[X]^n$ be quadratic homogeneous.
   (i) Show that $JH(X)Y = JH(Y)X$, where $Y = (Y_1, \ldots, Y_n)^t$ is a vector consisting of n new variables Y_i.
   (ii) Deduce from i) that if $(JH)^2 = 0$, then JH is strongly nilpotent. [Show first that $JH(X)JH(Y) + JH(Y)JH(X) = 0$. Then apply this to the vector Y and use i) and the Euler identity $JH(Y)Y = 2H(Y)$ to obtain $JH(X)H(Y) = 0$. Then differentiate with respect to Y.]
   (iii) Let $d \geq 4$. Show that $(JH)^d = 0$ does not imply that JH is strongly nilpotent. [Consider the map $H = (X_2X_5 - X_3X_4, X_1X_3, 0, X_1X_5, 0, \ldots, 0)$.]

6. Let $H \in k[X]^n$ be such that JH is strongly nilpotent. Put $F := X + H$. Show that F is **linearly triangulable**, i.e. that there exists $T \in Gl_n(k)$ such that $T^{-1}FT$ is of triangular form (Exercise 7 of the introduction of this book) and hence that F is tame.
7. Let H be as in 7.4.10. Show that $F := X + H$ is tame. [Multiply F from the right by suitable elementary maps.]
8. Let $F = X + H$ be defined by:

$$F := \left(\frac{1}{2}x_7^2 + x_1, x_8x_7 + x_2, \frac{1}{2}x_8^2 + x_3, -x_8x_2 + x_7x_3 + x_4,\right.$$

$$\left. x_7x_4 + x_8x_6 + x_5, x_8x_1 + x_6, x_7, x_8\right).$$

Then

$$JH := \begin{pmatrix} 0 & 0 & 0 & 0 & 0 & 0 & x_7 & 0 \\ 0 & 0 & 0 & 0 & 0 & 0 & x_8 & x_7 \\ 0 & 0 & 0 & 0 & 0 & 0 & 0 & x_8 \\ 0 & -x_8 & x_7 & 0 & 0 & 0 & x_3 & -x_2 \\ 0 & 0 & 0 & x_7 & 0 & x_8 & x_4 & x_6 \\ x_8 & 0 & 0 & 0 & 0 & 0 & 0 & x_1 \\ 0 & 0 & 0 & 0 & 0 & 0 & 0 & 0 \\ 0 & 0 & 0 & 0 & 0 & 0 & 0 & 0 \end{pmatrix}.$$

Show that $JH(Y_{(1)})^2 \cdot JH(Y_{(2)}) \neq 0$.

Notes

The first proof of 7.1.1 was given by Moh and Sathaye (unpublished). In fact they showed, using a computer, that in dimension 3 each polynomial map of degree ≤ 3 satisfying the Jacobian condition is invertible. Their computer output was about 100 pages. A conceptual proof of this result using methods of algebraic geometry was given by Vistoli in [288]. The origin of the class $H(n, A)$ was the observation, made by several authors, that in all cases known until then the rows of nilpotent Jacobian matrices (over a field k) are linearly dependent over k. For cubic homogeneous maps in four variables this was shown to be true by Hubbers [156], 7.1.4. For cubic homogeneous maps in any dimension this dependence property was explicitly stated as a conjecture by Meisters in [210], Conjecture 1, page 80. The same conjecture was made by Olech in [236], Conjecture B, page 135, where he offered a bottle of Polish vodka for either a proof or a counterexample of his conjecture. For quadratic linear maps the dependence property was stated as Conjecture 11.3 by Rusek in [255]. The generalized dependence problem was stated as a conjecture by Cima, Gasull and Mañosas in [56]. The first part of Theorem 7.1.7 is taken from [29], while the second part of this theorem is a slightly generalized version of the author's paper [110].

The results of §2 and 3 are taken from the papers [114] and [116] of Hubbers and the author. Exercise 9 at the end of §2 include the main results of Peretz' paper [241]. His proof relies on the study of the rational solutions of the so-called Burger's equation with no viscosity, which models the behaviour of gas (see [132]).

The proof of Theorem 7.3.4 is a simplified version of the one given in [116]. The simplified version of Theorem 7.3.21 given at the end of §3 is due to Hubbers and Wright [158].

The notion of strong nilpotence was introduced by Meisters and Olech in [213] and generalized by Hubbers and the author in [113] where Theorem 7.4.4 was proved. Another proof of this result, based on a result of Levitzki in [192] was given by Yu in [302]. Example 7.4.14 comes from the author's paper [109]. The quadratic homogeneous counterexample 7.4.15 is taken from Furter's thesis [134] in which he studies quadratic homogeneous automorphisms in connection with Problem 7.4.17.

Part II
Applications

8 Applications of polynomial mappings to dynamical systems

In the previous chapters we described several results concerning polynomial automorphisms and in particular in Chapter 7 we presented a large class of such automorphisms.

In this and the next chapter we will describe two topics to which results on polynomial automorphisms can be applied: dynamical systems and group actions. The aim of this section is to describe the impact of polynomial automorphisms on some problems concerning dynamical systems.

Stability theory of autonomous systems of differential equations started with the pioneering work of Lyapunov in his thesis around 1892. Later, several attempts were made to globalize his (local) results. One of those attempts is the Markus-Yamabe Conjecture, which was formulated in 1960. In the same spirit LaSalle in 1976 made an analogous conjecture for discrete dynamical systems. The Markus-Yamabe Conjecture in dimension two was finally solved affirmatively in 1993 independently by Gutierrez and Fessler. Furthermore, in dimension ≥ 4 analytic counterexamples were given by Bernat and Llibre in 1994 (cf [33]). It was observed by Fournier and Martelli in an unpublished manuscript that a positive solution of the Markus-Yamabe Conjecture for polynomial vector fields would imply the Jacobian Conjecture. So the interest in the Markus-Yamabe Conjecture extended beyond the interest of dynamical systems. However in 1995 the author in collaboration with Hubbers, Cima, Gasull and Mañosas found polynomial counterexamples to the Markus-Yamabe Conjecture in any dimension ≥ 3. These examples were based on results of the previous chapter. The complete story of how these examples were found is described in §3 of this chapter. In §1 we supply the necessary background concerning dynamical systems and in §2 we study the two dimensional case of both the Markus-Yamabe Conjecture and the LaSalle Conjecture. Finally in §4 we discuss Meisters' cubic-linear linearization conjecture, which is a refinement of an earlier conjecture of Meisters which played a crucial role in the final solution of the Markus-Yamabe conjecture as described in §3. Based on results of Chapter 6 we are able to give a counterexample to Meisters cubic-linear linearization conjecture (in dimension 17). As a byproduct we find quadratic and cubic homogeneous counterexamples to the Markus-Yamabe Conjecture in any dimension ≥ 5.

8.1 The Markus-Yamabe conjecture and a problem of LaSalle; some background

We start by recalling some well-known facts concerning stability of equilibria in dynamical systems.

Let $F : \mathbb{R}^n \to \mathbb{R}^n$ be a C^1-vector field with $F(0) = 0$. Consider the system of ordinary differential equations

$$\begin{cases} \dot{x_1}(t) &= F_1(x_1(t), \dots, x_n(t)), \\ & \vdots \\ \dot{x_n}(t) &= F_n(x_1(t), \dots, x_n(t)), \end{cases}$$

for short

$$\dot{x}(t) = F(x(t)). \tag{8.1.1}$$

Observe that $x(t) = 0$ is a solution of 8.1.1, which we call an **equilibrium** of 8.1.1. We say that the origin is **locally asymptotically stable** or that 0 is a **local attractor** of 8.1.1 if every solution of 8.1.1 which starts close enough to 0 is defined for all $t > 0$ and tends to 0 if t tends to infinity. More precisely, for every $\varepsilon > 0$ there exists $\delta > 0$ such that if $||x(0)|| < \delta$, then $||x(t)|| < \varepsilon$ for all $t > 0$ and $\lim_{t\to\infty} x(t) = 0$. The set of all $x_0 \in \mathbb{R}^n$ for which the solution satisfying $x(0) = x_0$ tends to 0 if t tends to infinity is called the **basin of attraction** of 8.1.1. If this is all of $\mathbb{R}^n$, then 0 is called **globally asymptotically stable** or a **global attractor**. In other words, 0 is a global attractor of 8.1.1 if every solution of this system tends to 0 if t tends to infinity.

Example 8.1.2 Let $F : \mathbb{R}^n \to \mathbb{R}^n$ be the linear map given by the matrix $A \in M_n(\mathbb{R})$. Then every solution of 8.1.1 is of the form

$$x(t) = e^{tA}x(0).$$

One deduces easily that 0 is a global attractor of 8.1.1 if and only if the real parts of all eigenvalues of A are negative.

This result was extended by Lyapunov to a local stability result. A weak version of his result is

Proposition 8.1.3 (Lyapunov). *Let $F : \mathbb{R}^n \to \mathbb{R}^n$ be a C^1-vector field such that $F(0) = 0$. If the real parts of all eigenvalues of $JF(0)$ are negative, then 0 is a local attractor of* 8.1.1.

We refer to [155] for a proof of this result.

Example 8.1.4 (The pendulum). Let us consider a pendulum moving in the vertical plane with constant mass m hanging on a rod of constant length l whose mass is neglected. If $\theta(t)$ is the counterclockwise angle from the vertical to the rod at time t, then one can deduce (see for example [155], page 183) that θ satisfies the following non-linear differential equation

$$\theta^{(2)} = -\frac{k}{m}\theta^{(1)} - \frac{1}{l}sin(\theta)$$

where k is some positive constant due to friction. So introducing a new variable ω, which we can interpret as angular velocity, we get the following first order system

$$\theta^{(1)} = \omega,\ \omega^{(1)} = -\frac{1}{l}sin(\theta) - \frac{k}{m}\omega.$$

So in the notation used above

$$F_1(\theta, \omega) = \omega,\ F_2(\theta, \omega) = -\frac{1}{l} sin(\theta) - \frac{k}{m}\omega.$$

One readily verifies that the real parts of the eigenvalues of $JF(0,0)$ are negative. So by Lyapunov's result (0, 0) is locally asymptotically stable. This of course agrees with what we know from experience: a pendulum which is moving will finally go to the equilibrium (0, 0), except if we start for example in the vertical position with angular velocity 0, i.e. at the point $(\pi, 0)$. This point is also an equilibrium of our system, but not a stable one. If we inspect the Jacobian matrix of F at $(\pi, 0)$ we find that one of its eigenvalues is positive. These and other considerations led Markus and Yamabe in 1960 to make their famous global stability conjecture (see [201]).

Markus-Yamabe Conjecture (MYC). Let $F : \mathbb{R}^n \to \mathbb{R}^n$ be a C^1-vector field with $F(0) = 0$ satisfying the so-called **Markus-Yamabe Assumption**, i.e. (MYA): for all $x \in \mathbb{R}^n$ the real parts of all eigenvalues of $JF(x)$ are negative then 0 is a global attractor of 8.1.1.

In their paper they proved the conjecture for $n = 2$ in case one of the four entries of JF is identically zero, i.e. one of the coordinates of F depends only on one variable. In particular this is the case when 8.1.1 is equivalent to a second order autonomous ODE.

Before the paper of Markus and Yamabe was published Hartman [153], generalizing a result of Krasovski, proved for arbitrary n that 0 is a global attractor of 8.1.1 if **the symmetric part of** $JF(x)$, i.e. the matrix $JF(x) + JF(x)^t$, has all its eigenvalues negative for all $x \in \mathbb{R}^n$.

Let us assume for a moment that the MYC is true. Then any C^1-vector field $F : \mathbb{R}^n \to \mathbb{R}^n$ satisfying MYA must be injective: namely suppose that $F(a) = F(b)$ for some $a \neq b$ in $\mathbb{R}^n$. Then put $G(X) := F(X+a) - F(a)$. Observe that $G(0) = 0$ and that G satisfies MYA. So our assumption implies that 0 is a global attractor of $\dot{x} = G(x)$, contradicting that the constant solution $x(t) = b - a \neq 0$ does not tend to 0 if t tends to infinity! Summarizing we showed: if MYC is true, then the following conjecture is true as well.

Weak Markus-Yamabe Conjecture. Let $F : \mathbb{R}^n \to \mathbb{R}^n$ be a C^1-vector field satisfying MYA. Then F is injective.

It was C. Olech who in [235] showed that in case $n = 2$ the weak MYC is in fact equivalent to MYC. He showed that a kind of "weak properness" of the map F implies global asymptotic stability. More precisely

Theorem 8.1.5 (Olech). Let $F : \mathbb{R}^2 \to \mathbb{R}^2$ be a C^1-vector field with $F(0) = 0$ satisfying MYA. If there exist constants $\varepsilon > 0$ and $r > 0$ such that

$$||F(x)|| \geq r \text{ if } ||x|| \geq \varepsilon$$

then 0 is a global attractor of 8.1.1.

We refer to [235] for a proof of this result.

Corollary 8.1.6 *Let $n = 2$. Then the weak MYC implies MYC.*

Proof: The MYA implies that $\det JF(x) > 0$ for all $x \in \mathbb{R}^2$. So by the local inverse function theorem it follows that F is a local homeomorphism. In particular there exists $\varepsilon > 0$ such that the open ball $||x|| < \varepsilon$ is mapped bijectively by F on an open neighborhood of $F(0) = 0$ in $\mathbb{R}^2$, whence it therefore contains an open ball, say $||y|| < r$. Since F is injective this means that no point outside the ball $||x|| < \varepsilon$ can be mapped into the ball $||y|| < r$. In other words, if $||x|| \geq \varepsilon$ then $||F(x)|| \geq r$. Then use Olech's theorem. □

Using Olech's theorem, Meisters and Olech in [212], in 1987 proved the two dimensional MYC for **polynomial vector fields**. We will give a version of their proof based on 1.1.32. As already remarked in the introduction of this chapter the two-dimensional C^1-case was proved independently by Fessler in [123] and Gutierrez in [147], both in 1993. See [140] for yet another proof due to Glutsuk.

Apart from the importance of the MYC for the field of dynamical systems, and hence for its applications to physics, chemistry, biology, economics etc., there was another reason for studying the MYC. A surprising relationship with the Jacobian Conjecture was discovered by Fournier and Martelli in an unpublished manuscript. They showed that if the MYC is true for all $n \geq 2$ and all polynomial vector fields, then the Jacobian Conjecture is true. To prove this result we need the following lemma, the proof of which is already given above just before the formulation of the Weak Markus-Yamabe Conjecture, but now restricted to polynomial maps of degree ≤ 3.

Lemma 8.1.7 *Let $n \geq 1$. If the MYC is true for all polynomial vector fields of $\mathbb{R}^n$ with degree ≤ 3, then every such vector field satisfying the MYA is injective.*

Proposition 8.1.8 (Fournier, Martelli). *If the MYC is true for all polynomial vector fields of $\mathbb{R}^n$ with degree ≤ 3 and all $n \geq 2$, then the Jacobian Conjecture is true.*

Proof: By 6.3.1 it suffices to prove the Jacobian Conjecture for all $n \geq 2$ and all polynomial maps $F : \mathbb{C}^n \to \mathbb{C}^n$ of the form

$$F = (-X_1 + H_1, \ldots, -X_n + H_n)$$

where each H_i is either zero or homogeneous of degree 3 and JH is nilpotent. As in 6.3.13 consider the polynomial map $\tilde{F} : \mathbb{R}^{2n} \to \mathbb{R}^{2n}$ defined by

$$\tilde{F} = (ReF_1, ImF_1, \ldots, ReF_n, ImF_n).$$

So we have $\tilde{F} = -\tilde{X} + \tilde{H}$, where $\tilde{H}$ is homogeneous of degree 3. Since JH is nilpotent, $JF = -I + JH$ is invertible by 6.2.11 and $\det J\tilde{F} = |\det JF|^2 = 1$, whence $J\tilde{F}$ is invertible. So by 6.2.11, $J\tilde{H}$ is nilpotent and consequently for each $x \in \mathbb{R}^{2n}$ all eigenvalues of $J\tilde{F} = -I + J\tilde{H}(x)$ are equal to -1 (Exercise 1). Hence the vector field $\tilde{F}$ satisfies the MYA. So by 8.1.7, $\tilde{F} : \mathbb{R}^{2n} \to \mathbb{R}^{2n}$ is injective and hence so is $F : \mathbb{C}^n \to \mathbb{C}^n$. But then it follows from 4.2.1 that F is invertible, as desired. □

In fact the argument above even gives

Corollary 8.1.9 *If the weak MYC is true for all $n \geq 2$ and all polynomial maps of degree ≤ 3, then the Jacobian Conjecture is true.*

A discrete version of MYC was formulated by LaSalle in 1976, [191].

LaSalle problem. Let $F : \mathbb{R}^n \to \mathbb{R}^n$ be a C^1-map satisfying the so-called **Discrete Markus-Yamabe Assumption (DMYA)**, i.e. (DMYA): for all $x \in \mathbb{R}^n$ all eigenvalues of $JF(x)$ have absolute value smaller than one.

Does $F(0) = 0$ imply that 0 is a **global attractor of the map** F, i.e. for each $x \in \mathbb{R}^n$ $F^k(x) \to 0$ if $t \to \infty$?

This problem was independently posed and studied by Cima, Gasull and Mañosas in 1995 [56], which finally appeared in [59]. They called it the **Discrete Markus-Yamabe problem** (DMYP). At that time W. Szlenk, working with them in Barcelona, found a rational counterexample to DMYP. Therefore they restricted their attention to polynomial maps. They made the trivial, but useful, observation that if the answer to DMYP for polynomial maps is affirmative, then DMYA implies that 0 is the only fixed point. This and the result of Exercise 2 below has led them to the following.

Fixed point conjecture. For all $n \geq 1$ every polynomial map $F : \mathbb{R}^n \to \mathbb{R}^n$ satisfying DMYA has a unique fixed point. In fact it is not difficult to verify (Exercise 3) that the fixed point conjecture is equivalent to its complex version, i.e.

Fixed point conjecture. For all $n \geq 1$ every polynomial map $F : \mathbb{C}^n \to \mathbb{C}^n$ satisfying DMYA has a unique fixed point.

Now the remarkable point is

Theorem 8.1.10 (Cima, Gasull, Mañosas). *The fixed point conjecture is equivalent to the Jacobian Conjecture.*

Proof: i) Assume that the Jacobian Conjecture holds. We show that the fixed point conjecture holds. Therefore let $F : \mathbb{C}^n \to \mathbb{C}^n$ be a polynomial map satisfying DMYA. Then by 8.1.11 below the characteristic polynomial of $JF(x)$ is independent of $x \in \mathbb{C}^n$. Consequently if we put $G(X) := F(X) - X$, then also the characteristic polynomial of $JG(x)$ is independent of $x \in \mathbb{C}^n$. Consequently $\det JG(x)$ is independent of x, i.e. $\det JG \in \mathbb{C}$. Since the eigenvalues of $JF(x)$ are smaller than 1 in absolute value, the eigenvalues of $JG(x)$ $(= JF(x) - I)$ are all non-zero, so $\det JG \in \mathbb{C}^*$. So by our assumption G is invertible and hence surjective. So $G(x_0) = 0$ for some $x_0 \in \mathbb{C}^n$, i.e. $F(x_0) = x_0$ and x_0 is unique since G is injective.

ii) Assume that the fixed point conjecture holds. By 6.3.1 and 4.2.1 it suffices to prove that for all $n \geq 2$ all maps of the form $F = X + H$ with JH nilpotent and H cubic homogeneous are injective. Therefore suppose that $F(y) = F(z)$ for some $z \neq y \in \mathbb{C}^n$. Put

$$G(X) := \frac{1}{2}(X - H(X) + F(y)).$$

The nilpotency of JH implies that for each x in $\mathbb{C}^n$ all eigenvalues of $JG(x)$ $(= \frac{1}{2}(I - JH(x))$ are equal to $\frac{1}{2}$. So from our assumption we deduce that G has only one fixed point. However $G(y) = y$ and $G(z) = \frac{1}{2}[z - H(z) + (z + H(z))] = z$ (since $F(y) = F(z)$). So G has at least two fixed points, a contradiction. □

Lemma 8.1.11 *Let $k = \mathbb{R}$ or $\mathbb{C}$ and $F : k^n \to k^n$ a polynomial map such that for all $x \in k^n$ the absolute values of all eigenvalues of $JF(x)$ are bounded by some positive number independent of x. Then the characteristic polynomial of $JF(x)$ does not depend on x.*

Proof: The coefficients of the characteristic polynomial of $JF(x)$ are the elementary symmetric polynomials in the eigenvalues of $JF(x)$ and hence they are bounded by some positive number independent of x. But the only bounded polynomials are constants. □

Remark 8.1.12 For a slightly different fixed point conjecture equivalent to the Jacobian Conjecture see Exercise 4 below.

Exercises for §8.1

1. Let k be any field, $\lambda \in k$ and $N \in M_n(k)$ a nilpotent matrix. Show that the eigenvalues of $\lambda I_n + N$ are all equal to λ.
2. Let $F : \mathbb{R}^2 \to \mathbb{R}^2$ be a C^1-map satisfying DMYA. Show that F has at most one fixed point. [Use that the weak MYC is true in case $n = 2$ and consider $G(X) := F(X) - X$.]
3. (i) Let $A, B \in M_n(\mathbb{R})$. Show that every eigenvalue λ of $\begin{pmatrix} A & -B \\ B & A \end{pmatrix}$ is also an eigenvalue of $A + iB$.
   (ii) Show that the complex version of the fixed point conjecture follows from the real version. [Let $F : \mathbb{C}^n \to \mathbb{C}^n$ be a polynomial map satisfying DMYA. Consider $\tilde{F} : \mathbb{R}^{2n} \to \mathbb{R}^{2n}$ as in 6.3.13. Use i) to show that every eigenvalue of $J\tilde{F}(x)$ is also eigenvalue of $JF(z)$, where for $z \in \mathbb{C}^n$ the corresponding point in $\mathbb{R}^{2n}$ is denoted by x.]
4. (Kulikov) Consider the following conjecture: $FC(n)$: Every polynomial map $H : \mathbb{C}^n \to \mathbb{C}^n$ such that JH is nilpotent has at most one fixed point.
   (i) Show that if $JC(n)$ is true, then $FC(n)$ is true.
   (ii) Show that if $FC(n)$ is true for all n, then $JC(n)$ is true. [If the Jacobian Conjecture is false there exists n and $F : \mathbb{C}^n \to \mathbb{C}^n$ of the form $X - H$ with JH nilpotent such that F is not injective, say $F(a) = F(b)$ for some $a, b \in \mathbb{C}^n$ with $a \neq b$. Consider $\tilde{H} := H + F(a)$.]

8.2 The Markus-Yamabe conjecture and the LaSalle problem in dimension two

In this section we give a proof of the two-dimensional Markus-Yamabe Conjecture for polynomial vector fields. The main ingredients are Olech's theorem and 1.1.32.

For the proof in the C^1-case we refer to [123], [147], [140] and [236]. In the second half of this section we show that the answer to the LaSalle problem is negative for rational maps and affirmative for polynomial maps in dimension two.

Theorem 8.2.1 (Meisters, Olech). *The two dimensional MYC is true for polynomial vector fields.*

Proof: i) Let $F : \mathbb{R}^2 \to \mathbb{R}^2$ be a polynomial map satisfying MYA. It suffices to prove that F is injective (by 8.1.6). From MYA it follows that $\det JF(x) \neq 0$ for all $x \in \mathbb{R}^2$, so $d := |\mathbb{R}(X) : \mathbb{R}(F)|$ is finite by 1.1.31 and $\#F^{-1}(x) \leq d$ for all $x \in \mathbb{R}^2$ by 1.1.32. Let $y_0 \in \mathbb{R}^2$ be such that $\#F^{-1}(y_0)$ is maximal, say N. Choose $x_0 \in F^{-1}(y_0)$, so $F(x_0) = y_0$. Put $G(X) := F(X + x_0) - y_0$. Then G also satisfies MYA and $\#G^{-1}(y) \leq N$ for all $y \in \mathbb{R}^2$ and $\#G^{-1}(0) = N$.

ii) Now we claim

(8.2.2) $\qquad G$ satisfies the hypothesis of Olech's theorem.

Let us assume 8.2.2 for a moment. Then it follows from 8.1.5 that 0 is a global attractor of $\dot{x} = G(x)$. Consequently $\#G^{-1}(0) = 1$ (for if $G(a) = 0$ with $a \neq 0$, then the constant solution $x(t) = a$ of $\dot{x} = G(x)$ cannot tend to 0 if t tends to infinity). So $N = 1$, whence $\#G^{-1}(y) \leq 1$ for all $y \in \mathbb{R}^2$. So G is injective and hence so is F. Then the result follows from 8.1.6.

iii) To prove 8.2.2 let $G^{-1}(0) = \{x_1, \ldots, x_N\}$. By the local inverse function theorem there exist an open ball with radius $r > 0$ around 0, denoted $B(r, 0)$, and bounded open neighborhoods $V_1, \ldots, V_N$ of the points $x_1, \ldots, x_N$ in $\mathbb{R}^2$ respectively such that $G_{|V_i} : V_i \to B(r, 0)$ is a homeomorphism for all i and $V_i \cap V_j$ is empty for all pairs (i, j) with $i \neq j$. We deduce now that

(8.2.3) $$G^{-1}(B(r, 0)) = V_1 \cup \cdots \cup V_N.$$

The inclusion "$\supset$" is obvious. Furthermore, suppose that there exists a point $x \in G^{-1}(B(r, 0))$ which does not belong to $V_1 \cup \cdots \cup V_N$. Since $G(x) \in B(r, 0) = G_{|V_i}(V_i)$ for all i, there exists $v_i \in V_i$ for all $1 \leq i \leq N$ with $G(x) = G(v_i)$, i.e. $v_i \in G^{-1}(G(x))$ for all i. So $G^{-1}(G(x))$ contains at least $N + 1$ points, namely $x, v_1, \ldots, v_N$ (use that $V_i \cap V_j$ is empty if $i \neq j$), contradicting the choice of N. So 8.2.3 holds.

iv) Finally, choose $\varepsilon > 0$ such that $V_i \subset B(\varepsilon, 0)$ for all i. Then it follows from 8.2.3 that if $||x|| \geq \varepsilon$, then $x \notin G^{-1}(B(r, 0))$, i.e. $||G(x)|| \geq r$, which proves 8.2.2. □

Now we consider the LaSalle problem. First we will show that the answer to this problem is negative for rational maps.

Example 8.2.4 (Szlenk). Let $F : \mathbb{R}^2 \to \mathbb{R}^2$ be given by

$$F(x, y) = \left(\frac{-ky^3}{1 + x^2 + y^2}, \frac{kx^3}{1 + x^2 + y^2} \right) \text{ where } k \in \left(1, \frac{2}{\sqrt{3}}\right).$$

Then F is injective and gives a negative answer to the LaSalle problem.

Proof: One easily verifies that F is injective and that $F^4(\frac{1}{\sqrt{k-1}}, 0) = (\frac{1}{\sqrt{k-1}}, 0)$. So 0 cannot be a global attractor of the map F. Finally we show that F satisfies the DMYA: therefore observe that

$$\det JF(x, y) = \frac{3k^2x^2y^2(x^2+y^2+3)}{(1+x^2+y^2)^3} \text{ and } \operatorname{Tr} JF(x, y) = \frac{-2kxy(x^2-y^2)}{(1+x^2+y^2)^2}.$$

So, using polar coordinates we get

$$\begin{aligned} \det JF(r,\theta) &= \frac{3k^2 sin^2(\theta)cos^2(\theta)r^4(r^2+3)}{(1+r^2)^3} \\ &= \frac{3k^2 sin^2(2\theta)(r^6+3r^4)}{4(r^6+3r^4+3r^2+1)} < \frac{3k^2}{4} < 1. \end{aligned} \tag{8.2.5}$$

Furthermore the discriminant of the characteristic polynomial of $JF(x, y)$ denoted $\Delta(x, y)$ is equal to

$$\Delta(x, y) = -4k^2x^2y^2 \frac{2x^4 + 8x^2y^2 + 2y^4 + 12x^2 + 12y^2 + 9}{(1+x^2+y^2)^4} \leq 0.$$

If $xy = 0$, then $\det JF(x, y) = \operatorname{Tr} JF(x, y) = 0$, so all eigenvalues of $JF(x, y)$ are zero. Finally, if $xy \neq 0$ then $\Delta(x, y) < 0$, hence all eigenvalues of $JF(x, y)$ are not real. Consequently if λ is such an eigenvalue, it follows that

$$|\lambda| = \sqrt{\det JF(x, y)} < \sqrt{\frac{3k^2}{4}} < 1$$

which concludes the proof. □

To conclude this section we show

Theorem 8.2.6 (Cima, Gasull, Mañosas). *Let $F : \mathbb{R}^2 \to \mathbb{R}^2$ be a polynomial map satisfying DMYA. Then F has a unique fixed point which is a global attractor of the map F. In particular for these maps the fixed point conjecture is true and the answer to the LaSalle problem is affirmative.*

Proof: Let $X = (X_1, X_2)$ and write x, y instead of X_1 respectively X_2. Let $F : \mathbb{R}^2 \to \mathbb{R}^2$ be a polynomial map satisfying DMYA.

i) We will show that there exists $T \in Gl_2(\mathbb{R})$ such that

$$T^{-1}FT = (c_1x + a(y), c_2y + c)$$

with $a(y) \in \mathbb{R}[y]$, $c_1, c_2 \in \mathbb{R}$ and $c \in \mathbb{R}$. Since $J(T^{-1}FT) = T^{-1}JF(TX)T$ we get that c_1, c_2 are eigenvalues of $JF(0)$, so $|c_1|, |c_2| < 1$. Then it follows from Exercise 1 that the map $T^{-1}FT$ has a global attractor, say $p \in \mathbb{R}^2$. Consequently $T(p)$ is a global attractor of the map F. Then use Exercise 3.

ii) Now we show that T as in i) exists. Since F satisfies DMYA it follows from 8.1.11 that both $\det JF$ and $\operatorname{Tr} JF$ are constants, say $\operatorname{Tr} JF = t \in \mathbb{R}$. Then $G := F - \frac{1}{2}tX$ satisfies $\operatorname{Tr} JG = 0$. So $\partial_1 G_1 + \partial_2 G_2 = 0$, whence $G_1 = \partial_2 P$ and $G_2 = -\partial_1 P$ for some $P \in k[x, y]$ by 1.3.53. Furthermore $d := \det JG \in \mathbb{R}$ implies that

$$d = -(\partial_1\partial_2 P)^2 + (\partial_1^2 P)(\partial_2^2 P) \in \mathbb{R}.$$

Then by 8.2.7 below there exist $T \in Gl_2(\mathbb{R})$, $g(y) \in \mathbb{R}[y]$, $a, b \in \mathbb{R}$ and $c \in \mathbb{R}^*$ such that

$$\begin{aligned} P(TX) &= cxy + g(y) + ax \quad \text{in case } d \in \mathbb{R}^*, \\ P(TX) &= ax + by + g(y) \quad \text{in case } d = 0. \end{aligned}$$

From this it follows in any case that

$$T^{-1}G(TX) = (c_1 x + a(y), c_2 y + c_3)$$

for some c_i in $\mathbb{R}$ and $a(y) \in \mathbb{R}[y]$. Then a similar equality holds for F, so the result follows from i). □

Proposition 8.2.7 (Dillen). *Let k be a field of characteristic zero and $P \in k[x, y]$. Let*

$$H(P) := -(\partial_1\partial_2 P)^2 + (\partial_1^2 P)(\partial_2^2 P)$$

*be the **Hessian** of P. Suppose that $H(P) \in k$. Then there exists $T \in Gl_2(k)$ such that $P(TX) = cxy + g(y) + ax$ for some a, c in k and $g(y)$ in $k[y]$. Furthermore $c \in k^*$ if and only if $H(P) \in k^*$.*

The proof of this proposition is based on the following lemma.

Lemma 8.2.8 *Let $P = \sum_{i=0}^{n} P_i(x)y^{n-i} \in k[x, y]$ with $P_0 \neq 0$. If $n \geq 2$ and $H(P) \in k$, then* $\deg P_i \leq i$ *for all i. In particular* $\deg_x P \leq \deg_y P$ $(= n)$.

Proof: Since $H(P) \in k$ the coefficient of y^{2n-2} in $H(P)$ is zero. So

$$n(n-1)P_0 P_0^{(2)} = n^2(P_0')^2.$$

If $d := \deg P_0 \geq 1$ then, looking at the highest x-term in this equation gives $n(n-1)d(d-1) = n^2 d^2$, whence $1 = n + d$, a contradiction. So $d = 0$, i.e. $\deg P_0 = 0$. The coefficient of y^{2n-3} in $H(P)$ is also zero. This gives

$$n(n-1)P_0 P_1^{(2)} = 0$$

so $P_1^{(2)} = 0$, i.e. $\deg P_1 \le 1$. Now we show that $\deg P_r \le r$ for all $2 \le r \le n$ by induction on r. So we assume that we already know that $\deg P_i \le i$ for all $0 \le i \le r-1$. Consider the coefficient of y^{2n-2-r} in $H(P)$. This coefficient is equal to

$$n(n-1)P_0P_r^{(2)} + \sum_{j=1}^{r-2}(n-j)(n-(j+1))P_jP_{r-j}^{(2)}$$
$$-\sum_{j=1}^{r-1}(n-j)(n-(r-j))P_j'P_{r-j}'.$$

Apart from the term $n(n-1)P_0P_r^{(2)}$ all other terms in this expression have degree $\le r-2$ by the induction hypothesis. Since $H(P) \in k$ we have $\deg H(P) = 0 \le r-2$. So we deduce that $\deg P_r^{(2)} \le r-2$, whence $\deg P_r \le r$, as desired. □

Corollary 8.2.9 *Let $n \ge 2$. If P satisfies $H(P) \in k$, $\deg_y P = n$ and $\deg_x P < n$, then $\deg_x P \le 1$.*

Proof: Suppose that $\deg_x P \ge 2$. Then by 8.2.8, interchanging the roles of x and y, we get $n = \deg_y P \le \deg_x P < n$, a contradiction. So $\deg_x P \le 1$. □

Proof of Proposition 8.2.7: i) The second part of the statement follows from the first one observing that if $\tilde{P} := P(TX)$, then $H(\tilde{P}) = (\det T)^2 H(P)$ (Exercise 2).

ii) Let $\deg P = n$. If $n \le 2$ then obviously $H(P) \in k$. So we may assume $n \ge 3$. Writing $P = \sum_{i=0}^n P_i$, its homogeneous decomposition, we deduce that $H(P_n) = \det J(JP_n) = 0$, so by 1.2.9 we get that $\partial_1 P_n$ and $\partial_2 P_n$ are algebraically dependent over k, whence $h(\partial_1 P_n, \partial_2 P_n) = 0$ for some homogeneous polynomial $h \in k[Y_1, Y_2]$. Over an algebraic closure $\overline{k}$ of k we can decompose h into linear factors. So there exist $p, q \in \overline{k}$ not both zero, such that $p\partial_1 P_n + q\partial_2 P_n = 0$. Since the coefficients of P_n all belong to k we may assume that p and q are in k. Then $(p\partial_1 + q\partial_2)P_n = 0$ implies that $P_n \in k[qx - py]$ and since P_n is homogeneous we get $P_n = c(qx-py)^n$, for some $c \in k^*$. So there exists $T \in Gl_2(k)$ such that $\tilde{P} := P(TX)$ satisfies $\deg \tilde{P} = n$, $\tilde{P}_n = cy^n$ for some $c \in k^*$ and $H(\tilde{P}) \in k$. Then it follows from 8.2.9 that $\deg_x \tilde{P} \le 1$, whence $\tilde{P} = b(y)x + g(y)$ for some $b(y), g(y)$ in $k[y]$. Finally using that $H(\tilde{P}) \in k$ we get the desired result.

Exercises for §8.2

1. Let $F : \mathbb{R}^2 \to \mathbb{R}^2$ be given by $F(x_1, x_2) = (c_1x_1 + a_1(x_2), c_2x_2 + a_2)$ where $c_1, c_2, a_2 \in \mathbb{R}$ and $a_1(x_2) \in \mathbb{R}[x_2]$. Suppose that $|c_1|, |c_2| < 1$.

   (i) Show that $(\frac{1}{1-c_1}a_1(\frac{a_2}{1-c_2}), \frac{a_2}{1-c_2})$ is a global attractor of the map F.
   (ii) Generalize i) to triangular maps $F : \mathbb{R}^n \to \mathbb{R}^n$.

2. Let R be any commutative ring, $T \in M_n(R)$ and $P \in R[X] := R[X_1, \ldots, X_n]$. Let $H(P) := \det J(JP)$. Put $\tilde{P} := P(TX)$. Show that $H(\tilde{P}) = (\det T)^2 H(P)$.

3. Let $F : \mathbb{R}^n \to \mathbb{R}^n$ be a continuous vector field such that $p \in \mathbb{R}^n$ is a global attractor of the map F.

   (i) Show that $F(p) = p$.
   (ii) Show that p is the only fixed point of F.

8.3 The story of the solution of the Markus-Yamabe Conjecture

In the previous section we saw that in dimension 2 both the MYC and the LaSalle problem were solved affirmatively for polynomial vector fields. Now we will show that for $n \geq 3$ the situation is completely different: we give polynomial counterexamples to both problems for all $n \geq 3$. These examples were found by the author and Hubbers (Nijmegen) in collaboration with Cima, Gasull and Mañosas (Barcelona) in November 1995. In this section I will describe the **story** which finally led to the discovery of these examples and hence to the complete solution of the 35 year old MYC. The reason I describe the whole story is that I believe that it is not only stimulating for doing research but also since it contains a lesson which I learned from it and which I want to share with the reader.

The starting point of the story was the discovery of the class $H(n, A)$ at the end of 1993 by the author in collaboration with Hubbers, inspired by his classification of the four-dimensional cubic homogeneous case, which started in September 1993 and was completed in February 1994.

Then in March 1994, Deng, Meisters and Zampieri proposed a new strategy to attack the Jacobian Conjecture based upon the following observation: let $F : \mathbb{C}^n \to \mathbb{C}^n$ be a polynomial map satisfying $F(0) = 0$ and $JF(0) = I$. Suppose that for some $s \in \mathbb{C}\backslash\{0, 1\}$ there exists an analytic automorphism $h_s : \mathbb{C}^n \to \mathbb{C}^n$ such that $h_s^{-1} \circ sF \circ h_s = sX$, then obviously $sF (= h_s \circ sX \circ h_s^{-1})$ is injective and hence so is F. Consequently by 4.2.1 F is an automorphism of $\mathbb{C}^n$. So their aim was to show that if F satisfies Keller's condition, i.e. $\det JF = 1$, then for some $s \in \mathbb{C}\backslash\{0, 1\}$ such an automorphism h_s would exist. In this case we say that sF is **globally analytically linearizable**. In fact the following well-known result, due to Poincaré and Siegel, together with Exercise 1 show that such an h_s exists **locally** and is unique if $| s | \neq 0$ or 1.

Theorem 8.3.1 (Poincaré, Siegel). *Let F be an analytic map from a neighborhood of $0 \in \mathbb{C}^n$ to a neighborhood of $F(0) = 0$ in $\mathbb{C}^n$. If the eigenvalues $s_1, \ldots, s_n$ of $JF(0)$ are* ***non-resonant*** *of type (C, r) for some $C > 0$ and $r > 0$, i.e.*

$$|s_j - s_1^{m_1} \ldots s_n^{m_n}| \geq C/|m|^r$$

for all $m = (m_1, \ldots, m_n)$ with $|m| \geq 2$ and all $1 \leq j \leq n$, then there exists a uniquely determined analytic map h defined in a neighborhood of $0 \in \mathbb{C}^n$ such that $h(0) = 0$, $Jh(0) = I$ and $h^{-1}Fh = JF(0)X$.

For a proof of this result we refer to [21], Chapter 5.

In [74] Deng, Meisters and Zampieri proved that for all s large enough h_s^{-1} is indeed entire. They could not prove such a result for h_s. In fact it was realized later that it is not difficult to see that for the map

$$F = (X + (Y + X^2)^2, Y + X^2)$$

h_s cannot be entire, for every $s \in \mathbb{C}\backslash\{0, 1\}$ (Exercise 2). However in the introduction of [74] the authors make the following conjecture.

DMZ-Conjecture. *Let $F = X + H$ with H homogeneous of degree $d \geq 2$. If* $\det JF = 1$, *then for all s sufficiently large there exists a global analytic automorphism h_s of $\mathbb{C}^n$ such that*

$$h_s^{-1} \circ sF \circ h_s = sX.$$

Meisters was very sceptical about this new approach and decided to do several computer experiments and computed h_s for various polynomial automorphisms of the form $X + H$, with H cubic homogeneous. To his own surprise, in all these cases the maps h_s turned out to be much nicer than expected; they were polynomial automorphisms! His scepticism changed into optimism and at the Curaçao conference, July 1994, he presented the following conjecture ([210]).

Meisters' Linearization Conjecture. *Let $F = X + H$ be a cubic homogeneous polynomial map with* $\det JF = 1$. *Then for almost all $s \in \mathbb{C}$ apart from a finite number of roots of unity, there exists a polynomial automorphism h_s of $\mathbb{C}^n$ such that $h_s^{-1} \circ sF \circ h_s = sX$.*

During the conference he offered a prize of \$100 to the first person to find a counterexample to his conjecture.

About one month after the conference Hubbers and I showed that Meisters' conjecture is true if we replace the condition "JH nilpotent", which in the case that H is homogeneous is equivalent to $\det JF = 1$, by the stronger condition "JH is strongly nilpotent", but H need not be homogeneous. More precisely we showed

Theorem 8.3.2 (van den Essen, Hubbers). *Let k be a field, $k(s)$ the field of rational functions in one variable and F of the form $X + H$ with $F(0) = 0$ and JH strongly nilpotent (H need not be homogeneous). Then there exists a polynomial automorphism $h_s \in \mathrm{Aut}_{k(s)}\, k(s)[X]$ such that*

$$h_s^{-1} \circ sF \circ h_s = sJF(0)X.$$

Furthermore, the zeros of the denominators of the coefficients of the X-monomials appearing in h_s are roots of unity.

The proof of this result will be given at the end of this section in order not to disturb the main line of the story. As an immediate consequence of 8.3.2 we get

Corollary 8.3.3 *If $n \leq 3$ Meisters' linearization conjecture is true.*

Proof: Apply 7.4.11 and 8.3.2. □

Now let us continue the story. During the Curaçao conference Harm Derksen made the following useful observation: let $F : \mathbb{C}^n \to \mathbb{C}^n$ be a polynomial map with $F(0) = 0$. If F is polynomially linearizable, i.e. if there exists a polynomial automorphism $G : \mathbb{C}^n \to \mathbb{C}^n$ such that $G^{-1} \circ F \circ G = L$, where L is the linear part of F, then there exists a positive integer N such that

$$\deg F^k \leq N, \quad \text{for all } k \geq 1.$$

Indeed, if we put $d := \max\{\deg G, \deg G^{-1}\}$ it follows from $F = G \circ L \circ G^{-1}$ that $F^k = G \circ L^k \circ G^{-1}$, whence $\deg F^k \leq d^2$ for all $k \geq 1$. Returning to Nijmegen after a beautiful stay at the island of Curaçao and a trip through Brazil, I started to look at Meisters' conjecture. My intention was to prove it for the nice class of polynomial automorphisms of the form $X + H$ with $H \in H(n, \mathbb{C})$. In trying to prove the conjecture for these maps I finally realized that taking F of the form

$$F = (X_1 + a_2 f(a_1X_1 + a_2X_2),\ X_2 - a_1 f(a_1X_1 + a_2X_2),$$
$$X_3 + c_3,\ X_4 + c_4)$$

where $f \in \mathbb{C}[X_3, X_4][T]$, $a_1, a_2 \in \mathbb{C}[X_3, X_4]$, $c_3 \in \mathbb{C}[X_4]$ and $c_4 \in \mathbb{C}$, the degrees of the powers of sF, where s is any complex number $\neq 0, 1$, could not be bounded. In fact taking almost the simplest choices, i.e. $f = T$, $a_1 = X_3$, $a_2 = X_4$, $c_3 = X_4^3$, to obtain a cubic homogeneous map and $c_4 = 0$, I had already found a counterexample. So we get

Proposition 8.3.4 (van den Essen). *Let $n \geq 4$ and $d := X_3X_1 + X_4X_2$. Then*

$$F = (X_1 + X_4d,\ X_2 - X_3d,\ X_3 + X_4^3,\ X_4, \ldots, X_n)$$

is a counterexample to Meisters' conjecture.

For the proof we refer to Exercise 3 (or [106]). I sent this example to Meisters on September 19, 1994 and one month later I received the promised \$100 prize!

Nevertheless the story had just begun. Immediately the question was raised: is this example a counterexample to the DMZ-conjecture? The answer turned out to be NO! In July 1995 I received a preprint of Gorni and Zampieri [144] in which they showed that, for each $s \in \mathbb{C}$ with $|s| \neq 1$, the map sF with F as in 8.3.4 is globally analytically linearizable!

So the DMZ-conjecture and hence that approach to the Jacobian Conjecture remained open. In their proof they essentially used that the F above depends linearly on X_1 and X_2. Then in September 1995 another event happened. I received a preprint of Bo Deng [73] in which he also showed that the map F is not a counterexample to the DMZ-conjecture. His proof is based on the following result.

Proposition 8.3.5 *Let $F : \mathbb{C}^n \to \mathbb{C}^n$ be an analytic map with $F(0) = 0$. Put $A := JF(0)$. Suppose that the eigenvalues of A are non-resonant and smaller than 1 in absolute value. Then there is equivalence between*

i) *F is globally analytically linearizable to its linear part, and*

ii) *F is an analytic isomorphism of $\mathbb{C}^n$ and 0 is a global attractor of the map F.*

Proof: First observe that since the eigenvalues of A are non-resonant they are all non-zero, so A is invertible.

i) $\to$ ii). If $h^{-1}Fh = A$ for some global analytic automorphism of $\mathbb{C}^n$, then $F = hAh^{-1}$, so F is an automorphism of $\mathbb{C}^n$. Furthermore, since the eigenvalues of A are smaller than 1 in absolute value, it follows easily that for every $x \in \mathbb{C}^n$, $A^k x \to 0$ if k tends to infinity. Consequently $F^k(x) = hA^k h^{-1}(x)$ tends to zero if k tends to infinity.

ii) $\to$ i). Since the eigenvalues of A are non-resonant, there exist neighborhoods U of 0 in $\mathbb{C}^n$ and V of $F(0) = 0$ in $\mathbb{C}^n$ and a local analytic isomorphism $h : U \to V$ such that $h^{-1}Fh = A$ on U (by 8.3.1). Now we show that h can be extended uniquely to a global analytic isomorphism of $\mathbb{C}^n$ satisfying $h^{-1}Fh(x) = Ax$ for all x of $\mathbb{C}^n$. Therefore let $x \in \mathbb{C}^n$. Since the eigenvalues of A are smaller than 1 in absolute value, there exists an integer k such that $A^p x \in U$ for all $p \geq k$. Define

$$h(x) := F^{-p}hA^p(x), \text{ where } p \geq k.$$

This definition does not depend on $p \geq k$ and h is one-to-one on $\mathbb{C}^n$ (Exercise 4). Furthermore, h is onto: in fact since 0 is a global attractor of the map F there exists k such that $F^p(x) \in V$ for all $p \geq k$. Then the inverse of h is defined by

$$h^{-1}(x) = A^{-p}h^{-1}F^p(x) \text{ where } p \geq k.$$

Again this definition is independent of $p \geq k$. Finally we show that $Fh(x) = hAx$ for all $x \in \mathbb{C}^n$: if $x \in \mathbb{C}^n$ and $A^p x \in U$ for all $p \geq k$, then

$$hAx = F^{-p}hA^p(Ax) = FF^{-(p+1)}hA^{p+1}(x) = Fh(x)$$

as desired. □

Corollary 8.3.6 *Let $F : \mathbb{C}^n \to \mathbb{C}^n$ be an analytic map such that $F(0) = 0$ and $JF(0) = I$. Let $s \in \mathbb{C}^*$ with $|s| < 1$. Then sF is globally analytically linearizable if and only if F is an analytic isomorphism of $\mathbb{C}^n$ and 0 is a global attractor of the map sF.*

Proof: This follows directly from 8.3.5 since the eigenvalues of $JsF(0) = sI$ are all equal to s and one easily verifies (Exercise 1) that the hypothesis $|s| < 1$ implies that they are non-resonant. □

Remark 8.3.7 Proposition 8.3.5, which I learned about from Deng's preprint [73] had already been proved several years earlier by Rosay and Rudin in [253].

When I saw 8.3.6 in Deng's paper I realized immediately that we had an effective tool to investigate the DMZ-conjecture by doing some computer experiments, namely just take a polynomial automorphism of the form $X+H$ with $H \in H(n, \mathbb{C})$ for some $n \geq 4$, take some $0 < s < 1$ and check if 0 is a global attractor of the map sF, by verifying that $(sF)^k$ tends to zero if k tends to infinity for randomly chosen points x.

I was convinced that for such F's this would be the case. So I said to my Ph.D student Engelbert Hubbers, just take some complicated $H \in H(5, \mathbb{C})$, form $F = X + H$ and check if for sF with $0 < s < 1$ the iterations $(sF)^k(x)$ tend to zero, and if this indeed is the case start proving! However, a few days later he came to me with the following example.

Let $s = (1/100)$ and $F = (X_1 + 3X_1X_2^2 + X_4^4 + X_1^3 - 3X_4^2X_1X_2^2 - 3X_4^2X_1X_2^2 - 2X_1^3X_4^3 - 3X_4^3X_1^2X_2 - X_4X_2^3 - X_4^4X_1^3 + 3X_1^2X_4X_2 + 2X_4X_1^3 - X_4^3 + X_2^3 + 3X_1^2X_2, -3X_4^2X_1X_2^2 - 6X_4^3X_1^2X_2 - 3X_4^4X_1^2X_2 - 3X_4^3X_1X_2^2 + X_2 + 3X_1X_4X_2^2 - X_4^3 + 3X_1X_2^2 - X_4^5X_1^3 - X_4^2X_2^3 + X_2^3 + 6X_1^2X_2X_4 - 3X_4^4X_1^3 + X_1^3 - 2X_1^3X_4^3 + 3X_1^2X_2 + 2X_4^2X_1^3 + 3X_1^3X_4 + X_4^5, 3X_4^2X_1X_2^2 + 3X_4^2X_1^2X_2 + 3X_4^3X_1^2X_2 + X_3 - 2X_1X_2^2 + X_4X_2^3 - X_4^4 - X_2^4 - X_2^3 - X_1^2X_4X_2 + X_4^4X_1^3 + 2X_1^3X_4^3 - X_1^2X_2 + X_4^2X_1^3, X_4, X_5)$.

So if one starts with the point $x = (0, 0, 0, 3.6314, 0)$ then the computations indicated that $F^k(x)$ tends to zero if k tends to infinity, but if one starts with $x = (0, 0, 0, 3.6315, 0)$ then the computations indicated that $F^k(x)$ tends to infinity if k tends to infinity! The same phenomenon happens if we take two such points with the first nine decimals equal and change the tenth decimal etc. Of course this was a fantastic discovery, which not only showed that we were on the right track showing that the DMZ-conjecture is false, but also that the answer to the LaSalle problem is negative!

The next day I went to Poland for a two week visit. In the airport waiting for my plane I started to think about how to prove this phenomenon. Hubbers' example was far too complicated to prove anything. So I had to find an easier example. Then I remembered that both in the Gorni-Zampieri paper and Deng's paper they essentially used that X_1 and X_2 in the example of 8.3.4 appear linearly! So I had to find an automorphism F in which both X_1 and X_2 do not appear linearly. Remembering that the map of 8.3.4 was of the form $X + H$ with $H \in H(4, \mathbb{C})$, namely

$$H = (X_4 f(X_3X_1 + X_4X_2), -X_3 f(X_3X_1 + X_4X_2), X_3^3, 0)$$

with $f = T$, it was not difficult to construct an automorphism where X_1 and X_2 do not appear linearly, namely take $f = T^2$, i.e.

$$F = (X_1 + X_4d^2, X_2 - X_3d^2, X_3 + X_4^3, X_4)$$

where $d := X_3X_1 + X_4X_2$. By 7.2.12 the corresponding F is an automorphism and by 7.2.6 JH is nilpotent, so all eigenvalues of $J(sF) = sI + sJH$ are equal to s. So to see that this F gives a counterexample to the LaSalle problem, I needed only to show that for some $0 < s < 1$ there exists a point x in $\mathbb{R}^4$ such that $(sF)^k(x)$ tends to infinity. During my stay at the University of Toruń, visiting Andrzej Nowicki, I succeeded in proving this: in fact the proof showed that if $0 < s < 1$ and $a \in \mathbb{R}$ such that $as > 1$, then $(sF)^k(a, a, a, a)$ tends to infinity if k tends to infinity. So the LaSalle problem had a negative answer if $n \geq 4$. Finally in the next week, visiting Ludwik Drużkowski at Krakow, using the same technique I could also prove that the DMZ-conjecture is false for all $n \geq 4$. In fact in the counterexamples to the LaSalle problem described above, the term X_4^3 could be replaced by any power X_4^m, with $m \geq 1$. Specializing to $m = 5$ this gives a quintic homogeneous counterexample to the DMZ-conjecture.

Since we will give easier counterexamples below, we refer to [115] for the proofs of these original results. We mention them for the sake of completeness of our story.

Then I sent an e-mail to Hubbers in Nijmegen, asking him to verify my proof by checking some iterations by computer. Now there really was chaos: in the beginning $F^k(x)$ indeed tended to infinity, however after 130,000 iterations the computer told us that $F^k(x)$ tended to zero!! Fortunately my proof was correct and not the computer this time.

Nevertheless, the LaSalle problem being answered in the negative or putting it differently, having found a negative answer to the Discrete Markus-Yamabe problem, it was of course very natural to ask: Is the system

$$(8.3.8) \qquad \begin{cases} \dot{x}_1 &= -x_1 + x_4 d(x)^2 \\ \dot{x}_2 &= -x_2 - x_3 d(x)^2 \\ \dot{x}_3 &= -x_3 + x_4^m \\ \dot{x}_4 &= -x_4 \end{cases}$$

for some m a counterexample to the Markus-Yamabe Conjecture or more precisely, can we find solutions which tend to infinity if t tends to infinity?

So in order to answer this question let us try to solve 8.3.8. First, solving the last equation we get $x_4 = c_4 e^{-t}$ for some $c_4 \in \mathbb{R}$. If $c_4 = 0$ then $x_4 = 0$ and one easily verifies that every solution x of 8.3.8 (having $x_4 = 0$) tends to zero if t tends to infinity. So from now on we may assume that $x_4 = c_4 e^{-t}$ with $c_4 \neq 0$. Substituting this in the third equation gives

$$x_3(t) = \frac{1}{1-m} c_4^m e^{-mt} + c_3 e^{-t} \text{ for some } c_3 \in \mathbb{R}.$$

Multiplying the first equation by x_3, the second by x_4 and adding the resulting equations we get

$$x_3 \dot{x}_1 + x_4 \dot{x}_2 = -d$$

which implies that $\dot{d} = -2d + x_4^m x_1$. So we get

$$(8.3.9) \qquad \dot{x}_1 = -x_1 + c_4 e^{-t} d^2, \; \dot{d} = -2d + c_4^m e^{-mt} x_1.$$

Solving the second equation of 8.3.9 for x_1 we get

$$x_1 = c_4^{-m} e^{mt} (\dot{d} + 2d).$$

Substituting this in the first equation of 8.3.9 we obtain that $d = x_3 x_1 + c_4 e^{-t} x_2$ satisfies

(8.3.10) $$d^{(2)} + (m+3)d^{(1)} + (2m+2)d = c_4^{m+1} e^{-(m+1)t} d^2.$$

Conversely, if d satisfies 8.3.10 then it is left to the reader to verify, Exercise 5, that

$$x_4(t) := c_4 e^{-t},\ x_3(t) := \frac{1}{1-m} c_4^m e^{-mt},\ x_1(t) := c_4^{-m} e^{mt}(\dot{d} + 2d)$$

and $x_2(t)$ defined by $x_3(t)x_1(t) + x_2(t)c_4 e^{-t} = d$ is a solution of 8.3.8.

Furthermore, if 8.3.10 admits a solution d tending to infinity if t does, then the equation $x_3(t)x_1(t) + x_2(t)c_4 e^{-t} = d$ and the definition of x_3 imply that not both x_1 and x_2 can tend to zero if t tends to infinity, so this would give a counterexample to the MYC. Consequently we would like to know if 8.3.10 admits a solution which tends to infinity if t tends to infinity. Here again we asked the help of our computer. However this time it was unable to help us to plot any solution of 8.3.10.

A crucial point was that we did not even try to find an exact solution of 8.3.10 since I believed, from experience with non-linear equations, that it would be impossible to find an exact solution! Nevertheless one readily verifies that in **our case** such a solution **does** exist and even better a solution which tends to infinity if t tends to infinity. For example if we consider the case $m = 3$, so we have

$$d^{(2)} + 6d^{(1)} + 8d = c_4^4 e^{-4t} d^2$$

and try a solution of the form $d(t) = ae^{bt}$ for some $a, b \in \mathbb{R}$, then one obtains

$$a(b^2 + 6b + 8)e^{bt} = c_4^4 a^2 e^{(2b-4)t}.$$

Equating the exponential powers we get $b = 4$ and hence $48a = c_4^4 a^2$, so $d(t) = 48c_4^{-4} e^{4t}$ is a solution which tends to infinity if t tends to infinity!

However, as already remarked above the computer could not help us and we did not even try to find an exact solution of 8.3.10, so we got stuck at this point! Fortunately, in the second week of November 1995, Anna Cima from Barcelona visited the University of Nijmegen and we discussed the situation and the possible counterexamples to MYC. Two weeks later I received the following e-mail from her: "As you thought, the counterexample to the Discrete Markus-Yamabe problem is also a counterexample to the continuous one. Take $m = 4$ then $x_1(t) = 490e^{9t}$, $x_2(t) = \frac{700}{3}e^{6t}$, $x_3(t) = -\frac{1}{3}e^{-4t}$ and $x_4(t) = e^{-t}$ is a solution of 8.3.8 which clearly goes to infinity if t increases".

Finally a week later I received another e-mail from Barcelona in which the above $n \geq 4$ counterexample was modified to a counterexample which works for all $n \geq 3$, thereby completely solving the Markus-Yamabe Conjecture, since the case $n \leq 2$ was solved already as described in the previous section. The precise result is

Theorem 8.3.11 (Cima, van den Essen, Gasull, Hubbers, Mañosas). *Let $n \geq 3$ and $F : \mathbb{R}^n \to \mathbb{R}^n$ be given by*

$$F(x_1, \ldots, x_n) = (-x_1 + x_3 d(x)^2, -x_2 - d(x)^2, -x_3, \ldots, -x_n)$$

where $d(x) = x_1 + x_3x_2$. Then F is a counterexample to the Markus-Yamabe Conjecture. More precisely, there exists a solution of $\dot{x} = F(x)$ which tends to infinity if t tends to infinity.

Proof: One easily verifies that for all $x \in \mathbb{R}^n$ all eigenvalues of $JF(x)$ are equal to -1 and that $x_1(t) = 18e^t, x_2(t) = -12e^{2t}, x_3(t) = \cdots = x_n(t) = e^{-t}$ is a solution of $\dot{x} = F(x)$ which tends to infinity if t does. □

In our paper [54] we also showed

Theorem 8.3.12 *Let $n \geq 3$ and $F : \mathbb{R}^n \to \mathbb{R}^n$ given by*

$$F(x_1, \ldots, x_n) = \left(\frac{1}{2}x_1 + x_3d(x)^2, \frac{1}{2}x_2 - d(x)^2, \frac{1}{2}x_3, \ldots, \frac{1}{2}x_n\right)$$

where $d(x) = x_1 + x_3x_2$. Then F gives a negative answer to the LaSalle problem. More precisely, there exists an initial condition $x^{(0)}$ such that the sequence $x^{(n+1)} = F(x^{(n)})$ tends to infinity if n tends to infinity.

Proof: One easily verifies that for all $x \in \mathbb{R}^n$ the eigenvalues of $JF(x)$ are equal to $\frac{1}{2}$. Taking

$$x^{(0)} = \left(\frac{147}{32}, -\frac{63}{32}, 1, 0, \ldots, 0\right)$$

it is easy to verify by induction that

$$x^{(n)} = \left(\frac{147}{32}.2^n, -\frac{63}{32}.2^{2n}, \frac{1}{2}^n, 0, \ldots, 0\right)$$

which obviously tends to infinity if n tends to infinity. □

(8.3.13) **A lesson**

Now let me conclude this part of the story (in fact the story had a continuation one year later, see §4) by describing what was for me the most important lesson I learned. At the point we arrived at equation 8.3.10 we got stuck; in fact I am convinced that I never would have found that there was such a simple solution as $d(t) = 48c_4^{-4}e^{4t}$. Why not? The main reason is that I was **psychologically blocked** by the fact that I **knew** from experience that "one cannot solve non-linear equations exactly" and certainly not if they are related to such a longstanding conjecture as the MYC. So this block made it **impossible for me to even think about the possibility** of such a **simple solution** as ae^{bt}. So it never came to my mind to even try such a solution!!

Let me give you one more example of such a psychological block which happened during the course of this story.

When the four-dimensional counterexample was found, I sent an e-mail to many people in the field, including Olech who had been working for many years on this problem. When he received the counterexample he was very enthusiastic. Of course he immediately tried to extend the ideas of our example to the missing three-dimensional case. After trying several days he gave up: he was convinced that at least the three-dimensional MYC was true after all. Then one week after the first e-mail I sent him a second e-mail, notifying him that now there was also a counterexample to the three-dimensional MYC. However I did not send him the example, only its existence. Within one day he sent me an e-mail with the following content:

"Is this the counterexample you have

$$\begin{aligned} \dot{x} &= -x + z(xz^2 + zy)^2, \\ \dot{y} &= -y - z^2(xz^2 + zy)^2, \\ \dot{z} &= -z? \end{aligned}$$

An unbounded solution is

$$x(t) = 150e^{5t}, y(t) = -180e^{4t}, z(t) = e^{-t}.$$

I discovered this example **after** I received your message that you have one in dimension 3. I tried to find one last week when I read your note. This shows how strongly I believed that a polynomial example is impossible." So the lesson learned is this: however difficult and impossible a problem may seem **just try it! Nothing is impossible unless you think it is!!**

To conclude this section we prove 8.3.2. Before we can prove this result we need a definition and some lemmas.

Definition 8.3.14 i) We say that $X_1^{i_1} \dots X_n^{i_n} > X_1^{i'_1} \dots X_n^{i'_n}$ if and only if $\sum_{j=1}^n i_j > \sum_{j=1}^n i'_j$ or if there is equality and there exists some $1 \le l \le n$ such that $i_j = i'_j$ for all $j < l$ and $i_l > i'_l$.

ii) The **rank** of a monomial is the index of this monomial in the ascending ordered list of monomials according i).

Example 8.3.15 The rank of X_1X_2 is 8 in the list of monomials of three variables X_1, X_2, X_3. Namely the ordered list starts as follows

$$X_3 < X_2 < X_1 < X_3^2 < X_2X_3 < X_2^2 < X_1X_3 < X_1X_2 < \cdots.$$

Lemma 8.3.16 *Let $F \in k(s)[X]^n$ be a polynomial map of the form*

$$F = (X_1 + a(X_2, \dots, X_n) + l_1, X_2 + l_2, \dots, X_{n-1} + l_{n-1}, X_n)$$

where each l_i is a linear form in the variables $X_{i+1}, \dots, X_n$ with coefficients in $k(s)$ and $a := a(X_2, \dots, X_n) \in k(s)[X_2, \dots, X_n]$ with leading monomial $cX_2^{i_2} \dots X_n^{i_n}$ with respect to the ordering of 8.3.14 *and $i_1 + \cdots + i_n \ge 2$. Then there exists a (triangular) polynomial automorphism $h \in k(s)[X]^n$ such that*

$$h^{-1} \circ sF \circ h = s(X_1 + \tilde{a} + l_1, X_2 + l_2, \dots, X_{n-1} + l_{n-1}, X_n)$$

where the leading monomial of $\tilde{a}$ is strictly smaller than that of a.

Proof: The letters a', a^* below denote elements of $k(s)[X_2, \ldots, X_n]$ which leading monomials are smaller than that of a. So $a = cX_2^{i_2} \ldots X_n^{i_n} + a'$. Furthermore one easily verifies that

$$(8.3.17) \qquad (X_2 + l_2)^{i_2} \ldots (X_{n-1} + l_{n-1})^{i_{n-1}} X_n^{i_n} = X_2^{i_2} \ldots X_n^{i_n} + a^*$$

Put $\mu := c/(s^{i_2+\cdots+i_n} - 1)$ and define $\tilde{a}$ by $s\tilde{a} = s^{i_2+\cdots+i_n}\mu a^* - sa'$. Now we will show that $h = (X_1 + \mu X_2^{i_2} \ldots X_n^{i_n}, X_2, \ldots, X_n)$ satisfies

$$(8.3.18) \qquad sF \circ h = h(s(X_1 + \tilde{a} + l_1, X_2 + l_2, \ldots, X_{n-1} + l_{n-1}, X_n))$$

which of course completes the proof. Therefore observe that for all $i \geq 2$ the i-th component of both sides of 8.3.18 are equal. So it remains to verify the equality for the first components.

The first component of the left-hand side of 8.3.18 is equal to

$$sX_1 + s\mu X_2^{i_2} \ldots X_n^{i_n} + scX_2^{i_2} \ldots X_n^{i_n} + sa' + sl_1.$$

Using 8.3.17 we get that the first component of the right-hand side of 8.3.18 is equal to

$$sX_1 + s\tilde{a} + sl_1 + s^{i_2+\cdots+i_n}\mu X_2^{i_2} \ldots X_n^{i_n} + s^{i_2+\cdots+i_n}\mu a^*$$

from which 8.3.18 follows. □

Now we are able to give

Proof of Theorem 8.3.2: By 7.4.4 we may assume that F is of triangular form. We use induction on n. The case $n = 1$ is obvious. So let $n \geq 2$. Write $F = (F_1, \tilde{F})$, where $\tilde{F} := (F_2, \ldots, F_n)$, $\tilde{X} := (X_2, \ldots, X_n)$ and $X = (X_1, \tilde{X})$. By the induction hypothesis there exists a (triangular) polynomial automorphism $\tilde{h} \in \mathrm{Aut}_{k(s)} k(s)[\tilde{X}]$ such that

$$\tilde{h}^{-1} \circ s\tilde{F} \circ \tilde{h} = sJ_{\tilde{X}}\tilde{F}(0).$$

Finally let $h := (X_1, \tilde{h})$. Then obviously h is a triangular automorphism of $k(s)[X]$ and if we write $F_1 = X_1 + a(X_2, \ldots, X_n) + l_1$, where l_1 is a linear form in the $X_2, \ldots, X_n$, then we get

$$h^{-1} \circ sF \circ h = s(X_1 + \tilde{a} + l_1, X_2 + l_2, \ldots, X_{n-1} + l_{n-1}, X_n)$$

where each l_i is a linear form in $X_{i+1}, \ldots, X_n$ and $\tilde{a}$ an element of $k(s)[X_2, \ldots, X_n]$ all of whose monomials have degree ≥ 2. So we only need to make the first component linear. Therefore let r be the rank of the leading monomial of $\tilde{a}$. Then by 8.3.16 there exists h_r such that

$$h_r^{-1} \circ (h^{-1} \circ sF \circ h) \circ h_r = s(X_1 + \tilde{a}_r + l_1, X_2 + l_2, \ldots, X_{n-1} + l_{n-1}, X_n)$$

where the rank of the leading monomial of $\tilde{a}_r$ is less than r. Proceeding in this way we get the desired result. □

Exercises for §8.3

1. Let $s \in \mathbb{C}$ with $|s| \neq 0, 1$. Show that the eigenvalues of sI_n are non-resonant of type $(C, 1)$, where $C := ||s| - |s|^2|$.
2. Let $F = (X + (Y + X^2)^2, Y + X^2)$ and $s \in \mathbb{C}\backslash\{0, 1\}$. Show that sF has three fixed points outside the origin. Deduce that sF is not globally analytically linearizable. [Use i)→ ii) of 8.3.5.]
3. Let F be as in 8.3.4 and for $s \in \mathbb{C}$ denote by $(sF)_1^m$ the first component of the iterated map $sF \circ \cdots \circ sF$.

   (i) Show by induction on m that for each $m \in \mathbb{N}$, $m \geq 1$ there exist positive integers p, q and polynomials $a(X_4), b(X_4)$ in $\mathbb{C}[X_4]$ satisfying the following properties:

   a) $(sF)_1^m = a(X_4)d + b(X_4)X_1$.

   b) $a(X_4) = s^p(X_4^4)^{m-1}X_4 + \text{l.o.t.}$

   c) $b(X_4) = s^q(X_4^4)^{m-1} + \text{l.o.t.}$

   (where l.o.t. means lower order terms in X_4).

   (ii) Deduce from i) that for any $s \in \mathbb{C}\backslash\{0, 1\}$ there does not exist a polynomial automorphism h_s such that $h_s^{-1} \circ sF \circ h_s = sX$. [Use the fact that if such a map does exist, then $\deg(sF)^m$ is bounded independent of m, contradicting b) and c).]

4. Complete the proof of i)→ ii) in the proof of 8.3.5 by showing that h does not depend on $p \geq k$ and h is one-to-one.
5. Let d satisfy 8.3.10. Show that $x_1(t) = c_4^{-m}e^{mt}(\dot{d} + 2d)$, $x_3(t) = \frac{1}{1-m}c_4^m e^{-mt}$, $x_4(t) = e^{-t}$ and $x_2(t)$ defined by $x_3(t)x_1(t) + x_2(t)c_4e^{-t} = d$ is a solution of 8.3.8.
6. Let $F = (X_1 + 3X_2^2X_3, X_2 + X_3^2, X_3)$. Show that for every $\varepsilon > 0$ the map $(1 + \varepsilon)F$ is polynomially linearizable to its linear part, i.e. there exists a polynomial automorphism h such that $h^{-1} \circ (1 + \varepsilon)F \circ h = (1 + \varepsilon)(X_1, X_2, X_3)$. Show that F is not polynomially linearizable.

8.4 Meisters' cubic-linear linearization conjecture and the MYC revisited

In the previous section we saw that the DMZ-conjecture is false. Nevertheless Meisters conjectured that for cubic-linear polynomial maps the DMZ-conjecture is true. This conjecture will be denoted by CLLC from now on. Observe that the truth of CLLC would still imply the truth of the Jacobian Conjecture by Drużkowski's reduction theorem. The problem of finding possible counterexamples is the fact that it is not easy to find cubic linear mappings for which one can prove something, since by the method of Chapter 6 the dimensions start to become big rather quick!

However in August 1996 I received a preprint of the very elegant paper [145] of Gorni and Zampieri in which they established the nice correspondence between cubic homogeneous and cubic linear polynomial mappings, as described in §4 of Chapter 6. This result reduced the study of Meisters' CLLC to the study of the DMZ-Conjecture for cubic homogeneous polynomial maps. So in order to show that CLLC is false one needs to find a cubic homogeneous counterexample to DMZ. I found such an example for all $n \geq 5$ when I was writing up Chapter 6 of this book. By using the Gorni-Zampieri theory it led to a counterexample of CLLC in dimension 17. Furthermore the example could be used to give quadratic and cubic-homogeneous counterexamples to both the LaSalle problem and MYC for all $n \geq 5$.

Now let us start describing the main results of this section.

Theorem 8.4.1 (van den Essen). *Let $n \geq 5$ and $F : \mathbb{C}^n \to \mathbb{C}^n$ be defined by*

$$\begin{aligned}F = (&X_1 + X_2X_5^2, X_2 + X_1^2X_5 - X_4X_5^2, X_3 + X_2^2X_5, X_4\\ &+ 2X_1X_2X_5 - X_3X_5^2, X_5, \ldots, X_n).\end{aligned}$$

Then F is invertible and for every $s \in \mathbb{C}\backslash\{0\}$, $|s| \neq 1$ sF is not globally analytically linearizable.

Proof: It is easy to verify by computer that F is invertible and to find the inverse (use 3.2.1). Observe that s is the only eigenvalue of $J(sF)(0) = sI_n$.

i) Let $0 < |s| < 1$. To show that sF is not globally analytically linearizable, it is sufficient to show that 0 is not a global attractor of the map sF (by 8.3.5). To do this we define

$$\begin{aligned}x_k(s) \;:=\; &(c_1(s)s^{-3k}, c_2(s)s^{-5k}, c_3(s)s^{-9k}, c_4(s)s^{-7k},\\ &s^k, \ldots, s^k), \text{ for all } k \geq 0\end{aligned}$$

where

$$\begin{aligned}c_1(s) &:= \varphi_3(s^2)\varphi_5(s^2)(s^4+1)(s^2-1)^2/s^{10},\\ c_2(s) &:= -\varphi_3(s^2)\varphi_5(s^2)(s^4+1)(s^2-1)^3(s^2+1)/s^{14},\\ c_3(s) &:= -(\varphi_3(s^2))^2\varphi_5(s^2)(s^4+1)^2(s^2-1)^5(s^2+1)^2/s^{18},\\ c_4(s) &:= (\varphi_3(s^2))^2\varphi_5(s^2)(s^4+1)(s^2-1)^4(s^8+s^6\\ &\quad+2s^4+2s^2+2)/s^{16}\end{aligned}$$

(φ_n denotes the n-th cyclotomic polynomial). Then a computation, which can be done for example by MAPLE, gives

$$(sF)(x_k(s)) = x_{k+1}(s), \text{ for all } k \geq 0.$$

So

$$(sF)^k(x_0(s)) = x_k(s), \text{ for all } k \geq 0.$$

Now observe that since $|s| \neq 1$ and $s \neq 0$ we have that $c_i(s) \neq 0$ for all $1 \leq i \leq 3$. So the first three components of $(sF)^k(x_0(s))$ tend to infinity if k tends to infinity. So 0 is not a global attractor of the map sF.

ii) Let $|s| > 1$. Put $G := F^{-1}$ and $t := s^{-1}$. So $0 < |t| < 1$. Suppose that sF is globally analytically linearizable. Then so is $(sF)^{-1} = G \circ t$ and hence, observing that t is the only eigenvalue of $J(G \circ t)(0)$, it follows that 0 is a global attractor of the map tG (since $(tG)^k(x) = t(G \circ t)^{k-1}(G(x))$. Now again we arrive at a contradiction by showing that 0 is not a global attractor of the map tG. To see this we define a point $x_0(t)$ such that $(tG)^k(x_0(t))$ tends to infinity if k tends to infinity. In fact define

$$\begin{aligned}
c_1(t) &:= \varphi_3(t^2)\varphi_5(t^2)(t^4+1)(t^2-1)^2/t^6,\\
c_2(t) &:= \varphi_3(t^2)\varphi_5(t^2)(t^4+1)(t^2-1)^3(t^2+1)/t^4,\\
c_3(t) &:= (\varphi_3(t^2))^2\varphi_5(t^2)(t^4+1)^2(t^2-1)^5(t^2+1)^2/t^{10},\\
c_4(t) &:= (\varphi_3(t^2))^2\varphi_5(t^2)(t^4+1)(t^2-1)^4(2t^8+2t^6\\
&\qquad +2t^4+t^2+1)/t^{12},
\end{aligned}$$

and put

$$x_k(t) := (c_1(t)t^{-3k}, c_2(t)t^{-5k}, c_3(t)t^{-9k}, c_4(t)t^{-7k}, t^k, \ldots, t^k), \text{ for all } k \geq 0.$$

Then as above we get $(tG)(x_k(t)) = x_{k+1}(t)$ and hence $(tG)^k(x_0(t)) = x_k(t)$ for all $k \geq 0$. It follows again that 0 is not a global attractor of the map tG. □

As a consequence we show that the CLLC is false in dimension 17.

Theorem 8.4.2 *Let* $\tilde{F} : \mathbb{C}^{17} \to \mathbb{C}^{17}$ *be defined by*

$$\begin{aligned}
\tilde{F} := \Bigg(& x_1 + \left(\frac{1}{6}x_4 + \frac{1}{6}x_5 - \frac{1}{3}x_6 - \frac{1}{6}x_7 - \frac{1}{6}x_8 + \frac{1}{3}x_9 + x_{17}\right)^3,\\
& x_2 + \left(\frac{1}{6}x_4 + \frac{1}{6}x_5 - \frac{1}{3}x_6 - \frac{1}{6}x_7 - \frac{1}{6}x_8 + \frac{1}{3}x_9 - x_{17}\right)^3,\\
& x_3 + \left(\frac{1}{6}x_4 + \frac{1}{6}x_5 - \frac{1}{3}x_6 - \frac{1}{6}x_7 - \frac{1}{6}x_8 + \frac{1}{3}x_9\right)^3,\\
& x_4 + \left(\frac{1}{6}x_1 + \frac{1}{6}x_2 - \frac{1}{3}x_3 + x_{17}\right)^3,\\
& x_5 + \left(-\frac{1}{6}x_1 - \frac{1}{6}x_2 + \frac{1}{3}x_3 + x_{17}\right)^3,\\
& x_6 + {x_{17}}^3,\\
& x_7 + \left(-\frac{1}{6}x_{10} - \frac{1}{6}x_{11} + \frac{1}{3}x_{12} + \frac{1}{12}x_{13} + \frac{1}{12}x_{14} - \frac{1}{12}x_{15} - \frac{1}{12}x_{16} + x_{17}\right)^3,\\
& x_8 + \left(-\frac{1}{6}x_{10} - \frac{1}{6}x_{11} + \frac{1}{3}x_{12} + \frac{1}{12}x_{13} + \frac{1}{12}x_{14} - \frac{1}{12}x_{15} - \frac{1}{12}x_{16} - x_{17}\right)^3,
\end{aligned}$$

$$x_9 + \left(-\frac{1}{6}x_{10} - \frac{1}{6}x_{11} + \frac{1}{3}x_{12} + \frac{1}{12}x_{13} + \frac{1}{12}x_{14} - \frac{1}{12}x_{15} - \frac{1}{12}x_{16}\right)^3,$$

$$x_{10} + \left(\frac{1}{6}x_1 - \frac{1}{6}x_2 - \frac{1}{3}x_6 + x_{17}\right)^3,$$

$$x_{11} + \left(\frac{1}{6}x_1 - \frac{1}{6}x_2 - \frac{1}{3}x_6 - x_{17}\right)^3,$$

$$x_{12} + \left(\frac{1}{6}x_1 - \frac{1}{6}x_2 - \frac{1}{3}x_6\right)^3,$$

$$x_{13} + \left(\frac{1}{6}x_1 + \frac{1}{6}x_2 - \frac{1}{3}x_3 + \frac{1}{6}x_4 + \frac{1}{6}x_5 - \frac{1}{3}x_6 - \frac{1}{6}x_7 - \frac{1}{6}x_8 + \frac{1}{3}x_9 + x_{17}\right)^3,$$

$$x_{14} + \left(\frac{1}{6}x_1 + \frac{1}{6}x_2 - \frac{1}{3}x_3 - \frac{1}{6}x_4 - \frac{1}{6}x_5 + \frac{1}{3}x_6 + \frac{1}{6}x_7 + \frac{1}{6}x_8 - \frac{1}{3}x_9 - x_{17}\right)^3,$$

$$x_{15} + \left(\frac{1}{6}x_1 + \frac{1}{6}x_2 - \frac{1}{3}x_3 + \frac{1}{6}x_4 + \frac{1}{6}x_5 - \frac{1}{3}x_6 - \frac{1}{6}x_7 - \frac{1}{6}x_8 + \frac{1}{3}x_9 - x_{17}\right)^3,$$

$$x_{16} + \left(\frac{1}{6}x_1 + \frac{1}{6}x_2 - \frac{1}{3}x_3 - \frac{1}{6}x_4 - \frac{1}{6}x_5 + \frac{1}{3}x_6 + \frac{1}{6}x_7 + \frac{1}{6}x_8 - \frac{1}{3}x_9 + x_{17}\right)^3,$$

$$x_{17}\Big)$$

Then $\tilde{F}$ is a counterexample to the CLLC.

Proof: Let F be as in 8.4.1 with $n = 5$. We follow the procedure and notation described in Chapter 6, §4. So we write each $F_i - X_i$ as a $\mathbb{C}$-linear combination of the following 16 cubic powers of linear forms

$$(X_2 + X_5)^3, (X_2 - X_5)^3, X_2^3, (X_1 + X_5)^3, (X_1 - X_5)^3,$$
$$X_5^3, (X_4 + X_5)^3, (X_4 - X_5)^3, X_4^3, (X_1 + X_2 + X_5)^3,$$
$$(X_1 - X_2 - X_5)^3, (X_1 + X_2 - X_5)^3, (X_1 - X_2 + X_5)^3,$$
$$(X_3 + X_5)^3, (X_3 - X_5)^3, X_3^3.$$

The matrix B_0 obtained in this way has rank 4 and we get that the matrix B, obtained by adding as 17-th column the fifth standard basis vector e_5, has full rank 5. So we can find a 17×5 matrix C satisfying $BC = I_5$. The matrix D_0 has already rank 5, so we get D by adding, as 17-th row, a row consisting of only zeros. Then let M be the matrix whose columns form a basis of the kernel of B. Then $(C|M)$ is invertible. Finally put $A := (D|0)(C|M)^{-1}$ and define

$$\tilde{F} : \mathbb{C}^{17} \to \mathbb{C}^{17} \text{ by } \tilde{F}(x) := x - (Ax)^{*3}.$$

Then $\tilde{F}$ is as above and it follows from 6.4.2 that $\tilde{F}$ is paired to F. Now let $s \in \mathbb{C}\backslash\{0\}, |s| \neq 1$. Then it follows from 6.4.11 that if $s\tilde{F}$ is globally analytically linearizable, then also sF is globally analytically linearizable, contradicting 8.4.1. So for every $s \in \mathbb{C}\backslash\{0\}, |s| \neq 1$ $s\tilde{F}$ is not globally analytically linearizable. □

Now we will show that the example of 8.4.1 also can be used to give quadratic and cubic homogeneous counterexamples to the MYC. To see this let F be as in 8.4.1. Put $H := F - X$. Then one readily verifies that $H = X_5 Q$, where Q is quadratic homogeneous. In fact

$$Q = (X_2X_5, X_1^2 - X_4X_5, X_2^2, 2X_1X_2 - X_3X_5, 0, \ldots, 0).$$

Theorem 8.4.3 *Let $n \geq 5$ and $F = -X + X_5Q$ (resp. $F = -X + Q$). Then F gives a counterexample to the MYC. More precisely*

$$x(t) = (120e^{3t}, 480e^{5t}, 23040e^{9t}, 11520e^{7t}, e^{-t}, \ldots, e^{-t})$$

(resp. $x(t) = (30e^t, 60e^{2t}, 720e^{4t}, 720e^{3t}, e^{-t}, \ldots, e^{-t})$) is a solution of $\dot{x} = F(x)$ which tends to infinity if t tends to infinity.

Proof: One readily verifies that for all $x \in \mathbb{R}^n$ all eigenvalues of $JF(x)$ are equal to -1 since $J(X_5Q)$ (resp. JQ) is nilpotent. That $x(t)$ is a solution of $\dot{x} = F(x)$ is left to the reader to verify. □

To conclude this section we describe how the map of 8.4.1 was found. The main idea is to construct F's of the form $X + H$ with JH nilpotent and satisfying the additional property that the H_i are linearly independent over $\mathbb{C}$. To find such maps one can proceed as follows: start with an automorphism of $\mathbb{C}^n$ of the form $X + F_{(2)} + F_{(3)}$, where $F_{(i)}$ is homogeneous of degree i. Then it follows from 6.2.12 that if $Y = (Y_1, \ldots, Y_n)$ are n new variables, then the map $H = (F_{(2)} + Y, -F_{(3)})$ has a nilpotent Jacobian matrix, i.e. $J_{X,Y}(H)$ is nilpotent. Of course if one starts with an automorphism of the form $X + F_{(2)} + \cdots + F_{(d)}$ with $d \geq 3$, one can use the formulae of 6.2.13. Then verify if the components of H are linearly independent over $\mathbb{C}$. If not, make a linear coordinate change $T \in GL_{2n}(\mathbb{C})$ such that $T^{-1}HT$ has its last component zero. Denote the new map $T^{-1}HT$ again by H and write $X_1, \ldots, X_{2n}$ instead of $X_1, \ldots, X_n, Y_1, \ldots, Y_n$. Then $H_{2n} = 0$. View the elements H_i, $1 \leq i \leq 2n - 1$ in $\mathbb{C}(X_{2n})[X_1, \ldots, X_{2n-1}]$. Then the Jacobian matrix $J_{X_1,\ldots,X_{2n-1}}(H_1, \ldots, H_{2n-1})$ is also nilpotent. Replace each H_i, $1 \leq i \leq 2n - 1$ by $H_i - H_i(0, \ldots, 0, X_{2n})$, i.e. remove the constant coefficient and verify if they are linearly independent over

$k := \mathbb{C}(X_{2n})$. If not, repeat the process above making again a coordinate transformation T with coefficients in k, etc.

If the H_i are linearly independent over k, specify some values in $\mathbb{C}$ for X_{2n} to obtain a map H such that its H_i are linearly independent over $\mathbb{C}$ (instead of k). We will illustrate this procedure with the following.

Example 8.4.4 Let $F = (X + Y^2, Y + (X + Y^2)^2)$. So we can write $F = (X, Y) + F_{(2)} + F_{(3)} + F_{(4)}$, where $F_{(2)} = (Y^2, X^2)$, $F_{(3)} = (0, 2XY^2)$ and $F_{(4)} = (0, Y^4)$. Then by 6.2.13 we introduce two sets of two variables, namely (Z, W) and (S, T) and we obtain that

$$H = (F_{(2)} - (S, T), F_{(4)}, F_{(3)} - (Z, W))$$

is such that $J_{X,Y,Z,W,S,T}H$ is nilpotent, i.e.

$$H = (Y^2 - S, X^2 - T, 0, Y^4, -Z, -W + 2XY^2)$$

has a nilpotent Jacobian matrix. Since $H_3 = 0$ we deduce that

$$J_{X,Y,W,S,T}(H_1, H_2, H_4, H_5, H_6) \text{ is nilpotent.}$$

We view the remaining five H_i's in $\mathbb{C}(Z)[X, Y, W, S, T]$. Since $H_5 = -Z$ we see that the fourth row of the 5×5 Jacobian matrix is zero. Consequently $J_{X,Y,W,T}$ (H_1, H_2, H_4, H_6) is nilpotent and we view H_1, H_2, H_4, H_6 in $\mathbb{C}(Z, S)[X, Y, W, T]$. Since $H_1 = Y^2 - S$ and $S \in \mathbb{C}(Z, S)$ we can replace H_1 by Y^2 and get that $J_{X,Y,W,T}(Y^2, X^2 - T, Y^4, -W + 2XY^2)$ is nilpotent. So if we rename the variables and the components as follows: $X_1 := X$, $X_2 := Y$, $X_3 := W$, $X_4 := T$, $h_1 := Y^2 = X_2^2$, $h_2 := X^2 - T = X_1^2 - X_4$, $h_3 := Y^4 = X_2^4$ and $h_4 := -W + 2XY^2 = -X_3 + 2X_1X_2^2$ we get that

$$J_{X_1,X_2,X_3,X_4}(X_2^2, X_1^2 - X_4, X_2^4, -X_3 + 2X_1X_2^2)$$

is nilpotent. Finally one readily verifies that h_1, h_2, h_3, h_4 are linearly independent over $\mathbb{C}$.

Exercises for §8.4

1. Let Q be as above, i.e. $Q = (X_2X_5, X_1^2 - X_4X_5, X_2^2, 2X_1X_2 - X_3X_5, 0, \ldots, 0)$. Put $F := X + Q$ and let $s \in \mathbb{C}$ with $0 < |s| < 1$. Show that sF gives a counterexample to the LaSalle problem. [Consider

$$\begin{aligned}
c_1(s) &:= \varphi_3(s)\varphi_5(s)(s^2+1)(s-1)^2/s^5,\\
c_2(s) &:= -\varphi_3(s)\varphi_5(s)(s^2+1)(s-1)^3(s+1)/s^7,\\
c_3(s) &:= -\varphi_3(s)^2\varphi_5(s)(s^2+1)^2(s-1)^5(s+1)^2/s^9,\\
c_4(s) &:= \varphi_3(s)^2(s^4+s^3+2s^2+2s+2)(s-1)^4\\
&\qquad \varphi_5(s)(s^2+1)/s^8.
\end{aligned}$$

[Let $x_0(s) := (c_1(s), \ldots, c_4(s), 0, \ldots, 0)$ and show that $(sF)^n(x_0(s)) = (c_1(s)s^{-n}, c_2(s)s^{-2n}, c_3(s)s^{-4n}, c_4(s)s^{-3n}, s^n, \ldots, s^n)$.]

2. Consider the automorphism

$$F = (X_1 + X_3(X_1X_4 + X_2X_3), X_2 - X_4(X_1X_4 + X_2X_3),\\ X_3 + X_4^2 + X_4^3, X_4).$$

Use the procedure described in 8.4.4 to arrive at a map $H : \mathbb{C}^4 \to \mathbb{C}^4$ with nilpotent Jacobian matrix and with rows that are linearly independent over $\mathbb{C}$. [This map F was used by the author to arrive at the map $(X_2, X_1^2 - X_4, X_2^2, 2X_1X_2 - X_3)$. After homogenization this gives the map

$$(X_2X_5, X_1^2 - X_4X_5, X_2^2, 2X_1X_2 - X_3X_5, 0),$$

i.e. the map Q which is used in both 8.4.1, $F = X + X_5Q$, and 8.4.3.]

Notes

Both Theorems 8.1.10 and 8.2.6 were obtained by Cima, Gasull and Mañosas in [56] and appeared in [59]. Also Example 8.2.4, which is due to Szlenk, is taken from their paper. Proposition 8.2.7 is taken from the paper [88] of Dillen. After the counterexample in Theorem 8.3.11 was found, Cima, Gasull and Mañosas in a subsequent paper [58] analyzed the basin of attraction of this example and also gave a class of counterexamples of the same type.

Theorem 8.3.2 was obtained by Hubbers and the author in [113]. The proof given here closely follows the proof given in that paper. An application of this result to polynomial automorphisms of finite order was given by Ivanenko in [160]. In particular he shows that a triangular automorphism of finite order is linearizable. See also the paper [183] of Kraft. Meisters' cubic-linear linearization conjecture appeared explicitly in [211] and as a question in [210], page 85. All results obtained in §4 are taken from the author's paper [111].

9 Group actions

In the previous chapter we saw how polynomial automorphisms played a role in the study of dynamical systems. In this chapter we show that polynomial automorphisms arise naturally in the study of group actions on affine varieties and discuss the interplay between group actions, polynomial automorphisms and locally finite, in particular locally nilpotent, derivations.

A fundamental theme throughout this chapter will be the study of the ring of invariants. After giving the basic definitions in §1 we prove Hilbert's finiteness theorem in §2, i.e. we show that the ring of invariants $A(X)^G$ of a reductive group action on any affine variety X over k is a finitely generated k-algebra. Hilbert's proof was criticized at the end of last century since it is not constructive. However, recently Harm Derksen in his thesis modified Hilbert's proof to make it constructive! In §3 we discuss Derksen's algorithm.

In §4 we consider a linearization conjecture due to Kambayashi. This conjecture asserts that every reductive group action on k^n is linearizable. We will prove this conjecture for $n = 2$. In fact we will show a stronger result, 9.4.2, namely that every algebraic group action on k^2 is conjugate to a subgroup of $\mathrm{Aff}(k, 2)$ or $J(k, 2)$. On the other hand, it was shown by Schwarz in 1989 that Kambayashi's conjecture is false in general. We refer the reader to the paper [262], where an O_4-action on k^4 is given which is not linearizable.

At the end of §4 we discuss a relationship between the still open linearization conjecture for $G := C_2$, the cyclic group of two elements, and the Cancellation problem and show how the positive solution of the cancellation problem due to Fujita in dimension two gives rise to the existence of so-called exotic $\mathbb{C}^n$'s for all $n \geq 3$.

In the last two sections we consider non-reductive group actions. More precisely in §6 we will show that the ring of invariants of a G_a-action need not be finitely generated, thereby giving a negative solution to Hilbert's fourteenth problem! In §5 we describe an algorithm which computes the ring of invariants of a given G_a-action when it is finitely generated over k. This algorithm is based on the kernel algorithm of Chapter 1 and the fact that every G_a-action arises as the exponential of a locally nilpotent derivation whose kernel equals the ring of invariants of the corresponding G_a-action.

At the end of §5 we discuss several open problems concerning locally nilpotent derivations.

9.1 Algebraic group actions: an introduction

In this section we recall some elementary facts and definitions concerning algebraic groups. For a more detailed treatment we refer to the textbooks [273] and [182].

Throughout this section k denotes an algebraically closed field of characteristic zero. By a **variety** we will always mean an affine variety over k. For several notions and results of algebraic geometry we refer to Appendix B.

Definition 9.1.1 i) Let G be a variety. If G is also a group such that both group operations, i.e. the multiplication map $\mu : G \times G \to G$ sending (x, y) to $x.y$ and the inversion map $i : G \to G$ sending x to x^{-1} are regular maps, then G is called a **linear algebraic group** or for short an **algebraic group**.

ii) Two algebraic groups G and G' are called **isomorphic** if there is an isomorphism $\varphi : G \to G'$ of varieties which is also an isomorphism of groups.

Example 9.1.2 1) The **linear group** $Gl_n(k)$, i.e. the set of invertible n by n matrices with elements in k. It is a variety (Appendix B) and

$$A(GL_n(k)) = k[T_{i,j}, 1 \leq i, j \leq n][d^{-1}]$$

where $d = \det(T_{i,j})$. For $n = 1$ we get the group $Gl_1(k) = k^*$, also denoted by G_m, which is called the **multiplicative group**.

2) The **special linear group** $Sl_n(k)$ consisting of the n by n matrices with determinant 1. Obviously $A(Sl_n(k)) = k[T_{i,j}, 1 \leq i, j \leq n]/(d-1)$.

3) $G_a := k$, the **additive group**. So $A(G_a) = k[T]$. The map

$$j : t \to \begin{pmatrix} 1 & t \\ 0 & 1 \end{pmatrix}$$

gives an isomorphism between G_a and the closed subgroup of $Sl_2(k)$ consisting of the matrices $j(t)$, where t runs through k.

4) The group $UT_n(k)$, i.e. the set of all $n \times n$ upper triangular matrices with only 1s on the main diagonal. So $A(UT_n(k)) = k[T_{i,j}, 1 \leq i < j \leq n]$.

5) The **Torus group** T_n, i.e. the set of all invertible diagonal matrices. One easily verifies that

$$\begin{aligned} A(T_n) &= k[T_{1,1}, \ldots, T_{n,n}][T_{1,1}^{-1}, \ldots, T_{n,n}^{-1}] \\ &= k[T_{1,1}, \ldots, T_{n,n}][(T_{1,1} \cdots T_{n,n})^{-1}]. \end{aligned}$$

6) The **orthogonal group** $O_n(k) = \{A \in Gl_n(k) | A^t A = I_n\}$.

7) All finite groups. Every finite group can be embedded in a permutation group and the permutation group S_n can be viewed as a closed subgroup of $Gl_n(k)$ by identifying a permutation σ of S_n with the matrix $\sum_{i=1}^n E_{i,\sigma(i)}$, where $E_{i,j}$ denotes the elementary matrix which has a 1 in position (i, j) and 0s everywhere else.

All examples above are closed subgroups of some $Gl_n(k)$ and hence algebraic groups, since one easily verifies that closed subgroups of algebraic groups are algebraic groups. In fact one can show that, up to an isomorphism, each algebraic group is a closed algebraic subgroup of some $Gl_n(k)$, i.e.

Proposition 9.1.3 *Let G be an algebraic group. Then G is isomorphic to a closed subgroup of some $Gl_n(k)$.*

We refer to [273] for a proof. See also Exercise 2.

Definition 9.1.4 Let G be an algebraic group and X a variety. We say that G **acts algebraically** on X, or that we have an **algebraic group action of G on X**, if there exists a regular map $\varphi : G \times X \to X$ denoted for brevity by $\varphi(g, x) = gx$, such that

1) $g_1(g_2x) = (g_1g_2)x$, for all $g_1, g_2 \in G$, all $x \in X$.
2) $1.x = x$, for all $x \in X$.

In this case we also say that X is a **G-variety**.

Example 9.1.5 i) Let $G = Gl_n(k)$ and $X = k^n$. Then the usual action of $Gl_n(k)$ on k^n is an algebraic group action.
ii) Let $X = k^n$ and F be an automorphism satisfying $F^2 = I$. Let $C_2 = \{-1, 1\}$ be the cyclic group of two elements. Then $-1.x = F(x)$ defines an algebraic group action of C_2 on X.

As already remarked in the beginning of this chapter, group actions give many examples of polynomial automorphisms, namely for each $g \in G$ the induced map $\varphi_g^* : A(X) \to A(X)$ defined by $\varphi_g^*(f)(x) = f(gx)$ is an automorphism of $A(X)$ with inverse $\varphi_{g^{-1}}^*$.

Now let X be a G-variety. For $g \in G$ and $f \in A(X)$ we define the function $g.f \in A(X)$ by the formula

$$(g.f)(x) = f(g^{-1}x), \text{ for all } x \in X.$$

This gives an action of G on the ring $A(X)$, i.e. a map $G \times A(X) \to A(X)$ which satisfies the properties 1) and 2) of 9.1.4. This map is called the **regular representation** of G on $A(X)$. Observe that $g.f = \varphi_{g^{-1}}^*(f)$.

A function $f \in A(X)$ is called an **invariant** of the group action if $g.f = f$ for all $g \in G$ or equivalently $\varphi_g^*(f) = f$ for all $g \in G$. The set of all invariant functions is denoted by $A(X)^G$. It is a k-subalgebra of $A(X)$ and is called the **ring of invariant functions** of the group action.

Example 9.1.6 Let $X = k^n$ and G be the symmetric group S_n. View S_n as a subgroup of $Gl_n(k)$, as described in 9.1.2. For the action of G on X we take the restriction to G of the standard action of $Gl_n(k)$ on X. This action induces, as described above, an action of G on the polynomial ring $k[X_1, \ldots, X_n]$ and it is well-known that $k[X_1, \ldots, X_n]^G$, consisting of all symmetric polynomials, is generated as a k-algebra by the elementary symmetric polynomials $p_k = \sum X_{i_1} \ldots X_{i_k}$ where the sum runs over all sequences $1 \le i_1 < \cdots < i_k \le n$. In fact $k[X_1, \ldots, X_n]^G$ is isomorphic to a polynomial ring in n variables since $p_1, \ldots, p_n$ are algebraically independent over k.

This and other results have led to one of the most important problems in **invariant theory**, namely

Problem 9.1.7 Let X be a G-variety. Is the ring $A(X)^G$ a finitely generated k-algebra?

The case $X = \mathbb{C}^n$ was studied first. It was Emmy Noether who showed that for finite groups the answer is yes. Also for other groups like Sl_2 the answer turned out to be affirmative. This was proved by Gordan in [142]. The real breakthrough was obtained by Hilbert who showed that for a large class of groups, including the ones just mentioned, the answer is affirmative (9.2.1). These are the so-called reductive groups. See 9.1.8 for the definition.

On the other hand it is not difficult to show that there exist irreducible affine varieties X such that $A(X)^{G_a}$ is not a finitely generated $\mathbb{C}$-algebra (Exercise 9, §5).

However for $X = \mathbb{C}^n$ the general case remained open until 1956. Then Nagata in [222] constructed an example of a group action on $\mathbb{C}^{32}$ whose ring of invariants is not finitely generated. This example also gives a counterexample to Hilbert's 14-th problem (1.2.22). More recently in [252] Roberts gave a new counterexample to Hilbert's 14-th problem. Deveney and Finston in [82] used this result to construct a G_a-action on $\mathbb{C}^7$ whose ring of invariants is not finitely generated. We will discuss this example in §6. See also 9.6.17 for a more recent five dimensional counterexample due to Daigle and Freudenburg. Now we turn to the reductive groups mentioned above. To describe these groups we need some more terminology.

Let V be a finite dimensional k-vector space with k-basis $(f_1, \ldots, f_d)$ and $\rho : G \to Gl(V)$ a homomorphism of abstract groups. So for each $g \in G$ we have

$$\rho(g)f_i = \sum c_{i,j}(g)f_j, \; c_{i,j}(g) \in k \text{ for all } i, j.$$

We say that ρ is a **linear representation** of G in V or a **regular representation** or a **rational representation** if $c_{i,j} \in A(G)$ for all i, j. One readily verifies that this definition does not depend on the k-basis (f) of V. We also say that V **is a** G**-module**. A linear representation is a special case of a group action, namely G acts by linear transformations on V, $g.v := \rho(g)v$ for all $g \in G$ and all $v \in V$.

A G-module $V \neq 0$ is called **irreducible**, or **simple** if it has no proper non-zero G-submodules. For example, the standard action of $Gl_n(k)$ on k^n is simple (Exercise 3). Furthermore V is called **completely reducible** or **semisimple** if for every G-submodule W of V there exists a complementary G-submodule W' of V such that $V = W \oplus W'$.

Definition 9.1.8 An algebraic group G is called **(linearly) reductive** if every finite dimensional G-module is completely reducible.

Example 9.1.9 (See [182] for a proof). The groups $Gl_n(k)$, $Sl_n(k)$, $T_n(k)$, $O_n(k)$ and all finite groups are reductive. The group G_a is not reductive, Exercise 4.

To conclude this section we give some useful results which will be used below.

Proposition 9.1.10 *Let X be a G-variety. Then the regular representation of G on $A(X)$ is locally finite and rational, i.e. for each finite dimensional linear subspace W of $A(X)$ the space $GW := \sum_{g \in G} g.W$ is finite dimensional and the induced representation of G on GW is rational.*

Proof: i) The action $\rho : G \times X \to X$ induces an algebra homomorphism $\rho^* : A(X) \to A(G \times X)$ which satisfies $\rho^*(f)(g, x) = f(g.x)$, for all $f \in A(X)$, $g \in G$

and $x \in X$. Furthermore, as described in Appendix A.2, the ring $A(G \times X)$ can be identified with $A(G) \otimes_k A(X)$: an element $p \otimes f$ corresponds to the regular function $(g, x) \to p(g)f(x)$.

ii) Let W be a finite dimensional linear subspace of $A(X)$. Then there exist $f_1, \ldots, f_n$ in $A(X)$ such that $\rho^*(W) \subset \sum_{i=1}^n A(G) \otimes f_i$. First we show that $G.W \subset \sum kf_i$, which implies that $G.W$ is finite dimensional. Therefore let $f \in W$. Then $\rho^*(f) = \sum p_i \otimes f_i$, for some $p_i \in A(G)$. Furthermore by i) we get

$$\begin{aligned}(g.f)(x) &= f(g^{-1}.x) = (\rho^* f)(g^{-1}, x) \\ &= \sum p_i(g^{-1}) f_i(x) = \left(\sum p_i(g^{-1}) f_i\right)(x).\end{aligned}$$

So

$$g.f = \sum p_i(g^{-1}) f_i \in \sum kf_i. \tag{9.1.11}$$

iii) Finally we show that the representation of G on GW is rational. Choose a basis $(f_1, \ldots, f_d)$ of $G.W \subset A(X)$ and extend it to a basis $(f_1, \ldots, f_d, \ldots, g_j, \ldots)$ of $A(X)$. Let $f \in GW$. Then

$$\rho^*(f) = \sum r_i \otimes f_i + \sum s_j \otimes g_j$$

with $r_i, s_j \in A(G)$. So, similarly to formula 9.1.11, we obtain

$$g.f = \sum r_i(g^{-1}) f_i + \sum s_j(g^{-1}) g_j, \text{ for all } g \in G.$$

Since $g.f \in GW$ we conclude that $s_j(g^{-1}) = 0$ for all $g \in G$, so $s_j = 0$ for all j. Consequently $g.f = \sum r_i(g^{-1}) f_i$. Since $g \to r_i(g^{-1})$ is regular for each i, the desired result follows. □

Corollary 9.1.12 *Let X be a G-variety. If G is reductive, then every finite dimensional $k[G]$-submodule W of $A(X)$ is a G-module and $A(X)$ is a semisimple $k[G]$-module.*

Proof: i) The first part follows from the last part of 9.1.10.

ii) To prove that $A(X)$ is a semisimple $k[G]$-module it suffices to show that each $f \in A(X)$ is contained in a sum of simple $k[G]$-submodules of $A(X)$, for then $A(X)$ is a sum of simple $k[G]$-submodules of $A(X)$ and hence semisimple, by §3 of Chapter 1. So let $f \in A(X)$. By 9.1.10 $W := k[G]f$ is finite dimensional and by i) it is a G-module. Since G is reductive, every G-submodule of W has a G-complement. So again by i) each $k[G]$-submodule of W has a $k[G]$-complement. So W is a semisimple $k[G]$-module. Consequently W, and hence f, is contained in a sum of simple $k[G]$-modules. □

Remark 9.1.13 From 9.1.12 we deduce: if G is reductive then $A(G)$ is a semisimple $k[G]$-module. The converse also holds ([182]).

So we get

(9.1.14) G is reductive if and only if $A(G)$ is a semisimple $k[G]$-module.

From this one immediately deduces that T_n is reductive, for $A(T_n)$ is a direct sum of the one-dimensional T_n-modules $kT_{1,1}^{i_1}\dots T_{n,n}^{i_n}$, where $i_1,\dots,i_n \in \mathbb{Z}$.

Exercises for §9.1

1. Show that the coordinate rings of the groups described in 1)-5) of 9.1.2 are as given there.
2. Let G be an algebraic group. Show that G is isomorphic to a closed subgroup of $Gl_n(k)$. [First show that there exists a finite dimensional k-vector space $E \subset A(G)$ with $GE \subset E$: just take a finite set of k-generators of $A(G)$ and apply 9.1.10. Let $(f_1,\dots,f_n)$ be a k-basis of E. So $g.f_i = \sum c_{j,i}(g)f_j$, for some $c_{i,j}(g) \in k$. Define $\varphi(g) = (c_{i,j}(g))_{1\le i,j\le n}$ and show that φ is a group isomorphism between G and $\varphi(G) \subset Gl_n(k)$. To see that $\varphi(G)$ is a closed subgroup of $Gl_n(k)$ we refer to the proof of 2.3.5, page 42 in [273].]
3. Show that the standard action of $Gl_n(k)$ on k^n is irreducible. [Observe that for any two non-zero vectors in k^n there exists an invertible linear map sending one vector to the other.]
4. Show that G_a is not reductive. [Use the description of G_a as a subgroup of $Sl_2(k)$ and show that the G_a-invariant subspace ke_1 of k^2 has no G_a-invariant complement, where e_1 is the first standard basis vector of k^2.]
5. Let G be an algebraic group and V an arbitrary k-vector space. A group homomorphism $\rho : G \to Gl(V)$ is called a **locally finite representation of** G if every vector $v \in V$ is contained in a finite dimensional k-linear subspace W of V such that $GW \subset W$ and such that the restriction $\rho_{|W} : W \to Gl(W)$ is a rational representation as in §1.

   (i) Show that every finite dimensional $k[G]$-submodule of V is a G-module.
   (ii) Prove that V is a semisimple $k[G]$-module if G is reductive. [Look at the proof of 9.1.12.]

9.2 Hilbert's finiteness theorem

The main aim of this section is to give an affirmative answer to 9.1.7 in case the group G is reductive. This result is due to Hilbert in [154]. In fact Hilbert's original result was for Sl_n, however his argument could easily be generalized to the version described in 9.2.1 below. The proof consists of two parts: first we show that we may assume that $X = k^n$ and that G acts linearly on X, i.e. as a subgroup of $Gl_n(k)$. Then we prove this special case in 9.2.3. Its proof is based on two facts. The first fact is that the polynomial ring $k[X_1,\dots,X_n]$ is Noetherian. This result, which is classical now, was first proved by Hilbert in [154] in order to use it as an essential ingredient in the proof of his finiteness theorem. The second fact is the existence of the so-called Reynolds operator (9.2.4).

Theorem 9.2.1 (Hilbert). *Let G be a reductive group and X a G-variety. Then $A(X)^G$ is a finitely generated k-algebra.*

Proof: i) Let $f_1, \ldots, f_r$ generate the k-algebra $A(X)$. By 9.1.10 the k-subspace W of $A(X)$ generated by the elements gf_i, where $g \in G$ and $1 \leq i \leq r$, is finite dimensional and the representation of G on W is rational. Let $x_1, \ldots, x_n$ be a k-basis of W. Then the x_i generate the k-algebra $A(X)$ and G acts linearly on the x_i, i.e.

$$g.x_i = \sum_j c_{i,j}(g)x_j, \text{ for some } c_{i,j}(g) \in k.$$

Consider the polynomial ring $A := k[T_1, \ldots, T_n]$ and let G act as a subgroup of $Gl_n(k)$ on A by the formula

$$g.T_i = \sum_j c_{i,j}(g)T_j.$$

Then one readily verifies that the surjective ring homomorphism $\varphi : A \to A(X)$ sending T_i to x_i respects the G-actions on both rings, so $\varphi(A^G) \subset A(X)^G$. However we even have

$$\varphi(A^G) = A(X)^G. \tag{9.2.2}$$

In fact this readily follows from Exercise 4 ii) below and Proposition 9.1.10, which guarantees that the locally finiteness hypothesis in Exercise 4 is satisfied. Using 9.2.2 the finiteness of $A(X)^G$ follows readily from the next result. □

Proposition 9.2.3 *Let G be reductive and act on $A := k[T_1, \ldots, T_n]$ as a subgroup of $Gl_n(k)$, i.e. for each $g \in G$ $g.T_i = \sum_j c_{i,j}(g)T_j$ for some $c_{i,j}(g) \in k$. Then A^G is a finitely generated k-algebra.*

The proof is based on the following result, which we will use only for the case $X = k^n$.

Proposition 9.2.4 *Let X be a G-variety and G a reductive group. Then there exists a unique G-invariant linear projection $R : A(X) \to A(X)^G$, i.e. $R(g.f) = R(f)$ for all $g \in G$, all $f \in A(X)$ and $R(f) = f$ for all $f \in A(X)^G$. It has the following properties:*

1) *R is $A(X)^G$-linear.*
2) *If $W \subset A(X)$ is a G-invariant linear subspace, then $R(W) \subset W$.*

Proof: Follows from Exercises 2, 3 and 5 at the end of this section: just apply 2 ii), 2 iii), 2 iv) to $V := A(X)$ and use Exercise 5 ii). □

The uniquely determined operator R in Proposition 9.2.4 is called the **Reynolds operator**.

Proof of Proposition 9.2.3: The polynomial ring A is a graded ring with the standard grading. Since G acts linearly on A, it follows that if f is an invariant, then all its homogeneous components are invariants too. Let I_N be the ideal of A generated by all invariants in $A_+ := \oplus_{n>0} A_n$. Then, by the remark just made and the fact that A is Noetherian, we deduce that I_N is generated by a finite number of homogeneous invariants, say $f_1, \ldots, f_r$ each of positive degree. Suppose now that h is a homogeneous invariant of degree d. Using induction on d we will prove that $h \in k[f_1, \ldots, f_r]$, which of course completes the proof. If $d = 0$ then $h \in k$. Otherwise write $h = \sum a_i f_i$ with $a_1, \ldots, a_r \in A$. Without loss of generality we may assume that a_i is homogeneous of degree $d - \deg f_i < d$ for all i. Applying the Reynolds operator to the case $X = k^n$, i.e. $A(X) = A$, yields

$$h = R(h) = R\left(\sum a_i f_i\right) = \sum R(a_i) f_i.$$

Now observe that by Exercise 2 $\deg R(a_i) \leq \deg a_i < d$ for all i. By the induction hypothesis we deduce that $R(a_1), \ldots, R(a_r) \in k[f_1, \ldots, f_r]$ whence $h \in k[f_1, \ldots, f_r]$.

Exercises for §9.2

1. Notation as in 9.2.4. Let G be a finite group. Show that the Reynolds operator R is given by $R(f) = \frac{1}{\#G} \sum_{g \in G} g.f$, for all $f \in A$.
2. Let $A = k[T_1, \ldots, T_n]$ and R be the Reynolds operator on A. Let A_d be the set of homogeneous polynomials of A of degree $d \geq 0$. Show that $R(A_d) \subset A_d$. [Use that A_d is G-invariant and property 2) of Proposition 9.2.4.]
3. Let G be a reductive group and $\rho : G \to Gl(V)$ a locally finite representation of G, Exercise 5, §1.

   (i) Show that there exists a surjective linear map $R : V \to V^G$ such that $R^2 = R$ and $R(gv) = R(v)$, for all $g \in G$, $v \in V$. [By Exercise 5, §1 V^G has a G-invariant complement W, i.e. $V = V^G \oplus W$ and $GW \subset W$. Define $R : V \to V^G$ to be the projection on V^G.]

   (ii) Show that V^G has a uniquely determined G-invariant complement, namely the subspace E_0 generated by all elements $g.v - v$, with $g \in G$ and $v \in V$. [Let $E := \ker R$. Show that $V^G = \ker(R - I)$ and deduce that $V = V^G \oplus E$. Furthermore show that $E_0 \subset E$ and, using Exercise 5, §1 once more, deduce that $E_0 = E$. Finally show that if C is any G-invariant complement of V^G then $E_0 \subset C$, whence $E_0 = C$.]

   (iii) Show that if R_1 is another operator satisfying the properties as in i) then $R_1 = R$, i.e. R is uniquely determined. [Use ii).]

   (iv) Show that R has the following property: if W is a G-invariant subspace of V then $R(W) \subset W$.

4. Let G be a reductive group, V, W locally finite G-modules and $\phi : V \to W$ G-linear. Let R_V, R_W denote the corresponding operators of Exercise 3.
   (i) Show that $\phi \circ R_V = R_W \circ \phi$. [Use the description of the uniquely determined G-invariant complement of V^G resp. W^G given in Exercise 3 ii).]
   (ii) Deduce that $\phi(V^G) \subset W^G$ and in case ϕ is surjective that $\phi(V^G) = W^G$.

5. Let G be a reductive group and V a locally finite G-module. Consider on $V^G \otimes_k V$ the locally finite G-action given by $g \cdot (v_0 \otimes v) = v_0 \otimes gv$, for all $v_0 \in V^G$ and all $v \in V$.
   (i) Show that $R_{V^G \otimes V} = 1 \otimes R_V$. [Use the uniqueness of the operator R in Exercise 3.]
   (ii) Let X be a G-variety and consider on $W := A(X)$ the regular representation of G. Show that the corresponding operator R_W on $A(X)$ is $A(X)^G$-linear. [Consider the G-linear map $\phi : A(X)^G \otimes_k A(X) \to A(X)$ given by $\phi(f \otimes a) = fa$ and use both i) and Exercise 4 i).]

9.3 Constructive invariant theory: Derksen's algorithm to compute the invariants for reductive groups

In the previous section we showed that $A(X)^G$ is a finitely generated k-algebra in case X is a G-variety and G is reductive. The proof Hilbert gave in 1890 was criticized because it was non-constructive. When Gordan, who obtained the first finiteness result for $A(k^n)^{Sl_2}$ in 1868, saw Hilbert's proof he made his famous exclamation: "Das ist Theologie und nicht Mathematik", i.e. "That is theology and not mathematics".

However, recently, more than a hundred years after Hilbert, Harm Derksen in his thesis was able to modify Hilbert's proof to make it constructive, i.e. he gave a constructive proof of 9.2.3, which by 9.2.2 gives a constructive proof of 9.2.1 as well, based on a clever description of the ideal I_N and a Gröbner basis computation.

Now let us start to describe the algorithm: from the argument given in the proof of 9.2.3 we see that it suffices to give a constructive method to find homogeneous invariants $f_1, \ldots, f_r$ which generate the ideal $I_N \subset A := k[X_1, \ldots, X_n]$. The next lemma shows that it suffices to find homogeneous generators of I_N, namely

Lemma 9.3.1 *If $h_1, \ldots, h_r \in A$ are homogeneous generators of I_N, then*

$$A^G = k[R(h_1), \ldots, R(h_r)].$$

Proof: Let $f_1, \ldots, f_s \in A$ be homogeneous invariants which generate the ideal I_N, we saw in the proof of 9.2.3 that they exist, and let $\mathfrak{m}$ be the maximal ideal $(X_1, \ldots, X_n)$ of A. The images of $f_1, \ldots, f_s$ in $I_N/\mathfrak{m}I_N$ span this vector space and the same holds for the images of the h_i. So we get

$$h_i = \sum_j c_{i,j} f_j + q_i, \; c_{i,j} \in k, \; q_i \in \mathfrak{m}I_N. \tag{9.3.2}$$

Since G acts linearly on A and $I_N = \sum Af_j$ is G-invariant, we deduce that $\mathfrak{m}I_N$ is G-invariant, so by property 2) of the Reynolds operator in 9.2.4 we get

$$(9.3.3) \qquad R(\mathfrak{m}I_N) \subset \mathfrak{m}I_N.$$

Furthermore $R(f_j) = f_j$. By 9.3.2 and 9.3.3 this gives

$$h_i - R(h_i) = q_i - R(q_i) \in \mathfrak{m}I_N.$$

So the images of $R(h_1), \ldots, R(h_r)$ in $I_N/\mathfrak{m}I_N$ span this vector space. Then the homogeneous Nakayama lemma implies that these homogeneous polynomials (Exercise 2, §2), which are also invariants, generate the ideal I_N. Then the argument given in the proof of 9.2.3 implies that $A^G = k[R(h_1), \ldots, R(h_r)]$. □

So it suffices to find homogeneous generators of the ideal I_N. Therefore let $V := k^n$ and define

$$\mathfrak{b} := \{h \in A(V \times V) | h(v, g.v) = 0, \text{ for all } v \in V, \text{ all } g \in G\}.$$

We will identify $A(V \times V)$ with $A(V) \otimes_k A(V)$ and with the polynomial ring $k[X, Y] := k[X_1, \ldots, X_n, Y_1, \ldots, Y_n]$.

Since G acts linearly on V it is not difficult to verify that $\mathfrak{b}$ is a homogeneous ideal in $k[X, Y]$, i.e. if f belongs to $\mathfrak{b}$ then all its homogeneous components with respect to the standard grading on $k[X, Y]$ belong to $\mathfrak{b}$ (Exercise 1). Now the key point is

Proposition 9.3.4 $I_N = ((Y_1, \ldots, Y_n) + \mathfrak{b}) \cap k[X]$.

Let us assume this result first. Then we claim that it suffices to find a finite set of generators of $\mathfrak{b}$. Namely if we have such a set of generators, then we get a finite set of homogeneous generators of $\mathfrak{b}$: since $\mathfrak{b}$ is homogeneous, just take all homogeneous components of all the finite number of generators, say $f_1(X, Y), \ldots, f_r(X, Y)$. Then by 9.3.4 $f_1(X, 0), \ldots, f_r(X, 0)$ are homogeneous generators of I_N, which as we saw above is sufficient!

So apart from 9.3.4 it remains to show how to construct generators for $\mathfrak{b}$. This can be done as follows: suppose that $A(G)$ is isomorphic to $k[Z_1, \ldots, Z_s]/J$, where J is an ideal in $k[Z_1, \ldots, Z_s]$ generated by $a_1, \ldots, a_t$. The linear action on A is given by

$$g.X_i = c_{i,1}(g)X_1 + \cdots + c_{i,n}(g)X_n, \text{ for all } g \in G$$

where $c_{i,j} \in k[Z_1, \ldots, Z_s]$. These elements are determined up to elements of J. Let $\Gamma \subset G \times V \times V$ be the Zariski-closed subset

$$\Gamma := \{(g, v, g.v) | g \in G, v \in V\}.$$

Then the coordinate ring of Γ is given by $k[Z_1, \ldots, Z_s, X_1, \ldots, X_n, Y_1, \ldots, Y_n]/I_\Gamma$, where I_Γ is generated by J and the elements

$$Y_i - \sum_j c_{i,j}X_j, \, 1 \leq i \leq n.$$

One easily verifies (Exercise 2) that

$$\mathfrak{b} = I_\Gamma \cap k[X, Y].$$

Finally choose an admissible ordering on the terms of $k[X, Y, Z]$ such that each Z_i is bigger than every monomial in X and Y and compute a Gröbner basis S of I_Γ with respect to this ordering. Then it follows from the elimination algorithm of Appendix C that $S \cap k[X, Y]$ is a Gröbner basis of $I_\Gamma \cap k[X, Y] = \mathfrak{b}$, as desired. So it remains to give the following.

Proof of Proposition 9.3.4: Under the identification $A(V \times V) = A(V) \otimes_k A(V) = k[X, Y]$, $A(V) \otimes k$ is identified with $k[X]$, $k \otimes A(V)$ with $k[Y]$ and we have to prove that

$$(A(V) \otimes \mathfrak{m} + \mathfrak{b}) \cap (A(V) \otimes k) = I_N$$

where $\mathfrak{m}$ is the maximal ideal $(Y_1, \ldots, Y_n)$ in $k[Y]$.

i) Let f be an invariant in $A(V)_+$. So $1 \otimes f \in A(V) \otimes \mathfrak{m}$. Furthermore

$$(f \otimes 1 - 1 \otimes f)(v, g.v) = f(v) - f(gv) = 0$$

since $f \in A(V)^G$, i.e. $f \otimes 1 - 1 \otimes f \in \mathfrak{b}$. Consequently

$$f \otimes 1 = 1 \otimes f + (f \otimes 1 - 1 \otimes f) \in (A(V) \otimes k) \cap (A(V) \otimes \mathfrak{m} + \mathfrak{b})$$

which implies the inclusion "$\supset$".

ii) Conversely, if $f \otimes 1 \in A(V) \otimes \mathfrak{m} + \mathfrak{b}$ we can write

$$f \otimes 1 = \sum c_i \otimes f_i + b_0 \tag{9.3.5}$$

with $c_i \in A(V)$, $f_i \in \mathfrak{m}$ and $b_0 \in \mathfrak{b}$. We view $V \times V$ as a G-variety where G acts only on the second factor. Using the identification $A(V \times V) = k[X, Y]$ one easily verifies that under this action $A(V \times V)^G = A(V) \otimes A(V)^G$, Exercise 3. So we get the corresponding Reynolds operator

$$\tilde{R} : A(V \times V) = A(V) \otimes A(V) \to A(V \times V)^G = A(V) \otimes A(V)^G.$$

Since $id \otimes R$ is also a Reynolds operator we get $\tilde{R} = id \otimes R$ because of its uniqueness (9.2.4). We apply $\tilde{R} = id \otimes R$ to 9.3.5. This gives

$$f \otimes 1 = \sum c_i \otimes R(f_i) + \tilde{R}(b_0). \tag{9.3.6}$$

Observe that $\mathfrak{b}$ is G-stable: if $h \in A(V \times V)$ and $g \in G$ then $g.h(v_1, v_2) = h(v_1, g^{-1}v_2)$ for all $v_1, v_2 \in V$. Consequently if $h \in \mathfrak{b}$ then $(g.h)(v, g'v) = h(v, g^{-1}g'v) = 0$ for all $g' \in G$ and all $v \in V$. So $gh \in \mathfrak{b}$. Hence property 2) of the Reynolds operator in 9.2.4 implies that $\tilde{R}(\mathfrak{b}) \subset \mathfrak{b}$.

iii) Finally, evaluating 9.3.6 on the elements (v, v) gives

$$f(v) = \sum c_i(v)R(f_i)(v), \text{ for all } v \in V$$

since $\tilde{R}(b_0) \in \mathfrak{b}$ and any element $h \in \mathfrak{b}$ satisfies $h(v, gv) = 0$ for all $g \in G$, i.e. in particular for $g = 1$ we have $h(v, v) = 0$. In other words

$$f = \sum c_i R(f_i).$$

Since $f_i \in \mathfrak{m}$ the same holds for $R(f_i)$, again using that $\mathfrak{m}$ is G-invariant and property 2) of the Reynolds operator, i.e. $R(f_i) \in A(V)_+$. Also $R(f_i) \in A(V)^G$. So by definition of I_N we deduce that $f \in I_N$. □

To conclude this section we summarize Derksen's algorithm.

1) Write $A(G) = k[Z_1, \ldots, Z_s]/(a_1(Z), \ldots, a_t(Z))$. Put $J := (a_1(Z), \ldots, a_t(Z))$.
2) Write $g.X_i = c_{i,1}(g)(X_1) + \cdots + c_{i,n}(g)X_n$, for some $c_{i,j} \in k[Z]$.
3) Let $I_\Gamma \subset k[X, Y, Z]$ be the ideal generated by J and the elements

$$Y_i - \sum_j c_{i,j}(Z)X_j, 1 \leq i \leq n.$$

Choose an admissible ordering on the terms of $k[X, Y, Z]$ such that $Z_i > X^\alpha Y^\beta$ for all i, α, β. Compute a Gröbner basis S of I_Γ.
4) Compute $S \cap k[X, Y]$. Let $f_1(X, Y), \ldots, f_r(X, Y)$ be the set of all homogeneous components of these Gröbner basis elements.
5) Then $A^G = k[R(f_1(X, 0)), \ldots, R(f_r(X, 0))]$.

Exercises for §9.3

1. Let $f = \sum f_d$ be the homogeneous decomposition of $f \in A$. Show that if $f \in \mathfrak{b}$ with $\mathfrak{b}$ as above, then $f_d \in \mathfrak{b}$ for all $d \geq 0$. [Choose $g \in G$ fixed. Then $f(x, g.x) = 0$ for all $x \in k^n$. Let $g.X_i = \sum_j c_{i,j}(g)X_j$, then $\tilde{f}_d := f_d(X, g.X)$ is also homogeneous of degree d and $\tilde{f}_d(x) = 0$ for all $x \in k^n$.]
2. Prove that $\mathfrak{b} = I_\Gamma \cap k[X, Y]$.
3. Let G act linearly on $V \times V$ only on the second component, i.e. $g.(v_1, v_2) = (v_1, g.v_2)$. Show that $A(V \times V)^G = A(V) \otimes_k A(V)^G$. [Write an element of $A(V \times V) = k[X, Y]$ in the form $\sum X^\alpha f_\alpha(Y)$ and use that for $g \in G$, $g.Y_i = \sum_j c_{i,j}(g)Y_j$ for some $c_{i,j}(g) \in k$.]
4. Let $G := G_m$ act on $V := k^4$ as follows: $\lambda.x = (\lambda^{-5}x_1, \lambda^{-3}x_2, \lambda^2 x_3, \lambda^4 x_4)$, for all $\lambda \in k^*$ and all $x \in k^4$. Compute $A(V)^G$. [Write $A(G) = k[Z_1, Z_2]/(Z_1Z_2 - 1)$. The matrix $(c_{i,j}(Z))$ representing the action on V is given by $c_{i,j} = 0$ if $i \neq j$, $c_{1,1} = Z_2^5$, $c_{2,2} = Z_2^3$, $c_{3,3} = Z_1^2$ and $c_{4,4} = Z_1^4$.]

9.4 A linearization conjecture for reductive group actions

Throughout this section k denotes an algebraically closed field of characteristic zero and we will consider group actions on $V := k^n$. The ultimate goal of the investigation of such group actions is to obtain a complete classification. For $n \geq 3$ very little is known, however if $n = 2$ rather precise statements can be made. In fact the main result of this section (9.4.2), which is due to Kambayashi [174], asserts that in case $n = 2$ every such action is conjugate to a subgroup of $\mathrm{Aff}(2, k)$ or $J(2, k)$. This implies that every reductive group action on k^2 becomes linear after a suitable coordinate change, 9.4.8. Kambayashi conjectured therefore in [174] that this last statement is true for all $n \geq 2$. This conjecture became known as the Linearization Conjecture for reductive group actions, see 9.4.1 below. It turned out that the conjecture is false in higher dimension: in 1989 Schwarz discovered the first counterexamples, namely non-linearizable actions of the orthogonal group O_2 on k^4 and of Sl_2 on k^7 (we refer to [262]). The first non-linearizable actions of finite groups e.g. for dihedral groups of order ≥ 10 on k^4 were found by Masuda, Moser-Jauslin and Petrie (see [196] and [202]). The stronger problem, namely whether there exist automorphisms of finite order which are non-linearizable, is still open even for automorphisms of order 2. In fact if all such automorphisms are linearizable, then the so-called cancellation problem would have an affirmative answer. At the end of this section we briefly discuss this problem and its relation to the linearization conjecture.

Before we formulate and prove the main result of this section let us first formulate Kambayashi's conjecture.

(9.4.1) **Kambayashi's Linearization Conjecture.**

Let G be a reductive group acting on k^n. Then there exists a polynomial coordinate change such that the action becomes linear.

In more detail: let $\rho : G \times V \to V$ be the given group action. Then for any $g \in G$ we get a polynomial automorphism ρ_g^* of $A(V)$ (§1). The subgroup of $\mathrm{Aut}\, A(V)$ consisting of all these elements ρ_g^* we denote by $\rho^*(G)$. Then the conjecture asserts that there exists $F \in \mathrm{Aut}\, A(V)$ such that

$$F^{-1}\rho^*(G)F \subset Gl_n(k)$$

the subgroup of $\mathrm{Aut}\, A(V)$ consisting of invertible linear automorphisms.

We will show that the conjecture is true if $n = 2$. In fact this follows from the following stronger result.

Theorem 9.4.2 *Let G be an algebraic group action on k^2. Then there exists $F \in$* $\mathrm{Aut}\, A(k^2)$ *such that either $F^{-1}\rho^*(G)F \subset \mathrm{Aff}(k, 2)$ or $F^{-1}\rho^*(G)F \subset T(k, 2)$.*

Proof: i) Let $A(k^2) = k[X_1, X_2]$ and $W := kX_1 + kX_2$. By 9.1.10 $G.W$ is a finite dimensional subspace of $k[X_1, X_2]$. Let $d := \max \deg f$, where f runs through $G.W$. So d is finite. Now let $g \in G$, then $g.X_j = \rho_g^*(X_j)$. Consequently

$$\deg \rho_g^* \le d, \text{ for all } g \in G. \tag{9.4.3}$$

ii) From 5.1.11 we know that $\mathrm{Aut}_k\, k[X_1, X_2]$ is the amalgamated free product of $\mathrm{Aff}(k, 2)$ and $J(k, 2)$ over their intersection. By 5.1.1 we can write every element H of $\mathrm{Aut}_k\, k[X_1, X_2]$ in the form

$$H = \lambda_0 \tau_1 \lambda_1 \dots \tau_l \lambda_l \tag{9.4.4}$$

with $\lambda_i \in \mathrm{Aff}(k, 2) \backslash J(k, 2)$, for all $1 \le i \le l-1$ and $\tau_i \in J(k, 2) \backslash \mathrm{Aff}(k, 2)$ for all $1 \le i \le l$. It was shown in 5.1.2 that

$$\mathrm{bideg}(\tau_1 \lambda_1 \dots \tau_l \lambda_l) = (\Pi_{j=1}^{l} \deg \tau_j, \Pi_{j=2}^{l} \deg \tau_j).$$

From this it follows that

$$\deg H \ge \Pi_{j=2}^{l} \deg \tau_j \ge 2^{l-1}. \tag{9.4.5}$$

Now apply 9.4.5 to the elements ρ_g^* which by 9.4.3 all have degree $\le d$. We deduce that $2^{l-1} \le d$, i.e.

$$l \le {}^2\log(2d). \tag{9.4.6}$$

iii) Finally define the length of an element of $\mathrm{Aut}_k\, k[X_1, X_2]$ as the minimal number of elements of $\mathrm{Aff}(k, 2) \cup J(k, 2)$ needed to express it as a product. From 9.4.4 we obtain that length $H \le 2l + 1$. So 9.4.6 implies

$$\text{length } \rho_g^* \le 2({}^2\log(2d)) + 1, \text{ for all } g \in G. \tag{9.4.7}$$

Now our theorem follows from the following general result about subgroups of amalgamated free products ([266], I.4.3, Theorem 8): if G_1 is an abstract group which is the amalgamated free product of its subgroups H_1 and H_2 over their intersection, and if furthermore the elements of G_1 have bounded length, then each subgroup of G_1 is conjugate to a subgroup of either H_1 or H_2. □

Corollary 9.4.8 *Every reductive group action of k^2 is linearizable.*

The proof of this result is left to the reader (Exercise 5). To conclude this section let us make some comments on the linearization conjecture.

If an action of G on V is linearizable, then the **Fixed point set**, i.e.

$$V^G := \{v \in V | g.v = v, \text{ for all } g \in G\},$$

is isomorphic to some k^d (Exercise 1). So this leads to

(9.4.9) **Fixed point problem.** *Let G be a reductive group action on V. Is the fixed point set V^G isomorphic to some k^d or at least non-empty?*

Although, as remarked in the beginning of this section, there exist counterexamples to the linearization conjecture, there do not yet exist counterexamples to the fixed point problem. The importance of the fixed point problem is illustrated by the following famous problem (we refer to [177] and [178] for more background).

(9.4.10) **Cancellation problem.** *Let Y be an affine variety such that $Y \times k \simeq k^n$ as algebraic varieties where $n \geq 2$. Does it follow that $Y \simeq k^{n-1}$?*

Before we give some comments on the Cancellation problem let us first show its relationship with the fixed point problem.

Proposition 9.4.11 *If the fixed point problem has an affirmative answer for the group $C_2 := \{-1, 1\}$, then the Cancellation problem has an affirmative answer too!*

Proof: Let Y be an affine variety satisfying $Y \times k \simeq k^n$, with $n \geq 2$. Consider the C_2-action on $Y \times k$ given by

$$-1.(y, t) := (y, -t).$$

Then $Y \times \{0\}$ is the fixed point set of this C_2-action and hence our hypothesis implies that $Y \times \{0\} \simeq k^d$ for some d. Obviously $d = n - 1$, whence $Y \simeq k^{n-1}$, as desired. □

So this leads to the question if the fixed point problem is true for C_2-actions on k^n, or stronger if the linearization conjecture is true for C_2-actions on k^n. Since a C_2-action on k^n is given by a polynomial automorphism of k^n satisfying $F^2 = I$, these questions can be reformulated as

Problem: Let $F \in \mathrm{Aut}_k\, k[X]$ with $F^2 = I$.

i) Does it follow that F is linearizable?
ii) Does it follow that the fixed point set of F is isomorphic to some k^d?

Now let us return to the Cancellation problem whose algebraic formulation we already saw in Chapter 2, §2, Exercise 3.

Suppose that Y is an affine variety such that $Y \times k \simeq k^n$. Then $A(Y)[T] \simeq k[X_1, \ldots, X_n]$. So the derivation $\frac{d}{dT}$ on $A(Y)[T]$ gives a locally nilpotent derivation on $k[X]$ which has kernel isomorphic to $A(Y)$. Since $k[X]$ is factorial, it follows from 1.3.36 that $A(Y)$ is factorial. Obviously the map $Y \times k \to Y$ is surjective, so composing with the isomorphism $k^n \simeq Y \times k$ we get a dominant morphism $k^n \to Y$. Consequently the following result, whose proof can be found in [218], implies the solution of the Cancellation problem in case $n = 3$.

Theorem 9.4.12 (Miyanishi, Sugie). *Let Y be a smooth affine surface such that $A(Y)$ is factorial and such that there exists a dominant morphism $k^N \to Y$ for some N. Then $Y \simeq k^2$ as affine varieties.*

Corollary 9.4.13 (Fujita). *The cancellation problem has an affirmative answer if $n = 3$, i.e. if Y is a surface.*

An analogous result of 9.4.12 is false if $\dim Y > 2$. The following counterexample is due to Russell in [258].

Example 9.4.14 (Russell). Let Y be the hypersurface in $\mathbb{C}^4$ given by

$$x + x^2y + z^3 + t^2 = 0.$$

Then Y is smooth and diffeomorphic to $\mathbb{C}^3$. Furthermore $A(Y)$ is factorial and there exist a dominant morphism $\mathbb{C}^3 \to Y$. However, Y and $\mathbb{C}^3$ are not isomorphic as algebraic varieties. This was first proved by Makar-Limanov in [199]. For a simplified proof we refer to Derksen's paper [77].

Hypersurfaces of the type described in the above example appeared in a list of Koras and Russell (see [181]) in connection with the linearization conjecture for $\mathbb{C}^*$-actions on $\mathbb{C}^3$. Finally by work of Koras, Russell, Makar-Limanov and Kaliman ([173]) this linearization conjecture has been solved affirmatively, i.e.

Theorem 9.4.15 (Kaliman, Koras, Makar-Limanov, Russell). *Every $\mathbb{C}^*$-action on $\mathbb{C}^3$ is linearizable.*

To conclude this section we mention one application of 9.4.12, namely the existence of **exotic algebraic $\mathbb{C}^n$'s**. For the rest of this section all algebraic varieties will be over the ground field $\mathbb{C}$.

Definition 9.4.16 A smooth affine variety X is called an **exotic algebraic structure on** $\mathbb{C}^n$ if X and $\mathbb{R}^{2n}$ are diffeomorphic as real manifolds but X and $\mathbb{C}^n$ are not isomorphic as algebraic varieties.

Now we will show how 9.4.12 can be used to construct exotic algebraic structures on $\mathbb{C}^n$ for all $n \geq 3$. Therefore we recall a result of Ramanujam (see [246], [53] and [172] for the proof of the next theorem and more about exotic $\mathbb{C}^n$'s).

Theorem 9.4.17 (Ramanujam). i) *If X is a smooth affine surface homeomorphic to $\mathbb{C}^2$, then X and $\mathbb{C}^2$ are isomorphic as algebraic varieties.*

ii) *If X is a contractible affine algebraic manifold of dimension ≥ 3, then X is diffeomorphic to $\mathbb{R}^{2n}$.*

Furthermore it was noticed by Ramanujam that in case X is a surface it is not enough to require that X be contractible. The following example given by tom Dieck and Petrie in [285] shows this. We refer to this paper for a proof: let a, d be positive integers such that $1 < a < d$ and $\gcd(a, d) = 1$. Let $X_{d,a}$ be the surface in $\mathbb{C}^3$ given by

$$((xz + 1)^d - (yz + 1)^a)/z = 0.$$

Then $X_{d,a}$ is a smooth contractible surface but not algebraically isomorphic to $\mathbb{C}^2$. Now we are able to give the announced construction of exotic algebraic structures on $\mathbb{C}^n$.

Proposition 9.4.18 *Let $X_{d,a}$ be one of the surfaces defined above. Then $E_{d,a} := \mathbb{C}^n \times X_{d,a}$ is an exotic algebraic structure on $\mathbb{C}^{n+2}$, for all $n \geq 1$.*

Proof: It follows from 9.4.17 ii) that $E_{d,a}$ is diffeomorphic to $\mathbb{R}^{2(n+2)}$. However $E_{d,a}$ and $\mathbb{C}^{n+2}$ are not isomorphic as algebraic varieties, for if $\mathbb{C}^n \times X_{d,a} \simeq \mathbb{C}^{n+2}$, then as observed above $A(X_{d,a})$ is factorial and we have a dominant morphism from $\mathbb{C}^{n+2} \to X_{d,a}$. So by 9.4.12 $X_{d,a} \simeq \mathbb{C}^2$ as algebraic varieties, a contradiction. □

Exercises for §9.4

1. Show that if an action of G on V is linearizable, then $V^G \simeq k^d$ as algebraic varieties, for some $d \geq 0$.
2. Show that the Cancellation problem has an affirmative answer if $n = 2$, i.e. show that $Y \times k \simeq k^2$ implies that $Y \simeq k$. [$A(Y)[T] = k[X_1, X_2]$ and the derivation $\frac{d}{dT}$ is locally nilpotent on $k[X_1, X_2]$ and its kernel is equal to $A(Y)$. Then apply 1.2.25 and 1.3.32 i).]
3. Let $F : k^4 \to k^4$ be given by
$$F(x, y, z, w) = \left(y, x, (w, z)\begin{pmatrix} 1 + xy + (xy)^2 & x^3 \\ -y^3 & 1 - xy \end{pmatrix}\right).$$
(i) Show that $F^2 = I$.
(ii) Show that $\operatorname{Fix} F \simeq k^2$.
(iii) Show that F is linearizable.
4. (Danielewski) Let $f := XY - Z^2 + 1$ and $g := X^2Y - Z^2 + 1$ in $\mathbb{C}[X, Y, Z]$. Let V resp. W be the algebraic varieties in $\mathbb{C}^3$ defined by f resp. g. It was shown by Fieseler that the first homology group at infinity of V resp. W is cyclic of order 2 resp. 4 and hence V and W are not isomorphic as algebraic varieties. However show that $V \times \mathbb{C} \simeq W \times \mathbb{C}$ as algebraic varieties. [Show that $\mathbb{C}[X, Y, Z, T]/(f) \simeq \mathbb{C}[X, Y, Z, T]/(g)$ by using the following identities. Let A be the quotient of the polynomial ring $\mathbb{C}[X, Y_1, Y_2, Z_1, Z_2, U]$ and the ideal I generated by $XY_1 - Z_1^2 + 1$, $X^2Y_2 - Z_2^2 + 1$ and $XU - Z_1 + Z_2$. Put $x_1 := X_1 + I$, $y_1 := Y_1 + I$ etc. Then $A = \mathbb{C}[x, y_2, z_2][u]$ since $z_1 = xu + z_2$ and $y_1 = u^2x + 2uz_2 + xy_2$. Similarly $A = \mathbb{C}[x, y_1, z_1][w]$ where $w = u^3x + 3u^2z_2 + 3uxy_2 + z_2y_2$ since $xw = y_1z_1 - 2u$ and $wz_1 = y_2 - u^2 + y_1^2$. So $2u = -xw + z_1y_1$, $y_2 = wz_1 + u^2 - y_1^2$ and $2z_2 = x^2w - z_1^3 + 3z_1$.]
5. Give a proof of 9.4.8.

9.5 G_a-actions

In the previous three sections we discussed reductive group actions on k^n and considered the following questions

1. Is the ring of invariants finitely generated?

2. Can we give an algorithm to compute the ring of invariants in case it is finitely generated?
3. Can we linearize all such actions?
4. Can we classify all such actions on k^2?

In this and the next section we investigate these questions for G_a-actions on k^n. Therefore we first show in 9.5.2 that there is a one-to-one correspondence between G_a-actions on $V := k^n$ and locally nilpotent derivations on $A(V)$, via exponentiation. See also Exercise 8 for a generalization to affine varieties. Under this correspondence the ring of invariants $A(V)^{G_a}$ equals the ring $A(V)^D$ of constants of the corresponding locally nilpotent derivation on $A(V)$. In the next section we will show that the ring of invariants need not be finitely generated, in contrast to 9.2.1. See also Exercise 9 where a G_a-action on a 3-dimensional irreducible affine variety $V \subset \mathbb{C}^4$ is given such that $A(V)^{G_a}$ is not a finitely generated $\mathbb{C}$-algebra. However if the ring $A(V)^{G_a}$ is finitely generated, then the kernel algorithm of Chapter 1 enables us to compute this ring. Concerning the third question above: the example of the G_a-action $x \to x+tv$ on k^n, where v is a fixed non-zero vector, shows that a G_a-action need not be linearizable, for the fixed point set of this action is empty, whereas the fixed point set of any linearizable algebraic group action on k^n is isomorphic to some k^d and hence non-empty. On the other hand Rentschler's theorem, 1.3.48, shows that any G_a-action on k^2 is triangulable (see 9.5.9 below). So one could ask if any G_a-action on k^n is triangulable. As we will see it turned out that the answer to this question is negative, for all $n \geq 3$. We will discuss several delicate questions and examples related to triangulability of G_a-actions and conclude this section with a list of challenging open problems.

Throughout this section k denotes a field of characteristic zero, $V := k^n$ and D is a locally nilpotent derivation on $A(V)$. By 1.2.14 and 1.3.19 we get a ring homomorphism

$$\exp TD : A(V) \to A(V)[T] = A(V \times k).$$

This induces a regular map $k \times V \to V$ given by

(9.5.1) $$t.v = (\exp tD)(X)_{|X=v}, \text{ where } X := (X_1, \ldots, X_n).$$

Since $\exp(0.D)(X) = X$ and $\exp(t_1+t_2)D(X) = \exp(t_1 D)(\exp(t_2 D)(X))$ it follows that 9.5.1 defines a G_a-action on V. Now we will show that the converse is true, i.e. each G_a-action on V arises in this way and D is uniquely determined by it (Exercise 1).

Proposition 9.5.2 *Let $\sigma : G_a \times V \to V$ be a G_a-action on V and denote by $\sigma^* : A(V) \to A(V)[T]$ the corresponding ring homomorphism. Then*

$$D(f) := \left(\frac{\sigma^*(f) - f}{T}\right)_{|T=0}, \text{ for all } f \in A(V)$$

defines a locally nilpotent derivation on $A(V)$ and $\sigma^* = \exp TD$.

Proof: i) Let $f \in A(V)$. Then $\sigma^*(f)(0, v) = f(\sigma(0, v)) = f(0.v) = f(v)$, so

$$\sigma^*(f)_{|T=0} = f. \tag{9.5.3}$$

So $D(f)$ is a well-defined element of $A(V)$ and we have

$$\sigma^*(f) = f + D(f) + \text{ higher order } T\text{-terms in } A(V)[T]. \tag{9.5.4}$$

From 9.5.4 it follows easily that D is a derivation on $A(V)$ (Exercise 2).

ii) Observe that $f(h.(t.v)) = f((h + t).v)$, so $\sigma^*(f)(h, t.v) = \sigma^*(f)(h + t, v)$ for all $t, h \in k$ and all $v \in V$. Consequently

$$\sigma^*(f)(H, t.v) = \sigma^*(f)(H + t, v), \text{ for all } t \in k \text{ and } v \in V \tag{9.5.5}$$

where H is a variable. So

$$\begin{aligned} \sigma^*(Df)(t, v) &= (Df)(t.v) \\ &= \left(\frac{\sigma^*(f)(H,t.v)-f(t.v)}{H}\right)_{|H=0} \\ &= \left(\frac{\sigma^*(f)(H+t,v)-\sigma^*(f)(t,v)}{H}\right)_{|H=0} \\ &= \left(\frac{d}{dT}\sigma^*(f)\right)(t, v). \end{aligned}$$

Since this holds for all $t \in k$ and all $v \in V$ we get

$$\sigma^*(Df) = \frac{d}{dT}(\sigma^*(f)), \text{ for all } f \in A(V). \tag{9.5.6}$$

From 9.5.3 and 9.5.6 it follows that $\sigma^*(f) = (\exp TD)(f)$ for all $f \in A(V)$, Exercise 3. Finally, since $(\exp TD)(f) = \sigma^*(f) \in A(V)[T]$ for all $f \in A(V)$ it follows from 1.3.19 that D is locally nilpotent on $A(V)$, which completes the proof. □

So instead of considering G_a-actions on V we may as well consider locally nilpotent derivations on $A(V)$. In the next lemma we therefore translate some of the geometric properties in algebraic terms.

Lemma 9.5.7 *Let $\sigma : G_a \times V \to V$ be a G_a-action defined by the locally nilpotent derivation D on $A(V)$. Then*

i) $A(V)^{G_a} = A(V)^D$.

ii) *The fixed point set of G_a, denoted* Fix G_a, *is equal to the set of common zeros of the ideal $I := (D(X_1), \dots, D(X_n))$.*

Proof: i) Let $f \in A(V)$. Then $f \in A(V)^{G_a}$ if and only if $t.f = f$ for all $t \in k$ if and only if $(\exp tD)(f) = f$ for all $t \in k$ if and only if $(\exp TD)(f) = f$ if and only if $D(f) = 0$.

ii) Observe that $D(f) \in I$ for all $f \in A(V)$. So for all $k \geq 1$ and each j we have in particular that $D^k(X_j)(v) = 0$ if v is a zero of I. Then the desired result follows easily from 9.5.1. □

Corollary 9.5.8 *The kernel algorithm of §4, Chapter 1 gives an algorithm to compute the ring of invariants of a given G_a-action, if it is a finitely generated k-algebra.*

(9.5.9) **Triangulable G_a-actions**

Now we address some specific questions concerning G_a-actions on V. So let σ be a G_a-action and D the corresponding locally nilpotent derivation. For each $t \in k$ we get an automorphism $\sigma_t{}^*$ of $A(V)$ which by 9.5.2 is equal to $\exp tD$.

We say that the action is **triangulable** if there exists $\varphi \in \operatorname{Aut} A(V)$ such that $\varphi^{-1}\sigma_t{}^*\varphi$ is of triangular form for all $t \in k$. It is left to the reader to verify (Exercise 4) that this is equivalent to $\varphi^{-1}D\varphi$ is of triangular form, i.e. D is triangulable.

Now this raises the following

Question 9.5.10 *Is every G_a-action on V triangulable?*

If $n = 2$ the answer is yes, due to Rentschler's theorem (1.3.48). However in [27] Bass gave the following example of a G_a-action on k^3 which is not triangulable.

Proposition 9.5.11 (Bass). *Let $D = (X_1X_3+X_2^2)(X_1\partial_2-2X_2\partial_3)$ on $k[X_1, X_2, X_3]$. Then D is locally nilpotent and the corresponding G_a-action on k^3 is not triangulable.*

The proof we will give is due to Popov, [245] in [60]. In his paper he extends Bass' example to any dimension $n \geq 3$ (Exercise 5).

Lemma 9.5.12 *Let k be an algebraically closed field. If a G_a-action σ on k^n is triangulable, then for each $t \in k$ the fixed point set of σ_t is a* ***cylindrical variety****, i.e. isomorphic to $k \times Z$ for some variety Z which depends on t.*

Proof: Let D be the locally nilpotent derivation corresponding to σ and let $\varphi \in \operatorname{Aut} A(V)$ be such that $\varphi^{-1}D\varphi$ is of triangular form. Then there exist polynomials $c_i(T) \in k[T]$ and $h_i(T, X_{i+1}, \ldots, X_n) \in k[T, X_{i+1}, \ldots, X_n]$ for all $1 \leq i \leq n$ such that

$$\varphi^{-1}\sigma_t{}^*\varphi(X_i) = c_i(t)X_i + h_i(t, X_{i+1}, \ldots, X_n), \quad \text{for all } t \in k.$$

Since $\varphi^{-1}\sigma_t{}^*\varphi$ is an automorphism, this implies that c_i does not vanish on k, so $c_i \in k^*$. Furthermore $\sigma_0{}^* = id$ which implies that $c_i = 1$ for all i. Consequently if we write $\varphi = F^*$, with $F \in \operatorname{Aut} V$, then we get

$$\operatorname{Fix} F\sigma_t F^{-1} = \{x \in k^n | h_i(t, x_{i+1}, \ldots, x_n) = 0, \text{ for all } i\}.$$

Since the variable X_1 does not appear in the polynomials h_i, it follows that $\operatorname{Fix} F\sigma_t F^{-1}$ is of the form $k \times Z$. Then the result follows observing that $\operatorname{Fix} \sigma_t = F(\operatorname{Fix} F\sigma_t F^{-1})$. □

Proof of Proposition 9.5.11: One readily verifies that the zero-set of $X_1X_3 + X_2^2$, which we denote by W, is contained in $\operatorname{Fix}(\exp tD)$ for all $t \in k$. Since for $t \neq 0$ the latter set has dimension ≤ 2 it follows that W is a component of $\operatorname{Fix}(\exp tD)$.

Now assume that our action is triangulable. Then it follows from 9.5.12 that W is isomorphic to a set of the form $k \times Z$. However this is a contradiction since one easily verifies that W has an isolated singularity, which is impossible for a set of the form $k \times Z$. □

The above example and Popov's generalization to all dimensions ≥ 3 led Snow in [272] in [184] to the following.

Question 9.5.13 *Is the condition that for each $t \in k$ the fixed point set of σ_t^* is isomorphic to some set $k \times Z$ the only obstruction for a G_a-action to be triangulable?*

or even more likely

Question 9.5.14 *If a G_a-action is **fixed point free**, i.e.* Fix G_a *is empty, does it follow that the action is triangulable?*

In [67] Daigle showed that the answer to the first question is no. Before we describe his result in more detail we first mention, without proof, some results concerning locally nilpotent derivations in two variables over a UFD.

Theorem 9.5.15 (Daigle, Freudenburg). *Let R be a UFD containing $\mathbb{Q}$, K its quotient field and D a non-zero R-derivation on $R[X, Y]$.*

1) *D is locally nilpotent if and only if there exist $P \in R[X, Y]$, which is a coordinate in $K[X, Y]$ and a non-zero element $a \in R[P]$ such that $D = a(P_Y\partial_X - P_X\partial_Y)$ with $\gcd(P_X, P_Y) = 1$. Moreover, if these conditions are satisfied $\ker D = R[P]$.*
2) *Let D be a locally nilpotent derivation on $R[X, Y]$. Then D has a slice in $R[X, Y]$ if and only if 1 belongs to the ideal in $R[X, Y]$ generated by $D(X)$ and $D(Y)$.*

Now let k be a field of characteristic zero and D a non-zero locally nilpotent derivation on $k[X, Y, Z]$ satisfying $D(X) = 0$ and $\gcd(D(Y), D(Z)) = 1$. So by 9.5.15 $D = P_Z\partial_Y - P_Y\partial_Z$ for some $P \in k[X, Y, Z]$ which is a coordinate in $k(X)[Y, Z]$ and $\ker D = k[X][P]$. With these notations we have

Proposition 9.5.16 (Daigle's triangulability criterion in dimension 3). *There is equivalence between*

i) *D is triangulable.*
ii) *D is triangulable over $k[X]$.*
iii) *There exists a coordinate Q of $k[X, Y, Z]$ over $k[X]$ such that $k(X)[P, Q] = k(X)[Y, Z]$.*

For the proofs of both results we refer to [70] and [67]. See also the Notes at the end of this chapter for some generalizations. Now we can give

Example 9.5.17 (Daigle). Let $D_1 = -P_Z\partial_Y + P_Y\partial_Z$, where $P := Y + \frac{1}{4}(XZ + Y^2)^2$ and $D_2 = X\partial_Y + (1 + Y^3)\partial_Z$. Then D_1 and D_2 are locally nilpotent derivations which satisfy

i) D_1 is not triangulable and D_2 is of triangular form.

ii) The ideals $(D_1(X), D_1(Y), D_1(Z))$ and $(D_2(X), D_2(Y), D_2(Z))$ are equal, so in particular $\operatorname{Fix} D_1 = \operatorname{Fix} D_2 = V \times k$, where V is the zero-set of $X = 0, Y^3 + 1 = 0$.

Proof: The only non-trivial point is to show that D_1 is not triangulable. Therefore we first observe that

$$k(X)[P, XZ + Y^2] = k(X)[Y, Z]. \tag{9.5.18}$$

Suppose now that D_1 is triangulable. Then by 9.5.16 there exists a coordinate Q of $k[X, Y, Z]$ over $k[X]$ which satisfies

$$k(X)[P, Q] = k(X)[Y, Z]. \tag{9.5.19}$$

From 9.5.18 and 9.5.19 we deduce that $k(X)[P, XZ + Y^2] = k(X)[P, Q]$. So writing this equality as $A[h] = A[Q]$, where $A := k(X)[P]$ and $h := XZ + Y^2$ we see that $Q = ch + g$, for some $c \in A^*$ and $g \in A$. In other words we get that $Q = c(XZ + Y^2) + g(P)$, for some $c \in k(X)^*$ and $g(T) \in k(X)[T]$, where T is a new variable. Multiplying by a suitable non-zero element of $k[X]$ we get

$$a_0(X)Q = a_1(X)(XZ + Y^2) + G(P) \tag{9.5.20}$$

for some non-zero $a_0(X), a_1(X) \in k[X]$ and $G(T) = \sum_{i=0}^{d} r_i(X)T^i$, with all $r_i(X) \in k[X]$. We may also assume that

$$\gcd(a_0, a_1, r_0, \ldots, r_d) = 1. \tag{9.5.21}$$

Since Q is a coordinate in $k[X, Y, Z]$, $q := Q(0, Y, Z)$ is a coordinate in $k[Y, Z]$. Substituting $X = 0$ in 9.5.20 gives

$$a_0(0)q = a_1(0)Y^2 + G_0\left(Y + \frac{1}{4}Y^4\right) \tag{9.5.22}$$

where $G_0(T) = \sum r_i(0)T^i$. If $a_0(0) = 0$, then 9.5.22 gives that $a_1(0) = 0$, which in turn implies that $G_0 = 0$, i.e. all $r_i(0) = 0$. This contradicts 9.5.21. So $a_0(0) \neq 0$ and from 9.5.22 we get that $a_1(0)Y^2 + G_0(Y + \frac{1}{4}Y^4)$ is a coordinate of $k[Y, Z]$ which only depends on Y, therefore must be linear in Y, a contradiction. Hence D_1 is not triangulable. □

On the other hand 9.5.14 remained open for all $n \geq 3$. In fact here we reach the limit of our knowledge of locally nilpotent derivations, at present, and we enter a world full of open problems. So let us conclude this section by formulating some of them.

Problem 9.5.23 Is a locally nilpotent derivation with a slice triangulable?

(9.5.24) **Comment.** The hypothesis $Ds = 1$ for some $s \in k[X]$ obviously implies that the fixed point set of the corresponding G_a-action is empty. Consequently if 9.5.14 has an affirmative answer, so has 9.5.23. Observe that the hypothesis in 9.5.23 is stronger. This can be seen from

Example 9.5.25 Let $D = (1 + X_3^2)\partial_1 + X_3\partial_2 + X_4\partial_3$. Then the corresponding G_a-action is fixed point free and D does not have a slice in $k[X_1, X_2, X_3, X_4]$ (Exercise 6).

Problem 9.5.26 Cancellation problem. Let D be a locally nilpotent derivation on $k[X]$. Assume that D has a slice in $k[X]$. Is $k[X]^D$ a polynomial ring in $n - 1$ variables?

Comment. As we saw in Chapter 2, §2, Exercise 3 this problem is just a reformulation of the Cancellation problem 9.4.10. If $n = 2$ the answer is affirmative (Exercise 2, §4) and as we saw in §4 the case $n = 3$ was solved affirmatively by Fujita. The case $n \geq 4$ remains open. In fact even the following special case is still open.

Problem 9.5.27 Let D be a triangular derivation having a slice in $k[X]$. Is $k[X]^D$ a polynomial ring in $n - 1$ variables?

Comment Open for all $n \geq 4$.

Since any non-zero locally nilpotent derivation on $k[X]$ has a slice in $k(X)$, namely $s = p/d$, where $d = D(p)$ and p a preslice (§3, Chapter 1) and $k(X) = Q(k[X]^D)(s)$ by 1.3.30 with s transcendental over $Q(k[X]^D)$, it is natural in the light of 9.5.26, to pose

Problem 9.5.28 Let D be a non-zero locally derivation on $k[X]$. Is $Q(k[X]^D)$ a rational function field in $n - 1$ variables over k?

Comment. It was proved by Deveney and Finston in [81] and [83] that the answer is affirmative if $n \leq 4$. The case $n \geq 5$ remains open.

In fact the case $n \leq 3$ follows from the equality $k(X) = Q(k[X]^D)(s)$ and the affirmative answer of the **birational cancellation problem** which asserts that if k is an algebraically closed field of characteristic zero and K_1, K_2 are finitely generated field extensions of k such that $K_1(X_1, \ldots, X_d) = K_2(Y_1, \ldots, Y_d)$, for some $d \geq 1$, then $K_1 \simeq K_2$ over k if $\operatorname{trdeg}_k K_1 = \operatorname{trdeg}_k K_2 \leq 2$. However it turned out that the birational cancellation problem has a negative answer if $\operatorname{trdeg}_k K_1 = \operatorname{trdeg}_k K_2 \geq 3$ (see Example 9.5.29 below, which was proved in [31]). Therefore the result obtained by Deveney and Finston in the case $n = 4$ is rather remarkable!

Example 9.5.29 (Beauville, Colliot-Thélène, Sansuc, Swinnerton-Dyer). Let k be an algebraically closed field with char $k \neq 2$. Let

$$K = Q(k[X, Y, Z, W]/(X^2 - a(W)Y^2 - f(W, Z))$$

where $f(W, Z) \in k[W, Z]$ is irreducible of degree 3 in Z and $a(W) = \operatorname{disc}_Z f(W, Z) \in k[W]\backslash\{0\}$, the Z-discriminant of f, is square free of degree ≥ 5. Then

i) K is not rational over k.

ii) $K(T_1, T_2, T_3)$ is k-isomorphic to a rational function field in six variables over k.

Exercises for §9.5

1. Let A be a $\mathbb{Q}$-algebra and D_1, D_2 locally nilpotent derivations on A such that $\exp TD_1 = \exp TD_2$. Show that $D_1 = D_2$.
2. Let A be a $\mathbb{Q}$-algebra and $\varphi : A \to A[T]$ a ring homomorphism such that $\varphi(a)_{|T=0} = a$ for all $a \in A$. Define $D : A \to A$ by $D(a) = (\frac{d}{dT}\varphi(a))_{|T=0}$. Show that D is a derivation on A.
3. Let A be a $\mathbb{Q}$-algebra and $\varphi : A \to A[T]$ a ring homomorphism such that $\varphi(a)_{|T=0} = a$ for all $a \in A$ and $\varphi(D(a)) = \frac{d}{dT}(\varphi(a))$ for all $a \in A$. Show that $\varphi = \exp TD$.
4. Let $V := k^n$ and D a locally nilpotent derivation on $A(V)$. Show that the corresponding G_a-action is triangular if and only if D is triangular.
5. (Popov) Let $n \geq 3$ and define $h := \sum_{j=1}^{d}(-1)^j X_j X_{n+1-j}$, where $d = n$ if n is odd and $d = n/2$ if n is even. Furthermore let $D_0 := \sum_{i=1}^{n-1} X_{i+1}\partial_i$ if n is odd and $D_0 := \sum_{i=1, i\neq n/2}^{n-1} X_{i+1}\partial_i$ if n is even.

   (i) Show that $D := hD_0$ is a non-triangular locally nilpotent derivation on $k[X]$ for all $n \geq 3$. [Show that $D_0(h) = 0$ and that 0 is an isolated singularity of the hypersurface $h = 0$.]

   (ii) Show that in case $n = 3$ Bass' example is conjugate to hD_0.

6. Show that the derivation of 9.5.25 is fixed point free, but has no slice in $k[X_1, \ldots, X_4]$. [Show that 1 does not belong to Im D and use the image membership algorithm of Chapter 1.]
7. Let $k[X, Y] := k[X_1, \ldots, X_n, Y_1, \ldots, Y_m]$ and D a so-called **elementary derivation** on $k[X, Y]$, i.e. a derivation of the form $D = \sum a_j(X)\partial_{Y_j}$. Show that if $1 \in (a_1, \ldots, a_m)$ then $k[X, Y]^D = k[F_1, \ldots, F_{n+m-1}]$ for some F_i which are algebraically independent over k. [Since $(a_1, \ldots, a_m)$ is a unimodular row, the Quillen-Suslin theorem ([190], Chapter XIII, Theorem 11.5) implies that there exists $A \in Gl_m(k[X])$ whose first row is $a_1, \ldots, a_m$. Put $C := A^{-1}$ and for each $1 \leq j \leq m$ let C_j be the j-th row of C^T and $s_j := \langle C_j, Y\rangle$ $(=\sum C_{j,i}Y_i)$. Show that $k[X, Y]^D = k[X, s_2, \ldots, s_m]$ and $Ds_1 = 1$.]
8. Let $V \subset k^n$ be an affine variety with coordinate ring $A(V) = k[X]/I(V)$ and let $\overline{X}_i$ denote the i-th coordinate function on V. Let D be a locally nilpotent k-derivation on $A(V)$. So there exists $m \in \mathbb{N}$ such that $D^p(\overline{X}_i)$ vanishes on V for all i and all $p \geq m$. Consequently

$$\exp tD(\overline{X}_i)(v) := \sum_{k\geq 0} \frac{1}{k!} t^k D^k(\overline{X}_i)(v)$$

is a well-defined element of k for all $v \in V$. Define

$$t.v := (\exp tD(\overline{X}_1)(v), \dots, \exp tD(\overline{X}_n)(v)) \in k^n, \text{ for all } t \in k, v \in V.$$

(i) Show that $t.v \in V$ for all $t \in k$, all $v \in V$. So we get a map $k \times V \to V$.
(ii) Show that this map is a G_a-action on V.
(iii) Conversely show that every G_a-action on V is of the above form, i.e. comes from a locally nilpotent derivation on $A(V)$.
(iv) Let $\sigma : G_a \times V \to V$ be a G_a-action on V defined by a locally nilpotent derivation D on $A(V)$. Show that $A(V)^{G_a} = A(V)^D$.

9. Let $R = \mathbb{C}[T^2, T^3] \subset \mathbb{C}[T]$. Consider on $\mathbb{C}[T][X, Y]$ the locally nilpotent derivation $T^3\partial_X - T^2\partial_Y$. The restriction of this derivation to $R[X, Y]$ we denote by D. Put $A := R[X, Y]^D$.

(i) Show that $A = \mathbb{C} + T^2\mathbb{C}[T][X + TY]$. [Use that $\ker(T^3\partial_X - T^2\partial_Y, \mathbb{C}[T][X, Y]) = \mathbb{C}[T][X + TY]$.]
(ii) Deduce from i) that $\mathbb{C}[T][X + TY]$ is not finite over A. [Consider the quotients modulo $T^2\mathbb{C}[T][X + TY]$.]
(iii) Deduce that the ideal $T^2\mathbb{C}[T][X + TY]$ in A is not finite over A. So A is not Noetherian. In particular $R[X, Y]^D (= A)$ is not a finitely generated R-algebra and hence not a finitely generated $\mathbb{C}$-algebra.
(iv) Let L be the quotient field of $R[X, Y]^D$. Deduce that $L \cap \mathbb{C}[T^2, T^3, X, Y]$ is not a finitely generated $\mathbb{C}$-algebra.
(v) Let $V := \{(x, y, z, w) \in \mathbb{C}^4 | z^3 - w^2 = 0\}$. Consider the G_a-action on V given by the locally nilpotent derivation D on $A(V) (\simeq R[X, Y])$ (Exercise 8). Deduce from iii) that $A(V)^{G_a}$ is not a finitely generated $\mathbb{C}$-algebra.

10. Let R be a UFD containing $\mathbb{Q}$ and $f \in R[X, Y]$, the polynomial ring in two variables. Let $D := f_Y\partial_X - f_X\partial_Y$.

(i) Show that f is a coordinate in $R[X, Y]$ if and only if D is locally nilpotent and $1 \in (f_X, f_Y)$ in $R[X, Y]$.
(ii) Let $p \in R$ and $G \in R[Y]$. Show that $f := pX + G(Y)$ is a coordinate in $R[X, Y]$ if and only if $1 \in (p, G'(Y))$ in $R[Y]$. [Ad i) Use 9.5.15 and the proof of 3.2.11.] Remark. By the main result in [37], the condition "R is a UFD" can be dropped.

11. (Edo) Let A be a UFD containing $\mathbb{Q}$, $n \geq 1$ and $G, F \in A[Y] := A[Y_1, \dots, Y_n]$. Let $a, p \in A$ be such that $1 \in Aa + Ap$ and all coefficients of G are nilpotent modulo p. Show that if F is a coordinate in $A[Y]$, then $aF + G + px$ is a coordinate in $A[x, Y_1, \dots, Y_n]$. [Put $R := A[Y_1, \dots, Y_{n-1}]$ and $y := Y_n$. Let $\alpha \in \mathrm{Aut}_A A[Y]$ be such that $\alpha(y) = F$ and consider $f := ay + \alpha^{-1}(G) + px \in R[x, y]$. Use Exercise 10 to show that f is a coordinate in $R[x, y]$. So there exists an R-automorphism σ of $R[x, y]$ with $\sigma(y) = f$. Finally, extend α to $R[x, y]$ by $\alpha(x) = x$ and consider $\alpha\sigma(y)$] Remark. As in the previous exercise the condition "R is a UFD" is superfluous.

12. (Edo) Let A be a UFD containing $\mathbb{Q}$, $m \geq 1$, $p_i \in A[z_i]$ and $G_i \in A[x_0, \ldots, x_{i-1}, z_1, \ldots, z_i]$ for all $1 \leq i \leq m$. Let $r_i \in A[z_i]$ be nilpotent modulo $p_i A[z_i]$ for all i. Show that $x_0 + \sum_{i=1}^{m} r_i G_i + \sum_{i=1}^{m} x_i p_i$ is a coordinate in $A[x_0, \ldots, x_m, z_1, \ldots, z_m]$. [Use induction on m and the previous two exercises] Remark. Again "R is a UFD" can be dropped.

9.6 G_a-actions and Hilbert's fourteenth problem

In 1900 at the international congress of mathematicians in Paris Hilbert posed 21 problems which should be a challenge for 20-th century mathematicians. The fourteenth problem of his list is the following.

Hilbert's fourteenth problem. Let k be a field and L a subfield of the field of rational functions $k(X_1, \ldots, X_n)$ containing k. Is $L \cap k[X_1, \ldots, X_n]$ a finitely generated k-algebra?

For an excellent survey of this problem we refer the reader to [159] and [221]. The problem was motivated by the finiteness problem of invariant theory described in §1. At the time of his address in Paris Hilbert believed that this last problem was solved affirmatively by a recent publication of L. Maurer, which turned out to be false. So he posed the more general question above. Indeed a positive answer to Hilbert's fourteenth problem would imply finite generation of $k[X_1, \ldots, X_n]^G$, since one easily verifies (Exercise 1) that

$$k[X_1, \ldots, X_n]^G = L \cap k[X_1, \ldots, X_n]$$

where L is the quotient field of the ring of invariants. In 1953 Hilbert's problem was generalized by Zariski ([306]) in the following way.

Zariski's problem. *Let k be a field and $k[a_1, \ldots, a_n]$ an affine normal domain, i.e. integrally closed, and let L be a subfield of $k(a_1, \ldots, a_n)$ containing k. Is $L \cap k[a_1, \ldots, a_n]$ a finitely generated k-algebra?*

In case the domain $k[a_1, \ldots, a_n]$ is not normal, the intersection need not be a finitely generated k-algebra as can be seen from Exercise 9 iv) in the previous section. In the same paper Zariski solved his problem affirmatively for the following special case.

Theorem 9.6.1 *If* $\operatorname{trdeg}_k L \leq 2$, *then $k[a_1, \ldots, a_n] \cap L$ is a finitely generated k-algebra.*

However in 1957 D. Rees in [249] gave a counterexample to Zariski's problem with $\operatorname{trdeg}_k L = 3$. Finally in 1958 Nagata ([222] in [43]) gave a counterexample to the finiteness of $k[X_1, \ldots, X_n]^G$ and hence to Hilbert's problem and Zariski's problem for $n = 32$.

Recently Roberts in [252], 1990, gave a new counterexample to Hilbert fourteen. A little later in [82], 1994, Deveney and Finston observed that Roberts' example could be realized by a G_a-action on k^7 (Exercise 5). To describe this result we introduce some notation: k is a field of characteristic zero, P the polynomial ring $k[X, Y, Z, S, T, U, V]$, B the subring $k[S, T, U]$ and $A = B[V, X, Y, Z]$. On P consider the derivation

$$D = X^3\partial_S + Y^3\partial_T + Z^3\partial_U + (XYZ)^2\partial_V.$$

The remainder of this section is devoted to the proof of the following theorem.

Theorem 9.6.2 (Deveney, Finston). *D is locally nilpotent on P and P^D is not a finitely generated k-algebra.*

Corollary 9.6.3 *The ring of invariants of the G_a-action on k^7 corresponding to D is not a finitely generated k-algebra and $L := Q(P^D) \subset Q(P)$ is a counterexample to Hilbert's 14-th problem.*

Proof: The first part follows from 9.5.2 and 9.5.7 and the last part from the equality $Q(P^D) \cap P = P^D$. □

The proof of 9.6.2 which we will give below follows the one given by Stefan Maubach in his master's thesis [204]. The main line of his proof follows the paper [82] of Deveney and Finston. However an essential new ingredient in the proof is the use of so-called D-invariant gradings on P, which turns out to be a powerful tool in the study of derivations: if G is any additive group, $R = \oplus_{\sigma \in G} R_\sigma$ a graded ring and D a derivation on R, then we say that the grading is **D-invariant** if $DR_\sigma \subset R_\sigma$ for all $\sigma \in G$.

Example 9.6.4 Let $R := P$. To simplify the notation we write $X_1 := X, X_2 := Y, X_3 := Z, X_4 := S, X_5 := T, X_6 := U$ and $X_7 := V$. So $D = X_1^3\partial_4 + X_2^3\partial_5 + X_3^3\partial_6 + (X_1X_2X_3)^2\partial_7$. For each $w \in \mathbb{Q}^7$ and $d \in \mathbb{Z}$ we denote by $R_d(w)$ the k-vector space generated by all monomials X^i $(= X_1^{i_1} \dots X_7^{i_7})$ with $\langle w, i\rangle = d$, where $\langle w, i\rangle := \sum w_j i_j$. One easily verifies that $R = \oplus R_d(w)$ is a graded ring, which we call the **w-grading** on R. Now we determine all D-invariant w-gradings on R. Therefore apply D to the monomial X^i. Let $d := \langle w, i\rangle$. The condition $DX^i \in R_d(w)$ for all i gives

$$\begin{aligned} \langle w, i + 3e_1 - e_4\rangle &= \langle w, i + 3e_2 - e_5\rangle \\ &= \langle w, i + 3e_3 - e_6\rangle \\ &= \langle w, i + 2e_1 + 2e_2 + 2e_3 - e_7\rangle \\ &= \langle w, i\rangle \end{aligned}$$

where e_j denotes the j-th standard basis vector in $\mathbb{Q}^7$, i.e.

$$\begin{aligned} \langle w, 3e_1 - e_4\rangle &= \langle w, 3e_2 - e_5\rangle \\ &= \langle w, 3e_3 - e_6\rangle \\ &= \langle w, 2e_1 + 2e_2 + 2e_3 - e_7\rangle \\ &= 0. \end{aligned}$$

It follows easily that the set of all $w \in \mathbb{Q}^7$ such that the corresponding w-grading is D-invariant is a $\mathbb{Q}$-vector space of dimension 3 with basis

$$w_{(1)} := e_1 + 3e_4 + 2e_7,\ w_{(2)} := e_2 + 3e_5 + 2e_7,\ w_{(3)} := e_3 + 3e_6 + 2e_7.$$

These three basis vectors contain all information about all D-invariant w-gradings on R. This led Stefan Maubach in [204] to define the following $\mathbb{Z}^3$-grading on R: for $\sigma \in \mathbb{Z}^3$ let R_σ be the k-vector space generated by all monomials X^i such that

$$(\langle i, w_{(1)}\rangle, \langle i, w_{(2)}\rangle, \langle i, w_{(3)}\rangle) = \sigma,$$

i.e. $(i_1 + 3i_4 + 2i_7,\ i_2 + 3i_5 + 2i_7,\ i_3 + 3i_6 + 2i_7) = (\sigma_1, \sigma_2, \sigma_3)$.

This grading will play a crucial role in the proof of 9.6.2. To prove 9.6.2 we use two gradings on P. Let $m = X^\alpha Y^\beta Z^\gamma S^a T^b U^c V^d$ be a monomial in P. We define

$$\begin{aligned} \deg m &:= a + b + c + d \in \mathbb{Z}, \\ \mathrm{Ddeg}\, m &:= (\alpha + 3a + 2d, \beta + 3b + 2d, \gamma + 3c + 2d) \in \mathbb{Z}^3. \end{aligned}$$

Then for $n \in \mathbb{Z}$ resp. $\sigma \in \mathbb{Z}^3$ we define P_n (resp. P_σ) := the k-linear subspace of P generated by all monomials m with $\deg m = n$ (resp. $\mathrm{Ddeg}\, m = \sigma$).

Then $P = \oplus_{n\in\mathbb{Z}} P_n$, resp. $P = \oplus_{\sigma\in\mathbb{Z}^3} P_\sigma$, give on P two structures of a graded ring. The first one is the "usual" grading with respect to the variables S, T, U, V and the second structure is exactly the one described in 9.6.4.

Lemma 9.6.5 *$P^D \cap P_n \subset (X, Y, Z)P_n$ for all $n \geq 1$.*

Proof: Let $n \geq 1$ and suppose that $v \in P^D \cap P_n$ but v is not in $(X, Y, Z)P_n$. Then v contains a monomial, with respect to S, T, U, V, of the form $gS^aT^bU^cV^d$ with $g \in k[X, Y, Z]$ and $g(0) \neq 0$. Since $n \geq 1$, at least one of a, b, c, d is positive. Let's assume $a > 0$ (the other cases are treated similarly). Writing out the equation $Dv = 0$, taking into account all monomials which can cancel with $X^3\partial_S(gS^aT^bU^cV^d)$, we get that the coefficient of $S^{a-1}T^bU^cV^d$ in Dv, which has to be zero, gives

$$gaX^3 + (b+1)Yf_2 + (c+1)Z^3 f_3 + (d+1)(XYZ)^2 f_4 = 0$$

for some f_i in $k[X, Y, Z]$. Using that $a \neq 0$ and $\mathrm{char}\, k = 0$ we get $gX^3 \in (Y^3, Z^3, X^2Y^2Z^2)$. Finally, using that $g(0) \neq 0$, we can write $g = g(0) + \tilde{g}$ with $\tilde{g} \in (X, Y, Z)$, and we deduce that

$$X^3 \in (X^4, X^3Y, X^3Z, Y^3, Z^3, X^2Y^2Z^2),$$

a contradiction. So $v \in (X, Y, Z)P_n$ as desired. □

Now the crucial point to prove is

Theorem 9.6.6 *There exists a sequence $a_0 = X, a_1, a_2, \ldots \in A$ satisfying*

i) $a_n \in P_n$, *for all* n.
ii) $D(a_n) = (XYZ)^2 a_{n-1}$, *for all* $n \geq 1$.

Before we prove this result we show how it implies 9.6.2.

Proof of Theorem 9.6.2: i) Define $F_n := \sum_{i=0}^{n}(-1)^i \frac{n!}{(n-i)!} a_i V^{n-i}$. Then one easily verifies that $F_n \in P^D \cap P_n$. Furthermore the coefficient of V^n is $a_0 = X$.

ii) Suppose that P^D is finitely generated. Since $DP_n \subset P_{n-1}$ we may assume that each of the generators belongs to some P_i. Let d be the maximum of all such i and consider the element F_{d+1}. Then F_{d+1} is a k-linear combination of products of these generators and each of these products must contain at least two factors coming from a P_i with $i \geq 1$. So by 9.6.5 $F_{d+1} \in (X, Y, Z)^2 P_{d+1}$. Whence X, the coefficient of V^{d+1} in P_{d+1}, belongs to $(X, Y, Z)^2$, a contradiction. □

The proof of 9.6.6 will consist of a series of reductions. Let $\phi : A \to B[X^3, Y^3, Z^3]$ be the B-homomorphism sending X to X^3, Y to Y^3 and Z to Z^3. Put $\tilde{D} := X\partial_S + Y\partial_T + Z\partial_U$. Then

(9.6.7) $$D \circ \phi = \phi \circ \tilde{D}.$$

Proposition 9.6.8 *If there exists a sequence* $\tilde{a}_0 := 1, \tilde{a}_1, \tilde{a}_2, \ldots$ *in* A *such that* $\tilde{a}_n \in P_n$ *for all* $n \geq 0$ *and* $\tilde{D}(\tilde{a}_{3n}) = XYZ\tilde{a}_{3n-1}$, $\tilde{D}(\tilde{a}_{3n+1}) = X\tilde{a}_{3n}$, $\tilde{D}(\tilde{a}_{3n+2}) = YZ\tilde{a}_{3n+1}$, *then there exists a sequence* $a_0 = X, a_1, a_2, \ldots$ *in* A *as described in* 9.6.6.

Proof: Define

$$a_{3n} = X\phi(\tilde{a}_{3n}),\ a_{3n+1} = Y^2Z^2\phi(\tilde{a}_{3n+1}) \text{ and } a_{3n+2} = X^2YZ\phi(\tilde{a}_{3n+2}).$$

By 9.6.7 we get

$$D(a_{3n}) = X\phi(\tilde{D}(\tilde{a}_{3n})) = X^4Y^3Z^3\phi(\tilde{a}_{3n-1}) = (XYZ)^2 a_{3n-1}.$$

We obtain the other relations similarly. □

To describe the second reduction we need some preparation. Let $\Phi : A \to B$ be the B-homomorphism sending X, Y, Z to 1 and let D_1 be the derivation $D_1 := \partial_S + \partial_T + \partial_U$ on B. Then

(9.6.9) $$D_1 \circ \Phi = \Phi \circ \tilde{D}.$$

Furthermore for $\sigma \in \mathbb{Z}^3$ let $A_\sigma = A \cap P_\sigma$. So $A = \oplus_{\sigma \in \mathbb{Z}^3} A_\sigma$ and if $m = X^\alpha Y^\beta Z^\gamma S^a T^b U^c \in A_\sigma$, then $\operatorname{Ddeg} m = (\alpha + 3a, \beta + 3b, \gamma + 3c) = \sigma$. Also

(9.6.10) $$\tilde{D}A_\sigma \subset A_\sigma \text{ for all } a \in A_\sigma \text{ and all } \sigma.$$

Denoting the restriction of Φ to A_σ by Φ_σ, then 9.6.9 and 9.6.10 imply that

(9.6.11) $$\Phi_\sigma \circ \tilde{D}(a) = D_1 \circ \Phi_\sigma(a) \text{ for all } a \in A_\sigma.$$

Finally for $0 \neq b \in B$ put

$$\operatorname{Deg} b := (\deg_S b, \deg_T b, \deg_U b) \in \bar{\mathbb{N}}^3$$

and consider on $\mathbb{N}^3$ the partial ordering $\leq$ defined by $(a, b, c) \leq (a', b', c')$ if and only if $a \leq a'$, $b \leq b'$ and $c \leq c'$. With these notations we get

Lemma 9.6.12 i) *$\Phi_\sigma(A_\sigma) = \{b \in B \mid \operatorname{Deg} b \leq \sigma\}$, for all $\sigma \in \mathbb{N}^3$.*

ii) *$\Phi_\sigma : A_\sigma \to \Phi_\sigma(A_\sigma)$ is injective hence bijective.*

Proof: i) Let $\sigma = (p, q, r)$ and $m = X^\alpha Y^\beta Z^\gamma S^a T^b U^c \in A_\sigma$. Then $\alpha + a = p$, $\beta + b = q$ and $\gamma + c = r$. So $\operatorname{Deg} \Phi_\sigma(m) = \operatorname{Deg} S^a T^b U^c = (a, b, c) \leq \sigma$. Conversely if $g = S^a T^b U^c$ satisfies $(a, b, c) \leq \operatorname{Deg} g \leq \sigma = (p, q, r)$, then $p - a, q - b, r - c \geq 0$. So $f := X^{p-a} Y^{q-b} Z^{r-c} S^a T^b U^c$ satisfies $\Phi(f) = g$ and $\operatorname{Ddeg} f = \sigma$.

ii) Let $m = X^\alpha Y^\beta Z^\gamma S^a T^b U^c \in A_\sigma$ such that $\Phi_\sigma(m) = S^a T^b U^c$. Then $\alpha + a = p$, $\beta + b = q$, $\gamma + c = r$, so $(\alpha, \beta, \gamma) = \sigma - (a, b, c)$, whence m is uniquely determined. □

Theorem 9.6.13 *There exists a sequence $\tilde{a}_0 = 1, \tilde{a}_1, \tilde{a}_2, \ldots$ in A, as in 9.6.8, if and only if there exists a sequence $b_0 = 1, b_1, b_2, \ldots$ in B such that*

i) *$D_1(b_n) = b_{n-1}$, for all $n \geq 1$.*

ii) *$\operatorname{Deg} b_{3n} \leq (2n, 2n, 2n)$, $\operatorname{Deg} b_{3n+1} \leq (2n+1, 2n, 2n)$, $\operatorname{Deg} b_{3n+2} \leq (2n+1, 2n+1, 2n+1)$.*

iii) *$b_n \in P_n$, for all $n \geq 0$.*

Proof: 1) Assume a sequence $\tilde{a}_0 = 1, \tilde{a}_1, \ldots$ as above exists. Then, since $\tilde{a}_0 = 1$ is Ddeg-homogeneous ($\operatorname{Ddeg} \tilde{a}_0 = (0, 0, 0)$) we may assume that all $\tilde{a}_n$ are Ddeg-homogeneous by 9.6.10. Then, again by 9.6.10 we get

$$\operatorname{Ddeg} \tilde{a}_{3n} = \operatorname{Ddeg} \tilde{D}(\tilde{a}_{3n}) = (1, 1, 1) + \operatorname{Ddeg} \tilde{a}_{3n-1}.$$

Similarly

$$\begin{aligned} \operatorname{Ddeg} \tilde{a}_{3n+1} &= (1, 0, 0) + \operatorname{Ddeg} \tilde{a}_{3n} \text{ and} \\ &\operatorname{Ddeg} \tilde{a}_{3n+2} = (0, 1, 1) + \operatorname{Ddeg} \tilde{a}_{3n+1} \end{aligned}$$

which implies by induction on n that

$$\begin{aligned} \operatorname{Ddeg} \tilde{a}_{3n} &= (2n, 2n, 2n), \\ \operatorname{Ddeg} \tilde{a}_{3n+1} &= (2n+1, 2n, 2n), \\ \operatorname{Ddeg} \tilde{a}_{3n+2} &= (2n+1, 2n+1, 2n+1). \end{aligned}$$

So

$$\begin{aligned} \operatorname{Deg} \tilde{a}_{3n} &\leq (2n, 2n, 2n), \\ \operatorname{Deg} \tilde{a}_{3n+1} &\leq (2n+1, 2n, 2n), \\ \operatorname{Deg} \tilde{a}_{3n+2} &\leq (2n+1, 2n+1, 2n+1). \end{aligned}$$

Finally put $b_n := \Phi(\tilde{a}_n)$ for all $n \geq 1$. Then by 9.6.9 we get

$$D_1 b_n = \Phi(\tilde{D}(\tilde{a}_n)) = b_{n-1}.$$

2) Conversely, let $b_0 = 1, b_1, b_2, \ldots$ in B satisfy i), ii) and iii) above. Since $\operatorname{Deg} b_{3n} \leq (2n, 2n, 2n)$ we get by 9.6.12 that

$$b_{3n} = \Phi_{(2n,2n,2n)}(\tilde{a}_{3n})$$

for some $\tilde{a}_{3n} \in A_{(2n,2n,2n)}$. Similarly we have

$$\begin{aligned} b_{3n+1} &= \Phi_{(2n+1,2n,2n)}(\tilde{a}_{3n+1}) \text{ and} \\ b_{3n+2} &= \Phi_{(2n+1,2n+1,2n+1)}(\tilde{a}_{3n+2}). \end{aligned}$$

Now we show that $\tilde{D}(\tilde{a}_{3n}) = XYZ\tilde{a}_{3n-1}$. The other relations one gets similarly. Put $\sigma = (2n, 2n, 2n)$. By 9.6.10 also $\tilde{D}(\tilde{a}_{3n}) \in A_\sigma$. Furthermore $\operatorname{Ddeg} XYZ\tilde{a}_{3n-1} = (1,1,1)+(2n-1, 2n-1, 2n-1) = \sigma$. So both the elements $\tilde{D}(\tilde{a}_{3n})$ and $XYZ\tilde{a}_{3n-1}$ belong to A_σ, hence by 9.6.12 ii) we need to show that their images under Φ_σ are equal. This follows from 9.6.11, namely

$$\begin{aligned} \Phi_\sigma(\tilde{D}(\tilde{a}_{3n})) &= D_1 b_{3n} \\ &= b_{3n-1} \quad \text{and} \\ \Phi_\sigma(XYZ\tilde{a}_{3n-1}) &= \Phi(\tilde{a}_{3n-1}) \\ &= \Phi_{(2n-1,2n-1,2n-1)}(\tilde{a}_{3n-1}) \\ &= b_{3n-1}. \end{aligned}$$

□

So it remains to prove the existence of the sequence $b_0 = 1, b_1, b_2, \ldots$ in B as described in 9.6.13. We will construct such a sequence inductively. More precisely

Theorem 9.6.14 *There exists $H_0 = 1, H_1, H_2, \ldots$ in $\ker D_1$ with $H_n \in P_n$ for all $n \geq 0$ such that $b_n = \sum_{i=0}^{n} \frac{1}{i!} S^i H_{n-i}$ is as described in* 9.6.13.

Proof: Suppose that we have found $H_0, H_1, \ldots, H_n$. Consider

$$c_{n+1} = \sum_{i=0}^{n} \frac{1}{(i+1)!} S^{i+1} H_{n-i}$$

and observe that $D_1 c_{n+1} = b_n$.

i) Assume that $n \equiv 1 \pmod 3$, say $n = 3l + 1$. The cases $n \equiv 0, 2 \pmod 3$ can be treated similarly. In ii) below we will show that there exists $H_{n+1} \in \ker D_1 \cap P_{n+1}$ of the form

$$H_{3l+2} = (S-T)^{l+1}(S-U)^{l+1} \sum_{i=0}^{l} \gamma_i (S-T)^{l-i}(S-U)^i,$$

$$\text{(9.6.15)} \qquad \gamma_i \in k$$

such that $b_{n+1} := c_{n+1} + H_{n+1}$ satisfies $\deg_S b_{3l+2} \le 2l+1$, i.e. $\partial_S^{2l+2} b_{3l+2} = 0$. One easily verifies that b_{n+1} also satisfies the degree conditions

$$\deg_U b_{3l+2} \le 2l+1 \text{ and } \deg_T b_{3l+2} \le 2l+1.$$

So b_{n+1} is as desired.

ii) To find H_{n+1} as described in i) observe that $c_{n+1} = c_{3l+2} \in P_{3l+2}$, whence $\partial_S^{2l+2} c_{n+1} \in P_l$. Furthermore $\partial_S^{2l+2} c_{n+1} \in \ker D_1$, namely $D_1 \partial_S^{2l+2} c_{n+1} = \partial_S^{2l+2} D_1 c_{n+1} = \partial_S^{2l+2} b_n = \partial_S^{2l+2} b_{3l+1} = 0$. So

$$\partial_S^{2l+2} c_{n+1} \in \sum_{i=0}^{l} k(S-T)^{l-i}(S-U)^i.$$

Then 9.6.16 below implies that $\partial_S^{2l+2} c_{n+1} = \partial_S^{2l+2}(-H_{3l+2})$, for some H_{3l+2} of the form 9.6.15, which completes the proof. □

Lemma 9.6.16 *Let $G_i := (S-T)^{l-i}(S-U)^i$, for all $0 \le i \le l$. Let V_1 be the k-span of the G_i and W the k-span of the elements $(S-T)^{l+1}(S-U)^{l+1}G_i$. Then $\partial_S^{2l+2} : W \to V_1$ is surjective.*

Proof: Since $\dim V_1 = \dim W$ $(= l+1)$ it suffices to prove that $\ker \partial_S^{2l+2} = 0$. This follows readily, since each non-zero element of W has S-degree $\ge 2l+2$. □

During the process of finishing this book I received a preprint of Freudenburg in which he gives a new counterexample to Hilbert fourteen in dimension 6!! Remarkably enough his proof goes along similar lines as the one given above. Even more recently the following five-dimensional counterexample was deduced from it by Daigle and Freudenburg. Their result is

Theorem 9.6.17 (Daigle, Freudenburg). *Let*

$$D = X^3\partial_S + S\partial_T + T\partial_U + X^2\partial_V$$

on $A := k[X, S, T, U, V]$. Then D is triangular and A^D is not a finitely generated k-algebra.

Remark 9.6.18 The derivation in 9.6.2 is an example of a so-called elementary derivation (§5, Exercise 7), i.e. a derivation on the $n+m$ variable polynomial ring $k[X, Y]$ of the form $D = \sum a_i(X)\partial_{Y_i}$. Elementary derivations played a crucial role in the proof of the stably tameness result in Chapter 7. Furthermore it is shown in [117] that if $n \le 2$ or $m \le 2$ then $k[X, Y]^D$ is a finitely generated k-algebra. See also Exercise 4 and Exercise 3, §4, Chapter 1. We do not know if there exists an elementary derivation with $n = 3$ and $m = 3$ such that its kernel is not a finitely generated k-algebra.

Theorem 9.6.2 shows in particular that a subring of a polynomial ring $k[X]$ containing k need not be a finitely generated k-algebra. This in contrast with the "field case" as can be seen from the following result, which was first proved by E. Noether in 1915 in case the characteristic of k is zero and by Samuel in 1952 for the general case.

Theorem 9.6.19 *Let $K \subset k(X_1, \ldots, X_n)$ be a subfield containing k. Then $K = k(g_1, \ldots, g_t)$ for some $g_i \in k(X_1, \ldots, X_n)$.*

Proof: i) Let $r = \text{trdeg}_k K$. So $r \leq n$ and say $g_1, \ldots, g_r$ is a transcendence basis of K over k. Then one can renumber the X_i such that $(g_1, \ldots, g_r, X_{r+1}, \ldots, X_n)$ form a transcendence basis of $k(X_1, \ldots, X_n)$ over k ([190], Chapter X, Theorem 1.1). So in particular $|k(X_1, \ldots, X_n) : k(g_1, \ldots, g_r, X_{r+1}, \ldots, X_n)|$ is finite and hence $m := |K(X_{r+1}, \ldots, X_n) : k(g_1, \ldots, g_r, X_{r+1}, \ldots, X_n)|$ is finite. We will show

$$|K : k(g_1, \ldots, g_r)| \leq m. \tag{9.6.20}$$

It follows that K is a finite dimensional vector space over $k(g_1, \ldots, g_r)$, say with basis $g_{r+1}, \ldots, g_t$. Then obviously $K = k(g_1, \ldots, g_r, g_{r+1}, \ldots, g_t)$.

ii) To prove 9.6.20 assume that there exist $y_1, \ldots, y_s \in K$ linearly independent over $k(g_1, \ldots, g_r)$ but linearly dependent over $k(g_1, \ldots, g_r, X_{r+1}, \ldots, X_n)$. So there exist $b_1, \ldots, b_s$ not all zero with $b_1 y_1 + \cdots + b_s y_s = 0$, where

$$b_i = \sum_{j \in \mathbb{N}^{n-r}} a_{i,j} {X_{r+1}}^{j_{r+1}} \ldots {X_n}^{j_n}, a_{i,j} \in k(g_1, \ldots, g_r).$$

We can write the linear relation as

$$\sum_{j \in \mathbb{N}^{n-r}} {X_{r+1}}^{j_{r+1}} \ldots {X_n}^{j_n} \left(\sum_{i=1}^{s} a_{i,j} y_i \right) = 0. \tag{9.6.21}$$

Now observe that $X_{r+1}, \ldots, X_n$ are algebraically independent over K: otherwise

$$\text{trdeg}_K K(X_{r+1}, \ldots, X_n) \leq n - r - 1$$

and hence $\text{trdeg}_k K(X_{r+1}, \ldots, X_n) \leq r + (n - r - 1) = n - 1$. Since

$$k(g_1, \ldots, g_r, X_{r+1}, \ldots, X_n) \subset k(X_{r+1}, \ldots, X_n)$$

we get

$$n = \text{trdeg}_k k(g_1, \ldots, g_r, X_{r+1}, \ldots, X_n) \leq n - 1,$$

a contradiction. So 9.6.21 gives $\sum_i a_{i,j} y_i = 0$ for all j, hence $a_{i,j} = 0$ for all i, j by the assumption on the $y_1, \ldots, y_s$. Consequently $b_i = 0$ for all i, a contradiction.

□

To conclude this section we mention, without proof, a classical result due to Weitzenböck in [291], 1932.

Theorem 9.6.22 (Weitzenböck). *Let k be a field of characteristic zero and assume that G_a acts as a subgroup of $Gl_n(k)$ on k^n. Then $A(k^n)^{G_a}$ is a finitely generated k-algebra.*

For a proof we refer the reader to the paper [267] of Seshadri. He embeds G_a in $G := Sl_2(k)$, as in 9.1.2, and defines an action of G on $A := k[X_1, \dots, X_n]$ and an action of G on another finitely generated k-algebra A' having two more generators in such a way that A'^G maps naturally onto A^{G_a}. Since by 9.2.1, A'^G is a finitely generated k-algebra, using that $Sl_2(k)$ is reductive, it follows that A^{G_a} is a finitely generated k-algebra. Observe that in §2 we assumed that k is algebraically closed, however to prove our result we may assume this due to 9.5.7 and Exercise 2.

Exercises for §9.6

1. Let G be a group acting on $V := k^n$. Show that $A(V)^G = A(V) \cap Q(A(V)^G)$.
2. Let k be a field and $\bar{k}$ an algebraic closure. Let D be a k-derivation on $k[X]$. Extend D to a $\bar{k}$- derivation $\overline{D}$ on $k[X]$. Show that $k[X]^D$ is a finitely generated k-algebra if and only if $\bar{k}[X]^{\overline{D}}$ is a finitely generated $\bar{k}$-algebra.
3. Consider $D = X^2\partial_S + Y^2\partial_T + Z^2\partial_U + (XYZ)^2\partial_V$ on $A := k[X, Y, Z, S, T, U, V]$. Show that A^D is a finitely generated k-algebra.
4. Let $D = X^3\partial_S + Y^3\partial_T + (XY)^2\partial_U$ on $A := k[X, Y, S, T, U]$.

   (i) Show that A^D is a finitely generated k-algebra; find a complete set of k-generators without using a computer. [To compute i) use the method of the kernel algorithm.]
   (ii) Check your results by an implementation of the kernel algorithm from Chapter 1.

5. Let $m \geq 2$, $P := k[X, Y, Z, S, T, U, V]$ and R the k-subalgebra of P generated by $X, Y, Z, XV - Y^mZ^mS, YV - X^mZ^mT, ZV - X^mY^mU$. Roberts showed in [252] that $Q(R) \cap P$ is not a finitely generated k-algebra. Let $D = X^{m+1}\partial_S + Y^{m+1}\partial_T + Z^{m+1}\partial_U + (XYZ)^m\partial_V$. Show that $Q(A^D) = Q(R)$. [Use the algorithm of Chapter 1, §4. Take as preslice $p := V$. So $d = D(V) = (XYZ)^m$ and $s = V/(XYZ)^m$. Then compute C_0.]
6. (Maubach) Let $A = k[X, Y, Z]$, $B = A[T]$, $a, b, c \geq 1$ and $D = Y^a\partial_X + Z^b\partial_Y + T^c\partial_Z$.

   (i) Show that the $\mathbb{Q}$-vector space of all $w \in \mathbb{Q}^4$ such that the w-grading on B is D-invariant is generated by $e := (abc, bc, c, 1)$.
   (ii) Let Ddeg denote the e-degree, so $\text{Ddeg}\, X^\alpha Y^\beta Z^\gamma T^\delta = abc.\alpha + bc.\beta + c.\gamma + \delta$. Let $D_1 := Y^a\partial_X + Z^b\partial_Y + T\partial_Z$. Denote by $\varphi : B \to B$ the A-homomorphism sending T to T^c. Show $D \circ \varphi = \varphi \circ D_1$.
   (iii) Show that if B^{D_1} is generated as a k-algebra by some F_i, then B^D is generated by the $\varphi(F_i)$ and T. [Let $G \in B^D$. Write $G = \sum_{i=0}^{c-1} G_i$, where G_i is the sum of all monomials m appearing in G such that $\text{Ddeg}\, m \equiv i \pmod{c}$. Deduce that each $G_i \in B^D$. Furthermore if $m = rX^\alpha Y^\beta Z^\gamma T^\delta$

is a monomial in G_i, $r \in k^*$, then $\text{Ddeg}\, m \equiv \delta \pmod{c}$. So $\delta = i + pc$ for some $p \in \bar{\mathbb{N}}$. Hence $m = rT^i X^\alpha Y^\beta Z^\gamma (T^c)^p$. Deduce that $G_i = T^i \tilde{G}_i$ with $\tilde{G}_i \in A[T^c]$. Then use ii) to conclude the proof of iii).]

(iv) Deduce from iii) that B^D is a finitely generated k-algebra. [It suffices, by iii), to prove that B^{D_1} is a finitely generated k-algebra. Then follow the kernel algorithm of Chapter 1, §4 taking $p := Z$ as preslice.]

Notes

The proof of Proposition 9.2.4 is taken from the paper [274] of Springer. All results of §3 are taken from Derksen's Ph.D thesis [78]. The first solution of the Cancellation problem in dimension 3 (Corollary 9.4.13) was given by Fujita in [133]. The results at the end of §4 come from the paper [53] of Choudary and Dimca. For more results about exotic spaces we refer to the papers [172], [171] of Kaliman, [304] of Zaidenberg and [285] of tom Dieck and Petrie.

Theorem 9.5.15, which appeared in [70], was originally proved by Eggermont and the author in [98] in case R is a polynomial ring over a field. Recently Theorem 9.5.15 was generalized by Bhatwadekar and Dutta in [39]: they showed amongst other things that if R is a Noetherian domain containing $\mathbb{Q}$ and D a locally nilpotent derivation on $R[X, Y]$ such that 1 belongs to the ideal generated by DX and DY, then $R[X, Y]$ is a polynomial ring in one variable over $R[X, Y]^D$. This result was in turn generalized by Berson, Maubach and the author in [37] to locally nilpotent derivations with divergence zero on $R[X, Y]$, where R is an arbitrary $\mathbb{Q}$-algebra. For several other results concerning locally nilpotent derivations the reader is referred to the papers [67], [69], [68] of Daigle, [81], [82], [83], [84], [85] of Deveney and Finston, [86] of Deveney, Finston and Gehrke, [124] of Finston and Walcher, [102], [107], [103] of the author, [129], [127], [128], [130], [131] of Freudenburg and the papers [217] of Miyanishi and [309] of Zurkowski. In particular I wish to mention the paper [130] of Freudenburg in which he constructs a locally nilpotent derivation in dimension 3 whose kernel does not contain a coordinate (see also Appendix F, Example 9.8), thereby giving a negative solution to question A2 in [108].

The proof of the Deveney-Finston-Roberts result given in §6 is taken from Maubach's master's thesis [204]. A proof of this result along similar lines was given by Kojima and Miyanishi in [180]. Freudenburg's recent 6-dimensional counterexample to Hilbert's fourteenth problem can be found in [131]. The 5-dimensional example of Daigle and Freudenburg appeared in [71].

10 The Jacobian conjecture

As the reader certainly has noticed by now, the Jacobian Conjecture has been studied extensively and various partial results have been obtained. This chapter will not be a collection of all these results. Instead it will describe several new ways to attack the conjecture.

In §1 we give some invertibility results concerning maps from $\mathbb{R}^n$ to $\mathbb{R}^n$, such as Hadamard's theorem and we briefly discuss the Real Jacobian Conjecture by giving Pinchuk's two-dimensional counterexample and some partial results due to Cima, Gasull and Mañosas.

In §2 we study the two-dimensional Jacobian Conjecture using Newton polygons and obtain several classical results, mainly due to Abhyankar in the early seventies. In particular the two-dimensional Jacobian Conjecture is reduced to the study of Jacobian pairs whose Newton polygon has only edges with positive slopes. The technique developed in this section, which was heavily influenced by an unpublished paper of V. Zurkowski, opens a way to study Newton polygons having edges with positive slopes.

For completeness' sake we discuss several more or less well-known special cases of the two-dimensional Jacobian Conjecture at the end of §2.

In §3 we study polynomial maps having integer coefficients and formulate a characteristic $p > 0$ version of the Jacobian Conjecture due to Adjamagbo. We show that this characteristic $p > 0$ version, if true, implies the classical (complex) Jacobian Conjecture. The interesting point is that this characteristic $p > 0$ version is open for all $n \geq 2$ and all p, so including the case $n = 2$ and $p = 2$!

In §4 we bring A_n-modules into the picture. The Jacobian condition induces an A_n-module structure on the polynomial ring. Our main theorem asserts that this A_n-module is regular holonomic, a result whose impact on the Jacobian Conjecture is not at all understood.

Finally in §5 we describe one more approach to the Jacobian Conjecture based on the question: is every homomorphism of $\mathbb{C}[X]$ which sends every linear coordinate to a coordinate an automorphism? If the Jacobian Conjecture holds, the answer to this question should be positive. On the other hand a result due to Mikhalev, Yu and Zolotykh (10.5.7) asserts that if $n \geq 3$ and $\mathbb{C}$ is replaced by any non-algebraically closed field, then the answer to the above question is negative.

10.1 Injectivity and invertibility of differentiable maps and the real Jacobian conjecture

In this section we discuss some classical and new results concerning global invertibility problems for differentiable maps. The starting point of each consideration on this subject is the local inverse function theorem: let $U \subset \mathbb{R}^n$ be an open set and $F : U \to \mathbb{R}^n$ a C^1-map, then for each $x \in U$ we have that F is locally C^1-invertible in a neighborhood of x with C^1-inverse if and only if $\det JF(x) \neq 0$.

One wants to globalize this result. So from now on let $F : \mathbb{R}^n \to \mathbb{R}^n$ be a C^1-map. The local inverse function theorem implies that a necessary condition for F to have a global C^1-inverse is that $\det JF(x) \neq 0$ for all $x \in \mathbb{R}^n$. However the example $(x, y) \to (e^x cos(y), e^x sin(y))$ shows that this condition is not sufficient since $F(1, 0) = F(1, 2\pi)$. Of course if F has a global inverse, then its inverse, being continuous, maps compact sets into compact sets, in other words F is **proper**, i.e. $F^{-1}(K)$ is compact for each compact set $K \subset \mathbb{R}^n$. It turns out that adding properness gives a sufficient condition for global invertibility, i.e.

Theorem 10.1.1 (Hadamard). *Let $F : \mathbb{R}^n \to \mathbb{R}^n$ be a C^1-map. Then F is a diffeomorphism if and only if* $\det JF(x) \neq 0$ *for all $x \in \mathbb{R}^n$ and F is proper.*

This theorem was obtained by Hadamard in 1906 ([150]). We refer the reader to the nice paper [143] of Gordon, where two proofs of this result are given.

A number of global univalence (=injectivity) results have been developed since 1906. We refer to the book [238] of Parthasarathy. Although injective C^1-maps are not always surjective, as can be seen from the easy example e^x, surjectivity and hence global invertibility is guaranteed for a large class of maps, due to the following result (for the proof of which we refer to [186]).

Theorem 10.1.2 (Kurdyka, Rusek). *Let $F : \mathbb{R}^n \to \mathbb{R}^n$ be an injective continuous mapping such that its* ***graph****, i.e. the set of all $(x, y) \in \mathbb{R}^n \times \mathbb{R}^n$ with $y = F(x)$, is an algebraic variety, then F is surjective.*

One attempt to obtain an injectivity result was made by the economist and Nobel laureate A. Samuelson in [259], 1953. He conjectured the following.

Samuelson Conjecture. *Let $F : \mathbb{R}^n \to \mathbb{R}^n$ be a C^1-map such that*

$$|JF|_k := \det\left(\frac{\partial F_i}{\partial x_j}\right)_{1 \le i,j \le k}(x) \neq 0, \qquad \text{for all } x \in \mathbb{R}^n \text{ and all } 1 \le k \le n; \tag{10.1.3}$$

then F is injective.

However Gale and Nikaido in [138] produced the following counterexample

$$F(x, y) = (f(x, y), g(x, y)) = (e^{2x} - y^2 + 3, 4e^{2x}y - y^3).$$

Then $f_x = 2e^{2x}$, $\det JF = e^{4x}(4e^{2x} + 5y^2)$ and $F(0, 2) = F(0, -2) = (0, 0)$. In 1992 Parthasarathy and the author showed in [119] that Samuelson's conjecture is true if we replace C^1-maps from $\mathbb{R}^n$ to $\mathbb{R}^n$ by polynomial maps from $\mathbb{C}^n$ to $\mathbb{C}^n$. In fact a more general result was obtained (Exercise 1). This result was generalized by Campbell in [45]. Furthermore in a subsequent paper [46] he showed that Samuelson's conjecture is true for a large class of maps from $\mathbb{R}^n$ to $\mathbb{R}^n$. More precisely

Theorem 10.1.4 (Campbell). *Let $F = (F_1, \ldots, F_n) : \mathbb{R}^n \to \mathbb{R}^n$ be a C^1-map. If F satisfies* 10.1.3 *and each component F_i has an algebraic graph, i.e. the set $\{(x, F_i(x)) | x \in \mathbb{R}^n\}$ is an algebraic variety, then F is a diffeomorphism.*

For a proof of this result, and other interesting things, we refer the reader to the very elegant paper [46].

Since the graph of an everywhere defined rational function $f = p/q$, where p and q are polynomials with q nowhere vanishing on $\mathbb{R}^n$ is obviously algebraic, namely given by $yq - p = 0$, we get

Corollary 10.1.5 *Each everywhere defined rational map from $\mathbb{R}^n$ to $\mathbb{R}^n$ satisfying* 10.1.3 *is a diffeomorphism. In particular Samuelson's conjecture is true for polynomial maps.*

This result shows that polynomial maps are better behaved than C^1-maps. So it is therefore natural to ask if for polynomial maps from $\mathbb{R}^n$ to $\mathbb{R}^n$ the condition $\det JF(x) \neq 0$ for all $x \in \mathbb{R}^n$ is sufficient to guarantee injectivity and hence by 10.1.2 global invertibility. One cannot expect the inverse to be a polynomial map, as can be seen from the example $x \to x + x^3$. This problem circulated in the mathematical community since the seventies under the name

Real Jacobian Conjecture, RJC(n). *Let $F : \mathbb{R}^n \to \mathbb{R}^n$ be a polynomial map such that* $\det JF(x) \neq 0$ *for all $x \in \mathbb{R}^n$, then F is injective.*

Of course replacing $\mathbb{R}$ by $\mathbb{C}$ in this statement we get the usual Jacobian Conjecture. Furthermore it is easy to see that *RJC(2n)* implies *JC(n)*: namely arguing as in Chapter 6, §3 we see that if $F : \mathbb{C}^n \to \mathbb{C}^n$ is a polynomial map with $\det JF \in \mathbb{C}^*$, then $\tilde{F} := (ReF_1, ImF_1, \ldots, ReF_n, ImF_n) : \mathbb{R}^{2n} \to \mathbb{R}^{2n}$ is a real polynomial map with $\det J\tilde{F} = |\det JF|^2 \in \mathbb{R}^*$. So if *RJC(2n)* is true we conclude that $\tilde{F}$ is injective and hence so is F.

However in 1994 Pinchuk in [243] found the following counterexample to the two-dimensional real Jacobian Conjecture.

Pinchuk's counterexample to RJC(2)

Let $t := XY - 1$, $h := t(XY + 1)$ and $f := (Xt + 1)^2 \left(\frac{h+1}{X}\right)$. Furthermore let $p, q \in \mathbb{R}[X, Y]$ be defined by $p = f + h$ and

$$\begin{aligned} q = &-t^2 - 6th(h+1) - 170fh - 91h^2 - 195fh^2 \\ &-69h^3 - 75h^3 f - \tfrac{75}{4}h^4. \end{aligned}$$

Put $F := (p, q)$. Then $F(1, 0) = F(-1, -2) = (0, -1)$ and

$$\det JF = t^2 + (t + f(13 + 15h))^2 + f^2.$$

Finally observe that $\det JF(x, y) > 0$ for all $x, y \in \mathbb{R}$: namely if $t(x, y) = 0$, then $x \neq 0$ and $h(x, y) = 0$, whence $f(x, y) = \frac{1}{x} \neq 0$.

Remark 10.1.6 It was Ronen Peretz who, guided by his theory of asymptotic values (see [240]), observed that Pinchuk's counterexample is of a special form, namely both components p, q above belong to the ring $\mathbb{R}[X, XY, X^2Y - X]$. So it is natural to ask if there exist counterexamples to the Jacobian Conjecture of the form (P, Q), where both P and Q belong to the ring $\mathbb{C}[X, XY, X^2Y - X]$. In [239] he answered this question in the negative! This result was recently generalized by David Wright in his paper [296].

Remark 10.1.7 Applying the stabilization methods of Chapter 6 to this example Hubbers, using a computer, found a cubic homogeneous counterexample to the real Jacobian Conjecture in dimension 2033, [157]. At the end of 1999 he even obtained a counterexample in dimension 1999!

On the other hand a positive result was obtained by Cima, Gasull, Llibre and Mañosas in 1994 ([55] and [57]). It asserts that the real Jacobian Conjecture is true for a large class of polynomial maps.

To explain this result we need to introduce some notation. Let $m = (m_1, \ldots, m_n)$ where each m_i is a positive integer and denote by χ_m the set of polynomial maps $F = (F_1, \ldots, F_n) : \mathbb{R}^n \to \mathbb{R}^n$ such that $\deg F_i \leq m_i$ for all i. By identifying χ_m with $\mathbb{R}^M$, where M denotes the number of all coefficients in the F_i, we endow χ_m with the so-called **coefficient-topology**. Furthermore we denote by J_m the set of all polynomial maps F of χ_m such that $\det JF(x) \neq 0$ for all $x \in \mathbb{R}^n$ and by $\operatorname{Int} J_m$ the interior of J_m. Then one of the main results of [57] is

Theorem 10.1.8 (Cima, Gasull, Mañosas). *If $F \in \operatorname{Int} J_m$, then F is injective and hence a diffeomorphism.*

The proof given below is based on the following lemma, due to Andrew Campbell.

Lemma 10.1.9 (Campbell). *Let $F \in \chi_m$ be such that 0 is the only zero of $F_+ = 0$. Then F is proper.*

Proof: Observe that $t^{m_i} F_i(\frac{x}{t})$ is a polynomial in x and t, so we get a well-defined continuous map $\tilde{F} : [0, 1] \times S^{n-1} \to \mathbb{R}$ given by

$$\tilde{F}(t, x) = |t^{m_1} F_1\left(\frac{x}{t}\right), \ldots, t^{m_n} F_n\left(\frac{x}{t}\right)|^2.$$

If F is not proper there exists a sequence $x_1, x_2, \ldots$ in $\mathbb{R}^n$ converging to infinity and such that $F(x_1), F(x_2), \ldots$ converges to some point $w \in \mathbb{R}^n$. For k large enough we have that $|x_k| > 1$ and each $F_i(x_k)$ is close to w_i. For such k we put $t_k := |x_k|^{-1}$ and $y_k := t_k x_k \in S^{n-1}$. Then

$$\tilde{F}(t_k, y_k) = |t_k^{m_1} F_1(x_k), \ldots, t_k^{m_n} F_n(x_k)|^2$$

converges to 0, since each component tends to zero, either because $m_i > 0$ and $t_k \to 0$, with $F_i(x_k)$ bounded, or if $m_i = 0$ because then F_i is the constant 0, since $F_+(0) = 0$. Taking a subsequence, if necessary, we may assume that $(t_k, y_k) \to (0, y)$. So by the continuity of $\tilde{F}$ we get $\tilde{F}(t_k, y_k) \to \tilde{F}(0, y) = |F_+(y)|^2$. We already saw that $\tilde{F}(t_k, y_k) \to 0$, so $F_+(y) = 0$. However $y \in S^{n-1}$, so $y \neq 0$, contradicting our hypothesis that 0 is the only solution of $F_+ = 0$. So F is proper. □

Corollary 10.1.10 (Randall). *If $F \in J_m$ is such that 0 is the only solution of $F_+ = 0$, then F is injective and hence a diffeomorphism.*

Proof: Follows immediately from Hadamard's theorem and Campbell's lemma. □

Proof of Theorem 10.1.8: i) By 10.1.10 it suffices to prove that $F_+(x) \neq 0$ for all $x \neq 0$. So assume that $F_+(a) = 0$ for some $a \neq 0$. We will derive a contradiction. Then $F_+(ta) = 0$ for all $t \in \mathbb{R}$, since F_+ is homogeneous. Differentiating with respect to t gives $JF_+(ta).a = 0$ for all t. In particular $JF_+(a).a = 0$. Since $a \neq 0$ this implies $\det JF_+(a) = 0$.

ii) Let r :=rank $JF_+(a)$. So $r < n$. We may assume that the first r rows of this matrix are linearly independent. Furthermore since $F \in J_m$ we may assume that $\det JF(x) > 0$ for all $x \in \mathbb{R}^n$. Choose polynomials $G_{r+1}, \ldots, G_n$, where each G_i is homogeneous of degree m_i, such that

(10.1.11) $$\det J((F_1)_{m_1}, \ldots, (F_r)_{m_r}, G_{r+1}, \ldots, G_n)(a) < 0$$

and define $F_\varepsilon := F + \varepsilon G$ for each $\varepsilon > 0$, where $G := (0, \ldots, 0, G_{r+1}, \ldots, G_n)$. Since $F \in \text{Int}\, J_m$ we can choose $\varepsilon > 0$ such that $F_\varepsilon \in J_m$ and furthermore so that $\det JF_\varepsilon(a)$ is arbitrarily close to $\det JF(a) > 0$, whence $\det JF_\varepsilon(a) > 0$ and hence, since $F_\varepsilon \in J_m$

(10.1.12) $$\det JF_\varepsilon(x) > 0 \text{ for all } x \in \mathbb{R}^n.$$

However one easily verifies that 10.1.11 implies that $\det J(F_\varepsilon)_+(a) < 0$, whence $\det JF_\varepsilon(ta) < 0$ for all sufficiently large t, a contradiction with 10.1.12. So $F_+(a) \neq 0$ for all $a \neq 0$, as desired.

Remark 10.1.13 The main results, 10.1.8 and 10.1.10 can be extended in the following sense ([57]): instead of considering on $\mathbb{R}[X] := \mathbb{R}[X_1, \ldots, X_n]$ the standard grading $(1, 1, \ldots, 1)$ we can as well consider gradings with respect to n-tuples $w = (w_1, \ldots, w_n)$ of positive integers (Chapter 5, §1) and denote by F_+ the highest w-degree homogeneous part of F. All results described above hold in this generality (Exercise 2).

Remark 10.1.14 In the proof of 10.1.8 the essential statement proved is: "if $F \in \text{Int}\, J_m$, then $\det JF_+(x) \neq 0$ for all $x \neq 0$". In fact the converse also holds ([55], Proposition 3.4). So we get

(10.1.15) $$F \in \text{Int}\, J_m \text{ if and only if } \det JF_+(x) \neq 0 \text{ for all } x \neq 0.$$

Furthermore it was shown in [57] that $\text{Int}\, J_m$ is not empty if and only if all m_i are odd numbers (see also Exercise 3). So in particular this applies to the case that all m_i are equal to 3. It is therefore interesting to observe that nevertheless all F of the form $X + H$ with JH nilpotent and $\deg H_i = 3$ for all i do not belong to $\text{Int}\, J_m$: namely the nilpotence condition implies that $\det JH(x) = 0$ for all x, whence $\det JF_+(x)$ $(= \det JH(x)) = 0$ for all $x \in \mathbb{R}^n$. So F does not belong to $\text{Int}\, J_m$, by 10.1.15. In fact every non-linear polynomial automorphism F over $\mathbb{R}$ does not belong to $\text{Int}\, J_m$, where $m = (m_1, \ldots, m_n)$ and $m_i = \deg F_i$ (Exercise 5).

Exercises for §10.1

1. (van den Essen, Parthasarathy) Notation as in 10.1.3. Let A be a commutative $\mathbb{Q}$-algebra. An elementary polynomial map E is called of **type** i if $E_j = X_j$ for all $j \neq i$. Let $F = (F_1, \ldots, F_n) \in A[X]^n$.

   (i) Show that $|J(F \circ E)|_p = |JF|_p(E)$ for all $1 \le p \le n$ and all elementary polynomial maps of type 1.

   (ii) Assume that $|JF|_i$ is a unit in A for all i. Show that $F = D \circ E_{(n)} \circ \cdots \circ E_{(1)}$ where each $E_{(p)}$ is an elementary polynomial map of type p and $D := (d_1X_1, \ldots, d_nX_n)$ with $d_1 := |JF|_1$ and $d_i := |JF|_i/|JF|_{i-1})$ for all $2 \le i \le n$. [Replace F by $D^{-1} \circ F$. Then we may assume that $|JF|_i = 1$ for all i. So $F_1 = X_1 + f_1$ with $f_1 \in A[X_2, \ldots, X_n]$. Put $E_{(1)} := (X_1 + f_1, X_2, \ldots, X_n)$ and consider $G := F \circ (E_{(1)})^{-1}$. So $G = (X_1, G_2, \ldots, G_n)$ and $|JG|_p = 1$ for all $1 \le p \le n$ by i). View $G_2, \ldots, G_n$ as elements of $A'[X_2, \ldots, X_n]$, where $A' := A[X_1]$ and use induction on the number of variables.]

2. Let $w := (w_1, \ldots, w_n)$ be an n-tuple of positive integers and consider the w-grading on $\mathbb{R}[X]$ (Chapter 5, §1) and for $P \in \mathbb{R}[X]$ denote by P_+ the w-homogeneous part of P of highest degree. For $F = (F_1, \ldots, F_n) \in \mathbb{R}[X]^n$ put $F_+ := ((F_1)_+, \ldots, (F_n)_+)$. Generalize 10.1.10 to any w-grading. [To prove a generalized version of 10.1.9 put $w_0 := w_1 \ldots w_n$ and $\overline{w}_i := w_0/w_i$ for all i. Define $|x|_w^2 := (\sum_i x_i^{2\overline{w}_i})^{1/w_0}$ for all $x \in \mathbb{R}^n$ and let S_w^{n-1} denote the set of all $x \in \mathbb{R}^n$ such that $|x|_w = 1$. Finally define $\tilde{F}_w : [0, 1] \times S_w^{n-1} \to \mathbb{R}$ by the formula $\tilde{F}_w(t, x) = |t^{w(F_1)} F_1(x_1/t^{w_1}, \ldots, x_n/t^{w_n}), \ldots, t^{w(F_n)} F_n(x_1/t^{w_1}, \ldots, x_n/t^{w_n})|_w^2$ where $w(F_i)$ denotes the w-degree of F_i.]
3. Let $m_1, \ldots, m_n$ be odd positive integers and $m = (m_1, \ldots, m_n)$. Show that $\operatorname{Int} J_m$ is not empty. [Consider $F = (F_1, \ldots, F_n)$, where $F_i = X_i + X_i^{m_i}$ if $m_i > 1$ and $F_i = X_i$ if $m_i = 1$.]
4. Let $p, q \in \mathbb{R}[X, Y]$ with degree ≤ 3. For $\varepsilon > 0$ define
$$F_\varepsilon = (-X - X^3 - XY^2 + \varepsilon p, -Y + X^3 - X^2Y - Y^3 + \varepsilon q).$$
Show that for all ε small enough F_ε is injective. [Consider F_0 and use 10.1.15.]
5. Let $F = (F_1, \ldots, F_n) \in \operatorname{Aut}_{\mathbb{R}} \mathbb{R}[X]$ be non-linear. Let $m_i := \deg F_i$ and $m = (m_1, \ldots, m_n)$. Show that F does not belong to $\operatorname{Int} J_m$. [Since $\det JF \in \mathbb{R}^*$ deduce that $\det JF_+ = 0$.]
6. (Adjamagbo, Maset) A surjective, non-injective local homeomorphism of $\mathbb{R}^2$. Let
$$f(z) = \sum_{n \ge 0} \frac{z^{2n+1}}{n!(2n+1)}$$
for all $z \in \mathbb{C}$. Put
$$F(x, y) = (\operatorname{Re} f(x + iy), \operatorname{Im} f(x + iy))$$
for all $(x, y) \in \mathbb{R}^2$.

(i) Show that $f'(z) = e^{z^2}$. Deduce that $\det JF(x, y) > 0$ for all $(x, y) \in \mathbb{R}^2$.

(ii) Show that f is surjective. [If f is not surjective then by the little Picard theorem $\mathbb{C}\backslash f(\mathbb{C}) = \{a\}$ for some $a \in \mathbb{C}$. However $\mathbb{C}\backslash f(\mathbb{C})$ is symmetric with respect to 0 since f is odd, but $f(0) = 0$.]

(iii) Show that f is not injective. [Use the Big Picard theorem.]

10.2 The two-dimensional Jacobian conjecture

Throughout this section k denotes a field of characteristic zero and R the polynomial ring in the variables X and Y over k. So by a polynomial we will mean an element of R and if we speak about the Jacobian Conjecture we will mean the two-dimensional case.

A polynomial P is said to have a **Jacobian mate** Q if $[P, Q] := \det J(P, Q) \in k^*$. In this case P and Q are called **Jacobian polynomials** and the pair (P, Q) is called a **Jacobian pair**.

The aim of this section is to give partial results on the Jacobian Conjecture and describe some equivalent formulations of it. The main tool is the Newton polygon of a polynomial (Chapter 5) together with gradings of R according to the slopes of the Newton polygon.

Similarity of Newton polygons

The first result we will give is

Theorem 10.2.1 *If (P, Q) is a Jacobian pair with both $\deg P$ and $\deg Q > 1$, then the Newton polygons of P and Q are* **similar**, *i.e. one is obtained from the other by a similarity transformation of radius $r > 0$ and origin $(0, 0)$.*

Remark 10.2.2 The condition $\deg P, \deg Q > 1$ is necessary, for $(X, Y + X^2)$ is clearly a Jacobian pair whose corresponding Newton polygons are not similar.

Before we prove this theorem we will discuss several generalities concerning gradings. A non-zero element $w = (u, v) \in \mathbb{Z}^2$ is called a **direction** if $\gcd(u, v) = 1$ and $u > 0$ or $v > 0$. To each such a direction we consider its w-grading on R (Chapter 5). So $R = \oplus R_n$, where R_n is the k-vector space generated by the monomials $X^i Y^j$ with $ui + vj = n$. A non-zero element P of R_n is called a **w-homogeneous** element of R and n is called its w-degree, denoted $w\deg P$. If $w = (1, 1)$ we write $\deg P$ instead of $w\deg P$. The element of highest w-degree in the homogeneous decomposition of a non-zero polynomial P is called its **w-leading form** and is denoted by P_+. The **w-degree of** P is by definition $w\deg P_+$.

Let $P = \sum_{i \le n} P_i$ with $P_n \neq 0$ and $Q = \sum_{j \le m} Q_j$ with $Q_m \neq 0$ be their respective w-homogeneous decompositions. Put $|w| := u + v$. Then the w-homogeneous decomposition $\sum_s [P, Q]_s$ is given by

$$[P, Q]_s = \sum_{i+j=s+|w|} [P_i, Q_j].$$

In particular if $[P, Q] = c \in k$ then $[P, Q]_s = 0$ if $s \neq 0$. So since $\deg P, \deg Q > 1$

$$(10.2.3) \qquad \sum_{i+j=m} [P_i, Q_j] = 0 \text{ if } m \neq |w| \text{ and } \sum_{i+j=|w|} [P_i, Q_j] = c.$$

We will show below, 10.2.7, that if (P, Q) is a Jacobian pair, then $w\deg P + w\deg Q - |w| > 0$. So by 10.2.3 it follows that $[P_+, Q_+] = 0$. Therefore the following lemma turns out to be very useful.

Lemma 10.2.4 *Let f, g be w-homogeneous polynomials such that $[f, g] = 0$. Then*

i) *There exists a w-homogeneous polynomial h such that $f = c_1 h^{n_1}$ and $g = c_2 h^{n_2}$ for some integers $n_1, n_2 \geq 0$ and $c_1, c_2 \in k^*$.*
ii) *h is not of the form ch_0^s for some $c \in k^*$, $h_0 \in R$ and an integer $s > 1$.*

Proof: To simplify the notation we write homogeneous instead of w-homogeneous. Let $D := f_Y \partial_X - f_X \partial_Y$. Then the hypothesis $[f, g] = 0$ is equivalent to $Dg = 0$. By 1.2.25 $\ker D = k[h]$ for some $h \in R$. Since f is homogeneous, so is D which in turn implies that h is homogeneous. Observe that $Df = 0$ which gives $f \in k[h]$. Since $Dg = 0$ we also have $g \in k[h]$. Then i) follows since f, g and h are homogeneous. Finally, if $h = ch_0^s$ then h_0 is homogeneous and $h_0 \in \ker D = k[h]$. So $h_0 = c'h^t$ for some $t \geq 1$ and $c' \in k^*$. From both equations it follows that $s = t = 1$, contradicting $s > 1$. □

Corollary 10.2.5 *If (P, Q) is a Jacobian pair with $w\deg P_+ + w\deg Q_+ - |w| > 0$, then $P_+ = c_1 h^{n_1}$ and $Q_+ = c_2 h^{n_2}$ for some $c_1, c_2 \in k^*$, $n_1, n_2 \geq 0$ and an h as in* 10.2.4.

Proof: Apply 10.2.3 and 10.2.4. □

For a polynomial f we define

$$t_X(f) := \max\{0, \deg_X f(X, 0)\},$$
$$t_Y(f) := \max\{0, \deg_Y f(0, Y)\},$$

where by definition $\deg 0 = -\infty$.

Proposition 10.2.6 *If (P, Q) is a Jacobian pair such that both* $\deg P, \deg Q > 1$, *then all numbers $t_X(P), t_X(Q), t_Y(P)$ and $t_Y(Q)$ are all positive.*

Proof: It suffices to prove $t_X(P) > 0$. First observe that P must contain mixed terms, for if for example $P = P(Y)$ then $[P, Q] \in k^*$ implies that $\deg P = 1$ contradicting $\deg P > 1$. Suppose that $t_X(P) = 0$. Since the Newton polygon of P is not a line segment, there exists a direction $w = (u, v)$ such that at least one point $\neq (0, 0)$ of $\operatorname{supp} P_+$ lies on the line $uX + vY = 0$, $u > 0$, $v < 0$ and no point of $\operatorname{supp} P$ lies in the area $uX + vY > 0$. Since the linear part of the polynomial map (P, Q) is invertible and $t_X(P) = 0$, we get $(1, 0) \in \operatorname{supp} Q$. This implies that $w\deg Q_+ \geq u > u + v$, since $v < 0$. By our choice of w we have $w\deg P_+ = 0$, whence $w\deg P_+ + w\deg Q_+ - |w| > 0$. So by 10.2.5 $P_+ = c_1h^{n_1}$, $Q_+ = c_2h^{n_2}$. Since $w\deg Q_+ \geq u > 0$ we get $n_2 > 0$ and $w\deg h > 0$. Then $P_+ = c_1h^{n_1}$ and $w\deg P_1 = 0$ imply that $n_1 = 0$, so $P_+ = c_1$. This contradicts the fact that supp P_+ contains at least one point $\neq (0, 0)$. So $t_X(P) > 0$ as desired. □

Lemma 10.2.7 *If (P, Q) is a Jacobian pair with $\deg P$, $\deg Q > 1$ then for any direction w we have*

$$w\deg P + w\deg Q - |w| > 0.$$

Proof: We may assume that $u \geq v$. So $u > 0$. Hence, using 10.2.6 $w\deg P \geq ut_X(P) \geq u$ and $w\deg Q \geq ut_X(Q) \geq u$. So $w\deg P + w\deg Q \geq 2u$. Consequently if $u > v$ the assertion is proved. Therefore assume $u = v$. So $u = v = 1$ since $\gcd(u, v) = 1$. Hence $w = (1, 1)$. So $w\deg P = \deg P$ and $w\deg Q = \deg Q$ which implies that $w\deg P + w\deg Q = \deg P + \deg Q \geq 4 > 2 = |w|$. □

Proof of Theorem 10.2.1: Let $n = \deg P$, $m = \deg Q$ and $w = (1, 1)$. Since (P, Q) is a Jacobian pair it follows from 10.2.5 and 10.2.7 that there exist $c_1, c_2 \in k^*$, $n_1, n_2 \geq 0$ and a homogeneous polynomial h such that $P_+ = c_1h^{n_1}$ and $Q_+ = c_2h^{n_2}$. So supp P_+ is similar to supp Q_+ with the origin as center of similarity and ratio $m : n$. Take one end of supp P_+ and the corresponding end of supp Q_+, say they are the points (a, b), resp. (c, d). So they satisfy $m(a, b) = n(c, d)$. If the Newton polygons of P and Q have no other edges we are done. So let us assume that the Newton polygon of P has another edge. Then there exists a direction w' such that the w'-leading forms P', Q' of P, resp. Q, contain the points (a, b), resp. (c, d). Again by 10.2.3 and 10.2.7 we get that $[P', Q'] = 0$ and, as before, we deduce that supp P' and supp Q' are similar with the origin as center of similarity. Since the points (a, b) and (c, d) are on supp P', resp. supp Q', we see that the ratio of similarity is again $m : n$. This argument can be applied to any neighboring edge successively. □

Generalities on edges of the Newton polygon of a Jacobian polynomial

Since by 10.2.1 the Newton polygons of polynomials of degree > 1 forming a Jacobian pair are similar, we will in the remainder of this section study the possible structure of the Newton polygon of a Jacobian polynomial P of degree > 1. The first result we will prove is

Proposition 10.2.8 *Let P be a polynomial of degree > 1 having a Jacobian mate of degree > 1 and $w \in \mathbb{Z}^2$ a direction. Then there exists a w-homogeneous polynomial H and an integer $r \geq 0$ such that*

$$[P_+, H] = P_+^r. \qquad (10.2.9)$$

To prove this result we introduce several ring extensions of R. More precisely let P and w be as in 10.2.8. To such a pair (P, w) we associate a ring $\tilde{R}_{P_+}$, also denoted $\tilde{R}$ if no confusion is possible, as follows: $\tilde{R}$ is the set of formal sums $f = \sum_{i \leq N} f_i$, for some $N \in \mathbb{Z}$, where each f_i is a w-homogeneous element of $R[P_+^{-1}]$ of degree i, i.e. $f_i = 0$ or $f_i = \frac{H}{P_+^r}$ for some w-homogeneous polynomial H of w-degree $dr + i$ where $d = w\deg P_+$.

If $f \neq 0$, the highest degree f_i appearing in this decomposition is, as before, denoted by f_+ and the corresponding i is called the **w-degree of f**, denoted $w\deg f$.

From 10.2.7 and 10.2.5 we deduce that there exists a w-homogeneous polynomial h such that $P_+ = ch^n$ for some $c \in k^*$ and some integer $n \geq 0$. Furthermore h is not of the form $c_0 h_0^s$ for some $c_0 \in k^*$, $h_0 \in R$ and an integer $s \geq 1$. Observe that $n > 0$, for if $n = 0$ then $P_+ = c \in k^*$ which implies that $(0,0) \in \operatorname{supp} P_+$ which contradicts that $t_X(P)$ and $t_X(Q)$ are strictly positive. Of course replacing P by $c^{-1}P$, we may assume that $P_+ = h^n$. The importance of the ring $\tilde{R}$ comes from the fact that it contains elements $P^{\frac{j}{n}}$ for each $j \in \mathbb{Z}$ which satisfy $(P^{\frac{j}{n}})_+ = h^j$. To see this observe that since $P_+ = h^n$ we can write

$$P = h^n + P_{d-1} + \cdots + P_0$$

where $d := w\deg P_+$. So

$$P = h^n(1 + g_{-1} + \cdots + g_{-d})$$

where $g_{-i} := \frac{P_{d-i}}{h^n}$ is homogeneous of degree $-i$ in $\tilde{R}$. Put

$$g := g_{-1} + \cdots + g_{-d}.$$

Then for each $j \in \mathbb{Z}$

$$(1+g)^{\frac{j}{n}} := \sum_{p=0}^{\infty} \binom{\frac{j}{n}}{p} g^p$$

is a well-defined element of $\tilde{R}$ satisfying $((1+g)^{\frac{j}{n}})^n = (1+g)^j$. So if for each $j \in \mathbb{Z}$ we put,

$$P^{\frac{j}{n}} := h^j(1+g)^{\frac{j}{n}},$$

then $P^{\frac{j}{n}}$ is an element of $\tilde{R}$ satisfying $(P^{\frac{j}{n}})^n = P^j$ and $(P^{\frac{j}{n}})_+ = h^j$.

Lemma 10.2.10 *Notation as above. If $0 \neq Q \in \tilde{R}$ is w-homogeneous and satisfies $[P_+, Q] = 0$, then $Q = ch^m$ for some $c \in k^*$ and an integer m.*

Proof: Since $Q = \frac{H}{h^r}$ for some $r \geq 0$ and a w-homogeneous polynomial H, the fact that $P_+ = h^n$ with $n > 0$ and the condition $[P_+, Q] = 0$ imply that $[h, H] = 0$. Then by 10.2.4 i) and ii) it follows that $H = ch^t$ for some $c \in k^*$ and $t \geq 0$. This implies the lemma. □

Lemma 10.2.11 *Let P be as in* 10.2.10 *and $Q \in \tilde{R}$ satisfying $[P, Q] \in k^*$. If furthermore $w\deg P + w\deg Q - |w| > 0$, then there exists $j \in \mathbb{Z}$ and $c \in k^*$ such that*

$$w\deg(Q - cP^{\frac{j}{n}}) < w\deg Q.$$

Proof: Since $w\deg P_+ + w\deg Q_+ - |w| > 0$ the hypothesis $[P, Q] \in k^*$ implies that $[P_+, Q_+] = 0$. So by 10.2.10 $Q_+ = ch^j$ for some $c \in k^*$ and $j \in \mathbb{Z}$. Since $(P^{\frac{j}{n}})_+ = h^j$ we get the desired result. □

Proof of Proposition 10.2.8: Let (P, Q) be a Jacobian pair with $[P, Q] = 1$. From 10.2.7 and 10.2.5 it follows that $P_+ = c_1 h^n$ and $Q_+ = c_2 h^m$ for some $c_1, c_2 \in k^*$ and $n, m \geq 0$. As observed above we may assume that $c_1 = 1$ and we have that $n > 0$. Then

$$w\deg(Q - c_2 P^{\frac{m}{n}}) < w\deg Q.$$

Observe that $Q_1 := Q - c_2 P^{\frac{m}{n}} \in \tilde{R}\ (= \tilde{R}_{P_+})$ and $[P, Q_1] = [P, Q] = 1$. If $w\deg P + w\deg Q_1 - |w| > 0$ we can repeat the argument using 10.2.11, i.e. we subtract another term of the form $c_2' P^{\frac{j}{n}}$ from Q_1, etc. After a finite number of steps we arrive at an element $Q^* \in \tilde{R}$ which satisfies $w\deg P + w\deg Q^* - |w| = 0$ and $[P, Q^*] = 1$. Consequently $[P_+, Q_+^*] = 1$. Since $Q_+^* = \frac{H}{P_+^r}$ for some $r \geq 0$ and some w-homogeneous element H, the condition $[P_+, Q_+^*] = 1$ implies that $[P_+, H] = P_+^r$, as desired.

Our next aim is to study the equation 10.2.9. Therefore we need some technical lemmas.

Lemma 10.2.12 *Let P be a non-zero w-homogeneous polynomial. Put $z := X^v Y^{-u}$. Then there exists a uniquely determined element $f(z) \in k[z]$ and integers a, b such that $P = X^a Y^b f(z)$ and $f(0)$ is non-zero.*

Proof: Let $(i_0, j_0) \in \operatorname{supp} P$. Then any solution $(i, j) \in \mathbb{Z}^2$ of the equation $ui + vj = n := w\deg P$ is of the form $(i, j) = (i_0, j_0) + s(-v, u)$ for some integer s since $\gcd(u, v) = 1$. Since $\operatorname{supp} P$ is finite there exists an integer N such that $-vi_0 + uj_0 + s(u^2 + v^2) \leq N$. So $s \leq s_0$ for some integer s_0 and hence every solution of $ui + vj = n$ is of the form $(i, j) = (i_0, j_0) + s_0(-v, u) + (s_0 - s)(v, -u)$. Putting $(a, b) := (i_0, j_0) + s_0(-v, u)$ and $t := s_0 - s \geq 0$, we get for each such pair (i, j) that $X^i Y^j = X^a Y^b z^t$, which implies the lemma. □

The pair (a, b) is called the **w-vertex of P**.

Now we consider the equality 10.2.9 and from now on we additionally assume that for the direction w chosen P_+ contains at least two points. Then using 10.2.12 we can write

$$P_+ = c_1 X^a Y^b f(z),\ H = c_2 X^s Y^t h(z)$$

with

$$f(z) = \Pi(z - a_i)^{n_i},\ h(z) = \Pi(z - a_i)^{h_i}$$

where $c_1, c_2 \in k^*$, all $a_i \in \bar{k}^*$ are different and n_i, h_i are integers ≥ 0 such that $n_i + h_i > 0$ for all i. The number of different a_i we denote by L. So by our hypothesis on P_+ we have $\sum n_i > 0$ and hence $L \geq 1$. Furthermore put $c := \frac{c_1 c_2}{c_1^r}$ and

$$\rho(z) := \det\begin{pmatrix} a & s \\ b & t \end{pmatrix} fh + \det\begin{pmatrix} a & v \\ b & -u \end{pmatrix} zfh'$$
$$+ \det\begin{pmatrix} v & s \\ -u & t \end{pmatrix} zf'h.$$

Lemma 10.2.13 $[P_+, H] = P_+^r$ *if and only if*

(10.2.14) $$X^{ar}Y^{br}\Pi(z-a_i)^{rn_i} = cX^{a+s-1}Y^{b+t-1}\rho(z).$$

Proof: Left to the reader (Exercise 1). □

In particular we see that $\rho(z) \neq 0$. So we get

$$d := ord_z\rho(z) \geq 0.$$

Lemma 10.2.15 *If* $[P_+, H] = P_+^r$ *then*

i) $(s,t) = (r-1)(a,b) + (1,1) - d(v,-u)$ *and*

ii) $\frac{c^{-1}}{z^{1-d}\Pi(z-a_i)^{h_i-(r-1)n_i}} = \frac{(a-b)+dw(P_+)}{z} + \sum_i \frac{-w(P_+)[h_i-(r-1)n_i]+|w|n_i}{z-a_i}$

where $w(P_+) := w\deg P_+$.

Proof: i) follows directly from 10.2.14 by comparing w-vertices. To obtain ii) divide both sides of 10.2.14 by $cX^{a+s-1}Y^{b+t-1}zfh$. The left-hand side of 10.2.14 becomes the left-hand side of ii) and the right-hand side of 10.2.14 becomes

(10.2.16) $$\frac{1}{z}\det\begin{pmatrix} a & s \\ b & t \end{pmatrix} + \det\begin{pmatrix} a & v \\ b & -u \end{pmatrix}\frac{h'}{h} + \det\begin{pmatrix} v & s \\ -u & t \end{pmatrix}\frac{f'}{f}.$$

Since P_+ contains the monomial X^aY^b we have $w(P_+) = au + bv$. Using i) we deduce that

$$\begin{aligned} \det\begin{pmatrix} a & s \\ b & t \end{pmatrix} &= \det\begin{pmatrix} a & 1 \\ b & 1 \end{pmatrix} + d.\det\begin{pmatrix} a & -v \\ b & u \end{pmatrix} \\ &= (a-b) + dw(P_+). \end{aligned}$$

Similarly

$$\det\begin{pmatrix} v & s \\ -u & t \end{pmatrix} = (r-1)w(P_+) + (v+u).$$

Furthermore

$$\frac{h'}{h} = \sum \frac{h_i}{z-a_i} \text{ and } \frac{f'}{f} = \sum \frac{n_i}{z-a_i}.$$

Substituting these formulae in 10.2.16 gives the desired result. □

Edges of Newton polygons of Jacobian polynomials having positive w-degree

Until now the only restriction on the direction w was that P_+ contains at least two points. Now we will study the edges of $N(P)$ which satisfy $w(P_+) > 0$ and $|w| > 0$. This condition is for example satisfied if we consider edges with negative slope.

Lemma 10.2.17 *Assume* $|w| > 0$ *and* $w(P_+) > 0$. *Then*

i) $h_i - (r-1)n_i = 1$ *for all* i.

ii) $\frac{c^{-1}}{z^{1-d}\prod(z-a_i)} = \frac{(a-b)+dw(P_+)}{z} + \sum_i \frac{|w|n_i - w(P_+)}{z-a_i}$.

Proof: From 10.2.15 we get $h_i - (r-1)n_i \leq 1$ for all i. Suppose that for some i $h_i - (r-1)n_i \leq 0$. Then multiply ii) in 10.2.15 by $z - a_i$ and sending z to a_i we get

$$-w(P_+)[h_i - (r-1)n_i] + |w|n_i = 0.$$

So by our assumption $h_i - (r-1)n_i \leq 0$ this gives

$$0 \leq |w|n_i = w(P_+)[h_i - (r-1)n_i] \leq 0.$$

Consequently we have all equalities. Then using $|w| > 0$ we get $n_i = 0$ and using $w(P_+) > 0$ we get $h_i - (r-1)n_i = 0$. So also $h_i = 0$ which implies that $h_i + n_i = 0$, contradicting $h_i + n_i > 0$.

The second statement follows readily from i) and the formulae in ii) of 10.2.15. □

Newton polygons of Jacobian polynomials having an edge with a negative slope

The main result which we will show is that in order to study the Jacobian Conjecture we may assume that $N(P)$ has no edge with negative slope. The proof is based on a reduction theorem due to Abhyankar. To formulate his result we introduce for any non-zero polynomial f

$$\delta(f) := \deg_X(f) + \deg_Y(f).$$

Theorem 10.2.18 (Abhyankar's reduction theorem). *Let P be a Jacobian polynomial and assume that $N(P)$ has an edge with a negative slope. Then there exists $\varphi \in T(k,2)$ such that*

$$\delta(\varphi(P)) < \delta(P).$$

The proof of this theorem is based on

Proposition 10.2.19 *Let P be a Jacobian polynomial and let $w = (u, v)$ be a direction with $u, v \geq 1$.*

i) *If* $d > 0$ *and* $L = 1$, *then* $P_+ = Y^{b-un_1}(X - a_1Y^u)^{n_1}$.

ii) *If* $d > 0$ *and* $L \geq 2$, *then* $P_+ = (X - a_1Y)^{n_1}(X - a_2Y)^{n_2}$.

iii) *If* $d = 0$, *then* $P_+ = X^a(X^v - a_1Y)^{n_1}$.

Proof: i) Suppose $d > 0$. Then the residue at $z = 0$ of the left-hand side of ii) in 10.2.17 is zero since $L \geq 1$. So $(a - b) + dw(P_+) = 0$, i.e. $a - b + d(ua + vb) = 0$ or equivalently $(1 + ud)a + (dv - 1)b = 0$. Since $b = a + dw(P_+) > 0$ and $u, v \geq 1$ we deduce that $(1 + ud)a = (dv - 1)b = 0$. So $a = 0$ and $d = v = 1$, whence $w(P_+) = ua + vb = b$. If $L = 1$ this gives $P_+ = Y^b(XY^{-u} - a_1)^{n_1} = Y^{b-n_1u}(X - a_1Y)^{n_1}$, which proves i).

ii) Now assume $L \geq 2$ (and still $d > 0$). Multiply ii) in 10.2.17 by z and let $z \to \infty$. Using that $d = v = 1$, $a = 0$ and $w(P_+) = b$ we deduce that $0 = (1 + u)\sum n_i + L(-b)$. Since $X^aY^bz^{\sum n_i}$ appears in P_+ and $z = XY^{-u}$ we have in particular that $b - u\sum n_i \geq 0$. So

$$bL = (1 + u)\sum n_i \leq 2u\left(\sum n_i\right) \leq 2b$$

whence $L \leq 2$. So $L = 2$ and we have all equalities. In particular $1 + u = 2u$, whence $u = 1$ and $b = \sum n_i = n_1 + n_2$ since $L = 2$. Consequently

$$\begin{aligned} P_+ &= X^0Y^{n_1+n_2}(XY^{-1} - a_1)^{n_1}(XY^{-1} - a_2)^{n_2} \\ &= (X - a_1Y)^{n_1}(X - a_2Y)^{n_2} \end{aligned}$$

which proves ii).

iii) Finally assume that $d = 0$. Again multiply ii) in 10.2.17 by z and let $z \to \infty$. Since $L \geq 1$ and $d = 0$ we get

$$0 = a - b + |w|\sum n_i + L(-w(P_+)). \qquad (10.2.20)$$

Put $X^{u_0}Y^{v_0} := X^aY^bz^{\sum n_i}$. Using $z = X^vY^{-u}$ this implies $a - b + |w|\sum n_i = u_0 - v_0$. Combining this with 10.2.20 we get $w(P_+) = uu_0 + vv_0$. So $u_0 - v_0 = L(uu_0 + vv_0)$, whence $u_0(1 - Lu) = v_0(1 + Lv)$. Since obviously $u_0(1 - Lu) \leq 0 \leq v_0(1 + Lv)$ we have all equalities. So $v_0 = 0$ and hence $u_0 > 0$, for if $u_0 = 0$, then $w(P_+) = 0$. So $1 - Lu = 0$, i.e. $L = u = 1$. Since $v_0 = b - \sum n_i$ from the definition, we deduce using $L = 1$ that $b = n_1$. Consequently

$$P_+ = X^aY^{n_1}(X^vY^{-1} - a_1)^{n_1} = X^a(X^v - a_1Y)^{n_1}$$

which concludes the proof. □

Proof of Abhyankar's reduction theorem

Since $N(P)$ has an edge with negative slope there is such an edge with lowest vertex. The part of P supported by that edge is the leading homogeneous part of P with respect to a w-grading where $w = (u, v)$ is a direction with $u, v \geq 1$. By 10.2.19 there are three cases.

Case i) $P_+ = Y^{p_1}(X - a_1Y^u)^{n_1}$, where $p_1 := b - un_1 \geq 0$ since $P_+ \in R$. So $v = 1$. Also $n_1 > 0$ since P_+ contains at least two points. Let $\varphi \in T(2, k)$ be given by

$$\varphi(X) = X + a_1Y^u, \varphi(Y) = Y.$$

Observe that $\deg_X \varphi(P) = \deg_X P$. Now we compare $\deg_Y \varphi(Y)$ with $\deg_Y(P)$. Therefore write $P = P_+ + P_0$ with $w\mathrm{deg} P_0 < w\mathrm{deg} P_+$. This means that each monomial $X^i Y^j$ appearing in P_0 satisfies $w\mathrm{deg} X^i Y^j < w\mathrm{deg} P_+$, i.e. $ui + j < un_1 + p_1$ since $Y^{p_1} X^{n_1}$ appears in P_+. Consider $\varphi(P_0)$. Its terms come from the expressions $(X + a_1 Y^u)^i Y^j$ where $ui + j < un_1 + p_1$. Since $\deg_Y (X + a_1 Y^u)^i Y^j \le ui + j < un_1 + p_1$ we get

$$\deg_Y \varphi(P) < \deg_Y(P_+) \le \deg_Y P$$

where we also used that $\varphi(P_+) = Y^{p_1} X^{n_1}$. Since, as already observed $\deg_X \varphi(P) = \deg_X(P)$, we deduce that $\delta(\varphi(P)) < \delta(P)$ which gives the desired result for case i). Case iii) follows from case i) by interchanging X and Y. Case ii) is left to the reader (Exercise 2). □

Corollary 10.2.21 *Let P be a Jacobian polynomial. Then there exists $\varphi \in \mathrm{Aut}_k R$ such that either $\varphi(P) = X$ or $N(\varphi(P))$ has no edge with a negative slope and in that case there exist integers $1 \le a \le b$ such that*

$$(a, b) \in \operatorname{supp} \varphi(P) \subset \{(i, j) | 0 \le i \le a, 0 \le j \le b\}.$$

Proof: We use induction on $\delta(P)$. If $\delta(P) = 1$, P is linear and the result is obvious. So let $\delta(P) > 1$. If P has an edge with negative slope, then it follows from Abhyankar's reduction theorem that there exists $\varphi \in T(k, 2)$ with $\delta(\varphi(P)) < \delta(P)$. Obviously $\varphi(P)$ has a Jacobian mate, so the result follows from the induction hypothesis. Hence we remain with the case that $\varphi(P)$ has no edge with negative slope. This implies that in the standard (1,1)-grading $P_+ = cX^a Y^b$ for some $c \in k^*$, a, b integers ≥ 0 and $\operatorname{supp} P \subset \{(i, j) | 0 \le i \le a, 0 \le j \le b\}$. Interchanging X and Y if necessary we may assume that $a \le b$. If $a = 0$ then $P = P(Y)$ and since P has a Jacobian mate this implies that P is linear. Otherwise $a \ge 1$ which concludes the proof. □

Corollary 10.2.22 *If P is a Jacobian polynomial, then P has at most two points at infinity.*

Proof: Let $w = (1, 1)$. If P_+ consists of only one point, then $P_+ = cX^a Y^b$ and we are done. If P_+ contains at least two points apply 10.2.19. □

Some equivalent formulations of the Jacobian conjecture

Theorem 10.2.23 (Abhyankar). *The following statements are equivalent.*

1) *Every polynomial P which has a Jacobian mate has one point at infinity.*
2) *The Newton polygon $N(P)$ of every polynomial P which has a Jacobian mate is a triangle (which can be a line segment).*
3) *For every polynomial P which has a Jacobian mate Q one has that either* $\deg P$ *divides* $\deg Q$ *or* $\deg Q$ *divides* $\deg P$.
4) *The Jacobian Conjecture.*

Proof: 4) $\to$ 1) and 4) $\to$ 2) follow from 5.1.18 and 4) $\to$ 3) from 5.1.6 a) and 5.1.11. Now assume 1). So let (P, Q) be a Jacobian pair. By 10.2.21 there exists $\varphi \in \mathrm{Aut}_k R$ such that either $\varphi(P) = X$ or $N(\varphi(P))$ has no edge with negative slope, in which case $\varphi(P)$ has two points at infinity. However $\varphi(P)$ has a Jacobian mate, namely $\varphi(Q)$, so we get a contradiction with the hypothesis. So the second case cannot happen. Hence $\varphi(P) = X$. Since $(\varphi(P), \varphi(Q)) = (X, \varphi(Q))$ is a Jacobian pair it follows easily that $(X, \varphi(Q))$ is invertible and hence so is (P, Q), which proves 4). So 1)$\to$4) holds.

Now assume 2) and let (P, Q) be a Jacobian pair. Since by hypothesis $N(P)$ is a triangle we have either that P is linear, in which case (P, Q) is invertible or by Abhyankar's reduction theorem there exists $\varphi \in \mathrm{Aut}_k R$ such that $\delta(\varphi(P)) < \delta(P)$. Again $(\varphi(P), \varphi(Q))$ is a Jacobian pair, so by induction on $\delta(P)$ the map $(\varphi(P), \varphi(Q))$ is invertible and hence so is (P, Q). So 2)$\to$4) holds.

Finally assume 3). Let $w = (1, 1)$. Then by 10.2.5 and 10.2.7 $P_+ = c_1 h^n$, $Q_+ = c_2 h^m$ with $c_1, c_2 \in k^*$, $n, m \geq 1$ and h a homogeneous polynomial. Suppose that $\deg Q$ divides $\deg P$. Then $m|n$ say $n = em$. We may assume $c_2 = 1$. Consider $P_1 := P - c_1 Q^e$. Then $[P_1, Q] = [P, Q]$ and $\max(\deg P_1, \deg Q) < \max(\deg P, \deg Q)$. So 4) follows by induction on $\max(\deg P, \deg Q)$, which concludes the proof. □

Some classical results on the Jacobian conjecture

The oldest result goes back to 1955, when Magnus studied volume preserving transformations.

Theorem 10.2.24 (Magnus). *Let $F = (P, Q)$ be a Keller map. If* $\gcd(\deg P, \deg Q) = 1$ *then F is invertible over k and* $\deg P = 1$ *or* $\deg Q = 1$.

Proof: If one of $\deg P$ or $\deg Q$ is 1, then one easily verifies that F is invertible over k. So we may assume that $m := \deg P$ and $n := \deg Q$ are both bigger than one. We will derive a contradiction. Therefore write

$$P = P_m + \cdots + P_0 \text{ and } Q = Q_n + \cdots + Q_0,$$

their respective homogeneous decompositions. Then by 10.2.5 there exists a homogeneous polynomial h such that $P_m = ah^{m'}$ and $Q_n = bh^{n'}$ with $a, b \in k^*$ and $n', m' \in \mathbb{N}$. Since $\gcd(n, m) = 1$ we deduce that $\deg h = 1$. Obviously we may assume that $h = X$. By 10.2.1 we have $t_Y(P)n = t_Y(Q)m$. So $t_Y(P)$ is divisible by m. However by 10.2.6 $t_Y(P) > 0$. So $t_Y(P) \geq m$. On the other hand $P_m = aX^m$ since $h = X$, so $t_Y(P) \leq m - 1$, a contradiction. □

Corollary 10.2.25 (Magnus). *If $F = (P, Q)$ is a Keller map and* $\deg P$ *or* $\deg Q$ *a prime number, then F is invertible over k.*

Proof: Suppose $\deg P = p$ is a prime number. We use induction on $\deg Q$. In case $\gcd(\deg P, \deg Q) = 1$ we are done by 10.2.24. So we may assume that $p = \deg P$ divides $\deg Q$. Again we have $P_m = ah^{m'}$, $Q_n = bh^{n'}$ for some $h \in k[X, Y]$. So m' divides n', i.e. $n' = dm'$ for some $d \in \mathbb{N}$. Hence if we put $Q_1 := Q - cP^d$ where

$b - ca^d = 0$, then we have $\deg Q_1 < \deg Q$. Obviously $\det J(P, Q_1) = \det J(P, Q) \in k^*$. So $k[P, Q_1] = k[X, Y]$ by the induction hypothesis. Consequently $k[P, Q] = k[P, Q_1] = k[X, Y]$ as desired. □

Magnus' theorem was first generalized in 1977 by Nakai and Baba in [23] to the case that $\gcd(\deg P, \deg Q) \leq 2$. Then in 1985 Appelgate and Onishi in [20] generalized this result to the case when $\gcd(\deg P, \deg Q)$ is a prime number. Their paper contains several errors. Finally Nagata repaired the proof and generalized Appelgate and Onishi's result. More precisely we have

Theorem 10.2.26 (Appelgate, Onishi, Nagata). *Let $F = (P, Q)$ be a Keller map and $d := \gcd(\deg P, \deg Q)$. If $d \leq 8$ or d is a prime number then F is invertible over k.*

For a proof of this result we refer to Nagata's paper [220].

Corollary 10.2.27 *If $F = (P, Q)$ is a Keller map and $\deg P$ or $\deg Q$ is a product of at most two prime numbers, then F is invertible over k.*

Proof: By 10.2.25 we may assume that $\deg P = p_1 p_2$, the product of two prime numbers which may be equal. So $d := \gcd(\deg P, \deg Q)$ is either 1 or p_1 or p_2 or $p_1 p_2$. In the first three cases we are done by 10.2.24 and 10.2.26. So it remains to verify the case $d = p_1 p_2$. Now we can use induction on $\deg Q$. Namely, since d divides $\deg Q$ we get $\deg P = p_1 p_2 = d$ divides $\deg Q$, say $\deg Q = n \deg P$ for some $n \in \mathbb{N}$. Then for suitable $c \in k^*$ we have $\deg Q_1 < \deg Q$, where $Q_1 := Q - cP^n$. So the pair (P, Q_1) satisfies the induction hypothesis, whence $k[X, Y] = k[P, Q_1] = k[P, Q]$, i.e. F is invertible over k. □

Another classical result was obtained by David Wright in 1978. To formulate his result recall that for any commutative ring R, $GE_n(R)$ denotes the smallest subgroup of $Gl_n(R)$ containing all elementary and invertible diagonal matrices.

Theorem 10.2.28 (Wright). *A polynomial map $F : k^2 \to k^2$ is invertible if and only if $JF \in GE_2(k[X, Y])$.*

Remark 10.2.29 i) It was shown by Cohn in [61] that

$$C := \begin{pmatrix} 1 + XY & -Y^2 \\ X^2 & 1 - XY \end{pmatrix}$$

belongs to $Gl_2(k[X, Y])$ but is not in $GE_2(k[X, Y])$. So the Jacobian Conjecture comes down to showing that any invertible Jacobian matrix belongs to $GE_2(k[X, Y])$.

ii) On the other hand it was shown by Suslin in [282] that if $m \geq 3$, then $GE_m(k[X_1, \ldots, X_n]) = Gl_m(k[X_1, \ldots, X_n])$ for every $n \geq 1$. A constructive proof of this result was more recently given by Park and Woodburn in [237].

Finally we would like to mention the following result due to Moh (1983). His highly non-trivial proof reduces the problem to four difficult cases, namely where the degrees of the polynomials P and Q are equal to (64, 48), (84, 56), (75, 50) and (99, 66).

Theorem 10.2.30 (Moh). *Let $F : k^2 \to k^2$ be a Keller map with* $\deg F \le 101$. *Then F is invertible.*

Some recent results

One of the most striking results obtained more recently (1993) is

Theorem 10.2.31 (Gwoździewicz). *Let k be an algebraically closed field and $F : k^2 \to k^2$ a Keller map. If there exists one line $\ell \subset k^2$ such that $F_{|\ell} : \ell \to k^2$ is injective, then F is invertible.*

Proof: We may assume that ℓ has the equation $Y = 0$. Define $f : k \to k^2$ by $f(x) = F(x, 0)$. So f is injective and $\det JF \in k^*$ implies that $f'(x) \neq 0$ for all $x \in k$. So by B.2.6 and 5.3.5 there exists a polynomial automorphism H such that $f(x) = H(x, 0)$ for all $x \in k$. Put $G = H^{-1} \circ F$. Then $\det JG \in k^*$ and $G(x, 0) = (x, 0)$. Write $G = (G_1, G_2)$. If both $\deg G_1 > 1$ and $\deg G_2 > 1$, then by 10.2.6 $t_X(G_2) > 0$, contradicting $G_2(x, 0) = 0$. So either $\deg G_1$ or $\deg G_2 \le 1$ in which case one easily deduces that G is invertible and hence so is $F(= H \circ G)$. □

Also in 1993, Shulim Kaliman proposed a new strategy to attack the Jacobian Conjecture. He formulated the

Weak Jacobian Conjecture Let $F = (P, Q) : \mathbb{C}^2 \to \mathbb{C}^2$ be a Keller map such that for every $c \in \mathbb{C}$ the polynomial $P - c$ is irreducible. Then F is invertible.

Theorem 10.2.32 (Kaliman). *If the weak Jacobian Conjecture is true, then the Jacobian Conjecture is true.*

To prove this theorem he shows the following result, which clearly implies 10.2.32.

Proposition 10.2.33 (Kaliman). *Let $F = (P, Q) : \mathbb{C}^2 \to \mathbb{C}^2$ be a Keller map. Then there exists an automorphism $G = (G_1, G_2)$ of the form $L_2 \circ E \circ L_1$, where both L_i are affine automorphisms and E is an elementary automorphism of the form $E(X, Y) = (X + g(Y), Y)$ such that $p := G_1(P, Q)$, the first component of $G \circ F$, satisfies: $p - c$ is irreducible for all $c \in \mathbb{C}$.*

In his paper Kaliman suggests that one may try to show that a polynomial p without singular points having the property that $p - c$ is irreducible for all $c \in \mathbb{C}$ is in fact a coordinate, which by means of 10.2.32 would imply the Jacobian Conjecture. However a family of counterexamples to this idea was constructed by Artal Bartolo, Cassou-Noguès and Luengo Velasco in [25]. The simplest of these counterexamples is the following. Let

$$s := XY + 1, h := Xs + 1 \text{ and } u := s^2 + y.$$

Then

$$p := h^2 u - \frac{5}{3} hs - \frac{1}{3} s$$

has all the desired properties, i.e. $p - c$ is irreducible for all $c \in \mathbb{C}$ and p has no singularities, however p is not a coordinate.

Exercises for §10.2

1. Give a proof of 10.2.13.
2. Complete the proof of Abhyankar's reduction theorem.

10.3 Polynomial maps with integer coefficients and the Jacobian Conjecture in positive characteristic

In Chapter 1 we showed that the Jacobian Conjecture over $\mathbb{C}$ is equivalent to the Jacobian Conjecture over $\mathbb{Z}$ (1.1.12 and 1.1.19).

In this section we restrict to polynomial maps with integer coefficients. We show that if a polynomial map $F : \mathbb{Z}^n \to \mathbb{Z}^n$ is surjective, then it is invertible over $\mathbb{Z}$. Also we show that if $\det JF \in \mathbb{Z}^*$ and for all but finitely many primes p the induced map $\overline{F} : \mathbb{F}_p^n \to \mathbb{F}_p^n$ is injective, then F is invertible over $\mathbb{Z}$. Using this result we discuss the Jacobian Conjecture in positive characteristic proposed by Adjamagbo in [10]: if $F \in \mathbb{F}_p[X]^n$ with $\det JF = 1$ and $|\mathbb{F}_p(X) : \mathbb{F}_p(F)|$ is not divisible by p, then F is invertible over $\mathbb{F}_p$.

It will be shown that this positive characteristic version of the Jacobian Conjecture implies the Jacobian Conjecture. At the end of this section we describe a nice argument due to van den Dries and McKenna, see [92], which shows that if there exists a counterexample to the Jacobian Conjecture, then there exists a counterexample with integer coefficients and having a rather exotic behavior. Most of the results of this section are taken from the paper [14] by Adjamagbo, Derksen and the author. The main tool is 10.3.1 which asserts that every finite set of algebraic numbers can be embedded in the ring $\mathbb{Z}_p$ of p-adic integers, for almost all p.

Throughout this section we have the following notation:

$\overline{\mathbb{Q}}$ is the algebraic closure of $\mathbb{Q}$ in $\mathbb{C}$ and for every prime p we denote by $\mathbb{Z}_p$ the ring of **p-adic integers**, i.e. the completion of $\mathbb{Z}$ with respect to the topology induced by the filtration $p\mathbb{Z} \supset p^2\mathbb{Z} \supset p^3\mathbb{Z} \supset \cdots$. The following result is crucial.

Theorem 10.3.1 *Let $a_1, \ldots, a_m \in \overline{\mathbb{Q}}$. Then for infinitely many prime numbers p there exists an injective ring homomorphism $\varphi : \mathbb{Z}[a_1, \ldots, a_m] \to \mathbb{Z}_p$.*

The proof of this theorem will be given in several steps. First we prove

Lemma 10.3.2 *Let $f(X) \in \mathbb{Z}[X]$ be irreducible and let E be a finite set of primes. Then there exist a prime number p not in E and a p-adic integer $a \in \mathbb{Z}_p$ with $f(a) = 0$.*

Proof: i) By Hensel's lemma (see for example [281], Proposition 2.3.1) it suffices to prove that there exist a prime number p not in E and an integer m such that $\deg f(X) = \deg \overline{f}(X)$, $\overline{f}(\overline{m}) = 0$ and $\overline{f}'(\overline{m}) \neq 0$. Here the upper bar denotes reduction modulo p.

ii) Let d be the discriminant of f. So $0 \neq d \in \mathbb{Z}$, since f is irreducible and hence has no multiple factors. Write

$$f(X) = f_n X^n + f_{n-1} X^{n-1} + \cdots + f_1 X + f_0.$$

Then we may assume that $f_0 \neq 0$, otherwise $f = X$ or $f = -X$ and we are done. Put $e := 2d \Pi_{p \in E} p$ and $m_l := f_0 e^l$, for all $l \in \mathbb{N}$. Then all m_l are different, since $e \geq 2$ and furthermore

$$f(m_l) = f_n f_0^n (e^l)^n + \cdots + f_1 f_0 e^l + f_0 = \tilde{f}(e^l) f_0 \tag{10.3.3}$$

where

$$\tilde{f}(X) := f_n f_0^{n-1} X^n + \cdots + f_1 X + 1. \tag{10.3.4}$$

Since $\tilde{f}$ can take one of the values 0, 1 or -1 at only a finite number of integers, there exists an $l \in \mathbb{N}$ such that $f(e^l)$ does not belong to $\{0, 1, -1\}$. Choose such an l and let p divide $\tilde{f}(e^l)$. Then by 10.3.3 $f(\overline{m_l}) = 0$ in $\mathbb{F}_p$. However p does not divide e, for otherwise by 10.3.4 and the fact that p divides $\tilde{f}(e^l)$ it follows that p divides 1. So p does not belong to E and p does not divide d. Whence $\overline{d} \neq 0$, implying that $\overline{f_n} \neq 0$, $\deg f(X) = \deg \overline{f}(X)$ and the discriminant of $\overline{f}(X)$ is non-zero. Since $\overline{f}(\overline{m_l}) = 0$ it follows that $\overline{f}'(\overline{m_l}) \neq 0$, as desired. □

Corollary 10.3.5 *Let E be a finite set of primes. If $c \in \overline{\mathbb{Q}}$, then there exist a prime number p, which does not belong to E, and an injective ring homomorphism φ : $\mathbb{Z}[c] \to \mathbb{Z}_p$.*

Proof: i) Let $g(X) \in \mathbb{Q}[X]$ be the minimal polynomial of c over $\mathbb{Q}$. Then there exists an integer m such that $f(X) := mg(X) \in \mathbb{Z}[X]$ and $f(X)$ is primitive, i.e. the gcd of the coefficients of $f(X)$ is 1. Observe that

$$\mathbb{Q}[X]f(X) \cap \mathbb{Z}[X] = \mathbb{Z}[X]f(X), \tag{10.3.6}$$

for if $\frac{1}{r}a(X)f(X) = h(X) \in \mathbb{Z}[X]$ with $0 \neq r \in \mathbb{Z}$ and $a(X) \in \mathbb{Z}[X]$ primitive, then $a(X)f(X) = rh(X)$. By Gauss' lemma $a(X)f(X)$ is also primitive, which implies that $r \in \{-1, 1\}$. So $h(X) \in \mathbb{Z}[X]f(X)$.

ii) Consider the substitution homomorphism $\sigma : \mathbb{Z}[X] \to \mathbb{Z}[c]$ sending X to c. Then $\ker \sigma = \mathbb{Z}[X] \cap \mathbb{Q}[X]f(X) = \mathbb{Z}[X]f(X)$ by 10.3.6. So

$$\mathbb{Z}[X]/\mathbb{Z}[X]f(X) \simeq \mathbb{Z}[c]. \tag{10.3.7}$$

iii) By 10.3.2 there exists a prime number p outside E and an element $a \in \mathbb{Z}_p$ such that $f(a) = 0$. So we get a ring homomorphism $\psi : \mathbb{Q}[X]/\mathbb{Q}[X]f(X) \to \mathbb{Q}_p$ which is injective since $f(X)$ is irreducible. Finally define φ to be the composition

$$\varphi : \mathbb{Z}[c] \simeq \mathbb{Z}[X]/\mathbb{Z}[X]f(X) \to \mathbb{Q}[X]/\mathbb{Q}[X]f(X) \to \mathbb{Q}_p$$

where the middle arrow, denoted i, is induced by the inclusion map $\mathbb{Z}[X] \to \mathbb{Q}[X]$. Since by 10.3.6 i is injective, φ is injective, as desired. □

Proof of Theorem 10.3.1: Put $k := \mathbb{Q}(a_1, \dots, a_m)$. By the theorem of the primitive element there exists an element $c \in \overline{\mathbb{Q}}$ such that $k = \mathbb{Q}[c]$. So each a_i is a polynomial in c with coefficients in $\mathbb{Q}$. Hence there exists an integer $s \neq 0$ such that $sa_i \in \mathbb{Z}[c]$ for all i. So $\mathbb{Z}[a_1, \dots, a_m] \subset S^{-1}\mathbb{Z}[c]$, where S is the multiplicatively closed subset of $\mathbb{Z}$ generated by s. Let E_1 denote the set of prime divisors of s. Then by 10.3.5, applied to E_1, there exists a prime number p not dividing s and an injective ring homomorphism $\psi : \mathbb{Z}[c] \to \mathbb{Z}_p$. Consequently, localizing at S and observing that $s^{-1} \in \mathbb{Z}_p$, since p does not divide s, we get an injective ring homomorphism $\psi : S^{-1}\mathbb{Z}[c] \to \mathbb{Z}_p$. Denote the restriction of ψ to the subring $\mathbb{Z}[a_1, \dots, a_m]$ of $S^{-1}\mathbb{Z}[c]$ by φ_1. Then φ_1 is as desired.

It remains to see that we have such a map for infinitely many prime numbers. Therefore repeat the argument with $E_2 := E \cup \{p\}$ which gives a new prime p_2 and an injective ring homomorphism $\varphi_2 : \mathbb{Z}[a_1, \dots, a_m] \to \mathbb{Z}_{p_2}$. Then again with $E_3 := E_2 \cup \{p_2\}$ etc. □

After these preparations we are ready to prove several results concerning polynomial maps with integer coefficients.

Theorem 10.3.8 *Let $F \in \mathbb{Z}[X]^n$. If $F : \mathbb{Z}_p^n \to \mathbb{Z}_p^n$ is injective for all but finitely many primes p, then $d :=$ det $JF \in \mathbb{Z}\backslash\{0\}$ and F is invertible over $\mathbb{Z}[d^{-1}]$.*

Proof: We first show that $F : \overline{\mathbb{Q}}^n \to \overline{\mathbb{Q}}^n$ is injective. Suppose that $a, b \in \overline{\mathbb{Q}}^n$ satisfy $F(a) = F(b)$. Put $R := \mathbb{Z}[a, b]$. Then by 10.3.1 there exists a prime number p and an injective ring homomorphism $\varphi : R \to \mathbb{Z}_p$. Furthermore we may assume that $F : \mathbb{Z}_p^n \to \mathbb{Z}_p^n$ is injective. The induced map from R^n to $\mathbb{Z}_p^n$ we also denote by φ and one easily verifies that $F(\varphi(x)) = \varphi(F(x))$ for all $x \in R^n$. So from $F(a) = F(b)$ we get $F(\varphi(a)) = \varphi(F(a)) = \varphi(F(b)) = F(\varphi(b))$. So $\varphi(a) = \varphi(b)$ since $F : \mathbb{Z}_p^n \to \mathbb{Z}_p^n$ is injective. Hence $a = b$, since φ is injective. So $F : \overline{\mathbb{Q}}^n \to \overline{\mathbb{Q}}^n$ is injective.

ii) From 4.2.1 it follows that F is invertible over $\overline{\mathbb{Q}}$. So det $JF \in \overline{\mathbb{Q}}^* \cap \mathbb{Z}[X] = \mathbb{Z}\backslash\{0\}$. By the formal inverse function theorem the inverse of F has all its components in $\mathbb{Z}[d^{-1}][[X]]$, hence in $\overline{\mathbb{Q}}[X] \cap \mathbb{Z}[d^{-1}][[X]] = \mathbb{Z}[d^{-1}][X]$. □

Corollary 10.3.9 *Let $F \in \mathbb{Z}[X]^n$. If $F : \mathbb{Z}^n \to \mathbb{Z}^n$ is surjective, then F is invertible over $\mathbb{Z}$.*

Proof: We may assume that $F(0) = 0$. So $F(p^k\mathbb{Z}) \subset p^k\mathbb{Z}$ for all primes p and all $k \in \mathbb{N}$. i) Since $F : \mathbb{Z}^n \to \mathbb{Z}^n$ is surjective it follows that $\overline{F} : (\mathbb{Z}/p^k\mathbb{Z})^n \to (\mathbb{Z}/p^k\mathbb{Z})^n$ is surjective, hence injective for all $k \geq 1$ and each prime p. So for each p $F : \mathbb{Z}_p^n \to \mathbb{Z}_p^n$ is injective. By 10.3.8 $d \in \mathbb{Z}\backslash\{0\}$ and it remains to verify that $d \in \{-1, 1\}$.

ii) To see this first observe that $F : (p\mathbb{Z})^n \to (p\mathbb{Z})^n$ is surjective for all p: namely if $g \in (p\mathbb{Z})^n$, then by the surjectivity of F there exists $h \in \mathbb{Z}^n$ with $F(h) = g$, implying that $\overline{F}(\overline{h}) = 0$ in $(\mathbb{Z}/p\mathbb{Z})^n$. Since $\overline{F} : (\mathbb{Z}/p\mathbb{Z})^n \to (\mathbb{Z}/p\mathbb{Z})^n$ is injective it follows that $\overline{h} = 0$ in $(\mathbb{Z}/p\mathbb{Z})^n$, i.e. $h \in (p\mathbb{Z})^n$ as desired.

iii) Consider the map $\overline{F} : (p\mathbb{Z}/p^2\mathbb{Z})^n \to (p\mathbb{Z}/p^2\mathbb{Z})^n$. This is just the $\mathbb{F}_p$-linear map $y \to JF(0) \pmod{p} y$. So if d does not belong to $\{-1, 1\}$ then for each p dividing d we get $\det JF(0) \pmod{p} = 0$ in $\mathbb{F}_p$, so $\overline{F}$ is not surjective. However by ii) $\overline{F}$ is surjective, a contradiction. □

In order to apply the above result to the Jacobian Conjecture we need the following.

Proposition 10.3.10 (Connell, van den Dries). *Let R be a domain and $F \in R[X]^n$ with $F(0) = 0$ and* $\det JF \in R^*$ *and $I \subset R$ a proper ideal. Put $\overline{R} := R/I$. If the induced map $\overline{F} : \overline{R}^n \to \overline{R}^n$ is injective, then $F : R^n \to R^n$ is injective.*

The proof is based on the following.

Lemma 10.3.11 *Let R be a domain, $F \in R[X]^n$ and $a = (a_1, \dots, a_n) \in R^n$, $b = (b_1, \dots, b_n) \in R^n$ distinct points with $F(a) = F(b)$.*

i) *If* $\det JF(0) \in R^*$, *then $a_1, \dots, a_n, b_1, \dots, b_n$ generate the unit ideal in R.*
ii) *If* $\det JF \in R^*$, *then $a_1 - b_1, \dots, a_n - b_n$ generate the unit ideal in R.*

Proof: First we show i). We may assume that R is Noetherian: let S be the subring of R generated by the a_i, b_j, the coefficients of each F_i and the coefficients of $\det JF(0)$. Since S is Noetherian the a_i, b_j generate the unit ideal in S and hence in R.

We may also assume that $F(0) = 0$. So F has a formal inverse G. Suppose now that the a_i, b_j do not generate the unit ideal in R. Then they are contained in a maximal ideal I. Let $\hat{R}$ be the I-adic completion of R. So, since R is a Noetherian domain, $R \to \hat{R}$ is injective. Since each a_i, b_j belongs to I so do the components of $F(a)$ and $F(b)$. Consequently the elements $G(F(a))$ and $G(F(b))$ are well-defined in $\hat{R}$. Finally, applying G to the equality $F(a) = F(b)$ we deduce that $a = b$, a contradiction. So i) is proved.

To prove ii) put $F_b(X) := F(X+b)$ and apply i) to the map F_b and the distinct points $a - b$ and 0, observing that $\det JF_b(0) = \det JF(b) \in R^*$ since $\det JF \in R^*$. □

Proof of Proposition 10.3.10: Suppose $F(a) = F(b)$ with $a \neq b$ in R^n. Then $1 \in (a_1 - b_1, \dots, a_n - b_n)$, by 10.3.10. So reducing modulo I we get

$$1 \in (\overline{a_1} - \overline{b_1}, \dots, \overline{a_n} - \overline{b_n}). \tag{10.3.12}$$

On the other hand $F(a) = F(b)$ implies that $\overline{F}(\overline{a}) = \overline{F}(\overline{b})$, so $\overline{a} = \overline{b}$ since $\overline{F}$ is injective. Consequently $\overline{a_i} - \overline{b_i} = 0$ for all i, contradicting 10.3.12.

As a consequence of the preceding results we get

Theorem 10.3.13 *Let $F \in \mathbb{Z}[X]^n$. If $\overline{F} : \mathbb{F}_p^n \to \mathbb{F}_p^n$ is injective for all but finitely many primes p and* $\det JF \in \mathbb{Z}^*$, *then F is invertible over $\mathbb{Z}$.*

Proof: By 10.3.8 it suffices to prove that $F : \mathbb{Z}_p^n \to \mathbb{Z}_p^n$ is injective for all but finitely many primes p. By 10.3.10 it therefore suffices to prove that $\overline{F} : (\mathbb{Z}_p/p\mathbb{Z}_p)^n \to (\mathbb{Z}_p/p\mathbb{Z}_p)^n$ is injective for all but finitely many primes p. However this is just the hypothesis! □

This result enables us to give a new (equivalent) formulation of the Jacobian Conjecture, namely

Proposition 10.3.14 *The Jacobian Conjecture is equivalent to the following statement: for all $n \geq 2$ and all polynomial maps $F \in \mathbb{Z}[X]^n$ with* $\det JF \in \mathbb{Z}^*$ *the induced map $\overline{F} : \mathbb{F}_p^n \to \mathbb{F}_p^n$ is injective for all p.*

Proof: Apply 10.3.13 and 1.1.19. □

Remark 10.3.15 It is interesting to observe that the condition "$\det JF \in \mathbb{Z}^*$" is equivalent to "$\det J\overline{F} \in \mathbb{F}_p{}^*$ for all p". So by 10.3.14, it can be written as

$$\forall_p[\det J\overline{F} \in \mathbb{F}_p^*] \Rightarrow \forall_p[\overline{F} : \mathbb{F}_p^n \to \mathbb{F}_p^n \text{ is injective}].$$

However the classical example $F = X - X^p$ shows that we can not change the order of the quantifier $\forall$, i.e. the statement

$$\forall_p[\det J\overline{F} \in \mathbb{F}_p^* \Rightarrow \overline{F} : \mathbb{F}_p^n \to \mathbb{F}_p^n \text{ is injective}]$$

is false.

(10.3.16) The Jacobian Conjecture in positive characteristic

As observed above the Jacobian Conjecture is false in characteristic $p > 0$ due to the example $F = X - X^p$. However in this case $|\mathbb{F}_p(X) : \mathbb{F}_p(F)| = p$, so in particular p divides the geometric degree of F ($= |\mathbb{F}_p(X) : \mathbb{F}_p(F)|$). This and other considerations led Adjamagbo in [10] to the following formulation of the Jacobian Conjecture in positive characteristic.

Jacobian Conjecture in positive characteristic JC($\mathbb{F}_p, n, p$).

Let $F \in \mathbb{F}_p[X]^n$. If $\det JF \in \mathbb{F}_p^*$ *and p does not divide $|\mathbb{F}_p(X) : \mathbb{F}_p(F)|$, then F is invertible over $\mathbb{F}_p$.*

Proposition 10.3.17 *If JC($\mathbb{F}_p, n, p$) is true for all $n \geq 1$ and all primes p, then the Jacobian Conjecture is true.*

Proof: Let $F \in \mathbb{Z}[X]^n$ with $\det JF \in \mathbb{Z}^*$ and put $d := |\mathbb{Q}(X) : \mathbb{Q}(F)|$. Then one easily verifies (Exercise 1) that for all but finitely many p's $|\mathbb{F}_p(X) : \mathbb{F}_p(\overline{F})| \leq d^n$. So for all p large enough p certainly cannot divide the geometric degree. Also $\det J\overline{F} \in \mathbb{F}_p^*$, so by $JC(\mathbb{F}_p, n, p)$ we deduce that for all such p $\overline{F} : \mathbb{F}_p^n \to \mathbb{F}_p^n$ is injective. Then apply 10.3.13. □

To conclude this section we describe a nice result of van den Dries.

Suppose that the Jacobian Conjecture is false. Then by Proposition 1.1.19 there exists a counterexample $F : \mathbb{C}^n \to \mathbb{C}^n$ with $\det JF = 1$ and integer coefficients. This F will be fixed from now on. Since F is dominant there exists a hypersurface $V := f^{-1}(0) \subset \mathbb{C}^n$ such that

$$\#F^{-1}(z) = |\mathbb{C}(X) : \mathbb{C}(F)| \text{ for all } z \in \mathbb{C}\backslash V$$

(Appendix B). This implies that $F : \mathbb{C}^n\backslash F^{-1}(V) \to \mathbb{C}^n\backslash V$ is a covering map. Finally let $\operatorname{dist}(v, w) := \max_i |v_i - w_i|$ denote the "distance" between two vectors v, w of $\mathbb{C}^n$ and $\operatorname{dist}(z, V) := \inf_{v\in V} \operatorname{dist}(z, v)$, for all $z \in \mathbb{C}^n$.

Proposition 10.3.18 (van den Dries). *For all* $a \in \mathbb{Z}[i]^n$ $\operatorname{dist}(F(a), V) \leq 1$.

Proof: Let $a \in \mathbb{Z}[i]^n$. Put $d := \operatorname{dist}(F(a), V)$ and suppose that $d > 1$. Let

$$D := \{z \in \mathbb{C}^n \,|\, \operatorname{dist}(z, F(a)) < d\}.$$

Then $D \subset \mathbb{C}^n\backslash V$. Furthermore let

$$\overline{D_1} := \{z \in \mathbb{C}^n \,|\, \operatorname{dist}(z, F(a)) \leq 1\}.$$

Since D is simply connected, there exists a unique continuous lift $G = (G_1, \ldots, G_n)$ such that $G(F(a)) = a$ and $F \circ G = 1_D$. Moreover since the polynomials F_i can be written as polynomials in $z_1 - F_1(a), \ldots, z_n - F_n(a)$ with coefficients in $\mathbb{Z}[i]$ because $a \in \mathbb{Z}[i]^n$ and the coefficients of F belong to $\mathbb{Z}$, and since $\det JF = 1$, the local analytic inverse function theorem implies that the unique lift G is holomorphic. More precisely each G_i is a power series

$$G_i = \sum g^{(i)}_{j_1,\ldots,j_n}(z_1 - F_1(a))^{j_1} \ldots (z_n - F_n(a))^{j_n}$$

with coefficients $g^{(i)}_{j_1,\ldots,j_n}$ in $\mathbb{Z}[i]$, which converges absolutely on the disc $\overline{D_1} \subset D$. In particular if $g^{(i)}_{j_1,\ldots,j_n} \neq 0$, then $|g^{(i)}_{j_1,\ldots,j_n}| \geq 1$ and the absolute convergence of the G_i on $\bar{D}_1$ implies that $\sum |g^{(i)}_{j_1,\ldots j_n}| < \infty$. Hence only finitely many coefficients are non-zero, implying that G_i is a polynomial. So G is a polynomial inverse of F, a contradiction. So $d \leq 1$ as desired. □

Looking more carefully at the above arguments we see that the only fact of $\mathbb{Z}[i]$ which is used is that $|g| \geq 1$ if $g \in \mathbb{Z}[i]$ and $g \neq 0$. So more generally, let $R \subset \mathbb{C}$ be a subring such that $|r| \geq 1$ for all $r \in R, r \neq 0$. Then the above arguments give

$$\operatorname{dist}(F(a), V) \leq 1 \text{ for all } a \in R^n.$$

Now one can show (Exercise 2) that these rings R are exactly all subrings of the ring of integers $\mathfrak{O}_K$, where K is a quadratic number field of the form $\mathbb{Q}(i\sqrt{m})$, where $m \geq 1$ is some integer. So we get

Theorem 10.3.19 $\operatorname{dist}(F(a), V) \leq 1$ *for all* $m \in \mathbb{Z}$, $m \geq 1$ *and all* $a \in \mathfrak{O}^n_K$, *where* $K = \mathbb{Q}(i\sqrt{m})$.

One could think that there do not exist dominant polynomial mappings $F : \mathbb{C}^n \to \mathbb{C}^n$ which satisfy the statement of 10.3.19, which would imply that the Jacobian Conjecture is true. However the following example shows that they do exist.

Example 10.3.20 (Makar-Limanov). Let $F_1 := X^6Y^4 + 2X^2Y$, $F_2 := X^9Y^6 + 3X^5Y^3 + 3X$ and $V := \{X = 0\} \cup \{X^3 - Y^2 = 0\}$. Then $\operatorname{dist}(F(p), V) \leq 1$ for all $p \in R^2$ and all subrings R of $\mathbb{C}$ in which 0 is isolated.

Proof: Observe that $F(0, y) = (0, 0) \in V$, $F(x, 0) = (0, 3x) \in V$. So assume $p = (x, y) \in R^2$ with $xy \neq 0$. Put $s = x^3y^2 + (xy)^{-1}$. Since $F_1(x, y) = s^2 - (xy)^{-2}$ and $F_2(x, y) = s^3 - (xy)^{-3}$, we get $\text{dist}(F(x, y), (s^2, s^3)) \leq 1$, because $|(xy)^{-1}| \leq 1$ (where the last inequality comes from the fact that every non-zero element r in R satisfies $|r| \geq 1$, namely if $|r| < 1$ then the sequence $r, r^2, r^3 \ldots$ in R converges to 0, contradicting the fact that 0 is isolated in R). □

Exercises for §10.3

1. Let $F \in \mathbb{Z}[X]^n$ with $\det JF \in \{-1, 1\}$. Put $d := |\mathbb{Q}(X) : \mathbb{Q}(F)|$. Show that for almost all primes p one has the inequality $|\mathbb{F}_p(X) : \mathbb{F}_p(\overline{F})| \leq d^n$.
2. Let R be a subring of $\mathbb{C}$ such that $|r| \geq 1$ for all $r \in R$, $r \neq 0$. Show that there exists a quadratic number field K such that R is a subring of $\mathfrak{O}_K$, the ring of integers of K. Conversely show that each subring R of such a ring $\mathfrak{O}_K$ has the property that $|r| \geq 1$ for all non-zero elements of R.

10.4 $\mathfrak{D}$-modules and the Jacobian conjecture

In this section we discuss the Jacobian Conjecture from the point of view of $\mathfrak{D}$-modules. The main aim is to indicate some directions in which $\mathfrak{D}$-modules can be used to attack the conjecture.

A first relationship with rings of differential operators was found by Vaserstein and Kac. They observed that if every k-endomorphism of the n-th Weyl Algebra A_n is surjective, then $JC(k, n)$ follows! It was later remarked by the author that in fact $JC(k, n)$ is equivalent to the statement that every order preserving endomorphism of A_n is surjective (10.4.2). Another relationship between $\mathfrak{D}$-modules and the Jacobian Conjecture was found by Bass: he used the hypothesis $\det JF \in k^*$ to equip $k[X]$ with the structure of a left $A_n(Y)$-module, namely by defining

$$Y_i.g(X) = F_i g(X) \text{ and } \frac{\partial}{\partial Y_j}.g(X) = \frac{\partial}{\partial F_j}.g(X)$$

where $\frac{\partial}{\partial F_j}$ is the derivation on $k[X]$ introduced in Chapter 2. Using standard results on A_n-modules he showed that $k[X]$ is a so-called holonomic A_n-module (Appendix E). In 10.4.6 below we strongly improve on this result by showing that $k[X]$ is a holonomic A_n-module with regular singularities.

(10.4.1) **Endomorphisms of the Weyl Algebra**

Let k be a field of characteristic zero and denote by A_n the n-th Weyl Algebra, i.e.

$$k[X_1, \ldots, X_n, \partial_1, \ldots, \partial_n]$$

where $\partial_i = \frac{\partial}{\partial X_i}$ as usual. To indicate the dependence on the variables $X_1, \ldots, X_n$ we sometimes write $A_n(X)$.

Let h be a k-endomorphism of A_n. So $\ker h \neq A_n$. Since A_n is a simple ring (Appendix E) it follows that $\ker h = 0$. So every k-endomorphism of A_n is injective.

Dixmier Conjecture. $DC(k,n)$ *Every k-endomorphism of A_n is surjective and hence an isomorphism.*

This conjecture is still open for all $n \geq 1$. In fact the original conjecture, due to Dixmier, was formulated for $n = 1$ in [89], 1968.

Now we will show that the Jacobian Conjecture is equivalent to a weak version of the Dixmier Conjecture. Therefore we consider on A_n the so-called order filtration (Appendix E)

$$F(0) \subset F(1) \subset F(2) \subset \cdots \subset A_n.$$

We say that an endomorphism h of A_n is **order preserving** if $h(F(m)) \subset F(m)$ for all $m \geq 0$.

Weak Dixmier Conjecture. $WDC(k,n)$ *Every order preserving endomorphism of A_n is surjective.*

Theorem 10.4.2 *For every $n \geq 1$ JC(k, n) is equivalent to WDC(k, n).*

Proof: i) Suppose $WDC(k,n)$ is true and let $F = (F_1, \dots, F_n)$ be a Keller map. By 2.2.3 we get a k-endomorphism of A_n given by

$$h(X_i) = F_i \text{ and } h(\partial_j) = \frac{\partial}{\partial F_j}, \text{ for all } i, j.$$

So by our hypothesis h is surjective, since obviously h is order preserving. In particular if $g \in k[X] \subset A_n$, then $g = h(P)$ for some $P = \sum a_\alpha(X)\partial^\alpha$ in A_n. So $g = \sum a_\alpha(F)(\frac{\partial}{\partial F})^\alpha$. Applying the operator g to the element $1 \in k[X]$ we get $g = a_0(F) \in k[F]$, since $\frac{\partial}{\partial F_j}(1) = 0$ for all j. Consequently $k[X] \subset k[F]$, whence $k[X] = k[F]$, i.e. F is invertible. So $JC(k,n)$ holds.

ii) Now assume that $JC(k,n)$ is true and let h be an order preserving endomorphism of A_n. So $h(X_i) \in F(0) = k[X]$ and $h(\partial_j) \in F(1) = k[X] + \sum_p k[X]\partial_p$, whence

$$h(X_i) = F_i \text{ and } h(\partial_j) = \sum_p a_{j,p}\partial_p + b_j$$

for some $F_i, a_{j,p}, b_j \in k[X]$.

From $[h(\partial_j), h(X_i)] = \delta_{i,j}$ we get $\sum a_{j,p}\partial_p(F_i) = \delta_{i,j}$, hence

$$(a_{j,p})(JF)^t = I_n. \tag{10.4.3}$$

So $\det JF \in k^*$, whence $k[X] = k[F]$ by our hypothesis. Put $D_j := h(\partial_j) - b_j$ and write ∂, D, b instead of $\partial_1, \dots, \partial_n$, resp. $D_1, \dots, D_n$, resp. $b_1, \dots, b_n$.

We have to prove that h is surjective, i.e. $h(k[X,\partial]) = k[X,\partial]$. So we must show that $k[F, D+b] = k[X,\partial]$. From $k[X] = k[F]$ we deduce that $k[F, D+b] = k[X, D+b] = k[X, D]$. Finally $D_j = \sum_p a_{j,p}\partial_p$ for all j. Since by 10.4.3

$(a_{j,p}) \in Gl_n(k[X])$ we derive that $\partial_i \in \sum_p k[X]D_p$ for all i. Hence $k[X, D] = k[X, \partial]$, which gives $k[F, D + b] = k[X, \partial]$, as desired. □

(10.4.4) **The A_n-module $F_*k[X]$**

Let $F : k^n \to k^n$ be a Keller map. Then we have the derivations $\frac{\partial}{\partial F_i}$ on $k[X]$ which satisfy the relations $[\frac{\partial}{\partial F_i}, \frac{\partial}{\partial F_j}] = 0$ and $[\frac{\partial}{\partial F_i}, F_j] = 0$, for all i, j. Consequently $k[X]$ becomes a left $A_n(Y)$-module by defining

$$Y_j.g = F_j g \text{ and } \frac{\partial}{\partial Y_i}.g = \frac{\partial}{\partial F_i} g \text{ for all } g \in k[X].$$

This left $A_n(Y)$-module will be denoted by $k[X]^{(J)}$. Obviously $k[X]^{(J)}$ contains the simple $A_n(Y)$-module $k[F]$. So the Jacobian Conjecture can be reformulated as

Jacobian Conjecture. *If* $\det JF \in k^*$, *then* $k[X]^{(J)}$ *is a simple* $A_n(Y)$*-module.*

It is not at all obvious to see that $k[X]^{(J)}$ is a finitely generated $A_n(Y)$-module. Nevertheless we will show that it is an $A_n(Y)$-module of finite length. In fact we will show that it is a holonomic $A_n(Y)$-module (see Appendix E for the definition), which implies that it is an $A_n(Y)$-module of finite length, which in turn implies that it is even a cyclic $A_n(Y)$-module, i.e. generated by one element.

Using some more or less standard results concerning A_n-modules one can show that the holonomicity of $k[X]^{(J)}$ follows from the finiteness of the field extension $k(F) \subset k(X)$ (Exercise 1). We will not follow this approach here, but instead we will show that $k[X]^{(J)}$ appears as a direct image of a regular holonomic A_n-module and hence by a non-trivial result from $\mathfrak{D}$-module theory we deduce that $k[X]^{(J)}$ is a holonomic A_n-module with regular singularities.

In order to describe this result put $V := k^n$ and $W := k^n$. So we get $F : V \to W$. The coordinate rings $A(V)$ (resp. $A(W)$) we denote by $k[X_1, \ldots, X_n]$ (resp. $k[Y_1, \ldots, Y_n]$). On V we consider the standard left $A_n(X)$-module $k[X]$. According to Appendix E we get a left $A_n(Y)$-module $F_*k[X]$ on W. Now the main result is

Theorem 10.4.5 *Let* $F : k^n \to k^n$ *be a Keller map. Then* $F_*k[X] \simeq k[X]^{(J)}$ *as left* $A_n(Y)$*-modules.*

Corollary 10.4.6 $k[X]^{(J)}$ *is a regular holonomic left* $A_n(Y)$*-module.*

Proof: Since $k[X]$ with the standard $A_n(X)$-module structure is regular holonomic, hence so is the left $A_n(Y)$-module $F_*k[X]$ by [42], Theorem 12.2. Then apply 10.4.5. □

Proof of Theorem 10.4.5: i) As described above, $k[X]^{(J)}$ is a left $A_n(Y)$-module. So it becomes a right $A_n(Y)$-module by putting

$$g.\frac{\partial}{\partial Y_i} = -\frac{\partial}{\partial Y_i} g = -\frac{\partial}{\partial F_i} g.$$

Since as described in Appendix E one can freely move from left to right A_n-modules and vice versa, it suffices to prove that $k[X]^{(J)}$ equipped with the right $A_n(Y)$-module structure is isomorphic to the right $A_n(Y)$-module $F_*^R k[X]$, i.e. the module

$$F_*^R k[X] = k[X] \otimes_{A_n(X)} \mathfrak{D}_{V\to W}$$

where

$$\mathfrak{D}_{V\to W} = k[X] \otimes_{k[Y]} A_n(Y)$$

and the left $A_n(X)$-module structure on $\mathfrak{D}_{V\to W}$ is given by

$$\frac{\partial}{\partial X_i}(g \otimes Q) = \frac{\partial g}{\partial X_i} \otimes Q + \sum_{p=1}^{n} g \frac{\partial F_p}{\partial X_i} \otimes \frac{\partial}{\partial Y_p}.Q.$$

ii) Therefore we consider the map $h : k[X] \to F_*^R k[X]$ given by $h(g) = g \otimes (1 \otimes 1)$ for all $g \in k[X]$. By 10.4.7 below h is surjective, by 10.4.9 h is injective and by 10.4.13 h is an isomorphism of right $A_n(Y)$-modules, which completes the proof. □

Lemma 10.4.7 *Let $F : k^n \to k^n$ be a Keller map and M a right $A_n(X)$-module. Then the map $h : M \to F_*^R M$ defined by $h(m) = m \otimes (1 \otimes 1)$ is surjective.*

Proof: From the left $A_n(X)$-module structure on $\mathfrak{D}_{V\to W}$ described in i) of the proof of 10.4.5 and the definition of $\frac{\partial}{\partial F_i}$ we get

$$\frac{\partial}{\partial F_i}(g \otimes Q) = \frac{\partial}{\partial F_i}(g) \otimes Q + g \otimes \frac{\partial}{\partial Y_i}.Q,$$
$$\text{for all } g \in k[X],\ Q \in A_n(Y).$$

So in particular

$$\frac{\partial}{\partial F_i}(1 \otimes 1) = (1 \otimes 1)\frac{\partial}{\partial Y_i} \text{ for all } i. \tag{10.4.8}$$

Each element of $F_*^R M$ is a finite sum of elements of the form

$$v = m \otimes g \otimes a(Y)\partial_Y{}^{\alpha}$$

where $m \in M$, $g \in k[X]$, $a(Y) \in k[Y]$ and $\alpha \in \bar{\mathbb{N}}^n$. It suffices to show that v belongs to $h(M)$. Therefore observe that $v = m.g.a(F) \otimes (1 \otimes 1)\partial_Y{}^{\alpha}$. So from 10.4.8 we derive that $v = m' \otimes (\frac{\partial}{\partial F})^{\alpha}(1 \otimes 1) = m_0 \otimes (1 \otimes 1) = h(m_0)$, where $m' = m.g.a(F)$ and $m_0 = m'(\frac{\partial}{\partial F})^{\alpha}$. So $v \in h(M)$, as desired. □

Lemma 10.4.9 *Let $F : k^n \to k^n$ be a Keller map. Then $h : k[X] \to F_*^R k[X]$ is injective.*

Proof: i) Let T be the translation $X \to X + F(0)$. Then $F = T \circ \tilde{F}$ with $\tilde{F}(0) = 0$. Since $F_*^R = \tilde{T}_*^R \circ F_*^R$ and T is invertible, we may replace F by $\tilde{F}$ and hence we can assume that $F(0) = 0$.

ii) The canonical embedding of $k[X]$ in $A_n(X)$ induces an isomorphism of right $A_n(X)$-modules between $k[X]$ and $A_n(X)/\sum \partial_i A_n(X)$. Consequently we get an isomorphism of right $A_n(Y)$-modules between $F_*^R k[X]$ and $\mathfrak{D}_{V\to W}/\sum \partial_i \mathfrak{D}_{V\to W}$, where an element $g\otimes(1\otimes 1)$ corresponds with the class $g\otimes 1+\sum \partial_i \mathfrak{D}_{V\to W}$. Let now $g \in k[X]$ with $h(g) = 0$, or equivalently $g\otimes 1 \in \sum \partial_i \mathfrak{D}_{V\to W}$, say $g(1\otimes 1) = \sum \partial_i m_i$ for some $m_i \in \mathfrak{D}_{V\to W}$.

iii) Observe that $A_n(Y)$ is a free left $k[Y]$-module with basis given by the monomials $\partial_Y{}^\alpha$. So $\mathfrak{D}_{V\to W}$ is a free left $k[X]$-module with basis given by the elements $e_\alpha := 1\otimes \partial_Y{}^\alpha$. Write the elements m_i on this basis with coefficients in $k[X]$, say $m_i = \sum a_\alpha^{(i)} e_\alpha$ with $d \in \mathbb{N}$ such that $|\alpha| \le d$ for all i. Then comparing the coefficients of the e_α in the relation $g(1 \otimes 1) = \sum \partial_i m_i$ we get the equations

(10.4.10) $$g = \sum \partial_i a_0^{(i)},$$

(10.4.11) $$\sum_{i=1}^{n}\sum_{p=1}^{n} a^{(i)}_{\alpha_1,\dots,\alpha_p-1,\dots,\alpha_n} \frac{\partial F_p}{\partial X_i} = 0 \text{ if } |\alpha| = d+1,$$

(10.4.12) $$\sum_{i,p} a^{(i)}_{\alpha_1,\dots,\alpha_p-1,\dots,\alpha_n} \frac{\partial F_p}{\partial X_i} + \sum_i \frac{\partial a^{(i)}_{\alpha_1,\dots,\alpha_n}}{\partial X_i} = 0 \text{ if } 0 < |\alpha| < d+1.$$

So we have to show that if $g \in k[X]$ is such that there exists $a_\alpha^{(i)}$ in $k[X]$ satisfying the equations 10.4.10, 10.4.11 and 10.4.12, then $g = 0$.

iv) Therefore we observe that $\det JF \in k^*$ and $F(0) = 0$ imply that F has a formal inverse. Replacing in all definitions above $k[X]$ by $k[[X]]$, the ring of formal power series in X over k, $A_n(X)$ by $k[[X]][\partial]$ and the same with respect to Y, we consider

$$h_F : k[[X]] \to F_*^R k[[X]]$$

and

$$h_G : F_*^R k[[X]] \to G_*^R(F_*^R k[[X]]).$$

Here by $F_*^R k[[X]]$ we mean the left $k[[X]][\partial]$-module obtained by replacing in both definitions of $\mathfrak{D}_{V\to W}$ and $F_*^R k[X]$ the rings $k[X]$, $k[Y]$ etc. by $k[[X]]$, $k[[Y]]$ etc. Since $G \circ F = X$ it follows that $G_*^R(F_*^R k[[X]]) \simeq k[[X]]$, where the isomorphism is given by the map $h_G \circ h_F$. In particular we get that h_F is injective.

v) Let now $g \in k[X]$ be as in iii) with $h(g) = 0$. View g as an element in $k[[X]]$ and observe that the relations 10.4.10, 10.4.11 and 10.4.12 can also be considered as relations in $k[[X]]$. Then arguing as in iii) they can be interpreted as $h_F(g) = 0$ in $F_*^R k[[X]]$. Since h_F is injective this means that $g = 0$ in $k[[X]]$, as desired. □

Lemma 10.4.13 *Notation as in* 10.4.9. *Then h is a homomorphism of right $A_n(Y)$-modules.*

Proof: It suffices to prove that $h(g.\frac{\partial}{\partial Y_i}) = h(g).\frac{\partial}{\partial Y_i}$ for all $g \in k[X]$ and all i. First

$$h\left(g.\frac{\partial}{\partial Y_i}\right) = g.\frac{\partial}{\partial Y_i} \otimes 1 \otimes 1 = -\frac{\partial}{\partial F_i}(g) \otimes (1 \otimes 1).$$

Furthermore

$$\begin{aligned} h(g).\frac{\partial}{\partial Y_i} &= (g \otimes (1 \otimes 1))\tfrac{\partial}{\partial Y_i} \\ &= g \otimes \frac{\partial}{\partial F_i}(1 \otimes 1) \quad \text{by 10.4.8} \\ &= g.\frac{\partial}{\partial F_i} \otimes (1 \otimes 1) \\ &= \left(\frac{\partial}{\partial F_i}\right)^t (g) \otimes 1 \otimes 1 \end{aligned}$$

where $(\frac{\partial}{\partial F_i})^t$ is the adjoint of $\frac{\partial}{\partial F_i}$ (Appendix E). Then the result follows from 10.4.14 below. □

Lemma 10.4.14 *The adjoint of $\frac{\partial}{\partial F_i}$ in $A_n(X)$ equals $-\frac{\partial}{\partial F_i}$.*

Proof: By definition $\frac{\partial}{\partial F_i} = \sum_j a_{i,j}\partial_j$, where the $a_{i,j} \in k[X]$ are uniquely determined by $(a_{i,j})(JF)^t = I_n$. So we get

$$a_{i,j} = (-1)^{i+j} \det(JF)_{(i,j)}$$

where $(JF)_{(i,j)}$ denotes the matrix obtained from JF by deleting the i-th row and j-th column. Since $\frac{\partial}{\partial F_i} = \sum_j a_{i,j}\partial_j$ we get $(\frac{\partial}{\partial F_i})^t = \sum_j -\partial_j a_{i,j} = -\frac{\partial}{\partial F_i} - \sum_j \partial_j(a_{i,j})$. So it remains to show that

(10.4.15) $$\sum_{j=1}^{n}(-1)^{i+j}\partial_j(\det(JF)_{(i,j)}) = 0.$$

This follows from 10.4.17 below. □

Let R be a commutative ring $q \in \mathbb{N}$ and $A = (a_1, \ldots, a_q) \in M_q(R)$, a matrix whose i-th column is a_i. If τ is a derivation on R then it is well-known that

(10.4.16) $$\tau(\det(a_1, \ldots, a_q)) = \sum_{k=1}^{q} \det(a_1, \ldots, a_{k-1}, \tau(a_k), a_{k+1}, \ldots, a_q).$$

Now let $n \geq 2$ and $d_1, \ldots, d_n$ be n pairwise commuting derivations on R. For a column $G \in R^{n-1}$ we define

$$D(G) := \sum_{j=1}^{n}(-1)^j d_j(\det(d_1G, \ldots, \hat{d_j G}, \ldots, d_nG)).$$

Lemma 10.4.17 $D(G) = 0$ *for all* $G \in R^{n-1}$.

Proof: By 10.4.16 we get

$$D(G) = \sum_{j=1}^{n}(-1)^j \left[\sum_{k<j} \det(d_1 G, \ldots, d_j d_k G, \ldots, \hat{d_j} G, \ldots, d_n G)\right.$$

$$\left. + \sum_{k>j} \det(d_1 G, \ldots, \hat{d_j} G, \ldots, d_j d_k G, \ldots, d_n G)\right].$$

Since

$$\sum_{j=1}^{n}\sum_{k>j} a_{j,k} = \sum_{k=1}^{n}\sum_{j<k} a_{j,k} = \sum_{j=1}^{n}\sum_{k<j} a_{k,j}$$

for all $a_{j,k} \in R$ we get

$$D(G) = \sum_{j=1}^{n}\sum_{k<j}\Big[(-1)^j \det(d_1 G, \ldots, d_j d_k G, \ldots, \hat{d_j} G, \ldots, d_n G)$$

$$+ (-1)^k \det(d_1 G, \ldots, \hat{d_k} G, \ldots, d_j d_k G, \ldots, d_n G)\Big].$$

Finally one verifies that each summand in the above double sum is zero, which completes the proof. □

Exercises for §10.4

1. Let $F : k^n \to k^n$ be a Keller map.

   (i) Show that there exists a non-zero element $P \in k[F]$ such that $k[X][P^{-1}]$ is a finitely generated $k[F]$-module. [The hypothesis implies that $k(X)$ is finite over $k(F)$.]

   (ii) Deduce from i) that $k[X][P^{-1}]$ is a holonomic left $k[F][\frac{d}{dF}]$-module, and hence so is $k[X]$. [Use E.1.6.]

10.5 Endomorphisms sending coordinates to coordinates

Throughout this section k denotes a field of characteristic zero. If φ is a k-automorphism of the polynomial ring $k[X] := k[X_1, \ldots, X_n]$, then one easily verifies that $\varphi(f)$ is a coordinate of $k[X]$ for every coordinate f of $k[X]$. Conversely we have

Problem 10.5.1 Let φ be a k-endomorphism of $k[X]$ such that $\varphi(f)$ is a coordinate of $k[X]$ for every coordinate f of $k[X]$. Does it follow that φ is an automorphism?

This problem was posed by Shpilrain and the author in [120]. Before we look at it let us first point out its relevance for the Jacobian Conjecture.

Theorem 10.5.2 *If the Jacobian Conjecture is true in dimension $n-1$, $n \geq 2$, then* 10.5.1 *has an affirmative answer in dimension n, in case k is an algebraically closed field.*

The proof of this result is based on the following lemma which is due to Derksen.

Lemma 10.5.3 *Let φ be a k-endomorphism of $k[X]$ given by $\varphi(X_i) = F_i$ for all i. If φ sends linear coordinates to coordinates, i.e. $\varphi(\sum c_i X_i)$ is a coordinate for all non-zero n-tuples $(c_1, \ldots, c_n)$ in k^n, then* $\det JF(x) \neq 0$ *for all $x \in k^n$. In particular if k is algebraically closed, then* $\det JF \in k^*$.

Proof: Suppose that $\det JF(x) = 0$ for some x. Then the rows of $JF(x)$ are k-linearly dependent, so

$$\sum_i c_i(\partial_1 F_i(x), \ldots, \partial_n F_i(x)) = (0, \ldots, 0)$$

for some c_i in k, not all zero. So if we put $f := \sum c_i F_i$, then $\partial_1 f(x) = \cdots = \partial_n f(x) = 0$. However f is a coordinate, so its partial derivatives have no common zero, a contradiction. The last statement is obvious. □

Proof of Theorem 10.5.2: Let $\varphi(X_i) = F_i$ for all i. Since $F_1 (= \varphi(X_1))$ is a coordinate, there exists an automorphism ψ such that $\psi(X_1) = F_1$. Then, replacing φ by $\psi^{-1} \circ \varphi$, we may assume that $F_1 = X_1$. By 10.5.3 $\det JF \in k^*$ and since $F_1 = X_1$ we deduce that $\det J_{X_2,\ldots,X_n}(F_2, \ldots, F_n) \in k^*$. From the $n-1$-dimensional Jacobian Conjecture it follows that $k(X_1)[F_2, \ldots, F_n] = k(X_1)[X_2, \ldots, X_n]$, whence $k(F_1, \ldots, F_n) = k(X_1, F_2, \ldots, F_n) = k(X_1, \ldots, X_n)$. So $k[F] = k[X]$ by Keller's theorem. □

In the proof of 10.5.2 we only used that φ is an endomorphism sending **linear** coordinates to coordinates. So we formulate

Problem 10.5.4 Let φ be a k-endomorphism of $k[X]$ sending all linear coordinates to coordinates. Does it follow that φ is an automorphism?

If so, then we get

Corollary 10.5.5 *Let k be algebraically closed. If the Jacobian Conjecture is true in dimension $n-1$ ($n \geq 2$), then* (10.5.4) *has an affirmative answer in dimension n.*

Corollary 10.5.6 *Let k be algebraically closed. If $n = 2$, then* 10.5.4 *has an affirmative answer.*

Recently an affirmative answer for 10.5.4 in dimension two for any field of characteristic zero was given by Cheng and the author in [51]. On the other hand for $n \geq 3$ and k a non-algebraically closed field we have

Theorem 10.5.7 (Mikhalev, Yu, Zolotykh). *For any $n \geq 3$ and any non-algebraically closed field k, 10.5.4 has a negative answer.*

Proof: i) Since k is not algebraically closed there exists a non-zero polynomial $q(T) \in k[T]$ such that $1 + Tq(T)$ has no zeros in k. Define $\varphi \in \mathrm{End}_k\, k[X]$ by

$$\varphi(X_1) = X_1 + X_2 q(X_3), \varphi(X_2) = X_2 - X_1 X_3 \text{ and } \varphi(X_i) = X_i$$

for all $i \geq 3$. Then one readily verifies that $\det J\varphi = 1 + X_3 q(X_3)$ does not belong to k^*, so φ is not an automorphism.

ii) Now we show that $h := \varphi(\sum c_i X_i)$ is a coordinate for every non-zero n-tuple $(c_1, \ldots, c_n)$ in k^n. Namely if $c_i \neq 0$ for some $4 \leq i \leq n$ or $c_2 = 0$, the statement is obvious. So we may assume that $c_4 = \cdots = c_n = 0$ and furthermore that $c_2 = 1$. So $h = c_1 X_1 + c_3 X_3 - X_1 X_3 + X_2 q_0(X_3)$, where $q_0(T) = 1 + c_1 q(T)$. Since $1 + Tq(T)$ has no zeros in k, in particular $q_0(c_1) \neq 0$. Then the theorem follows from the next lemma. □

Lemma 10.5.8 *Let $c_1, c_3 \in k$ and $f(T) \in k[T]$ with $f(c_1) \neq 0$. Then $h := c_1 X_1 + c_3 X_3 - X_1 X_3 + X_2 f(X_3)$ is a coordinate in $k[X]$.*

Proof: We use induction on $r := \deg f(T)$. If $f(T) \in k^*$, the assertion is obvious. So let $r \geq 1$ and write $f(T) = g(T) + aT^r$ with $\deg g(T) \leq r - 1$. Define the elementary automorphism E by

$$E(X_1) = X_1 + aX_2 X_3^{r-1} \text{ and } E(X_i) = X_i \text{ for all } i \geq 2.$$

Then $E(h) = c_1 X_1 + c_3 X_3 - X_1 X_3 + X_2(g(X_3) + c_1 a X_3^{r-1})$. Now observe that $f_1(T) := g(T) + c_1 a T^{r-1}$ satisfies $\deg f_1(T) \leq r - 1$ and $f_1(c_1) = g(c_1) + ac_1^r = f(c_1) \neq 0$. So the induction hypothesis implies that $E(h)$ is a coordinate in $k[X]$, hence so is h. □

At present we do not know the answer to 10.5.4 in case k is algebraically closed. However in case $k = \mathbb{C}$, 10.5.1 was completely solved by Jelonek in [167].

Theorem 10.5.9 (Jelonek). *If $k = \mathbb{C}$, 10.5.1 has an affirmative answer in any dimension.*

In the remainder of this section we sketch Jelonek's proof which is based on several results of his in [163], [164] and [165]. For the complete proof we refer the reader to [167] and the papers just mentioned.

In order to give some idea of the proof we need to introduce some notions and recall some of the results on uniruled varieties.

Let X be an irreducible affine variety of dimension ≥ 1. Then X is called $\mathbb{C}$-**uniruled** if for every point x in X there is an affine parametric line l in X through x, where a curve l is called an **affine parametric line** if l is affine and there exists a dominant regular mapping $\varphi : \mathbb{C} \to l$. An arbitrary algebraic set X is called **$\mathbb{C}$-uniruled** if it has pure dimension ≥ 1, i.e. each irreducible component has the same dimension, and each irreducible component is $\mathbb{C}$-uniruled. The empty set is also called $\mathbb{C}$-uniruled.

Now let $F : \mathbb{C}^n \to \mathbb{C}^n$ be a polynomial map and $y \in \mathbb{C}^n$. We say that F **is proper at** y if there is an open neighborhood U of y such that the restriction $res_{F^{-1}(U)} : F^{-1}(U) \to U$ is proper, i.e. the inverse image of each compact set is compact. Put $S_F :=$ the set of all y in $\mathbb{C}^n$ at which F is not proper.

This set was studied by Jelonek in [163]. His main result is

Proposition 10.5.10 (Jelonek). *Let $F : \mathbb{C}^n \to \mathbb{C}^n$ be dominant. Then S_F is either empty or a $\mathbb{C}$-uniruled hypersurface. In particular if $n = 1$ S_F is always empty.*

Remark 10.5.11 The set S_F can be computed as follows: since F is dominant the field extension $\mathbb{C}(F) \subset \mathbb{C}(X)$ is algebraic, so for each $1 \le i \le n$ there exists an irreducible polynomial $h_i(Y_1, \dots, Y_n, T)$ in $k[Y_1, \dots, Y_n, T]$ such that $h(F_1, \dots, F_n, X_i) = 0$ (Chapter 3, §2, Exercise 7). Let $d_i := \deg_T h_i$ and denote by $a_i(Y)$ the coefficient of T_i^d in h_i. Then the set S_F is given by the equation $a_1(Y) \dots a_n(Y) = 0$.

The proof of 10.5.9 rests on the following technical lemma, for the proof of which we refer to [167].

Lemma 10.5.12 *Let $n \ge 2$ and $V \subset \mathbb{C}^n$ be an irreducible hypersurface. Then there exist uncountably many hypersurfaces $V(f)$ such that $V \cap V(f)$ is not $\mathbb{C}$-uniruled.*

Proof of Theorem 10.5.9: i) Let φ be a $\mathbb{C}$-endomorphism of $\mathbb{C}[X]$ sending each coordinate of $\mathbb{C}[X]$ to $\mathbb{C}[X]$. Denote by F the corresponding polynomial map from $\mathbb{C}^n$ to $\mathbb{C}^n$. Then by 10.5.3, $\det JF \in \mathbb{C}^*$. Consequently it suffices to prove that S_F is empty. Therefore assume that S_F is not empty and let V be an irreducible component of S_F. Then it follows from 10.5.10 and 10.5.12 that there exists a coordinate f such that $V \cap V(f)$ is not a $\mathbb{C}$-uniruled variety.

ii) On the other hand consider the restriction $F' : F^{-1}(V(f)) \to V(f)$ of F to $F^{-1}(V(f))$. Observe that by our hypothesis both $V(f)$ and $F^{-1}(V(f))$ are coordinate hypersurfaces, hence isomorphic to $\mathbb{C}^{n-1}$. So we can view F' as a map from $\mathbb{C}^{n-1}$ to $\mathbb{C}^{n-1}$ with non-vanishing Jacobian. Counting the number of points in the fibers of F' one can deduce that $V(f) \cap V \subset S'_F$. Since both varieties have the same dimension and S'_F is $\mathbb{C}$-uniruled, we conclude that $V(f) \cap V$ is $\mathbb{C}$-uniruled as well, contradicting i). □

Notes

Corollary 10.1.10 was first obtained by Randall in [247]. The proof of the similarity of the Newton polygons of Jacobian mates given here is due to Nagata in [220]. The method used to study the edges of Newton polygons and to prove Abhyankar's reduction theorem is strongly influenced by the unpublished paper [309] of Victor Zurkowski. Theorem 10.2.18 and Theorem 10.2.23 were first obtained by Abhyankar and can be found in his paper [3]. Magnus' theorem and Corollary 10.2.25 were first proved by Magnus in [197]. Theorem 10.2.31 is taken from Gwoździewicz's paper [149]. This paper improves an earlier result of Tutaj and the author in [121] in which they showed that injectivity on three lines implies invertibility. Moh's result can be found in [219] and Wright's result, Theorem 10.2.28, was proved in [292]. See also his paper [295] in [108]. Kaliman's results Theorem 10.2.32 and Proposition 10.2.33 are obtained in [170].

A promising new approach to the two-dimensional Jacobian Conjecture was initiated by Y. Stein in [279] where he establishes an interesting relationship between Jacobian polynomials and so-called Toeplitz sequences.

Corollary 10.3.9 was obtained by McKenna and van den Dries. In fact in [92] they obtained a more general result. Proposition 10.3.10, Lemma 10.3.11 as well as Proposition 10.3.18 are taken from the paper [63] of Connell and van den Dries.

The equivalence of the Jacobian Conjecture and the weak Dixmier Conjecture comes from the author's paper [101]. The A_n-module $k[X]^{(J)}$ was introduced by Bass in [28] where he also showed that this module is holonomic. In [101] the author proved Theorem 10.4.5 and deduced that $k[X]^{(J)}$ is regular holonomic.

Theorem 10.5.7 was obtained by Mikhalev, Yu and Zolotykh in [214]. Theorem 10.5.9 also holds if we replace $\mathbb{C}$ by any algebraically closed field of characteristic zero. This was communicated to me by Jelonek.

A topic not treated in this book is the study of test polynomials: a polynomial $p \in k[X]$ is called a **test polynomial** if every k-endomorphism φ of $k[X]$ which satisfies $\varphi(p) = p$ is a k-automorphism of $k[X]$. For several results about test polynomials we refer to the papers [91] and [122].

Part III
Appendices

Appendix A Some commutative algebra

A.1 Rings

For a more detailed treatment of commutative algebra we refer to the books [22], [190], [203], [307] and [308].

Throughout this chapter A denotes a commutative ring containing 1 and $0 \neq 1$. Each such ring contains a maximal ideal and hence a prime ideal. The set of all prime ideals (resp. all maximal) ideals of A is denoted by $Spec A$ (resp. $Max A$). Rings having precisely one maximal ideal are called **local** rings. A ring A is local if and only if the set of non-units is an ideal in A, which is the unique maximal ideal $\mathfrak{m}$ of A. The field $A/\mathfrak{m}$ is called the **residue field** of A. The intersection of all maximal ideals of a ring is called its **Jacobson radical** and the intersection of all prime ideals the **nilradical**, often denoted by η. The nilradical of A is equal to the set of nilpotent elements of A. If A has no nilpotent elements except zero, A is called a **reduced** ring. The ring $\overline{A} := A/\eta$ is reduced. One easily verifies that the nilradical of a polynomial ring $A[X]$ equals $\eta A[X]$.

A prime ideal $\wp$ is said to have **height** n, notation $ht\wp = n$, if there exists a strictly increasing chain of prime ideals $\wp_0 \subset \wp_1 \subset \cdots \subset \wp_n = \wp$ and this is the longest such chain. The **(Krull) dimension of** A, denoted $\dim A$, is the supremum of the heights of all prime ideals contained in A. So if $\dim A$ is finite, then it is just the length of the longest chain of prime ideals contained in A. If $\wp_1 \subset \wp_2$ are prime ideals of the same height, then they are equal. If $A = k[X_1, \ldots, X_n]$, the polynomial ring over a field k, then $\dim A = n$ and more generally if $\wp$ is a prime ideal then $\dim A/\wp = \dim A - ht\wp$. Also in this case $\dim A/\wp = \text{trdeg}_k\, Q(A/\wp)$.

A set $x_1, \ldots, x_r$ of elements of A is called a **regular sequence** of A if x_1 is not a zero-divisor of A and for all $2 \leq i \leq r$ x_i is not a zero-divisor of $A/(x_1, \ldots, x_{i-1})$. If A is a local ring with maximal ideal $\mathfrak{m}$, then the **depth** of A, denoted depth A is the maximum length of all regular sequences of A with all x_i in $\mathfrak{m}$.

A.2 Modules

An **A-module** M is an additive group equipped with a scalar multiplication $A \times M \to M$, $(a, m) \to am$ which satisfies

i) $a(m_1 + m_2) = am_1 + am_2$, for all $a \in A$, all $m_1, m_2 \in M$.
ii) $(a_1 + a_2)m = a_1 m + a_2 m$, for all $a_1, a_2 \in A$, all $m \in M$.
iii) $(a_1 a_2)m = a_1(a_2 m)$, for all $a_1, a_2 \in A$, all $m \in M$.
iv) $1.m = m$, for all $m \in M$.

Submodules, quotient modules, direct sums and A-module homomorphisms, also called A-linear maps, are defined in the obvious way.

Let M and N be A-modules, then there exists an A-module T together with an A-bilinear map $i : M \times N \to T$ which has the following universal property: if P is any A-module and $F : M \times N \to P$ an A-bilinear map, then there exists a uniquely determined A-linear map $f : T \to P$ such that $F \circ i = f$.

Furthermore if (T, i) and (T', i') are two pairs with this property, then there exists a unique isomorphism $j : T \to T'$ such that $j \circ i = i'$. The module T is called the **tensor product** of M and N and is denoted by $M \otimes_A N$ or just $M \otimes N$. The elements $i((m, n))$ are denoted by $m \otimes n$. Each element of $M \otimes N$ is a finite sum of such elements. The bilinearity of i implies that $am \otimes n = a(m \otimes n) = m \otimes an$ and $(m_1 + m_2) \otimes n = m_1 \otimes n + m_2 \otimes n$, for all m, m_1, m_2, a, n, etc.

If $f : M_1 \to M_2$ and $g : N_1 \to N_2$ are A-linear we get an A-linear map $f \otimes g : M_1 \otimes N_1 \to M_2 \otimes N_2$ given by $f \otimes g(m \otimes n) = f(m) \otimes g(n)$, for all $m \in M_1, n \in N_1$. One has the following isomorphisms.

$$\begin{aligned} A \otimes_A M &\simeq M, \\ (M_1 \otimes M_2) \otimes M_3 &\simeq M_1 \otimes (M_2 \otimes M_3), \\ M \otimes (\oplus_{i \in I} N_i) &\simeq \oplus_{i \in I} (M \otimes N_i). \end{aligned}$$

If $f : A \to B$ is a ring homomorphism, B becomes an A-module by defining $a.b = f(a)b$. Therefore to an A-module M we can associate $M_B := B \otimes_A M$, which gets a B-module structure by defining $b(b' \otimes m) = bb' \otimes m$. We say that M_B is the B-module obtained from the A-module M by **extension of scalars**.

If also $g : A \to C$ is a ring homomorphism, then we can form the A-module $B \otimes_A C$. It becomes a ring by defining $(b \otimes c)(b' \otimes c') = bb' \otimes cc'$. If $A = k$ is a field, $B = A(X)$ the coordinate ring of an affine variety $X \subset k^n$ (Appendix B) and $C := A(Y)$ the coordinate ring of some affine variety $Y \subset k^m$, then the rings $A(X) \otimes_k A(Y)$ and $A(X \times Y)$ are isomorphic. The isomorphism is given by

$$f \otimes g \to ((x, y) \to f(x)g(y)).$$

A.3 Localization

Let $S \subset A$ be a **multiplicatively closed subset** of A, i.e. $1 \in S$ and if $s, t \in S$, then also $st \in S$. To such a set we associate the so-called **localization of A at S**, denoted $S^{-1}A$, i.e. the set of equivalence classes under the equivalence relation $(s, a) \sim (s', a')$ if and only if there exists $t \in S$ such that $t(s'a - sa') = 0$. The class of (s, a) we denote by $s^{-1}a$ or $\frac{a}{s}$. In complete analogy to the case of the rational numbers we define an addition and a multiplication in $S^{-1}A$. So it becomes a ring. If for example A is a domain and $S = A \backslash \{0\}$ we have $S^{-1}A = Q(A)$, the quotient field of A. In a similar way, to an A-module M we can associate the set $S^{-1}M$. With the obvious definitions of addition and scalar multiplication with elements of $S^{-1}A$, $S^{-1}M$ becomes an $S^{-1}A$-module which is isomorphic to $M_{S^{-1}A}$. Important examples arise if we take $S = \{1, f, f^2, \ldots\}$ for $f \in A$, in which case we write M_f, resp. A_f, instead of $S^{-1}M$, resp. $S^{-1}A$, or $S = A \backslash \wp$, where $\wp$ is a prime ideal in A. In this case we write $A_\wp$, $M_\wp$ instead of $S^{-1}A$ resp. $S^{-1}M$. The ring $A_\wp$ is a local ring with maximal ideal $\wp A_\wp$. Localizations are useful tools since various problems can be reduced to the local ring case by the local-global principle. This method is based on the following easy fact: $M = 0$ if and only if $M_\wp = 0$ for all $\wp \in Spec A$ if and only if $M_\mathfrak{m} = 0$ for all $\mathfrak{m} \in Max A$. The prime ideals of $S^{-1}\wp$ are in one-to-one correspondence with the prime ideals of A which do not intersect S. If $f : A \to S^{-1}A$ denotes the canonical map given by $f(a) = \frac{a}{1}$, this correspondence is given by $\wp \to f(\wp)$ and $f^{-1}(\mathfrak{q}) \leftarrow \mathfrak{q}$. It follows that $\dim A_\wp = ht\wp$.

A.4 Completions

Let M be an A-module and I an ideal in A. Then the **I-adic topology** of M is the topology whose basis of neighborhoods of $m \in M$ is given by the sets $m + I^n M$, $n \geq 1$. The **I-adic completion of** M, denoted $\hat{M}$, is the completion of M with respect to the I-adic topology, i.e. the elements of $\hat{M}$ are the equivalence classes of Cauchy sequences under the relation that two Cauchy sequences $(m_n)_{n\geq 1}$ and $(m'_n)_{n\geq 1}$ are equivalent if their differences $m_n - m'_n$ tend to zero in the I-adic topology of M. Then $\hat{A}$ has a ring structure in the obvious way and similarly $\hat{M}$ is an $\hat{A}$-module, which is isomorphic to $M_{\hat{A}}$ in case M is a finite A-module and A is Noetherian (see below). Finally, if I is an ideal in A, then $\hat{M}/\hat{I}\hat{M} \simeq M/IM$ and in particular $\hat{A}/\hat{I} \simeq A/I$.

A.5 Finiteness conditions and integral extensions

We say that a module M is a **finite A-module** or a **finitely generated A-module** if there exist $m_1, \ldots, m_n \in M$ such that

$$M = Am_1 + \cdots + Am_n.$$

A ring A is called **Noetherian** if every ideal in A is a finite A-module or equivalently if every ascending chain of ideals in A stabilizes. If A is Noetherian, then every submodule of a finite A-module is again a finite A-module. If A is Noetherian, then also all quotient rings A/I, all localizations $S^{-1}A$, all I-adic completions and all polynomial and power series rings over A are Noetherian. Consequently, localizations of finitely generated algebras over $\mathbb{Z}$ or a field are Noetherian.

Let $A \subset B$ be a ring extension. An element b in B is called **integral over** A if it satisfies a monic polynomial with coefficients in A or equivalently if $A[b]$ is a finite A-module. We say that $A \subset B$ is an **integral extension** or that B **is integral over** A if each element in B is integral over A. If furthermore B is a finitely generated A-algebra, then B is a finite A-module. In case $A \subset B$ is an integral extension, then for each $\wp \in Spec A$ there exists $\mathfrak{q} \in Spec B$ which **lies over** $\wp$, i.e. satisfies $\mathfrak{q} \cap A = \wp$. Furthermore $\wp$ is maximal if and only if $\mathfrak{q}$ is maximal, and if both $\mathfrak{q}_1 \subset \mathfrak{q}_2$ lie over $\wp$, then $\mathfrak{q}_1 = \mathfrak{q}_2$.

If $A \subset B$ is a ring extension, the **integral closure of** A **in** B is the set of all elements of B which are integral over A. If A is a domain with quotient field $Q(A)$, then the

integral closure of A, denoted $\tilde{A}$, is the integral closure of A in $Q(A)$. If $A = \tilde{A}$, A is called **integrally closed.** A Noetherian domain of dimension 1 which is integrally closed is called a **Dedekind ring.**

A Noetherian local ring A with maximal ideal $\mathfrak{m}$ and residue field k is called a **regular local ring** (resp. a **Cohen-Macaulay ring,** abbreviated CM-ring) if $\dim_k \mathfrak{m}/\mathfrak{m}^2 = \dim A$ (resp. depth $A = \dim A$). A regular local ring of dimension 1 is the same as a Dedekind ring which is also a local ring. A regular local ring is an example of a CM-ring. A Noetherian ring A is called a **regular ring** (resp. a **CM-ring**) if $A_\mathfrak{m}$ is a regular local ring (resp. a CM-ring) for every maximal ideal $\mathfrak{m}$ of A. Hence a regular ring is a CM-ring. If A is a regular ring (resp. a CM-ring), then so are $A[X_1, \ldots, X_n]$, $A[[X_1, \ldots, X_n]]$ and $A_\wp$ for every $\wp$ in Spec A.

Proposition A.5.1 (Nullstellensatz, algebraic form). *Let k be any field and A a finitely generated k-algebra. If A is a field then A is algebraic over k.*

Corollary A.5.2 (Nullstellensatz, geometric form). *Let k be an algebraically closed field and I an ideal in $k[X] := k[X_1, \ldots, X_n]$. If $f \in k[X]$ is such that $f(x) = 0$ for every zero $x \in k^n$ of I, then $f^m \in I$ for some $m \in \mathbb{N}$.*

A.6 The universal coefficients method

To prove or disprove a statement which should hold for all rings it is often very useful to use what we will call **the universal coefficients method**: first describe a given problem in terms of elements of the ring and its relations and then consider the "universal ring" having these relations, i.e. the quotient of a polynomial ring in as many variables as there are elements involved, and the ideal of relations amongst these elements. Let us illustrate this method with two examples.

Problem 1 Let A be a ring. Suppose that both x and y generate the same ideal, i.e. $Ax = Ay$, so $x = zy$ and $y = wx$ for some $z, w \in A$. Are z and w units in A?

Problem 2 Does the Cayley-Hamilton theorem hold for matrices with coefficients in an arbitrary ring?

Solution to Problem 1 The problem is completely determined by the elements x, y, z and w and the relations $x = zy$ and $y = wx$. Therefore we consider the "universal" ring having these properties, i.e. we consider the polynomial ring $\mathbb{Z}[X, Y, Z, W]$ and to obtain the relations above we define

$$A_U := \mathbb{Z}[X, Y, Z, W]/(X - ZY, Y - WX)$$

and put $x_U := \bar{X}$, $y_U := \bar{Y}$, $z_U := \bar{Z}$ and $w_U := \bar{W}$. So obviously $x_U = z_U y_U$ and $y_U = w_U z_U$. Furthermore it is necessary and sufficient to investigate Problem 1 for the universal ring A_U (the "sufficiency" follows using the substitution homomorphism $x_U \to x$, $y_U \to y$, $z_U \to z$ and $w_U \to w$).

Finally it is easy to solve our problem in A_U: namely if z_U is a unit in A_U, there exists $V \in \mathbb{Z}[X, Y, Z, W]$ such that $v = \bar{V}$ in A_U satisfies $v z_U = 1$, i.e. $VZ \equiv 1$

(mod $(X - ZY, Y - WX)$) in $\mathbb{Z}[X, Y, Z, W]$. Then substituting $X = Y = Z = W = 0$ we get $0 = 1$, a contradiction. So z_U is not a unit in A_U.

Solution to Problem 2 We assume the Cayley-Hamilton theorem for the field case, i.e. if $B \in M_n(k)$, k a field then $p_B(B) = 0$, where $p_B(X) = \det(XI_n - B)$ is the characteristic polynomial of B.

Now let A be any ring and $B = (b_{i,j}) \in M_n(A)$. Then again the problem is completely determined by the elements $b_{i,j}$. Therefore we consider the universal ring

$$A_U := \mathbb{Z}[B_{i,j}; 1 \leq i, j \leq n]$$

where $B_{i,j}$ are n^2 variables and the universal matrix $B_U := (B_{i,j}) \in M_n(A_U)$. Now observe that A_U is a domain, hence contained in its quotient field K. So we can view B_U in $M_n(K)$. Then applying the Cayley-Hamilton theorem for the field case we obtain $p_{B_U}(B_U) = 0$. Since obviously $p_{B_U}(X) \in A_U[X]$ we get the Cayley-Hamilton theorem for the ring A by applying the substitution homomorphism $B_{i,j} \to b_{i,j}$ to the relation $p_{B_U}(B_U) = 0$.

Appendix B Some basic results from algebraic geometry

B.1 Algebraic sets

Throughout this section k denotes an algebraically closed field. By $\mathbb{A}^n$ or k^n we denote the **n-dimensional affine space over k** whose points are of the form $x = (x_1, \ldots, x_n)$ with $x_i \in k$. An **algebraic subset of $\mathbb{A}^n$** is the common zero-set of a (finite) set of polynomials in $k[X_1, \ldots, X_n]$. One easily verifies that the intersection of any number of algebraic sets is again an algebraic set, and the union of two (and hence of a finite number) of algebraic sets is also algebraic. Since obviously the empty set and the whole affine space $\mathbb{A}^n$ are algebraic (being the zero-set of the polynomial 1 resp. 0) we get that the algebraic subsets of $\mathbb{A}^n$ form the closed subsets of a topology on $\mathbb{A}^n$, called the **Zariski-topology**. Its open sets are by definition the complements of the algebraic subsets. If V is an algebraic subset of $\mathbb{A}^n$, then on V we get the **induced Zariski topology**. The intersection of all closed subsets of V containing a given set $M \subset V$ is closed and is called the **Zariski-closure of M** and is denoted by $\overline{M}$. A subset M is called **dense in** V if $\overline{M} = V$.

An important result is that every algebraic set V is a finite union $V = V_1 \cup \cdots \cup V_s$, where each V_i is a so-called irreducible algebraic set: an algebraic set is called **reducible** if there exist closed subsets $V_1 \subset V$, $V_2 \subset V$ with $V_1 \neq V$, $V_2 \neq V$ such that $V = V_1 \cup V_2$. Otherwise V is called **irreducible**. If we assume that V_i is not contained in V_j for all $i \neq j$, then this so-called **irreducible decomposition** of V is unique up to permutation.

To study the algebraic set V we introduce the set of regular functions on it: a function $f : V \to k$ is called **regular** if there exists a polynomial function $F \in k[X_1, \ldots, X_n]$ such that $f(x) = F(x)$ for all $x \in V$. The set of regular functions on V is called the **coordinate ring on** V and is denoted by $A(V)$. A useful tool to study the coordinate ring is the **local ring of V at a point** $v \in V$, i.e. the ring $A(V)_v$ obtained from $A(V)$ by localizing at the maximal ideal $\mathfrak{m}_v$ of $A(V)$ consisting of all $f \in A(V)$ satisfying $f(v) = 0$. A point $v \in V$ is called a **singular point** of V or simply a **singularity** of V if the local ring $A(V)_v$ is not regular (Appendix A). V is called **regular** if V has no singular points, i.e. if $A(V)_v$ is regular for all v in V. Similarly V is called **Cohen-Macaulay** if $A(V)_v$ is Cohen-Macaulay (Appendix A) for all v in V, in other words if $A(V)$ is Cohen-Macaulay.

One easily verifies that the map $F \to F_{|V}$ induces an isomorphism between $k[X_1, \ldots, X_n]/I(V)$ and $A(V)$, where $I(V)$ is the ideal in $k[X_1, \ldots, X_n]$ consisting of all polynomials F vanishing at all points of V, i.e. $F_{|V} = 0$. Furthermore V is irreducible if and only if $I(V)$ is a prime ideal if and only if $A(V)$ is a domain. If V is irreducible then the **dimension of** V is defined as the transcendence degree of $k(V)$ over k, where $k(V)$ is the quotient field of $A(V)$, also called **the field of rational functions on** V. In general, if V is an arbitrary algebraic set then $\dim V := \max \dim V_i$, where $V = V_1 \cup \cdots \cup V_s$ is the irreducible decomposition of V.

Now let $V \subset \mathbb{A}^n$ and $W \subset \mathbb{A}^m$ be algebraic sets. A map $f : V \to W$ is called **regular** or **a morphism of algebraic varieties** if there exist m regular functions $f_1, \ldots, f_m$ on V such that $f(x) = (f_1(x), \ldots, f_m(x))$ for all $x \in V$. Such a regular map induces a k-homomorphism $f^* : A(W) \to A(V)$ between the coordinate rings given by $f^*(g) = g \circ f$, for all $g \in A(W)$. In fact regular mappings can be characterized as mappings carrying regular functions into regular functions: for suppose a mapping of algebraic sets $f = (f_1, \ldots, f_m) : V \to W$ is such that for any regular function g on W the function $f^*(g)$ is also regular. Then in particular this is so for the coordinate functions restricted to W, i.e. $f^*(Y_{i|W})$ is regular for all $1 \leq i \leq m$. But this means that each map $f_i : V \to k$ is regular. So f is regular. Consequently regularity can be expressed in terms of the coordinate rings, i.e. f is regular if and only if $f^*(g) \in A(V)$ for all $g \in A(W)$. Also one easily verifies that every k-homomorphism $\varphi : A(W) \to A(V)$ is of the form $\varphi = f^*$, where f is a regular map from V to W. So the mapping $f \to f^*$ gives a bijection between the regular mappings from V to W and the k-algebra homomorphisms from $A(W)$ to $A(V)$. Also one easily verifies that for each $v \in V$ a morphism $f : V \to W$ induces a k-algebra homomorphism $f^* : A(W)_{f(v)} \to A(V)_v$.

A regular map f is called an **isomorphism** if it has an inverse, i.e. there exists a regular map $g : W \to V$ such that $f \circ g = 1_{|W}$ and $g \circ f = 1_{|V}$. In this case V and W are called **isomorphic**. It is not difficult to verify that f is an isomorphism if and only if f^* is a k-algebra isomorphism between $A(W)$ and $A(V)$. A regular map $f : V \to W$ is called **finite** if the induced map $f^* : A(W) \to A(V)$ makes $A(V)$ into a finite $A(W)$-module. Since $A(V)$ is a finitely generated k-algebra this is equivalent to $A(V)$ being integral over $f^*(A(W))$.

B.2 Morphisms of irreducible affine algebraic varieties

Now we restrict to regular mappings between irreducible algebraic sets $f : V \to W$. Such a map is called **dominant** if $\overline{f(V)} = W$. This is equivalent to $f^* : A(W) \to A(V)$ is injective. So in this case f^* induces an embedding of $k(W)$ into $k(V)$ and if V and W have the same dimension the field extension $f^*(k(W)) \subset k(V)$ is finite, since $k(V)$ is a finitely generated field extension over k. The degree of this extension is called **the geometric degree of** f, denoted **degree** f or $d(f)$. If $d(f) = 1$, i.e. $k(V) = f^*(k(W))$, then f is called a **birational isomorphism** or we simply say that f is **birational**. The degree of f describes the number of points of a "generic" fiber of f. More precisely

Proposition B.2.1 *Assume char* $k = 0$. *Let* $f : V \to W$ *be a dominant map between irreducible sets of the same dimension. Then there exists a dense subset* U *of* W *such that* $\# f^{-1}(w) = d(f)$ *for all* $w \in U$.

The condition char $k = 0$ in this and the next proposition is necessary, namely if char $k = p > 0$, then the Frobenius automorphism $f : x \to x^p$ obviously satisfies $d(f) = p$ while $\# f^{-1}(x) = 1$ for all $x \in k$.

An irreducible algebraic set V is called **normal** if its coordinate ring $A(V)$ is normal, i.e. integrally closed in $k(V)$.

Proposition B.2.2 *Assume char* $k = 0$. *Let* $f : V \to W$ *be a dominant map between normal varieties of the same dimension. If* $\# f^{-1}(w) < \infty$ *for all* $w \in W$, *then*

$$d(f) = \max\{\# f^{-1}(w) | w \in W\}.$$

A **normalization** of an irreducible algebraic set V is a normal algebraic set $\tilde{V}$ together with a regular map $\pi : \tilde{V} \to V$ that is finite and a birational isomorphism.

Proposition B.2.3 *Let* V *be an irreducible algebraic set. Then there exists a normalization* $\pi : \tilde{V} \to V$ *which satisfies the following universal property: for every normal set* W *and any dominant map* $f : W \to V$ *there exists exactly one regular map* $\varphi : W \to \tilde{V}$ *such that* $f = \pi \circ \varphi$.

From the universal property it follows readily that the normalization $\pi : \tilde{V} \to V$ is uniquely determined up to an isomorphism.

Now we generalize the notion of algebraic subset of some affine space $\mathbb{A}^n$ to the coordinate free notion of **affine variety**.

Definition B.2.4 A set Z together with a k-algebra $A(Z)$ consisting of functions from Z to k is called an **affine variety** (over k) if there exist an algebraic subset $V \subset \mathbb{A}^n$ for some n and a bijection $\varphi : Z \to V$ which identifies $A(V)$ with $A(Z)$, i.e. for all $f \in A(V)$ the function $f \circ \varphi$ belongs to $A(Z)$ and the map $f \to f \circ \varphi$ is an isomorphism between $A(V)$ and $A(Z)$.

Example B.2.5 Let $P \in k[X] := k[X_1, \dots, X_n]$. Then the set $Z := \mathbb{A}^n \backslash P^{-1}(0)$, which obviously is not an algebraic subset of $\mathbb{A}^n$ since any polynomial vanishing on Z vanishes on $\mathbb{A}^n$, together with the k-algebra $k[X, P^{-1}]$ is an affine variety: one readily verifies that the restriction of the projection $p : \mathbb{A}^{n+1} \to \mathbb{A}^n$, given by $p(x, t) = x$ for all $x \in \mathbb{A}^n$ and all $t \in k$, to the algebraic set

$$V := \{(x, t) \in \mathbb{A}^{n+1} | tP(x) - 1 = 0\} \subset \mathbb{A}^{n+1}$$

gives a bijection between V and Z which induces an isomorphism between $k[X, P^{-1}]$ and $A(V) = k[X, T]/(TP(X) - 1)$.

As before we call $A(Z)$ the **coordinate ring of** Z or the **ring of regular functions on** Z. The Zariski-topology on Z is defined via the bijection φ. It follows that the Zariski-topology does not depend on the choice of φ.

A map $f : V \to W$ between two affine varieties is called **regular** or a **morphism** if $f^*(g) \in A(V)$ for all $g \in A(W)$. By our observations above this definition agrees with the one given for algebraic subsets of affine spaces. It is left to the reader to verify that all concepts introduced above and all results described above can be extended to the case of affine varieties.

A morphism $f : V \to W$ between affine varieties is called an **open immersion** if $f(V)$ is an open subset of W and the map $f : V \to f(V)$ is an isomorphism of affine varieties. For example the inclusion map $i : Z \to \mathbb{A}^n$, where Z is the set of B.2.5, is an open immersion.

Zariski's Main Theorem. *Assume char $k = 0$. Let $f : V \to W$ be a morphism of irreducible affine varieties, where W is normal. If f is birational and* ***quasi-finite,*** *i.e. f has finite fibers, then f is an open immersion.*

As a very special case of this result we have

Corollary B.2.6 *If $f := (f_1, \ldots, f_n) : k \to k^n$ is injective and $f'(t) \neq 0$ for all $t \in k$, then $k[f_1(t), \ldots, f_n(t)] = k[t]$.*

Finally we mention the "weak form of Bezout's theorem" due to S. Abhyankar in [1], 12.3.1 p. 272.

Proposition B.2.7 *Let $F_1, \ldots, F_d \in k[X_1, \ldots, X_n] \backslash k$. Then the number of isolated points in the set of common zeroes of the F_i's is at most* $\deg F_1 \ldots \deg F_d$.

The assumption made in the beginning of this Appendix, that k is algebraically closed, cannot be dropped in the last proposition. This can be seen from Example B.1 in Appendix F.

Appendix C Some results from Gröbner basis theory

C.1 Definitions and basic properties

For a nice introduction to Gröbner basis theory and its applications we refer the reader to the textbooks [32] and [65].

Throughout this section k denotes a field and $k[X] := k[X_1, \ldots, X_n]$. By T we denote the set of **terms** $X_1^{i_1} \ldots X_n^{i_n}$ with $i_1, \ldots, i_n \geq 0$. Let $<$ be a total ordering on T. We say that $<$ is **admissible** if $1 < t$ for all $t \in T$, $t \neq 1$ and if for all $t_1, t_2, t \in T$ the relation $t_1 < t_2$ implies that $tt_1 < tt_2$. One can show that such an ordering is a **well-ordering** on T, i.e. every non-empty subset of T has a smallest element under $<$ or equivalently every decreasing sequence in T, $t_1 \geq t_2 \geq \cdots$ eventually stabilizes.

Let $0 \neq f \in k[X]$. Then we can write f as a finite sum $f = \sum c_t t$, with all $c_t \in k^*$. The greatest t appearing in f is called the **leading term of** f, denoted $lt(f)$. The corresponding coefficient c_t is called the **leading coefficient of** f, denoted $lc(f)$. Furthermore we put $lt(0) = 0$ and $lc(0) = 0$. So we get maps

$$lt : k[X] \to T \cup \{0\} \quad \text{and} \quad lc : k[X] \to k.$$

Definition C.1.1 Let $G \subset k[X]$ be a finite subset and f a non-zero element of $k[X]$. We say that f **can be reduced modulo** G, if there exists a term t in f, i.e. with non-zero coefficient, such that $t = s.lt(g)$ for some $s \in T$ and $g \in G$.

Definition C.1.2 Let I be a non-zero ideal of $k[X]$, $<$ an admissible ordering and $G \subset I$ a finite subset.

i) G is called a **Gröbner basis** of I (with respect to $<$) if for every $f \in I$ there exist $s \in T$ and $g \in G$ such that $lt(f) = s.lt(g)$.

ii) A Gröbner basis G of I is called **minimal** if for every $g \in G$ the set $G\backslash\{g\}$ is not a Gröbner basis of I.

iii) A Gröbner basis G of I is called **reduced** if for every $g \in G$ we have: g cannot be reduced modulo $G\backslash\{g\}$ and $lc(g) = 1$ for all $g \in G$.

It follows easily that a reduced Gröbner basis is minimal. Furthermore one can prove that with respect to a chosen admissible ordering $<$, a reduced Gröbner basis is unique!

Now one of the main points of Gröbner basis theory is that given any finite set of generators of a non-zero ideal of $k[X]$ one can construct out of it a Gröbner basis, resp. a reduced Gröbner basis, of I. Also, as the word "basis" in Gröbner basis suggests, every Gröbner basis of I forms a set of generators of I. Now we describe some applications of Gröbner basis theory.

C.2 Applications: several algorithms

The normal form algorithm

Let I be a non-zero ideal in $k[X]$, $<$ an admissible ordering on $k[X]$ and G a Gröbner basis of I with respect to $<$. Suppose that $f = \sum c_t t$ contains terms which are reducible modulo G. Let t_1 be the largest of these terms. So $t_1 = s.lt(g)$ for some $g \in G$ and $s \in T$. Define $f_1 := f - (c_{t_1}/lc(g))sg \in I$. Then either f_1 has no terms which are reducible modulo G or each term of f_1 which is reducible modulo G is smaller than t_1. In the latter case define f_2, etc. So either the process stops or we get a sequence of terms $t_1 > t_2 > \cdots$ which is impossible since $<$ is a well-ordering. The element we obtain at the end turns out to be independent of G and is determined only by f and I and the ordering chosen. It is called the **normal form** of f and is denoted by f_+. Obviously the construction above shows that f_+ has the following properties: $f - f_+ \in I$ and f_+ is not reducible modulo G.

The ideal membership algorithm

Let I, $<$, G and f be as above. The **(ideal) membership algorithm** decides if f belongs to I, namely $f \in I$ if and only if $f_+ = 0$.

The elimination algorithm

Proposition C.2.1 *Let I be an ideal in $k[X, Y] := k[X_1, \ldots, X_n, Y_1, \ldots, Y_m]$ and $<$ an admissible ordering on the terms of $k[X, Y]$ such that $X_i > Y^\alpha$ for all i and $\alpha \in \overline{\mathbb{N}}^m$. If G is a Gröbner basis (resp. reduced Gröbner basis) of I, then $G \cap I$ is a Gröbner basis (resp. reduced Gröbner basis) of $k[Y] \cap I$.*

Proof: Let $0 \neq f \in k[Y] \cap I$. Since G is a Gröbner basis of I, there exist $s \in T$ and $g \in G$ with $lt(f) = s.lt(g)$. Since $f \in k[Y]$, $lt(f) \in k[Y]$, whence both s and $lt(g)$ belong to $k[Y]$. The ordering $<$ chosen implies that $g \in k[Y] \cap G$. Since $lt(f) = s.lt(g)$ with $s \in k[Y]$ it follows that $k[Y] \cap G$ is a Gröbner basis of $k[Y] \cap I$. That a reduced Gröbner basis gives a reduced Gröbner basis is obvious. □

The relation algorithm

Let $F_1, \ldots, F_m \in k[X]$ and I an ideal in $k[X]$ generated by $Q_1, \ldots, Q_s$. Let

$$J := \{P(Y) \in k[Y] | P(F_1, \ldots, F_m) \in I\}.$$

Observe that

(1) $P(Y) \equiv P(F) \pmod{(Y_1 - F_1, \ldots, Y_m - F_m)}$ for all $P(Y) \in k[Y]$.

Proposition C.2.2 *Let $<$ be an admissible ordering on the terms of $k[X, Y]$ such that $X_i > Y^\alpha$ for all i and all $\alpha \in \overline{\mathbb{N}}^m$. Let G be a Gröbner basis of*

$$(Y_1 - F_1(X), \ldots, Y_m - F_m(X), Q_1(X), \ldots, Q_s(X)).$$

Then $G \cap k[Y]$ is a Gröbner basis of J.

Proof: Let $P(Y) \in k[Y]$. Then $P(Y) \in J$ if and only if $P(F) \in I = (Q_1, \ldots, Q_s)$ if and only if $P(Y) \in (Y_1 - F_1, \ldots, Y_m - F_m, Q_1, \ldots, Q_s)$, by (1). So $J = k[Y] \cap (Y_1 - F_1, \ldots, Y_m, Q_1, \ldots, Q_s)$. Then apply C.2.1. □

The (algebra) membership algorithm

Let $k[x] := k[x_1, \ldots, x_n]$ be a finitely generated k-algebra whose kernel under the homomorphism $k[X] \to k[x]$ sending X_i to x_i is generated by the elements $Q_1, \ldots, Q_s \in k[X]$. Let $f_1(X), \ldots, f_m(X), g(X) \in k[X]$. The algebra membership algorithm decides if $g(x)$ belongs to the k-subalgebra $k[f] := k[f_1(x), \ldots, f_m(x)]$ of $k[x]$ and, if it does, computes a polynomial $P(Y) \in k[Y] := k[Y_1, \ldots, Y_m]$ such that $g(x) = P(f_1(x), \ldots, f_m(x))$. More precisely

Proposition C.2.3 *Let $k[X, Y] := k[X_1, \ldots, X_n, Y_1, \ldots, Y_m]$. Choose an admissible ordering $<$ on the terms of $k[X, Y]$ such that $X_i > Y^\alpha$ for all i, α. Let*

$$J := (Y_1 - f_1(X), \ldots, Y_m - f_m(X), Q_1(X), \ldots, Q_s(X))$$

and let $g(X)_+$ be the normal form of $g(X)$ with respect to J. Then $g(x) \in k[f]$ if and only if $g(X)_+ \in k[Y]$. Furthermore, in case $g(X)_+ = P(Y) \in k[Y]$ we have $g(x) = P(f_1(x), \ldots, f_m(x))$.

Proof: i) If $g(X)_+ \in k[Y]$, say $g(X)_+ = P(Y)$, then $g(X) - P(Y) \in J$. Making the substitutions $Y_j := f_j(x)$ and $X_i := x_i$ for all i, j gives $g(x) = P(f_1(x), \ldots, f_m(x))$.

ii) Conversely assume that $g(x) = P(f_1(x), \ldots, f_m(x))$ for some $P(Y) \in k[Y]$. Then

$$g(X) - P(f_1(X), \ldots, f_m(X)) \in (Q_1(X), \ldots, Q_s(X))$$

and since $Y_i \equiv f_i(X) \pmod J$ for all i we get that $g(X) - P(Y) \in J$. Hence $g(X)_+ - P(Y) \in J$. If $g(X)_+$ does not belong to $k[Y]$, then $lt(g(X)_+ - P(Y)) = lt g(X)_+$, by the choice of $<$. Furthermore since $g(X)_+ - P(Y) \in J$, its leading term, which as we just saw is the leading term of $g(X)_+$, is reducible modulo B, where B is a Gröbner basis of J with respect to $<$, contradicting the definition of $g(X)_+$. □

Computing degrees of some field extensions

Finally we describe an algorithm which is used in Chapter 3. For more general results on computing the degree of a field extension we refer to [284]. In fact we prefer to give our algorithm instead of a more general one, because the one given here, which is sufficient for our application, is much faster.

We have the following situation: $F_1, \ldots, F_m \in k[X] := k[X_1, \ldots, X_n]$ algebraically independent over k and X_1 algebraic over $k(F) := k(F_1, \ldots, F_m)$. Let $d := |k(F)(X_1) : k(F)|$. So d is finite. Our purpose is to calculate d. Therefore we proceed as follows. Introduce m new variables $Y_1, \ldots, Y_m$ and consider $k(Y)[X_1, \ldots, X_n]$. Choose on the terms X^α an admissible ordering $<$. Let J be the ideal in $k(Y)[X]$ given by

$$J = (Y_1 - F_1(X), \ldots, Y_m - F_m(X))$$

and let G be a Gröbner basis of J with respect to $<$.

Proposition C.2.4 *G contains an element of the form*

$$a_e(Y)X_1^e + \cdots + a_1(Y)X_1 + a_0(Y)$$

with $a_i \in k(Y)$ and $a_e \neq 0$. Furthermore d is equal to the minimal X_1-degree of all such elements in G.

Proof: Let $P(T) \in k(F)[T]$ be the minimal polynomial of X_1 over $k(F)$. So $\deg_T P(T) = d$. Multiplying the equation $P(X_1) = 0$ by a suitable non-zero element of $k[F]$ we find $a_i(F) \in k[F]$ with $a_d(F) \neq 0$ such that

$$a_d(F)X_1^d + \cdots + a_1(F)X_1 + a_0(F) = 0.$$

Since $a_i(F) \equiv a_i(Y) \pmod J$ we deduce that

$$a_d(Y)X_1^d + \cdots + a_1(Y)X_1 + a_0(Y) \in J.$$

So $X_1^d \in lt(J)$. Hence there exist $1 \leq e \leq d$ and $g \in G$ such that $X_1^e = lt(g)$. Observe that if $e = 0$ then $1 \in G \subset J$, a contradiction. This implies that $g = g_e X_1^e + \cdots + g_1 X_1 + g_0$ for some $g_i \in k(Y)$ with $g_e \neq 0$. Let now g be such that e is minimal under all such elements in G, then $e = d$, for if $e < d$ we get a contradiction with the degree d of the minimal polynomial $P(T)$. □

Appendix D Flatness

D.1 Flat modules and algebras

Throughout this section A and B denote commutative rings.

Definition D.1.1 i) An A-module M is called a **flat** A-module if for every injective A-module homomorphism $u : N_1 \to N_2$ the induced map $1 \otimes u : M \otimes N_1 \to M \otimes N_2$ is injective.

ii) A flat A-module M is called **faithfully flat** if $M \otimes N \neq 0$ for every non-zero A-module N.

Example D.1.2 i) The A-module A is faithfully flat. More generally every free A-module is faithfully flat. In particular the polynomial ring $A[X_1, \ldots, X_n]$ is a faithfully flat A-module.

ii) If S is a multiplicatively closed subset of A, then $S^{-1}A$ is a flat A-module ([22]).

iii) If A is a Noetherian ring and I an ideal in A, then the I-adic completion $\hat{A}$ of A is a flat A-module ([22]).

Let $f : A \to B$ be a ring homomorphism. We say that f is a **flat homomorphism**, or that B **is flat over** A if B is a flat A-module. It is left to the reader to prove the following result (Exercise 3).

Proposition D.1.3 *Let M be an A-module and N a B-module.*

i) *If M is flat (faithfully flat) over A, then $M_B := B \otimes_A M$ is flat (resp. faithfully flat) over B.*

ii) *If B is flat (faithfully flat) over A and N is a flat (resp. faithfully flat) B-module, then N is a flat (resp. faithfully flat) A-module.*

iii) *If N is faithfully flat over B and N is flat over A, then B is flat over A.*

iv) *If N is faithfully flat over both A and B, then B is faithfully flat over A.*

(D.1.4) **Warning.** Let N be a B-module which is faithfully flat over A, and furthermore assume that B is faithfully flat over A. Then N need not be flat over B (see Exercise 4, where we give an example where even both N and B are free A-modules!).

Another result which is not difficult to prove is ([203], Theorem 7.4).

Theorem D.1.5 *Let M be a flat A-module and N_1, N_2 two submodules of an A-module N. Then as submodules of $M \otimes N$ we have*

$$M \otimes_A (N_1 \cap N_2) = (M \otimes N_1) \cap (M \otimes N_2).$$

In particular, taking $N = A$, $M = B$ and using that for an ideal I of A the subset $B \otimes_A I$ of $B \otimes_A A = B$ coincides with BI we get

Corollary D.1.6 *If $A \to B$ is flat and I_1 and I_2 are ideals of A, then $B(I_1 \cap I_2) = BI_1 \cap BI_2$.*

This result can often be used to show that a given ring homomorphism $A \to B$ is not flat (Exercise 4). Another result, which plays a role in connection with the Jacobian Conjecture is

Proposition D.1.7 (Bass). *Let $A \subset B$ be a ring extension, where A and B are domains with quotient fields $Q(A)$, resp. $Q(B)$. Suppose that A is a UFD and that $B^* \cap A \subset A^*$. If B is flat over A then $Q(A) \cap B = A$.*

Proof: Let $b \in B \cap Q(A)$. So $b = p/q$ with $p, q \in A$, $q \neq 0$ and $\gcd(p, q) = 1$. Since $\gcd(p, q) = 1$ the map $.p : A/qA \to A/qA$ is injective. Since B is flat over A we deduce that the induced map

$$1 \otimes .p : B \otimes_A A/qA \to B \otimes_A A/qA$$

is also injective. So the map $.p : B/qB \to B/qB$ is injective. Since $p = bq$ it follows that this last map is the zero-map. The injectivity then implies that $B/qB = 0$, i.e. $B = qB$. In particular $q \in B^* \cap A \subset A^*$. So $q \in A^*$, whence $b = p/q \in A$, as desired. □

Flatness is a local property. This is expressed in the following result.

Proposition D.1.8 *Let N be a B-module. Then N is a flat A-module if and only if $N_{\mathfrak{q}}$ is flat over $A_{\mathfrak{p}}$ (where $\mathfrak{p} := f^{-1}(\mathfrak{q})$) for every prime ideal $\mathfrak{q}$ of B (or the same condition for every maximal ideal $\mathfrak{q}$ of B.)*

For the (easy) proof we refer to [203], Theorem 7.1.

Since this result reduces the study of flat modules to that over local rings, D.1.8 combined with the following result gives a powerful tool to study flatness.

Theorem D.1.9 *Let $(A, \mathfrak{m})$ and $(B, \mathfrak{n})$ be Noetherian local rings, $\hat{A}$ and $\hat{B}$ their respective completions and $f : A \to B$ a **local** homomorphism, i.e. $f(\mathfrak{m}) \subset \mathfrak{n}$. Let M be a finite B-module and $\hat{M} := \hat{B} \otimes_B M$. Then M is flat over A if and only if $\hat{M}$ is flat over A if and only if $\hat{M}$ is flat over $\hat{A}$.*

For a proof we refer to [203], Theorem 22.4.

For local rings flatness is more accessible, namely we have

Theorem D.1.10 *Let A be a local ring and M an A-module of **finite presentation**, i.e. of the form A^n/M_1 for some n and some finitely generated A-module M_1. Then M is flat over A if and only if M is a free A-module.*

The proof follows from [203], Corollary of Theorem 7.12 and [203] Theorem 2.5.

D.2 Flat morphisms between affine algebraic varieties

From now on k will denote an algebraically closed field and X and Y will be affine algebraic varieties over k.

Let $F : X \to Y$ be a regular map. So we get the induced map $F^* : A(Y) \to A(X)$. Then F is called **flat** if and only if for all $x \in X$, $A(X)_x$ is flat over $A(Y)_{f(x)}$. Using the Nullstellensatz and D.1.8 one deduces that F is flat if and only if F^* is flat.

Theorem D.2.1 *Let X, Y be irreducible varieties over k and $F : X \to Y$ regular. Assume that Y is regular and X Cohen-Macaulay. Then F is flat if and only if every fiber of F has dimension* $\dim X - \dim Y$.

For the proof we refer to [203], Theorem 15.1 and Theorem 23.1. In fact Matsumura formulates this result for irreducible k-schemes, where k is an arbitrary field. Now we will show how to construct non-trivial flat maps from k^n to k^n. Therefore we need one more result, for the proof of which we refer to [223].

Theorem D.2.2 (Nagata). *Let A be any commutative ring and $A[T]$ the polynomial ring in one variable T over A. Let $f(T) \in A[T]$ and denote by J the ideal in A generated by the coefficients of $f(T)$. Then $A[T]/(f(T))$ is flat over A if and only if J is generated by an idempotent, i.e. $J = (e)$ for some $e \in A$ with $e^2 = e$.*

Construction of flat maps from k^n to k^n

Let $(F_1, \ldots, F_{n+1}) \in \mathrm{Aut}_k\, k^{n+1}$ and put $f := F_{n+1}$. Then

$$k[X_1, \ldots, X_{n+1}]/(f) \simeq k[F_1, \ldots, F_n] \simeq A := k[X_1, \ldots, X_n].$$

So $V(f) \simeq k^n$. Expand f in powers of $T := X_{n+1}$, say $f = \sum_{i=0}^{d} f_i T^i$ with each f_i in A. By Nagata's theorem $A[T]/(f)$ is flat over A if and only if the f_i generate the unit ideal in A and the f_i have no common zero in k^n (Nullstellensatz). Consequently, since both A and $A[T]/(f)$ are isomorphic to $k[X_1, \ldots, X_n]$, polynomials f such that $V(f_0, \ldots, f_d)$ is empty give rise to flat maps from k^n to k^n. To illustrate this procedure we give

Example D.2.3 (A non-surjective flat map from k^2 to k^2).

Consider the Nagata map

$$(F_1, F_2, F_3) = (X - 2aY - a^2 Z, Y + aZ, Z)$$

where $a := XZ + Y^2$. Put $f := F_2 + 1 = XZ^2 + Y^2 Z + (Y + 1) \in k[X, Y][Z]$. Then $(f_2, f_1, f_0) = (1)$, so $k[X, Y, Z]/(f)$ is flat over $k[X, Y]$. Furthermore the corresponding map $V(f) \to k^2$, $(x, y, z) \to (x, y)$ is not surjective since $(0, 0)$ is not in the image (for $f(0, 0, z) = 1$ for all $z \in k$). Since $V(f) \simeq k^2$ we get a non-surjective flat map from k^2 to k^2.

To get an explicit description of this map, observe that the isomorphism between $k[X, Y, Z]/(f)$ and $k[F_1, F_3]$ is given by

$$\overline{X} \to H_1(F_1, 0, F_3),\ \overline{Y} \to H_2(F_1, 0, F_3),\ \overline{Z} \to H_3(F_1, 0, F_3)$$

where (H_1, H_2, H_3) is the inverse of (F_1, f, F_3). Since

$$H_1 = (X + 2aY - a^2 Z)_{|Y:=Y-1} \text{ and } H_2 = (Y - aZ)_{|Y:=Y-1}$$

one easily obtains that the desired description of our map from k^2 to k^2 is given by

$$(x, y) \to (x - 2(xy+1) - (xy+1)^2 y, -1 - (xy+1)y).$$

It is left to the reader to fill in the details. See also Exercise 7.

Final remark. For interesting results concerning testing of flatness we refer to the recent paper [287] of Vasconcelos.

Exercises for §D.2

1. Let S be a multiplicatively closed subset of a commutative ring A. Show that $A \to S^{-1}A$ is faithfully flat if and only if $S \subset A^*$.
2. Let M be an A-module. Show that M is faithfully flat over A if and only if $\mathfrak{m}M \neq M$ for all maximal ideals $\mathfrak{m}$ in A. Deduce that $A[[X_1, \ldots X_n]]$ is faithfully flat over A.
3. Give a proof of D.1.3. [Use $N \otimes_B (B \otimes_A M) \simeq N \otimes M$, for all A-modules M and B-modules N.]
4. Let $A = k[X^2]$, $B = k[X^2, X^3]$ and $N = k[X]$, where k is a field.

   (i) Show that both N and B are free A-modules of rank 2.
   (ii) Use D.1.6 to show that N is not flat over B. [Take $I_1 = X^2k[X]$ and $I_2 = X^3k[X]$.]

5. Let k be a field. Show that $k[X, XY] \subset k[X, Y]$ is not flat. [Use D.1.7.]
6. Let k be an algebraically closed field and $F : k^n \to k^n$ a Keller map. It is shown in Chapter 1 that $k[F] \subset k[X]$ is flat. Give a new proof of this flatness result based on the formal inverse function theorem. [Use D.1.9, D.1.8 and the Nullstellensatz.]
7. Fill out the details of D.2.3. Using the explicit description of the map given in D.2.3, show that it is a flat map. [Show that each fiber has at most two elements, whence all fibers have dimension zero. Then apply D.2.1.]

Appendix E $\mathfrak{D}$-modules

E.1 A_n-modules

In this section we give some basic definitions and results from the theory of $\mathfrak{D}$-modules. For the proofs of all results below we refer the reader to [64], [40], [242], [261] and [42].

We will describe results concerning differential operators with polynomial coefficients, although there are also many results concerning rings of differential operators whose coefficients are formal or convergent power series.

Starting point is the ring of differential operators over the polynomial ring $k[X] := k[X_1, \ldots, X_n]$, where k is a field of characteristic zero.

Definition E.1.1 The n-th **Weyl Algebra** over k, denoted $A_n(X)$ or A_n is the subring of $\operatorname{End}_k k[X]$ generated by the multiplication maps $a.$, where $a \in k[X]$ (which send each element g in $k[X]$ to ag) and the k-derivations on $k[X]$. Notation: $A_n = k[X_1, \ldots, X_n, \partial_1, \ldots, \partial_n]$, where $\partial_i := \frac{\partial}{\partial X_i}$.

One easily verifies the commutator relations

$$[\partial_i, X_j] = \delta_{i,j}, [\partial_i, \partial_j] = 0, [X_i, X_j] = 0, \text{ for all } i, j.$$

Due to these equations one can write every element $P \in A_n$ uniquely as a finite sum of the form

$$P = \sum_{|\alpha| \leq m} a_\alpha(X) \partial^\alpha$$

where $\partial^\alpha = \partial_1{}^{\alpha_1} \ldots \partial_n{}^{\alpha_n}$, $a_\alpha = a_{\alpha_1,\ldots,\alpha_n}(X) \in k[X]$ and $|\alpha| = \alpha_1 + \cdots + \alpha_n$. If $P \neq 0$, then the **order** of P is the maximum of all $|\alpha|$ with $a_\alpha \neq 0$. We define the order of 0 as 0. A first result one obtains is

Proposition E.1.2 *A_n is a **simple ring**, i.e. every two-sided ideal in A_n is either zero or the whole ring.*

The ring A_n is equipped with the so-called **order filtration**. Namely for each $r \in \mathbb{N}$ define $F(r)$ as the set of all $P \in A_n$ whose order is $\leq r$. Then one easily verifies that we have an ascending chain

$$F(0) \subset F(1) \subset F(2) \subset \cdots$$

of subgroups of A_n satisfying $1 \in F(0)$ and $F(p)F(q) \subset F(p+q)$ for all $p, q \geq 0$. Now suppose that M is a finitely generated left A_n-module, say $M = A_n m_1 + \cdots + A_n m_s$. Then by defining

$$F(r, M) := F(r)m_1 + \cdots + F(r)m_s$$

we get a **filtration on** M, i.e. the $F(r, M)$, $r \geq 0$ form an ascending chain

$$F(0, M) \subset F(1, M) \subset F(2, M) \subset \cdots$$

of M such that $F(p)F(q,M) \subset F(p+q,M)$ for all $p,q \geq 0$. To both filtrations described above one can associate their corresponding graded objects, namely

$$\begin{aligned} \mathrm{gr}_F A_n &:= \oplus_{p\geq 0} F(p)/F(p-1), \quad \text{where } F(-1) := 0. \\ \mathrm{gr}_F M &:= \oplus_{p\geq 0} F(p,M)/F(p-1,M). \end{aligned}$$

Then $\mathrm{gr}_F M$ is a graded module over the graded ring $\mathrm{gr}_F A_n$. Furthermore one easily verifies that $\mathrm{gr}_F A_n \simeq k[X_1,\ldots,X_n,Y_1,\ldots,Y_n]$ (the $2n$-variable polynomial ring), where Y_i corresponds to the class $\partial_i + F(0) \in F(1)/F(0) \subset \mathrm{gr}_F A_n$.

To define the notion of a holonomic A_n-module we need to introduce the characteristic variety of M. Therefore we consider

$$I_M := \mathrm{Ann}_{\mathrm{gr}_F A_n} \mathrm{gr}_F M = \{x \in \mathrm{gr}_F A_n | xm = 0, \text{ for all } m \in \mathrm{gr}_F M\}.$$

So, identifying $\mathrm{gr}_F A_n$ with $k[X,Y]$, we get that I_M is an ideal in $k[X,Y]$. Then the **characteristic variety of** M, denoted $V(M)$, is the affine variety $\subset k^{2n}$ defined by the ideal I_M, i.e.

$$V(M) := V(I_M) \subset k^{2n}.$$

This variety does not depend on the choice of the generators m_i. The starting point of the theory of A_n-modules, due to Bernstein in [34] and [35], is the following inequality, called the **Bernstein inequality**.

Proposition E.1.3 $\dim V(M) \geq n$ *for every non-zero finitely generated left A_n-module M.*

Definition E.1.4 A finitely generated non-zero left A_n- module M is called **holonomic** if $\dim V(M) = n$.

Example E.1.5 The left A_n-module $k[X]$ with the standard A_n-module structure $\partial_i.g = \partial_i(g)$, for all $g \in k[X]$, is a holonomic left A_n-module. Namely one easily verifies by writing $k[X] = A_n.1$ that $V(M) = V(Y_1,\ldots,Y_n)$.

A much less obvious result is

Theorem E.1.6 *Let $0 \neq P \in k[X]$. Then $k[X][P^{-1}]$ equipped with the standard left A_n-module structure, i.e.*

$$\partial_i.gP^{-r} = \partial(g)P^{-r} + (-r)g\partial_i(P)P^{-(r+1)}$$

for all $g \in k[X]$, $r \geq 0$ and all i, is a holonomic left A_n-module. More generally, if M is a left A_n-module which is a finitely generated $k[X][P^{-1}]$-module, then M is a holonomic left A_n-module.

Holonomic A_n-modules have all kinds of nice properties such as

Proposition E.1.7 i) *Let $0 \to M_1 \to M_2 \to M_3 \to 0$ be an exact sequence of finitely generated left A_n-modules. Then M_2 is holonomic if and only if M_1 and M_3 are holonomic.*

ii) *If M is a holonomic left A_n-module, then M is a left A_n-module of finite length.*
iii) *If M is an A_n-module of finite length, then M is a cyclic A_n-module, i.e. $M = A_n m$ for some $m \in M$.*

E.2 Direct and inverse images

A deeper part of the theory concerns the preservation of holonomy under various constructions. To describe these results we need to introduce the so-called direct and inverse images of A_n-modules.

The inverse image

Put $V = k^n$, $W = k^m$ and let $F : V \to W$ be a polynomial map. The coordinates on V we denote by $X_1, \dots, X_n$ and the coordinates on W by $Y_1, \dots, Y_m$. So $A(V) = k[X_1, \dots, X_n]$ and $A(W) = k[Y_1, \dots, Y_m]$. Let M be a left $A_m(Y)$-module on W. We will associate to M a left $A_n(X)$-module on V, denoted $F^*(M)$, as follows: first observe that in particular M is a left $k[Y] = A(W)$-module. Since the map $F : V \to W$ induces a ring homomorphism $F^* : A(W) \to A(V)$ we obtain a left $A(V)$-module by extension of scalars, i.e.

$$F^*(M) := A(V) \otimes_{A(W)} M (= k[X] \otimes_{k[Y]} M).$$

Rewriting gives

$$F^*(M) = (k[X] \otimes_{k[Y]} A_m(Y)) \otimes_{A_m(Y)} M.$$

So if we put

$$\mathfrak{D}_{V \to W} := k[X] \otimes_{k[Y]} A_m(Y)$$

we get

$$F^*(M) = \mathfrak{D}_{V \to W} \otimes_{A_m(Y)} M.$$

On $\mathfrak{D}_{V \to W}$ we get a left $A_n(X)$-module structure by defining

$$\frac{\partial}{\partial X_i}(g \otimes Q) = \frac{\partial g}{\partial X_i} \otimes Q + \sum_{p=1}^{m} g \frac{\partial F_p}{\partial X_i} \otimes \frac{\partial}{\partial Y_p} . Q.$$

This enables us to equip $F^*(M)$ with the structure of a left $A_n(X)$-module in the obvious way. The left $A_n(X)$-module $F^*(M)$ is called the **inverse image of M** under F.

Left versus right A_n-modules

Let M be a left A_n-module. Then M can be given the structure of a right A_n-module, denoted M^t, by putting

$$m.P = P^t m$$

where P^t is the **adjoint operator** of P. The adjoint of P is defined by the rules $(PQ)^t = Q^t P^t$, $(P+Q)^t = P^t + Q^t$, $X_i^t = X_i$ and $\partial_i^t = -\partial_i$. Obviously putting $P.m = m.P^t$ we can also turn any right A_n-module into a left A_n-module, denoted also by M^t. So we can freely move from right to left A_n-modules and vice versa.

The direct image of a right A_n-module

Let M be a right $A_n(X)$-module. Observe that $\mathfrak{D}_{V\to W}$ is an $A_n(X)$-$A_m(Y)$-bimodule. Therefore we put

$$F_*^R(M) := M \otimes_{A_n(X)} \mathfrak{D}_{V\to W}.$$

Using the right $A_m(Y)$-module structure of $\mathfrak{D}_{V\to W}$ we get that $F_*^R(M)$ is a right $A_m(Y)$-module, called the **direct image** of M under F.

The direct image of a left $A_n(X)$-module

Let M be a left $A_n(X)$-module. Then, as described above, M^t is a right $A_n(X)$-module. So we get the right $A_m(Y)$-module $F_*(M^t)$. Turning this into a left $A_m(Y)$-module we get

$$F_*(M) := (F_*^R(M^t))^t.$$

So

$$F_*(M) = (M^t \otimes_{A_n(X)} \mathfrak{D}_{V\to W})^t.$$

The module $F_*(M)$ is called the **direct image of** M under F.

One of the main theorems in the algebraic theory of $\mathfrak{D}$-modules is

Theorem E.2.1 *(Preservation of holonomy)*

i) *If M is a holonomic left $A_n(X)$-module, then $F_*(M)$ is a holonomic left $A_m(Y)$-module.*
ii) *If M is a holonomic left $A_m(Y)$-module, then $F^*(M)$ is a holonomic left $A_n(X)$-module.*

To conclude this section we mention that there is an important special subclass of the class of holonomic modules, namely the **regular holonomic A_n-modules** (also called A_n-modules with regular singularities).

An example of such a module is $k[X]$ with the standard A_n-module structure and more generally all its localizations $k[X][P^{-1}]$ also equipped with the standard A_n-module structure as described above, where P is any non-zero polynomial in $k[X]$.

If $0 \to M_1 \to M_2 \to M_3 \to 0$ is an exact sequence of A_n-modules, then M_2 is regular holonomic if and only if M_1 and M_3 are. So submodules and quotient modules of regular holonomic modules are regular holonomic.

However one of the highlights of the theory of $\mathfrak{D}$-modules is the fact that both the direct image and the inverse image of a regular holonomic A_n-module are again regular holonomic! We refer the reader to [42] for more details.

Appendix F Special examples and counterexamples

During several years of research I encountered many "strange" examples, i.e. examples which were counterexamples to conjectures made or questions posed. Very often these examples could also be used to provide counterexamples to other, completely different questions as well. Therefore I think that it is useful to collect various of these examples at the end of this book. Several appeared throughout the text, others are new. Some are rather easy while others are deep and form solutions to longstanding problems.

The order in which these examples are arranged is according to the chapter to which they are related.

Chapter 1

1.1 Let $k = \mathbb{F}_3$ and $F(X) = X + X^3 \in k[X]$. Then $\det JF = 1$ and $F \circ F(x) = x$ for all $x \in k$. However F is not invertible over k.

1.2 Let $k = \mathbb{R}$ and $F : \mathbb{R} \to \mathbb{R}$ be given by $F(x) = x + x^3$ for all $x \in \mathbb{R}$. Then $\det JF(x) \geq 1$ for all $x \in \mathbb{R}$ and F is bijective. However F has no polynomial inverse.

1.3 Let $R := \mathbb{Z}[Y]/(2Y)$, $a := \overline{Y}$ and $F(X) := X - aX^2$. Then $F'(X) = 1$ but F is not invertible over R (1.1.17).

1.4 (A non-flat polynomial map with rational inverse). Let $F : \mathbb{R}^3 \to \mathbb{R}^3$ be given by

$$F(X_1, X_2, X_3) = (X_1, X_2 - X_1X_3, X_3 + X_1X_2).$$

Then $\det JF(x) \neq 0$ for all $x \in \mathbb{R}^3$ and F has a rational inverse. However F is not flat (1.1.36).

1.5 Let $F : \mathbb{C} \to \mathbb{C}^2$ be given by

$$F(z) = (z^2(z-1), z(z-1)), \text{ for all } z \in \mathbb{C}.$$

Then JF has a left inverse but F is not injective.

1.6 Let $F : \mathbb{C}^2 \to \mathbb{C}$ be given by

$$F(x, y) = x + (x(1 + xy))^2 \text{ for all } (x, y) \in \mathbb{C}^2.$$

Then JF has maximal rank at each point of $\mathbb{C}^2$, however there does not exist a polynomial map $\varphi : \mathbb{C} \to \mathbb{C}^2$ such that $F(\varphi(z)) = z$ for all $z \in \mathbb{C}$.

1.7 Let $n \geq 3$, $r \geq n$ and

$$D = X_1\partial_1 + X_2\partial_2 + (r - n + 2)X_3\partial_3.$$

Then the minimal number of generators of $\mathbb{C}[X]^D$ is equal to r (1.2.28).

Chapter 2

2.1 Let $d \geq 2$ and R a ring containing an element $t \neq 0$ such that $t^2 = 0$. Let

$$F = (X + Y^d, Y) \circ (X, Y + tX^d).$$

Then $F \in \mathrm{Aut}_R\, R[X, Y]$ and $\det JF = 1$. However $\deg F^{-1} = d^2 > 2d - 1 = \deg F$ (2.3.5).

2.2 Let R be a ring containing an element t such that $t^6 \neq 0$ and $t^7 = 0$. Furthermore let

$$\begin{aligned} F_1 &= X - \tfrac{4}{3}t^3X^2 - 2tXY + \tfrac{64}{27}t^6X^3 + \tfrac{8}{3}t^4X^2Y + 4t^2XY^2 + Y^3, \\ F_2 &= Y + \tfrac{8}{3}t^3XY + tY^2. \end{aligned}$$

Then $F = (F_1, F_2) \in \mathrm{Aut}_R\, R[X, Y]$ with $\deg F = 3$ and $\deg F^{-1} = 9$.

Chapter 3

3.1 Let $n \geq 3$ and $F : \mathbb{C}^n \to \mathbb{C}^n$ be given by

$$F = (X_1 + X_1^2X_2, X_2 - X_1X_2^2, X_3, \ldots, X_n).$$

Then $\mathbb{C}[X]/(F_1, \ldots, F_{n-1}) \simeq \mathbb{C}[T]$. However $(F_1, \ldots, F_{n-1})$ is not extendible in the sense of Chapter 3.

3.2 Let R be a ring containing an element $t \neq 0$ such that $t^2 = 0$. Then

$$F = (X_1 + tX_1.X_2 \ldots X_n, X_2, \ldots, X_n)$$

is a polynomial automorphism over R which has the same face polynomials as the identity map (3.2.8).

Chapter 4

4.1 Let $k = \overline{\mathbb{F}_p}$ and $F(x) = x^p$ for all $x \in k$. Then k is algebraically closed, F is an automorphism of k, but F has no polynomial inverse. Furthermore for all $x \in k^n$ we have that $\#F^{-1}(x) = 1$ is strictly smaller than the geometric degree of F, which is p.

4.2 The map $f : \mathbb{C} \to W := V(X^2 - Y^3) \subset \mathbb{C}^2$ given by $f(x) = (x^3, x^2)$ is a homeomorphism between $\mathbb{C}$ and W. Furthermore the coordinate rings of $\mathbb{C}$ and W have the same units, however f is not an isomorphism of algebraic varieties (4.4.1).

Chapter 5

5.1 (A non-tame automorphism of $R[X,Y]$).
Let $R = \mathbb{C}[Z]$ and $F \in \mathrm{Aut}_R\, R[X,Y]$ given by

$$F = (X - 2aY - a^2 Z, Y + Za)$$

where $a = XZ + Y^2$. Then F is a non-tame automorphism of $R[X,Y]$, 5.1.9.

5.2 (A quadratic non-tame automorphism of $R[X,Y]$).
Let $R = \mathbb{C}[Z, W]$ and $F \in \mathrm{Aut}_R\, R[X,Y]$ given by

$$F = (X + W(ZX + WY)^2, Y - Z(ZX + WY)^2).$$

Then F is a non-tame automorphism (Chapter 5, §1, Exercise 6).

5.3 Let R be a commutative ring containing two non-zero elements a, b with $ab = 0$. Then $F = (F_1, F_2) = (X + bY^2, Y + aX^3)$ is an R-automorphism of $R[X,Y]$ such that $\deg F_1 (= 2)$ does not divide $\deg F_3 (= 3)$ (Chapter 5, §1, Exercise 3).

5.4 Let k be a field of characteristic $p > 0$ and $a \in \mathbb{N}$ with $\gcd(a, p) = 1$. Let $f(T) = T^{p^2}$ and $g(T) = T^{ap} + T$. Then $k[f(T), g(T)] = k[T]$, however $\deg f(T)$ does not divide $\deg g(T)$ and $\deg g(T)$ does not divide $\deg f(T)$.

5.5 (Shpilrain, Yu, [269]) Let $k \geq 1$ and $p_0, p_1, \ldots, p_k$ be different prime numbers such that $p_0 > p_1 \ldots p_k$. Let

$$f_i := (X^{p_0} - Y^{p_1 \cdots p_i})^{p_{i+1} \cdots p_k}, \text{ for all } 1 \leq i \leq k.$$

Then

i) $\mathbb{C}[X,Y]/(f_i) \simeq \mathbb{C}[X,Y]/(f_j)$ for all i, j and
ii) If $i \neq j$ there does not exist $\varphi \in \mathrm{Aut}_{\mathbb{C}}\, \mathbb{C}[X,Y]$ such that $f_i = \varphi(f_j)$.

5.7 Let $V \subset \mathbb{C}^3$ be the curve parameterized by $\rho_1(t) = (t^7, t^{11}, t^{13})$ and $W \subset \mathbb{C}^3$ the curve parameterized by $\rho_2(t) = (t^7 + t^{14}, t^{11}, t^{13})$. Let $G : \mathbb{C}^3 \to \mathbb{C}^3$ be given by $G(X,Y,Z) = (X + X^2, Y, Z)$. Then $g := G_{|V}$ induces an isomorphism between V and W which cannot be extended to an isomorphism of $\mathbb{C}^3$ (Example 7, §3, Chapter 5).

Chapter 6

6.1 Let $F = (X - YZ - X^3Y - X^2Z, Y + X^2, Z + X^3) \in \mathbb{F}_5[X,Y,Z]^3$. Then $\det JF = 1$ and $\deg F = 4 (< 5)$, however F is not invertible over $\mathbb{F}_5$.

Chapter 7

7.1 Let $H = (Y, X^2 - W, Y^2, 2XY - Z)$. Then JH is nilpotent and the rows of JH are linearly independent over $\mathbb{C}$.

7.2 Let $H = (W(ZX + WY), -Z(ZX + WY), W^3, 0)$. Then JH is nilpotent, but $J(H \circ H)$ is not nilpotent.

7.3 Let $n \geq 4$, $d := X_3X_1 + X_4X_2$ and $H = (X_4d, -X_3d, 0, \ldots, 0)$. Then JH is nilpotent but JH is not strongly nilpotent (7.4.11).

7.4 Let $n \geq 5$, $H = (H_1, \ldots, H_n)$, where $H_i = 0$ for all $i > 5$ and

$$\begin{aligned} H_1 &= 0, H_2 = X_1X_3, H_3 = X_1X_4 + \frac{1}{2}X_2^2, \\ H_4 &= X_1X_5 - X_2X_3, H_5 = \frac{1}{2}X_3^2. \end{aligned}$$

Then JH is nilpotent but JH is not strongly nilpotent (7.4.10).

7.5 Let $A = \mathbb{Z}[X, Y, Z]/(X^2 + YZ)$, $H_1 = xX_1 + yX_2$, $H_2 = zX_1 - xX_2$ in $A[X_1, X_2]$, where $x = \overline{X}$, $y = \overline{Y}$ and $z = \overline{Z}$. Then $H = (H_1, H_2) \in \overline{H}(2, A)$ but $H \notin H(2, A)$.

7.6 Let $n \geq 5$. Let $F = X + H$ be the cubic homogeneous map given by

$$H = (3X_4^2X_2 - 2X_3X_4X_5, X_4^2X_5, X_4^3, X_5^3, 0, \ldots, 0).$$

Then $\deg F^{-1} = 13 > (\deg F)^{\mathrm{ind}(JH)-1}$, where $\mathrm{ind}(JH)$ is the nilpotency index of JH (7.4.14).

7.7 Let $n \geq 6$. Let $F = X + H$ be the quadratic homogeneous map given by

$$\begin{aligned} H = (&2X_2X_6 - 2X_3^2 - X_4X_5, 2X_3X_5 - X_4X_6, X_5X_6, \\ &X_5^2, X_6^2, 0, \ldots, 0). \end{aligned}$$

Then $\deg F^{-1} = 6 > (\deg F)^{\mathrm{ind}(JH)-1}$ (7.4.15).

7.8 The derivation $D = (1 + YU - ZW)\partial_X + W\partial_Y + U\partial_Z$ on $A := k[X, Y, Z, U, W]$ is locally nilpotent (nice of order ≤ 2) and has no fixed points. However D has no slice in A (Chapter 1, §4, Exercise 2 and 7.3.13).

7.9 Let H be as in 7.1. Then JH is nilpotent, however $H^k \neq 0$ for all $k \geq 1$.

7.10 (Derksen) Consider on $\mathbb{C}^{[8]} := \mathbb{C}[P, X, Y, Z, A, B, C, Q]$ the derivation $D := R^2D_A + RQ^2D_B + Q^4D_C$, where $R := AX + BY + CZ - PQ$ and $D_A := X\partial_P + C\partial_Y - B\partial_Z + Q\partial_A$, $D_B := Y\partial_P + A\partial_Z - C\partial_X + Q\partial_B$ and $D_C := Z\partial_P + B\partial_X - A\partial_Y + Q\partial_C$. Then D is a locally nilpotent derivation on $\mathbb{C}^{[8]}$ satisfying $D^2P = D^2A = D^2B = D^2C = D^2X = D^2Y = D^2Z = DQ = 0$. However, D is not nice (of some order $\leq m$) with respect to $P, X, Y, \ldots, Q$ and $\Delta = \{\partial_P, \partial_X, \partial_Y, \ldots, Q\}$. (Exercise 6, Section 3, Chapter 7).

Chapter 8

8.1 Let $k \in (1, \frac{2}{\sqrt{3}})$ and let $F : \mathbb{R}^2 \to \mathbb{R}^2$ be given by

$$F(x, y) = \left(\frac{-ky^3}{1 + x^2 + y^2}, \frac{kx^3}{1 + x^2 + y^2} \right).$$

Then for every $(x, y) \in \mathbb{R}^2$ the eigenvalues of $JF(x, y)$ have absolute value smaller than 1, however 0 is not a global attractor: in fact $F^4(\frac{1}{\sqrt{k-1}}, 0) = (\frac{1}{\sqrt{k-1}}, 0)$ (8.2.4).

8.2 Let $n \geq 3$ and $F : \mathbb{R}^n \to \mathbb{R}^n$ be given by

$$F(x_1, \ldots, x_n) = \left(\frac{1}{2}x_1 + x_3 d(x)^2, \frac{1}{2}x_2 - d(x)^2, \frac{1}{2}x_3, \ldots, \frac{1}{2}x_n \right)$$

where $d(x) = x_1 + x_3 x_2$. Then for each $x \in \mathbb{R}^n$ all eigenvalues of $JF(x)$ are equal to $\frac{1}{2}$, however the point $x_0 := (\frac{147}{32}, -\frac{63}{32}, 1, 0, \ldots, 0)$ satisfies $\lim_{n\to\infty} ||F^n(x_0)|| = \infty$ (8.3.12).

8.3 Let $n \geq 3$ and $F : \mathbb{R}^n \to \mathbb{R}^n$ be given by

$$F(x_1, \ldots, x_n) = (-x_1 + x_3 d(x)^2, -x_2 - d(x)^2, -x_3, \ldots, -x_n)$$

where $d(x) = x_1 + x_3 x_2$. Then for each $x \in \mathbb{R}^n$ all eigenvalues of $JF(x)$ are equal to -1, however the solution $x(t) = (18e^t, -12e^{2t}, e^{-t}, \ldots, e^{-t})$ tends to infinity if t tends to infinity (8.3.11).

8.4 Let $n \geq 5$ and $F : \mathbb{C}^n \to \mathbb{C}^n$ be defined by

$$F = (X_1 + X_2 X_5^2, X_2 + X_1^2 X_5 - X_4 X_5^2, X_3 + X_2^2 X_5, \\ X_4 + 2X_1 X_2 X_5 - X_3 X_5^2, X_5, \ldots, X_n).$$

Then F is invertible and for every non-zero complex number s, $|s| \neq 1$, sF is not globally analytically linearizable (8.4.1).

Chapter 9

9.1 Let $D_1 = -P_Z \partial_Y + P_Y \partial_Z$, where $P = Y + \frac{1}{4}(XZ + Y^2)^2$ and $D_2 = X\partial_Y + (1 + Y^3)\partial_Z$ on $\mathbb{C}[X, Y, Z]$. Then both derivations have the same fixed point sets. Furthermore D_2 is of triangular form, however D_1 is not triangulable.

9.2 Let $D = (1 + Z^2)\partial_X + Z\partial_Y + W\partial_Z$. Then D is locally nilpotent, fixed point free, but has no slice in $\mathbb{C}[X, Y, Z, W]$ (9.5.25).

9.3 Let $D = (XZ + Y^2)(X\partial_Y - 2Y\partial_Z)$ on $\mathbb{C}[X, Y, Z]$. Then D is locally nilpotent and $F = \exp D$ is not triangulable. In particular the corresponding G_a-action on $\mathbb{C}^3$ is not triangulable (9.5.11).

9.4 Let $K = Q(\mathbb{C}[X, Y, Z, W]/(X^2 - a(W)Y^2 - f(W, Z)))$, where $f(W, Z) \in \mathbb{C}[W, Z]$ is irreducible of degree 3 in Z and $a(W) = \operatorname{disc}_Z f(W, Z) \in \mathbb{C}[W]\setminus\{0\}$ (the Z-discriminant of f) is square-free of degree ≥ 5. Then K is not rational over $\mathbb{C}$, however $\mathbb{C}(T_1, T_2, T_3)$ is $\mathbb{C}$-isomorphic to a rational function field in six variables (9.5.29).

9.5 Let D be the locally nilpotent derivation $T^3\partial_X - T^2\partial_Y$ on $\mathbb{C}[T][X, Y]$. Let D_0 be the restriction of D to $R[X, Y]$, where $R = \mathbb{C}[T^2, T^3]$. Then $R[X, Y]^{D_0}$ is not a finitely generated R-algebra. In particular both $R[X, Y]^{D_0}$ and $Q(R[X, Y]^{D_0}) \cap \mathbb{C}[T^2, T^3, X, Y]$ are not finitely generated as $\mathbb{C}$-algebras (Chapter 9, §5, Exercise 9).

9.6 (Daigle, Freudenburg) Let $D = X^3\partial_S + S\partial_T + T\partial_U + X^2\partial_V$ on $R = \mathbb{C}[X, S, T, U, V]$. Then D is locally nilpotent and R^D is not a finitely generated $\mathbb{C}$-algebra. In particular the ring of invariants of the corresponding G_a-action on $\mathbb{C}^5$ and the ring $Q(R^D) \cap R$ are both not finitely generated as $\mathbb{C}$-algebras ([71]).

9.7 Let $F : \mathbb{F}_p^2 \to \mathbb{F}_p^2$ be given by

$$F(X, Y) = (X, Y + X(X - 1)).$$

Then $F^p = I$ but F is not linearizable since $\operatorname{Fix} F = \{0\} \times \mathbb{F}_p \cup \{1\} \times \mathbb{F}_p$ is not isomorphic to a linear subspace of $\mathbb{F}_p^2$.

9.8 (Freudenburg) Let $F_1 = XZ + Y^2$, $F_2 = ZF_1^2 + 2X^2YF_1 - X^5$ and $D : \mathbb{C}[X, Y, Z] \to \mathbb{C}[X, Y, Z]$ be defined by $D(g) = \det J(F_1, F_2, g)$ for all $g \in \mathbb{C}[X, Y, Z]$. Then D is locally nilpotent and has no coordinate in its kernel ([130]).

9.9 (Danielewski) Let V (resp. W) be the zero set of $f := XY - Z^2 + 1$ (resp. $g := X^2Y - Z^2 + 1$) in $\mathbb{C}^3$. Then the algebraic varieties V and W are not isomorphic, however $V \times \mathbb{C} \simeq W \times \mathbb{C}$ as algebraic varieties (Exercise 4, Chapter 9, §4).

Chapter 10

10.1 Let $t = XY - 1$, $h = t(Xt + 1)$, $f = (Xt + 1)^2(\frac{h+1}{X})$,

$$\begin{aligned} p &= f + h \\ q &= -t^2 - 6th(h+1) - 170fh - 91h^2 - 195fh^2 \\ &\quad -69h^3 - 75h^3f - \frac{75}{4}h^4. \end{aligned}$$

Then $F = (p, q) : \mathbb{R}^2 \to \mathbb{R}^2$ satisfies $\det JF > 0$ on $\mathbb{R}^2$ and $F(1, 0) = F(-1, -2) = (0, -1)$, so F is not injective (Chapter 10, §1).

10.2 Let $n \geq 3$ and $\varphi \in \mathrm{End}_{\mathbb{R}}\, \mathbb{R}[X]$ given by

$$\begin{aligned}\varphi(X_1) &= X_1 + X_2X_3, \varphi(X_2) = X_2 - X_1X_3 \text{ and} \\ \varphi(X_i) &= X_i \text{for all } i \geq 3.\end{aligned}$$

Then φ sends each linear coordinate to a coordinate, however φ is not an automorphism (10.5.7).

10.3 Let $F : \mathbb{R}^2 \to \mathbb{R}^2$ be given by the analytic map

$$f(z) = \sum_{n \geq 0} z^{2n+1}/(n!(2n+1)), \text{ for all } z \in \mathbb{C}.$$

Then F is a surjective, non-injective local homeomorphism (Exercise 6, §1, Chapter 10).

10.4 (An open embedding of $\mathbb{C}^2$ in a two dimensional affine variety)

Let $V := \{(s, t, u) \in \mathbb{C}^3 | su - t^2 + t = 0\}$ and $F : \mathbb{C}^2 \to V$ be given by

$$F(x, y) := (y, xy, x^2y - x).$$

Then F is an open embedding of $\mathbb{C}^2$ in V. More precisely

$$\mathbb{C}^2 \simeq \{(s, t, u) \in V | s \neq 0\} \cup \{(s, t, u) \in V | t \neq 1\}.$$

10.5 Let $F : \mathbb{R}^2 \to \mathbb{R}^2$ be given by

$$F(x, y) = (f(x, y), g(x, y)) = (e^{2x} - y^2 + 3, 4e^{2x}y - y^3).$$

Then both f_x and $\det JF$ are positive on $\mathbb{R}^2$. However $F(0, 2) = F(0, -2) = (0, 0)$, so F is not injective (Chapter 10, §1).

Appendix B

B.1 (Drużkowski) Let

$$\begin{aligned}F = (((x-1)(x-2)\ldots(x-5))^2 + ((y+1)(y+2)\ldots \\ (y+5))^2, z(y+1), z).\end{aligned}$$

Then the fiber above $(0, 0, 0)$ has 25 points, however $\deg F_1 \cdot \deg F_2 \cdot \deg F_3 = 10 \cdot 2 \cdot 1 = 20$.

Appendix D

D.1 (A non-surjective flat map from $\mathbb{C}^2 \to \mathbb{C}^2$)

Let $F : \mathbb{C}^2 \to \mathbb{C}^2$ be given by

$$F(X, Y) = (X - 2(XY + 1) - (XY + 1)^2Y, -1 - (XY + 1)Y).$$

Then F is flat (or equivalently has finite fibers), however $(0, 0) \notin \mathrm{Im}\, F$.

D.2 Let $A \subset B \subset C$ be given by

$$A = \mathbb{C}[X^2], B = \mathbb{C}[X^2, X^3], C = \mathbb{C}[X].$$

Then B and C are free A-modules of rank 2, however C is not flat over B.

D.3 (Eisenbud, Huneke) Let $A = \mathbb{C}[X_1, X_2 + X_1X_2^2] \subset B = \mathbb{C}[X_1, X_2]$. Then B is faithfully flat over A but not a projective A-module.

Bibliography

[1] S. Abhyankar, *Resolution of singularities of Embedded Algebraic Surfaces*, Academic Press, New York, 1966

[2] ———, *Lectures in algebraic geometry*, Notes by Chris Christensen, 1974

[3] ———, *Expansion Techniques in Algebraic Geometry*, Tata institute of fundamental research, Tata Institute, 1977

[4] ———, *On the semigroup of a meromorphic curve*, Kinokuniya Book-Store, Tokyo, 1978

[5] ———, Algebraic geometry for scientists and engineers, *Mathematical Surveys*, Vol. 35, AMS, 1990

[6] S. Abhyankar and W. Li, On the Jacobian Conjecture: a new approach via Gröbner bases, *J. of Pure and Applied Algebra*, 61(1989), 211–222

[7] S. Abhyankar and T. Moh, Newton-Puisseux expansion and generalized Tschirnhausen transformation, *J. Reine Angew. Math.*, 260(1973), 47–83

[8] ———, Newton-Puisseux expansion and generalized Tschirnhausen transformation, *J. Reine Angew. Math.*, 261(1973), 29–54

[9] ———, Embeddings of the line in the plane, *J. Reine Angew. Math.*, 276(1975), 148–166

[10] K. Adjamagbo, On separable algebras over a U.F.D. and the Jacobian Conjecture in any characteristic, In Essen [108], *Proceedings of the conference 'Invertible Polynomial Maps'*, pp. 89–104

[11] ———, Sur les endomorphisms injectifs et les automorphisms des schemas, Tech. Report 112, *Publ. Math. de l'Univ. Pierre et Marie Curie*, 1995

[12] ———, Sur les morphisms injectifs et les isomorphisms des variétés algébriques affines, Tech. Report 111, *Publ. Math. de l'Univ. Pierre et Marie Curie*, 1995

[13] K. Adjamagbo and P. Boury, A resultant criterion and formula for the inversion of a rational map in two variables, *J. of Pure and Applied Algebra*, 79(1992), 1–13

[14] K. Adjamagbo, H. Derksen and A. van den Essen, *On polynomial maps in positive characteristic and the Jacobian Conjecture*, report 9208, Univ. of Nijmegen, 1992

[15] K. Adjamagbo and A. van den Essen, A resultant criterion and formula for the inversion of a polynomial map in two variables, *J. of Pure and Applied Algebra*, 64(1990), 1–6

[16] ———, A differential criterion and formula for the inversion of a polynomial map in several variables, *J. of Pure and Applied Algebra*, 65(1990), 97–100

[17] J. Alev, A note on Nagata's automorphism, In Essen [108], *Proceedings of the conference 'Invertible Polynomial Maps'*, pp. 215–222

[18] J. Alev and M. Chamarie, Dérivations et Automorphismes de quelques algèbres quantiques, *Comm. in Algebra*, 20(1992), No. 6, 1787–1802

[19] R. Alperin, Homology of the group of automorphisms of $k[X, Y]$, *J. of Pure and Applied Algebra*, 15(1979), 109–115

[20] H. Appelgate and H. Onishi, The Jacobian Conjecture in two variables, *J. of Pure and Applied Algebra*, 37(1985), 215–227

[21] V.I. Arnold, Geometrical Methods in the Theory of Ordinary Differential Equations, *Grundlehren Math. Wiss.*, Vol. 250, Springer-Verlag, New York Heidelberg Berlin, 1983, Chapter 5: Resonances, Poincaré's Theorem, and Siegel's Theorem

[22] M. Atiyah and I. MacDonald, *Introduction to commutative algebra*, Addison-Wesley Publishing Company, 1969

[23] K. Baba and Y. Nakai, A generalization of Magnus' theorem, *Osaka J. Math.*, 14(1977), 403–409

[24] B. Barkee, D.K. Dennis, and S.S.-S. Wang, Automorphisms are determined by their face polynomials, *Arch. Math.*, 55(1990), 429–430

[25] E. Artal Bartolo, P. Cassou-Noguès and I. Luengo Velasco, On polynomials whose fibers are irreducible with no critical points, *Math. Ann.*, 299(1994), 477–490

[26] H. Bass, The Jacobian Conjecture and inverse degrees, arithmetic and geometry, *Prog. Math.* (1983), No. 36, 65–75

[27] ———, A non-triangular action of G_a on A^3, *J. of Pure and Applied Algebra*, 33(1984), 1–5

[28] ———, Differential Structure of Etale Extensions of Polynomial Algebras, *Commutative Algebra* (M. Hochster, C. Huneke and J.D. Sally, eds.), Springer-Verlag, New York, 1989, Proceedings of a Microprogram Held, June 15–July 2, 1987

[29] H. Bass, E. Connell and D. Wright, The Jacobian Conjecture: Reduction of Degree and Formal Expansion of the Inverse, *Bulletin of the American Mathematical Society*, 7(1982), 287–330

[30] H. Bass and D. Wright, Localisation in the K-theory of invertible algebras, *J. of Pure and Applied Algebra*, 9(1976), 89–105

[31] A. Beauville, J.L. Colliot-Thélène, J.J. Sansuc and P. Swinnerton-Dyer, Varietes stablement rationelles non rationelles, *Annals of Mathematics*, 121(1985), 283–318

[32] T. Becker and V. Weispfennig, *Gröbner Bases*, Cambridge Univ. Press, Cambridge, 1993

[33] J. Bernat and J. Llibre, Counterexample to Kalman and Markus-Yamabe conjectures in dimension 4, *Dynamics of Continuous, Discrete and Impulsive systems*, 2(1996), 337–379

[34] J. Bernstein, Modules over a ring of differential operators: study of the fundamental solutions of equations with constant coefficients, *Funct. Anal. Appl.*, 5(1971), 89–101

[35] ———, The analytic continuation of generalized functions with respect to a parameter, *Funct. Anal. Appl.*, 6(1972), 273–285

[36] J. Berson, *Derivations on* $A[X]$, Master's thesis, University of Nijmegen, 1999

[37] J. Berson, A. van den Essen and S. Maubach, *Derivations having divergence zero on* $R[X, Y]$, To appear in Israel J. of Math.

[38] S. Bhatwadekar and A. Dutta, Linear planes over a D.V.R., *J. of Algebra*, 166(1994), 393–405

[39] ———, Kernel of locally nilpotent R-derivations of $R[X, Y]$, *Transactions of the American Mathematical Society*, 349(1997), 3303–3319

[40] J. Björk, Rings of differential operators, *North-Holland Math. Library Series*, Vol. 21, North-Holland, 1979

[41] S. Bochner and W. Martin, *Several complex variables*, Princeton Univ. Press, New Jersey, 1948

[42] A. Borel et al., Algebraic D-modules, *Perspectives in Math.*, Vol. 2, Academic Press, London-New York, 1987

[43] Cambridge University, *Proceedings Int. Congress Math. 1958*, New York, Cambridge Univ. Press, 1966

[44] L.A. Campbell, A condition for a polynomial map to be invertible, *Math. Ann.*, 205(1973), 243–248

[45] ———, Decomposing Samuelson maps, *Linear Algebra and its Applications*, 187(1993), 227–238

[46] L.A. Campbell, Rational Samuelson maps are univalent, *J. of Pure and Applied Algebra*, 92(1994), 227–240

[47] I. Canals and E. Lluis, Acerca de un resultado de Segre, *Anal. Inst. Math. Univ. Nacional Autonoma de Mexico*, 10(1970), 1–15

[48] M. Chardin, Un algorithm pour le calcul du résultant de trois polynomes homogènes en trois variables, Effective methods in algebraic geometry, *Progress in Math.*, Vol. 94, Birkhäuser, 1991

[49] Chau and Nga, *A remark on Yu's theorem*, preprint, 1998

[50] C.C. Cheng, Cubic linear Keller maps, To appear in *J. of Pure and Applied Algebra*

[51] C.C. Cheng and A. van den Essen, Endomorphisms of the Plane Sending Linear Coordinates to Coordinates, To appear in *Proceedings of the American Mathematical Society*, 2000

[52] C.C. Cheng and S.S.-S. Wang, An algorithm that determines whether a polynomial map is bijective, In Essen [108], *Proceedings of the conference 'Invertible Polynomial Maps'*, pp. 169–176

[53] A. Choudary and A. Dimca, Complex hypersurfaces diffeomorphic to affine spaces, *Kodai. Math. J.*, 12(1994), No. 2, 171–178

[54] A. Cima, A. van den Essen, A. Gasull, E.-M.G.M. Hubbers and F. Mañosas, A polynomial counterexample to the Markus-Yamabe Conjecture, *Advances in Mathematics*, 131(1997), No. 2, 453–457

[55] A. Cima, A. Gasull, J. Llibre and F. Mañosas, Global injectivity of polynomial maps via vector fields, In Essen [108], *Proceedings of the conference 'Invertible Polynomial Maps'*, pp. 105–124

[56] A. Cima, A. Gasull and F. Mañosas, *The discrete Markus-Yamabe problem*, Prepublicacions Núm. 26/1995, Universitat Autònoma de Barcelona, 1995

[57] ———, Injectivity of polynomial local homeomorphisms of $\mathbb{R}^n$, *Nonlinear Analysis, Theory, Methods & Applications*, 26(1996), No. 4, 877–885

[58] ———, A polynomial class of Markus-Yamabe counterexamples, *Publicacions Matematiques Univ.* Autonòma de Barcelona, 41(1997), 85–100

[59] ———, The discrete Markus-Yamabe problem, *Nonlinear Analysis*, 35(1999), 343–354

[60] A.M. Cohen et al. (eds.), Algebraic groups, *Lecture Notes in Mathematics*, Vol. 1271, Springer-Verlag, Berlin-Heidelberg-New York, 1987, Proceedings of a symposium in honour of T.A. Springer

[61] P.M. Cohn, On the structure of $G\ell_2$ of a ring, *Publ. I.H.E.S.*, 30(1966), 365–413

[62] E. Connell, On the K-theory of algebras of polynomial extensions, *J. of Pure and Applied Algebra*, 7(1976), 169–174

[63] E. Connell and L. van den Dries, Injective polynomial maps and the Jacobian Conjecture, *J. of Pure and Applied Algebra*, 28(1983), 235–239

[64] S. Coutinho, A primer of algebraic D-modules, *London Math. Soc. Student Texts*, Vol. 33, Cambridge Univ. Press, 1995

[65] D. Cox, J. Little and D. O'Shea, *An introduction to Computational Algebraic Geometry and Commutative Algebra*, Springer-Verlag, New York, Berlin, Heidelberg, 1996

[65a] P. Craighero, A result on m-Flats in $\mathbb{A}_k^n$, Rend. Sem. Mat. Univ. Padova, 75(1986), 39–46

[66] S. Cynk and K. Rusek, Injective endomorphisms of algebraic and analytic sets, *Annales Polonici Mathematici*, 56(1991), No. 1, 29–35

[67] D. Daigle, A necessary and sufficient condition for triangulability of derivations of $k[X, Y, Z]$, *J. of Pure and Applied Algebra*, 113(1996), 297–305

[68] D. Daigle, On some properties of locally nilpotent derivations, *J. of Pure and Applied Algebra*, 114(1997), 221–230

[69] ———, Homogeneous locally nilpotent derivations, *J. of Pure and Applied Algebra*, 128(1998), 109–132

[70] D. Daigle and G. Freudenburg, Locally nilpotent derivations over a UFD and an application to rank two locally nilpotent derivations on $k[X_1, \ldots, X_n]$, *J. of Algebra*, 204(1998), 353–371

[71] ———, A counterexample to Hilbert's fourteenth problem in dimension five, *J. of Algebra*, 221(1999), 528–535

[72] ———, A note on triangular derivations on $k[X_1, X_2, X_3, X_4]$, *To appear in Proceedings of the American Mathematical Society*, 2000

[73] B. Deng, *Automorphic conjugation, global attractor, and the Jacobian conjecture*, preprint 1995

[74] B. Deng, G.H. Meisters and G. Zampieri, Conjugation for polynomial mappings, *(ZAMP) Z. Angew. Math. Phys.*, 46(1995), 872–882

[75] H. Derksen, The kernel of a derivation, *J. of Pure and Applied Algebra*, 84(1993), 13–16

[76] ———, Inverse degrees and the Jacobian Conjecture, *Communications in Algebra*, 22(1994), No. 12, 4793–4794

[77] ———, *More on the hypersurface* $x + x^2y + z^2 + t^3 = 0$ *in* $\mathbb{C}^4$, preprint, 1995

[78] ———, *Constructive invariant theory and nonlinearizable holomorphic group actions*, Ph.D. thesis, Univ. of Basel, 1997

[79] H. Derksen, O. Hadas and L. Makar-Limanov, Newton polytopes of invariants of additive group actions, preprint, 1999, *To appear in J. of Pure and Applied Algebra*

[80] H. Derksen and F. Kutzschebauch, Nonlinearizable holomorphic group actions, *Math. Ann.*, 311(1998), 41–53

[81] J.K. Deveney and D.R. Finston, Rationally triangulable automorphisms, *J. of Pure and Applied Algebra*, 72(1991), 1–4

[82] ———, G_a-actions on $\mathbb{C}^3$ and $\mathbb{C}^7$, *Comm. Alg*, 22(1994), No. 15, 6295–6302

[83] ———, Fields of G_a invariants are ruled, *Canadian Mathematical Bulletin*, 37(1994), 37–41

[84] ———, Algebraic aspects of additive group actions on complex affine space, In Essen [108], *Proceedings of the conference 'Invertible Polynomial Maps'*, pp. 179–190

[85] ———, *A proper* G_a *action on* $\mathbb{C}^5$ *which is not locally trivial*, 1995

[86] J.K. Deveney, D.R. Finston and M. Gehrke, G_a actions on $\mathbb{C}^n$, *Comm. Alg.*, 22(1994), No. 12, 4977–4988

[87] W. Dicks, Automorphisms of the polynomial ring in two variables, *Publ. Sec. Mat. Univ. Autònoma Barcelona*, 27(1983), 155–162

[88] F. Dillen, Polynomials with constant Hessian determinant, *J. of Pure and Applied Algebra*, 71(1991), 13–18

[89] J. Dixmier, Sur les algèbres de Weyl, *Bull. Soc. Math. France*, 96(1968), 209–242

[90] V. Drensky, J. Gutierrez and J.-T. Yu, Gröbner bases and the Nagata automorphism, *J. of Pure and Applied Algebra*, 135(1999), 135–153

[91] V. Drensky and J.-T. Yu, Test polynomials for automorphisms of polynomial and free associate algebras, *J. of Algebra*, 207(1998), 491–510

[92] L. van den Dries and K. McKenna, Surjective polynomial maps and a remark on the Jacobian problem, *Manuscripta Math.*, 67(1990), 1–15

[93] L.M. Drużkowski, An Effective Approach to Keller's Jacobian Conjecture, *Math. Ann.*, 264(1983), 303–313

[94] L.M. Drużkowski, *The Jacobian Conjecture*, preprint 492, Institute of Mathematics, Polish Academy of Sciences, IMPAN, Śniadeckich 8, P.O. Box 137, 00-950 Warsaw, Poland, 1991

[95] ———, The Jacobian Conjecture in case of rank or corank less than three, *J. of Pure and Applied Algebra*, 85(1993), 233–244

[96] L.M. Drużkowski and K. Rusek, The formal inverse and the Jacobian conjecture, *Annales Polonici Mathematici*, 46(1985), 85–90

[97] E. Edo, Variables of $A[x, y]$ and transfert, *To appear in the proceedings of the conference Poly'99 held at Kraków*, 6–11 July 1999

[98] C. Eggermont and A. van den Essen, A class of triangular derivations having a slice, *J. of Pure and Applied Algebra*, 114(1997), 209–215

[99] A. van den Essen, *A criterion to decide if a polynomial map is invertible and to compute the inverse*, report 8653, University of Nijmegen, 1986

[100] ———, A criterion to decide if a polynomial map is invertible and to compute the inverse, *Comm. in Algebra*, 18(1990), No. 10, 3183–3186

[101] ———, *$\mathcal{D}$-modules and the Jacobian Conjecture*, Report 9108, University of Nijmegen, Toernooiveld, 6525 ED Nijmegen, The Netherlands, 1991

[102] ———, Locally finite and locally nilpotent derivations with applications to polynomial flows and polynomial morphisms, *Proceedings of the American Mathematical Society*, 116(1992), No. 3, 861–871

[103] ———, Locally nilpotent derivations and their applications III, *J. of Pure and Applied Algebra*, 98(1993), 15–23

[104] ———, *Conjectures and problems surrounding the Jacobian Conjecture*, Recent Results on the Global Asymptotic Stability Jacobian Conjecture (M. Sabatini, ed.), Matematica 429, Università di Trento, 1993, Extended version as Report 9345, University of Nijmegen, Toernooiveld, 6525 ED Nijmegen, The Netherlands, 1993

[105] ———, An algorithm to compute the invariant ring of a G_a action on an affine variety, *J. Symbolic Computation*, 16(1993), 531–555

[106] ———, A counterexample to a conjecture of Meisters, In Automorphisms of Affine Spaces [108], *Proceedings of the conference 'Invertible Polynomial Maps'*, pp. 231–234

[107] ———, Locally finite and locally nilpotent derivations with applications to polynomial flows and morphisms and G_a-actions II, *Proceedings of the American Mathematical Society*, 121(1994), No. 3, 667–678

[108] A. van den Essen (ed.), Automorphisms of Affine Spaces, Curaçao, Caribbean Mathematics Foundation, Kluwer Academic Publishers, July 4–8 1994, 1995, *Proceedings of the conference 'Invertible Polynomial Maps'*.

[109] ———, A counterexample to a conjecture of Drużkowski and Rusek, *Annales Polonici Mathematici*, LXII.2(1995), 173–176

[110] ———, *Nilpotent Jacobian matrices with independent rows*, Report 9603, University of Nijmegen, Toernooiveld, 6525 ED Nijmegen, The Netherlands, 1996

[111] ———, A counterexample to Meisters' Cubic-Linear Linearization Conjecture, *Indagationes Mathematicae*, New Series 9(1998), No. 3, 333–339

[112] ———, The exponential conjecture and the nilpotency subgroup of the automorphism group of a polynomial ring, *Prepublicacions, Univ. Autònoma de Barcelona*, April 1998

[113] A. van den Essen and E.-M.G.M. Hubbers, Polynomial maps with strongly nilpotent Jacobian matrix and the Jacobian Conjecture, *Linear Algebra and its Applications*, 247(1996), 121–132

[114] A. van den Essen and E.-M.G.M. Hubbers, A new class of invertible polynomial maps, *J. of Algebra*, 187(1997), 214–226

[115] ———, Chaotic Polynomial Automorphisms; counterexamples to several conjectures, *Advances in Applied Mathematics*, 18(1997), 382–388

[116] ———, $\mathcal{D}_n(A)$ for polynomial automorphisms and stable tameness, *J. of Algebra*, 192(1997), 460–475

[117] A. van den Essen and T. Janssen, *Kernels of elementary derivations*, Report 9548, University of Nijmegen, Toernooiveld, 6525 ED Nijmegen, The Netherlands, 1995

[118] A. van den Essen and M. Kwieciński, On the reconstruction of polynomial automorphisms from their face polynomials, *J. of Pure and Applied Algebra*, 80(1992), 327–336

[119] A. van den Essen and T. Parthasarathy, Polynomial maps and a conjecture of Samuelson, *Linear Algebra and its Applications*, 177(1992), 191–195

[120] A. van den Essen and V. Shpilrain, Some combinatorial questions about polynomial mappings, *J. of Pure and Applied Algebra*, 119(1997), 47–52

[121] A. van den Essen and H. Tutaj, A remark of the two-dimensional Jacobian conjecture, *J. of Pure and Applied Algebra*, 96(1994), 19–22

[122] K.-Q. Feng and J.-T. Yu, *New classes of testpolynomials of polynomial algebras*, preprint, 1998

[123] R. Fessler, A solution of the two dimensional Global Asymptotic Jacobian Stability Conjecture, *Annales Polonici Mathematici*, 62(1995), 45–75

[124] D. Finston and S. Walcher, Centralizers of locally nilpotent derivations, *J. of Pure and Applied Algebra*, 120(1997), 39–49

[125] E. Formanek, Two notes on the Jacobian Conjecture, *Arch. Math.*, 49(1987), 286–291

[126] F. Forstneric, J. Globevnik and J. Rosay, Nonstraightenable complex lines in $\mathbb{C}^2$, *Ark. Mat*, 34(1996), 97–101

[127] G. Freudenburg, *On tameness and triangulability of polynomial automorphisms*, preprint

[128] ———, Triangulability criteria for additive group actions on affine space, *J. of Pure and Applied Algebra*, 105(1995), 267–275

[129] ———, A note on the kernel of a locally nilpotent derivation, *Proceedings of the American Mathematical Society*, 124(1996), 27–29

[130] ———, Actions of G_a on $\mathbb{A}^3$ defined by homogeneous derivations, *J. of Pure and Applied Algebra*, 126(1998), 169–181

[131] ———, A counterexample to Hilbert's fourteenth problem in dimension six, *To appear in J. of Transformation Groups*, 2000

[132] J. Fritz, Partial differential equations, fourth ed., *Applied Mathematical Sciences*, Vol. 1, Springer, 1986

[133] T. Fujita, On Zariski problem, *Proc. Japan Acad. Ser. A, Math. Sci*, 55(1979), 106–110

[134] J. Furter, *Sur les automorphisms quadratiques de l'espace affine*, Ph.D. thesis, Univ. de Grenoble, 1995

[135] ———, On the variety of automorphisms of the affine plane, *J. of Algebra*, 195(1997), 604–623

[136] ———, On the degree of the inverse of an automorphism of the affine space, *J. of Pure and Applied Algebra*, 130(1998), 277–292

[137] P. Gabriel and Y. Nouazé, Idéax premiers de l'Algèbre Enveloppante d'une Algèbre de Lie Nilpotente, *J. of Algebra*, 6(1967), 77–99

[138] D. Gale and N. Nikaido, The Jacobian matrix and global univalence of mappings, *Math. Ann.*, 159(1965), 81–93

[139] R. Ganong, On plane curves with one place at infinity, *J. Reine Angew. Math.*, 307/308(1979), 173–193

[140] A.A. Glutsuk, The complete solution of the Jacobian problem for planar vector fields, *Uspeki Mat. Nauk*, 49(1994), No. 3, 179–180, In Russian

[141] K. Goodearl and R. Warfield, An introduction to noncommutative noetherian rings, *London Math. Soc. Student Texts*, Vol. 16, Cambridge University Press, 1989

[142] P. Gordan, Beweis dass jede Covariante und Invariante einer binären Form eine ganze Function mit numerischen Coefficienten solcher Formen ist, *J. Reine Angew. Math.*, 69(1868), 323–354

[143] W. Gordon, On the diffeomorphisms of euclidian space, *American Math. Monthly*, 79(1972), No. 7, 755–759

[144] G. Gorni and G. Zampieri, On the existence of global analytic conjugations for polynomial mappings of Yagzhev type, *J. of Math. Analysis and Applications*, 201(1996), 880–896

[145] ———, On cubic-linear polynomial mappings, *Indagationes Mathematicae*, New Series 8(1997), No. 4, 471–492

[146] A. Grothendieck and J. Dieudonne, Elements de Geométrie Algébrique IV, *Publ. Math. I.H.E.S.*, 24(1965)

[147] C. Gutierrez, A solution to the bidimensional Global Asymptotic Stability Conjecture, *Ann. Inst. H. Poincarré. Anal. Non Linéaire*, 12(1995), No. 6, 627–672

[148] A. Gutwirth, An inequality for certain pencils of plane curves, *Proceedings of the American Mathematical Society*, 12(1961), 631–638

[149] J. Gwoździewicz, Injectivity on one line, *Bull. Soc. Sci. Łódź*, 7(1993), 59–60, Série: Recherches sur les déformationes XV

[150] J. Hadamard, Sur les transformationes ponctuelles, *Bull. Soc. Math. France*, 34(1906), 71–84, Oeuvres pp. 349–363 & 383–384

[151] O. Hadas, On the vertices of newton polytopes associated with an automorphism of the ring of polynomials, *J. of Pure and Applied Algebra*, 76 (1991), 81–86

[152] O. Hadas and L. Makar-Limanov, Newton polytopes of constants of locally nilpotent derivations, preprint, 1999, *To appear in Communications in Algebra.*

[153] P. Hartman, On stability in the large for systems of ordinary differential equations, *Canadian Journal of Mathematics*, 13(1961), 480–492

[154] D. Hilbert, Über die Theorie der algebraische Formen, *Math. Ann.*, 36(1890), 473–534

[155] M. Hirsch and S. Smale, *Differential equations, dynamical systems and linear algebra*, Acad. Press, New York, 1974

[156] E.-M.G.M. Hubbers, *The Jacobian Conjecture: Cubic Homogeneous Maps in Dimension Four*, Master's thesis, University of Nijmegen, Toernooiveld, 6525 ED Nijmegen, The Netherlands, February 17 1994, directed by Essen, A. van den.

[157] ———, *Nilpotent Jacobians*, Ph.D. thesis, Univ. of Nijmegen, 1998

[158] E.-M.G.M. Hubbers and D. Wright, *Stably tame automorphisms*, Report 9817, University of Nijmegen, Toernooiveld, 6525 ED Nijmegen, The Netherlands, 1998

[159] J. Humphreys, Hilbert's fourteenth problem, *American Math. Monthly*, (1978), 341–353

[160] N. Ivanenko, Some classes of linearizable polynomial maps, *J. of Pure and Applied Algebra*, 126(1998), 223–232

[161] Z. Jelonek, The extension of regular and rational embeddings, *Math. Ann.*, 277(1987), 113–120

[162] Z. Jelonek, The extension of polynomial embeddings and the Jacobian Conjecture, *Math. Ann.*, 294(1992), 289–293

[163] ———, The set of points at which a polynomial map is not proper, *Annales Polonici Mathematici*, 58(1993), 259–266

[164] ———, Affine smooth varieties with finite group of automorphisms, *Math. Zeitschrift*, 216(1994), 575–591

[165] ———, *Testing sets for properness of polynomial mappings*, IMUJ preprint, 1996

[166] ———, A hypersurface which has the Abhyankar-Moh property, *Math. Ann.*, 308(1997), 73–84

[167] ———, A solution of the problem of van den Essen and Shpilrain, *J. of Pure and Applied Algebra*, 137(1999), 49–55

[168] H.W.E. Jung, Über ganze birationale Transformationen der Ebene, *J. Reine Angew. Math.*, 184(1942), 161–174

[169] S. Kaliman, Extensions of isomorphisms between affine algebraic subvarieties of k^n to automorphisms of k^n, *Proceedings of the American Mathematical Society*, 113(1991), No. 2, 325–334

[170] ———, On the Jacobian Conjecture, *Proceedings of the American Mathematical Society*, 117(1993), No. 1, 45–51

[171] ———, Smooth contractible hypersurfaces in $\mathbb{C}^n$ and exotic algebraic structures on $\mathbb{C}^3$, *Math. Zeitschrift*, 214(1993), 499–510

[172] ———, Exotic structures on $\mathbb{C}^n$ and $\mathbb{C}^*$-actions on $\mathbb{C}^3$, pp. 299–307 in Proc. on 'Complex Analysis and Geometry'. *Lecture Notes in Pure and Applied Mathematics*, 173, Marcel Dekker Inc., 1995

[173] S. Kaliman, M. Koras, L. Makar-Limanov and P. Russell, $\mathbb{C}^*$-actions on $\mathbb{C}^3$ are linearizable, *Electronic Research Announcements of the A.M.S.*, 3(1997)

[174] T. Kambayashi, Automorphism group of a polynomial ring and algebraic group actions on affine space, *J. of Algebra*, 60(1979), 439–451

[175] M. Kang, On Abhyankar-Moh's epimorphism theorem, *Amer. J. Math.*, 113(1991), 399–421

[176] ———, Injective morphisms of affine varieties, *Proceedings of the American Mathematical Society*, 119(1993), No. 1, 1–4

[177] M.-C. Kang, The biregular cancellation problem, *J. of Pure and Applied Algebra*, 45(1987), 241–253

[178] ———, A note on the birational cancellation problem, *J. of Pure and Applied Algebra*, 77(1992), 141–154

[179] O.H. Keller, Ganze Cremona-Transformationen, *Monatsh. Math. Phys.*, 47(1939), 299–306

[180] M. Kojima and M. Miyanishi, On Robert's counterexample to the fourtheenth problem of Hilbert, *J. of Pure and Applied Algebra*, 122(1998), No. 3, 277–292

[181] M. Koras and P. Russell, *Contractible threefolds and $\mathbb{C}^*$-actions on $\mathbb{C}^3$*, Tech. Report 5, Concordia-Laval-McGill, 1995, CICMA Reports.

[182] H. Kraft, *Geometrische Methoden in der Invariantentheorie*, Vieweg-Verlag, 1984

[183] ———, *A remark on equivariant affine line bundles*, preprint, 1996

[184] H. Kraft, T. Petrie and G.W. Schwarz (eds.), Topological methods in algebraic transformation groups, *Progress in Math.*, Vol. 80, Birkhäuser, Boston-Basel-Berlin, 1989, proceedings of the conference 'Topological Methods in Algebraic Transformation Groups' held at Rutgers University, 4–8 April, 1988

[185] W. van der Kulk, On polynomial rings in two variables, *Nieuw Archief voor Wiskunde*, 3(1953), No. 1, 33–41

[186] K. Kurdyka and K. Rusek, Surjectivity of certain injective semialgebraic transformations of $\mathbb{R}^n$, *Math. Zeitschrift*, 200(1988), 141–148

[187] K. Kwieciński, Automorphisms from face polynomials via Gröbner bases, *J. of Pure and Applied Algebra*, 82(1992), 65–70

[188] M. Kwieciński, A Gröbner basis criterion for isomorphisms of algebraic varieties, *J. of Pure and Applied Algebra*, 74(1991), 275–279

[189] J. Lang and S. Maslamani, Some results on the Jacobian Conjecture in higher dimension, *J. of Pure and Applied Algebra*, 94(1994), 327–330

[190] S. Lang, *Algebra*, Addison-Wesley Publishing Company, 1984

[191] J. LaSalle, The stability of dynamical systems, *CBMS-NSF Regional Conference Series in Applied Math.*, Vol. 25, 1976, second printing 1993

[192] J. Levitzki, Über nilpotente Unterringen, *Math. Ann.*, 105(1931), 620–622

[193] W. Li, On a problem about face polynomials, *J. of Pure and Applied Algebra*, 60(1989), 269–272

[194] W. Li and J.-T. Yu, Reconstructing birational maps from their face functions, *Manuscripta Mathematica*, 76(1992), 353–366

[195] ———, Computing minimal polynomials and the degree of unfaithfulness, *Communications in Algebra*, 21(1993), 3557–3569

[196] L. Moser-Jauslin M. Masuda and T. Petrie, The equivariant Serre Problem for abelian groups, *Topology*, 35(1996), No. 2, 329–334

[197] A. Magnus, On polynomial solutions of a differential equation, *Math. Scand.*, 3(1955), 255–260

[198] L. Makar-Limanov, *On automorphisms of certain algebras*, Ph.D. thesis, Moscow State University, 1970

[199] ———, On the hypersurface $x + x^2y + z^2 + t^3$ in $\mathbb{C}^4$ or a $\mathbb{C}^3$-like threefold which is not $\mathbb{C}^3$, *Israel J. Math.*, 96(1996), 419–429

[200] ———, *Locally nilpotent derivations, a new ring invariant and applications*, Lecture notes, 1998

[201] L. Markus and H. Yamabe, Global stability criteria for differential systems, *Osaka Math. Journal*, 12(1960), 305–317

[202] M. Masuda, L. Moser-Jauslin and T. Petrie, Equivariant algebraic vector bundles over representations of reductive groups: Applications, *Proc. Natl. Acad. Sci. USA*, 88(1991), 9065–9066

[203] H. Matsumura, *Commutative ring theory*, Cambridge University Press, Cambridge, 1989, Cambridge Studies in Advanced Mathematics 8

[204] S. Maubach, *Hilbert* 14 *and related subjects*, Master's thesis, Univ. of Nijmegen, 1998

[205] R. Mauldin (ed.), *The Scottish Book: Mathematics from the Scottish Café*, Birkhäuser, Boston-Basel-Stuttgart, 1979

[206] J. McKay and S.S.-S. Wang, An inversion formula for two polynomials in two variables, *J. of Pure and Applied Algebra*, 40(1986), 245–257

[207] J. McKay and S.S.-S. Wang, An elementary proof of the automorphism theorem for the polynomial ring in two variables, *J. of Pure and Applied Algebra*, 52(1988), 91–102

[208] J. McKay and S.S.-S. Wang, On the inversion formula for two polynomials in two variables, *J. of Pure and Applied Algebra*, 52(1988), 103–119

[209] J. McKay, S.S.-S. Wang and T. Moh, On face polynomials, *J. of Pure and Applied Algebra*, 52(1988), 121–125

[210] G.H. Meisters, Polyomorphisms conjugate to Dilations, In Essen [108], *Proceedings of the conference 'Invertible Polynomial Maps'*, pp. 67–88

[211] ———, *The Cubic-Linear Linearization Conjecture*, This paper is available by World-Wide-Web at `http://www.math.unl.edu/~gmeister`, November 1995

[212] G.H. Meisters and C. Olech, Solution of the Global Asymptotic Stability Jacobian Conjecture for the Polynomial Case, *Analyse Mathématique et Applications*, Gauthier-Villars, Paris, 1988, pp. 373–381

[213] ———, Strong Nilpotence Holds in Dimensions up to Five Only, *Linear and Multilinear Algebra*, 30(1991), 231–255

[214] A. Mikhalev, J.-T. Yu and A. Zolotykh, Images of coordinate polynomials, *Algebra Colloquium*, 4(1997), No. 2, 159–162

[215] M. Miyanishi, Analytic irreducibility of certain curves on a nonsingular affine rational surface, *Proceedings of the International Symposium on Algebraic Geometry*, Kyoto 1977 (Tokyo), Kinokuniya, 1978, pp. 575–587

[216] ———, *Curves on rational and unirational surfaces*, Tata Institute of Fundamental Research 60, Tata Institute, 1978

[217] ———, *Normal affine subalgebras of a polynomial ring*, Algebraic and Topological Theories - to the memory of Dr. Takehiko Miyata, Kinokuniya, Tokyo, 1985, pp. 37–51

[218] M. Miyanishi and T. Sugie, Affine surfaces containing cylinderlike open sets, *J. Math Kyoto Univ.*, 20(1980), 11–42

[219] T. Moh, On the global Jacobian Conjecture and the configuration of roots, *J. Reine Angew. Math.*, 340(1983), 140–212

[220] M. Nagata, a revised version of both [225] and [226]

[221] ———, *Lectures on the Fourteenth Problem of Hilbert*, Tata Institute of Fundamental Research 31, Tata Institute, 1965

[222] ———, On the fourteenth problem of Hilbert, In *Proceedings Int. Congress Math. 1958*, [43], pp. 459–462

[223] ———, Flatness of an extension of a commutative ring, *J. Math. Kyoto Univ.*, 9(1969), 439–448

[224] ———, On the automorphism group of $k[X, Y]$, *Kyoto Univ. Lectures in Math. 5*, Kyoto University, Kinokuniya – Tokyo, 1972

[225] ———, *Two dimensional jacobian conjecture*, Proceedings of the 3rd KIT Mathematics Workshop (Taejŏn) (M.H. Kim and K.H. Ko, eds.), Korean Institute of Technology, 1988, pp. 77–98

[226] ———, Some remarks on the two-dimensional Jacobian Conjecture, *Chin. J. Math.*, 17(1989), 1–7

[227] ———, *Field theory*, Marcel Dekker, Inc., New York and Basel, 1997

[228] M. Nagata and A. Nowicki, Rings of constants for k-derivations in $k[x_1, \ldots, x_n]$, *J. Math. Kyoto Univ.*, 28(1988), No. 1, 111–118

[229] P. Nousiainen and M. Sweedler, Automorphisms of polynomial and power series rings, *J. of Pure and Applied Algebra*, 29(1983), 93–97

[230] A. Nowicki, Commutative basis of derivations in polynomial and power series rings, *J. of Pure and Applied Algebra*, 40(1986), 279–283

[231] A. Nowicki, *Polynomial derivations and their rings of constants*, Univ. of Toruń, 1994

[232] A. Nowicki, Richman's proof of the Abhyankar-Moh theorem (in Polish), *Materialy XVI konferencji Szkoleniowej z Analizy i Geometrii Zespolonej, Łódź*, 49–71, 1995

[233] A. Nowicki and J.-M. Strelcyn, Generators of rings of constants for some diagonal derivations in polynomial rings, *J. of Pure and Applied Algebra*, 101(1995), 207–212

[234] S. Oda, *The Jacobian problem and the simply-connectedness of* $\mathbb{A}^n$ *over a field k of characteristic zero*, Osaka Univ. preprint, 1980

[235] C. Olech, On the global stability of an autonomous system on the plane, *Contributions to Diff. Eq.*, 1(1963), 389–400

[236] ———, *On Markus-Yamabe stability conjecture*, International Meeting on Ordinary Differential Equations and Their Applications, University of Florence, Florence, Italy, September 20–24 1993, 1995, pp. 127–137

[237] H. Park and C. Woodburn, An algorithmic proof of Suslin's stability theorem for polynomial rings, *J. of Algebra*, 178(1995), 277–298

[238] T. Parthasarathy, *On Global Univalence Theorems*, LNM, Vol. 977, Springer-Verlag, Berlin-Heidelberg-New York, 1983

[239] R. Peretz, *Polynomially Parametrizable Surfaces*, preprint, 1996

[240] ———, The variety of asymptotic values of a real polynomial etale map, *J. of Pure and Applied Algebra*, 106(1996), No. 1, 103–112

[241] ———, Mapping with composite part and with constant Jacobian, *Applied Math. Letters*, 3(1998), 39–43

[242] F. Pham, Singularités des systemes différentieles de Gauss-Manin, *Progress in Math*, Vol. 2, Birkhäuser, Boston, 1979

[243] S. Pinchuk, A counterexample to the real Jacobian Conjecture, *Math. Zeitschrift*, 217(1994), 1–4

[244] A. Płoski, Algebraic dependence of polynomial automorphisms, *Bulletin of the Polish Academy of Sciences*, 34(1986), No. 11, 653–659

[245] V.L. Popov, On actions of G_a on $\mathbf{A}^n$., In Cohen et al. [60], *Proceedings of a symposium in honour of T.A. Springer*

[246] C. Ramanujam, A topological characterization of the affine plane as an algebraic variety, *Annals of Mathematics*, 94(1971), 69–88

[247] J.D. Randall, The real Jacobian problem, *Proceedings of Symposia in Pure Mathematics*, 40(1983), 411–414

[248] M. Razar, Polynomial maps with constant jacobian, *Israel J. of Math.*, 32(1979), 97–106

[249] D. Rees, On a problem of Zariski, *Illinois J. of Math.*, 2(1958), 145–149

[250] R. Rentschler, Opérations du groupe additif sur le plan, *C.R.Acad.Sci.Paris*, 267(1968), 384–387

[251] D. Richman, On the computation of minimal polynomials, *J. of Algebra*, 103(1986), 1–17

[252] P. Roberts, An infinitely generated symbolic blow-up in a power series ring and a new counterexample to Hilbert's fourteenth problem, *J. of Algebra*, 132(1990), 461–473

[253] J.-P. Rosay and W. Rudin, Holomorphic maps from $\mathbb{C}^n$ to $\mathbb{C}^n$, *Transactions of the American Mathematical Society*, 310(1988), 47–86

[254] L. Rudolph, Embeddings of the line in the plane, *J. Reine Angew. Math.*, 337(1982), 113–118

[255] K. Rusek, Polynomial Automorphisms, preprint 456, Institute of Mathematics, *Polish Academy of Sciences*, IMPAN, Śniadeckich 8, P.O. Box 137, 00-950 Warsaw, Poland, May 1989

[256] K. Rusek and T. Winiarski, Polynomial automorphisms of $\mathbb{C}^n$, *Univ. Iagell. Acta Math.*, 24(1984), 143–149

[257] P. Russell, Simple birational extensions of two dimensional affine rational domains, *Compositio Mathematica*, 33(1976), 197–208

[258] ———, On a class of $\mathbb{C}^3$-like threefolds, *Preliminary Report*, 1992

[259] P.A. Samuelson, Prices of factors and goods in general equilibrium, *Rev. Econom. Stud.*, 21, 1953

[260] A. Sathaye, On linear planes, *Proceedings of the American Mathematical Society*, 56(1976), 1–7

[261] P. Schapira, *Microdifferential systems in the complex domain*, Grundlehren der mathematischen Wissenschaften, Vol. 269, Springer-Verlag, Berlin-Heidelberg, 1985

[262] G. Schwarz, Exotic algebraic group actions, *C.R. Acad. Sci. Paris*, 309(1989), 89–94

[263] B. Segre, Forme differentziali e loro integrali, Vol. II, *Docet, Roma*, 1956

[264] ———, Corrispondenze di mobius e transformazioni cremoniane intere, *Classe di Scienzo Fisiche. mat. e Naturel. 91, Atti della acc. della Scienze di Torino*, 1956–1957

[265] A. Seidenberg, Derivations and integral closure, *Pacific Journal*, 16(1966), 167–173

[266] J.-P. Serre, *Trees*, Springer-Verlag, Berlin-Heidelberg-New York, 1980

[267] C.S. Seshadri, Quotient Spaces modulo Reductive Algebraic Groups, *Annals of Mathematics*, 95(1972), 511–556

[268] I.R. Shafarevich, On some infinite dimensional groups, *Rend. Math. e Appl.*, 25(1966), 208–212

[269] V. Shpilrain and J.-T. Yu, Embeddings of curves in the plane, *J. of Algebra*, 217(1999), 668–678

[270] S. Smale, Mathematical problems for the next century, *Math. Intellingencer*, 20(1998), No. 2, 7–15

[271] M.K. Smith, Stably tame automorphisms, *J. of Pure and Applied Algebra*, 58(1989), 209–212

[272] D.M. Snow, *Unipotent actions on affine space*, In Kraft et al. [184], proceedings of the conference 'Topological Methods in Algebraic Transformation Groups' held at Rutgers University, 4–8 April, 1988

[273] T.A. Springer, *Linear Algebraic Groups*, Birkhäuser Verlag, 1981

[274] ———, *Aktionen Reduktiver Gruppen auf Varietäten*, Algebraische Transformationsgruppen und Invariantentheorie (H. Kraft, P. Slodowy and T.A. Springer, eds.), Birkhäuser Verlag, 1989

[275] V. Srinivas, On the embedding dimension of an affine variety, *Math. Ann.*, 289(1991), 125–132

[276] Y. Stein, On linear differential operators related to the Jacobian Conjecture, *J. of Pure and Applied Algebra*, 57(1989), 175–186

[277] ———, Linear differential operators related to the Jacobian Conjecture have a closed image, *J. D'Analyse Mathématique*, 54(1990), 237–245

[278] ———, Derivations generated by polynomials, their images and complements of the images, In Essen [108], *Proceedings of the conference 'Invertible Polynomial Maps'*, pp. 137–144

[279] ———, The Jacobian problem as a system of ordinary differential equations, *Israel Journal of Mathematics*, 89(1995), 301–319

[280] I. Stewart, *Galois theory*, Chapman and Hall, London, 1973

[281] J. Strooker, *Homological questions in local algebra*, LNS, Vol. 145, Cambridge Univ. Press, 1990

[282] A. Suslin, On the structure of the special linear group over polynomial rings, *Izv. Akad. Nauk. S.S.S.R., Ser. Mat.*, 41(1977), 235–252

[283] M. Suzuki, Propriétés topologiques des polynômes de deux variables complexes, et automorphismes algébriques de l'espace C^2, *J. Math. Soc. Japan*, 26(1974), No. 3, 241–257

[284] M. Sweedler, Using Gröbner basis to determine the algebraic and trancendental nature of field extensions: return of the killer tag variables, *Applied Algebra, Algebraic Algorithms and Error-Correcting Codes* (G. Cohen, T. Mora and O. Moreno, eds.), LNCS, Vol. 673, Springer, 1993, pp. 66–75

[285] T. tom Dieck and T. Petrie, Contractible affine surfaces of Kodaira dimension one, *Japan J. Math.*, 16(1990), 147–169

[286] W. Vasconcelos, Derivations on commutative noetherian rings, *Mathematische Zeitschrift*, 112(1969), 229–233

[287] ———, Flatness testing and torsionfree morphisms, *J. of Pure and Applied Algebra*, 122(1997), 313–321

[288] A. Vistoli, The Jacobian conjecture in dimension 3 and degree 3, *J. of Pure and Applied Algebra*, 142(1999), No. 1, 79–89

[289] S.S.-S. Wang, A Jacobian criterion for separability, *J. of Algebra*, 65(1980), 453–494

[290] Z. Wang, *Locally nilpotent derivations of polynomial rings*, Ph.D. thesis, University of Ottawa, 1999

[291] R. Weitzenböck, Über die invarianten Gruppen, *Acta Math.*, 58(1932), 231–293

[292] D. Wright, The amalgamated free product structure of $GL_2(K[X_1, \ldots, X_n])$ and the Weak Jacobian Theorem for two variables, *J. of Pure and Applied Algebra*, 12(1978), 235–251

[293] ———, On the Jacobian Conjecture, *Illinois J. of Math.*, 15 (1981), No. 3, 423–440

[294] ———, The Jacobian Conjecture: linear triangularization for cubics in dimension three, *Linear and Multilinear Algebra*, 34(1993), 85–97

[295] ———, Normal forms and the Jacobian Conjecture, In Essen [108], *Proceedings of the conference 'Invertible Polynomial Maps'*, pp. 145–156

[296] ———, *Affine surfaces fibered by affine lines over the projective line*, preprint, 1996.

[297] A.V. Yagzhev, On Keller's problem, *Siberian Math. J.*, 21(1980), 747–754

[298] J.-T. Yu, Face polynomials and inversion formula, *J. of Pure and Applied Algebra*, 78(1992), 213–219

[299] ———, Ph.D. thesis, Univ. of Notre Dame, 1993

[300] ———, Degree bounds for minimal polynomials and polynomial automorphisms, *J. of Pure and Applied Algebra*, 92(1994), 199–201

[301] ———, On the Jacobian Conjecture: reduction of coefficients, *J. of Algebra*, 171(1995), 515–523

[302] ———, On generalized strongly nilpotent matrices, *Linear and Multilinear Algebra*, 41(1996), 19–22

[303] ———, Complementary Conjecture Revisited, *J. of Pure and Applied Algebra*, 131(1998), 99–108

[304] M. Zaidenberg, *Affine acyclic curves and surfaces*, Tech. Report 90-52, Max-Planck Institut für Mathematik, Bonn, 1990, preprint

[305] A. Zaks, Dedekind subrings of $k[x_1, \ldots, x_n]$ are rings of polynomials, *Israel J. of Mathematics*, 9(1971), 285–289

[306] G. Zariski, Interprétations algébrico-géométriques du quatorzième problème de Hilbert, *Bull. Sci. Math.*, 78(1954), 155–168

[307] O. Zariski and P. Samuel, *Commutative algebra*, Vol. I, D. van Nostrand Comp. Inc., 1956

[308] ———, *Commutative algebra*, Vol. II, D. van Nostrand Comp. Inc., 1960

[309] V. Zurkowski, *Locally finite derivations*, preprint, 1993

Authors Index

Index

MIX
Papier aus verantwortungsvollen Quellen
Paper from responsible sources
FSC® C105338

If you have any concerns about our products,
you can contact us on
ProductSafety@springernature.com

In case Publisher is established outside the EU,
the EU authorized representative is:
Springer Nature Customer Service Center GmbH
Europaplatz 3, 69115 Heidelberg, Germany

Printed by Libri Plureos GmbH
in Hamburg, Germany